JOHN E. FREUND
Arizona State University

RONALD E. WALPOLE
Roanoke College

mathematical statistics

third edition

Prentice-Hall, Inc., *Englewood Cliffs, New Jersey 07632*

Library of Congress Cataloging in Publication Data

Freund, John E
Mathematical statistics.

Includes bibliographies and index.
1. Mathematical statistics. I. Walpole, Ronald E., joint author. II. Title.
QA276.F692 1980 519.5 79-16146
ISBN 0-13-562066-X

Editorial/production supervision by Karen J. Clemments
Interior design by Judy Winthrop
Cover design by Maurice A. Kruth
Manufacturing buyer: Ed Leone

Printed in the United States of America

10 9 8 7 6

PRENTICE-HALL INTERNATIONAL, INC., *London*
PRENTICE-HALL OF AUSTRALIA PTY., LIMITED, *Sydney*
PRENTICE-HALL OF CANADA, LTD., *Toronto*
PRENTICE-HALL OF INDIA PRIVATE LIMITED, *New Delhi*
PRENTICE-HALL OF JAPAN, INC., *Tokyo*
PRENTICE-HALL OF SOUTHEAST ASIA PTE. LTD., *Singapore*
WHITEHALL BOOKS LIMITED, *Wellington, New Zealand*

Affectionately dedicated to
our wives, Mickey and Norma

contents

PREFACE

INTRODUCTION

PROBABILITY

PROBABILITY DISTRIBUTIONS

3

MATHEMATICAL EXPECTATION

4

SPECIAL PROBABILITY DISTRIBUTIONS

5

SPECIAL PROBABILITY DENSITIES

6

INTERVAL ESTIMATION

11

HYPOTHESIS TESTING: THEORY

12

HYPOTHESIS TESTING: APPLICATIONS

13

REGRESSION AND CORRELATION

14

ANALYSIS OF VARIANCE

15

NONPARAMETRIC METHODS

16

APPENDIX: THE ALGEBRA OF EVENTS

I

APPENDIX: SUMS AND PRODUCTS

II

preface

Like its first and second editions, this book is designed for a two-semester or three-quarter calculus-based introduction to mathematical statistics. Most of the differences between this edition and the preceding ones reflect the changes that have taken place in recent years in statistical thinking, and in the teaching of statistics. Also, there have been extensive changes in format, which should make the book easier to read and easier to teach.

In addition to substantial changes in notation, the basic material on distribution theory has been reorganized, there is a new chapter combining the material on functions of random variables, the theoretical and applied aspects of estimation have been expanded and placed in two chapters, an expanded coverage is given to nonparametric statistics, the introduction to analysis of variance has been rewritten with more emphasis on the concepts of experimental design, the material on Boolean Algebra has been placed into an appendix, and there are many new exercises and illustrations.

The authors would like to express their appreciation for the many constructive comments which they have received from their colleagues; also, they are indebted to Harry Gaines for his efforts which led to their collaboration on this new edition of MATHEMATICAL STATISTICS, to Ms. Karen J. Clemments for her cooperation during the production stages of the book, to Doug Freund for his editorial assistance, and to Ms. Elizabeth L. Leonard for typing the first draft of the manuscript.

Finally, the authors would like to express their appreciation to the McGraw-Hill Book Company for their permission to reproduce in Table II material from their *Handbook of Probability and Statistics with Tables*, and to Professor E. S. Pearson and the *Biometrika* trustees for their permission to reproduce the material in Tables IV, V, and VI.

John E. Freund

Ronald E. Walpole

introduction

1

1.1 HISTORICAL BACKGROUND

In recent years, the growth of statistics has made itself felt in almost every phase of human activity. Statistics no longer consists merely of the collection of data and their presentation in charts and tables—it is now considered to encompass the science of basing inferences on observed data and the entire problem of making decisions in the face of uncertainty. This covers considerable ground since uncertainties are met when we flip a coin, when a dietician experiments with food additives, when an actuary determines life insurance premiums, when a quality control engineer accepts or rejects manufactured products, when a teacher compares the abilities of his students, when an economist forecasts trends, when a newspaper predicts an election, and so forth.

It would be presumptuous to say that statistics, in its present state of development, can handle all situations involving uncertainties, but new techniques are constantly being developed and modern statistics can, at least, provide the framework for looking at these situations in a logical and systematic fashion. In other words, statistics provides the models that are needed to study situations involving uncertainties, in the same way as calculus provides the models that are needed to describe, say, the concepts of Newtonian physics.

The beginnings of the mathematics of statistics may be found in mid-eighteenth-century studies in probability motivated by interest in games of chance. The theory thus developed for "heads or tails" or "red or black" soon found applications in situations where the outcomes were "boy or girl," "life or death," or "pass or fail," and scholars began to apply probability theory to actuarial problems and some aspects of the social sciences. Later, probability and statistics were introduced into physics by L. Boltzmann, J. Gibbs, and J. Maxwell, and in this century they have found applications in all phases of human endeavor which in some way involve an element of uncertainty or risk. The names which are connected most prominently with the growth of mathematical statistics in the first half of this century are those of R. A. Fisher, J. Neyman, E. S. Pearson, and A. Wald. More recently, the work of R. Schlaifer, L. J. Savage, and others, has given impetus to statistical theories based essentially on methods which date back to the eighteenth-century English clergyman Thomas Bayes.

The approach to statistics presented in this book is essentially the classical approach, with methods of inference based largely on the work of J. Neyman and E. S. Pearson. However, the more general decision-theory approach is introduced in Chapter 9 and some Bayesian methods are presented in Chapter 10.

1.2 MATHEMATICAL PRELIMINARY: COMBINATORIAL METHODS

In many problems of statistics we must list all the alternatives that are possible in a given situation, or at least determine how many different possibilities there are. In connection with the latter, we often use the following theorem, sometimes called the "multiplication rule" for possibilities or choices:

THEOREM 1.1 If an operation consists of two steps, of which the first can be made in n_1 ways and for each of these the second can be made in n_2 ways, then the whole operation can be made in $n_1 \cdot n_2$ ways.

Here, "operation" stands for any kind of procedure, process, or task.

To justify this theorem, let us define the ordered pair (x_i, y_j) to be the outcome which arises when the first step results in possibility x_i and the second step results in possibility y_j. Then, the set of all possible outcomes is composed of the following $n_1 \cdot n_2$ pairs:

$$(x_1, y_1), (x_1, y_2), \ldots, (x_1, y_{n_2})$$
$$(x_2, y_1), (x_2, y_2), \ldots, (x_2, y_{n_2})$$
$$\cdots$$
$$\cdots$$
$$\cdots$$
$$(x_{n_1}, y_1), (x_{n_1}, y_2), \ldots, (x_{n_1}, y_{n_2})$$

EXAMPLE 1.1

Suppose that someone wants to go by bus, by train, or by plane on a week's vacation to one of the five East North Central States. Find the number of different ways in which this can be done.

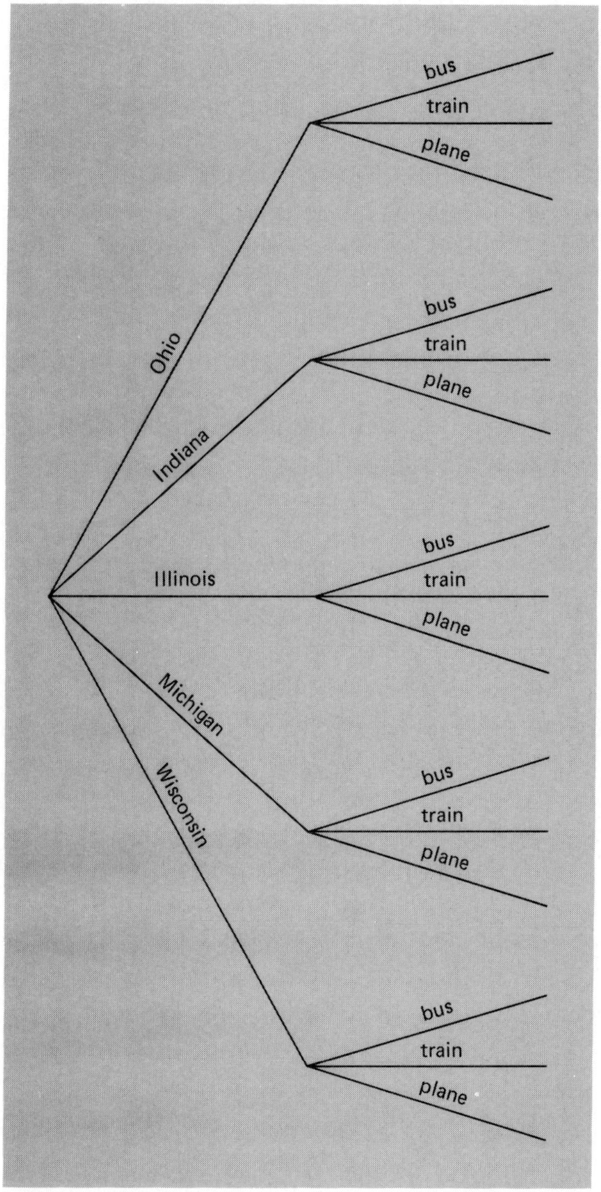

Figure 1.1 Tree diagram.

Solution

The particular state can be chosen in $n_1 = 5$ ways and the means of transportation can be chosen in $n_2 = 3$ ways. Therefore, the trip can be carried out in $5 \cdot 3 = 15$ possible ways. If an actual listing of all the possibilities is desirable, a **tree diagram** like that in Figure 1.1 provides a systematic approach. This diagram shows that there are $n_1 = 5$ branches (possibilities) for the number of states and for each of these branches there are $n_2 = 3$ branches (possibilities) for the different means of transportation. It is apparent that the 15 possible ways of taking the vacation are represented by the 15 distinct paths along the branches of the tree.

EXAMPLE 1.2

How many possible outcomes are there when a red die and a green die are thrown?

Solution

The red die can land in any one of six ways, and for each of these six ways the green die can also land in six ways. Therefore, the pair of dice can land in $6 \cdot 6 = 36$ ways.

Theorem 1.1 may be extended to cover situations where an operation consists of any fixed number of steps. The general case is stated in the following theorem:

THEOREM 1.2 If an operation consists of k steps, of which the first can be made in n_1 ways, for each of these the second step can be made in n_2 ways, for each of the first two the third step can be made in n_3 ways, and so forth, then the whole operation can be made in $n_1 \cdot n_2 \cdot \ldots \cdot n_k$ ways.

EXAMPLE 1.3

How many different lunches are possible consisting of a soup, a sandwich, a dessert, and a drink if one can select from 4 different soups, 3 kinds of sandwiches, 5 desserts, and 4 drinks?

Solution

The total number of lunches would be $4 \cdot 3 \cdot 5 \cdot 4 = 240$.

EXAMPLE 1.4

How many ways can one mark a true-false test consisting of 20 questions?

Solution

If a true-false test consists of 20 questions, there are

$$\underbrace{2 \cdot 2 \cdot 2 \cdot 2 \cdot \ldots \cdot 2 \cdot 2}_{20 \text{ factors}} = 1{,}048{,}576$$

different ways in which one can mark the test, and only one of these corresponds to the case where each answer is correct.

Frequently, we are interested in situations where the outcomes are the different *orders* or *arrangements* that are possible for a group of objects. For example, we might want to know how many different arrangements are possible for electing the president, vice-president, treasurer, and secretary from the 24 members of a club, or we might want to know how many different arrangements are possible for seating 6 persons around a table. Different arrangements like these are called **permutations**.

EXAMPLE 1.5

How many permutations are there of all three of the letters a, b, and c?

Solution

The possible arrangements are *abc*, *acb*, *bac*, *bca*, *cab*, and *cba*, so the number of distinct permutations is six. Using Theorem 1.2, we could have arrived at this answer without actually listing the different permutations. Since there are three choices to select a letter for the first position, then two for the second position, leaving only one letter for the third position, the total number of permutations is $3 \cdot 2 \cdot 1 = 6$.

Generalizing the argument used in this example, we find that n distinct objects can be arranged in $n(n-1)(n-2) \cdot \ldots \cdot 3 \cdot 2 \cdot 1$ ways. We represent this product by the symbol $n!$, which is read "n factorial." Thus, $1! = 1$, $2! = 2 \cdot 1 = 2$, $3! = 3 \cdot 2 \cdot 1 = 6$, and so on. By definition, $0! = 1$.

THEOREM 1.3 The number of permutations of n distinct objects is $n!$.

EXAMPLE 1.6

How many different orders are possible for introducing the 5 starting players of a basketball team to the public?

Solution

There are $5! = 5 \cdot 4 \cdot 3 \cdot 2 \cdot 1 = 120$ different orders for introducing the starting lineup.

EXAMPLE 1.7

The number of permutations of the four letters a, b, c, and d is 24, but what is the number of permutations if we take only two of the four letters, or as it is usually put, if we take the four letters two at a time?

Solution

Again using Theorem 1.2, we find that we have two positions to fill with four choices for the first and then three choices for the second for a total of $4 \cdot 3 = 12$ permutations.

Generalizing the argument used in this example, we find that n distinct objects taken r at a time can be arranged in $n(n-1) \cdot \ldots \cdot (n-r+1)$ ways. We represent this product by the symbol ${}_nP_r$.

THEOREM 1.4 The number of permutations of n distinct objects taken r at a time is

$${}_nP_r = n(n-1) \cdot \ldots \cdot (n-r+1)$$

or, in factorial notation,

$${}_nP_r = \frac{n!}{(n-r)!}$$

To obtain the second formula for ${}_nP_r$ we made use of the identity

$$n(n-1) \cdot \ldots \cdot (n-r+1) \cdot (n-r)! = n!$$

In applications, the first formula is generally easier to use, but the one in factorial notation is easier to remember and more easily programmed for solution on a digital computer.

EXAMPLE 1.8

Four names are drawn from the 24 members of a club for the offices of president, vice-president, treasurer, and secretary. In how many different ways can this be done?

Solution

The number of permutations of 24 distinct objects taken 4 at a time is

$$_{24}P_4 = 24 \cdot 23 \cdot 22 \cdot 21 = 255{,}024$$

EXAMPLE 1.9

In how many ways can a local chapter of the American Chemical Society schedule three speakers for three different meetings, if they are all available on any of five possible dates?

Solution

The number of permutations of 5 distinct objects taken 3 at a time is

$$_5P_3 = 5 \cdot 4 \cdot 3 = 60$$

Permutations that occur when objects are arranged in a circle are called **circular permutations**. Two circular permutations are not considered different if corresponding objects in the two arrangements are preceded and followed by the same objects as we proceed in a clockwise direction. For example, if four persons are playing bridge, we do not get a new permutation if they all move one position in a clockwise direction.

EXAMPLE 1.10

How many circular permutations are there of four persons playing bridge?

Solution

By considering one person in a fixed position and arranging the other three in 3! ways, we find that there are six different arrangements (circular permutations) of four persons playing bridge.

Generalizing the argument used in this example, we get the result stated in the following theorem:

THEOREM 1.5 The number of permutations of n distinct objects arranged in a circle is $(n - 1)!$.

Throughout our discussion it has been assumed that the n objects from which we select r objects and form permutations are all distinct. Thus, our results cannot be used, for example, to determine the number of ways in which we can arrange the letters in the word "book," or the letters in the word "receive."

EXAMPLE 1.11

How many permutations are there of the letters in the word "book?"

Solution

If we distinguish for the moment between the two o's by labeling them o_1 and o_2, there are $4! = 24$ different permutations of the symbols b, o_1, o_2, and k. However, if we drop the subscripts, then bo_1ko_2 and bo_2ko_1, for instance, both yield $boko$, and since each pair of permutations with subscripts yields but one arrangement without subscripts, the total number of arrangements of the letters in the word "book" is $\frac{24}{2} = 12$.

EXAMPLE 1.12

How many permutations are there of the letters in the word "receive"?

Solution

With subscripts on the e's there are 7! permutations of the letters in the word "receive," but since there are $3! = 6$ permutations of e_1, e_2, and e_3 which lead to the same arrangement of the letters in "receive," there are only $\frac{7!}{3!} = 7 \cdot 6 \cdot 5 \cdot 4 = 840$ different arrangements of the letters in the word "receive."

Generalizing the argument used in these two examples, we get the result stated in the following theorem:

THEOREM 1.6 The number of permutations of n objects of which n_1 are of one kind, n_2 are of a second kind, ..., n_k are of a kth kind, and $n_1 + n_2 + \ldots + n_k = n$, is

$$\frac{n!}{n_1! \cdot n_2! \cdot \ldots \cdot n_k!}$$

EXAMPLE 1.13

In how many ways can 2 oaks, 3 pines, and 2 maples be arranged in a straight line if one does not distinguish between trees of the same kind?

Solution

The total number of distinct arrangements is

$$\frac{7!}{2! \cdot 3! \cdot 2!} = 210$$

Often we are interested in determining the number of ways of selecting r objects from among n distinct objects without regard to the order in which they are selected. Such selections are called **combinations**.

EXAMPLE 1.14

In how many ways can a person gathering data for a market research organization interview 3 of the 20 families living in a certain apartment house?

Solution

If we cared about the order in which the families are interviewed, the answer would be

$${}_{20}P_3 = 20 \cdot 19 \cdot 18 = 6{,}840$$

but each set of 3 families would then be counted $3! = 6$ times. If we are not interested in the order in which the 3 families are interviewed, there are thus only $\frac{6{,}840}{6} = 1{,}140$ ways in which 3 of the 20 families can be selected.

Actually, "combination" means the same as "subset," and when we ask for the number of combinations of r objects selected from a set of n distinct objects, we are simply asking for the total number of subsets of r objects that can be selected from a set of n distinct objects. In general, there are $r!$ permutations of the objects in a subset of r objects, so that the ${}_nP_r$ permutations of r objects selected from a set of n distinct objects contain each subset $r!$ times. Dividing ${}_nP_r$ by $r!$ and denoting the result by the symbol $\binom{n}{r}$, we thus have:

THEOREM 1.7 The number of combinations of r objects selected from a set of n distinct objects is

$$\binom{n}{r} = \frac{n(n-1)(n-2)\cdot\ldots\cdot(n-r+1)}{r!}$$

or, in factorial notation,

$$\binom{n}{r} = \frac{n!}{r!\cdot(n-r)!}$$

Again, the first formula is generally easier to use, but the one in factorial notation is easier to remember and more easily programmed for solution on a digital computer.

EXAMPLE 1.15

In how many different ways can 6 tosses of a coin yield 2 heads and 4 tails?

Solution

This question is equivalent to asking in how many different ways one can select the 2 tosses on which heads is to occur; applying Theorem 1.7, we thus find the answer to be

$$\binom{6}{2} = \frac{6\cdot 5}{2!} = 15$$

This result could also have been obtained by the rather tedious process of enumerating the various possibilities, HHTTTT, TTHTHT, HTHTTT, . . . , where H stands for head and T for tail.

EXAMPLE 1.16

How many committees of two chemists and one physicist can be formed from four chemists and three physicists?

Solution

Since two of four chemists can be selected in $\binom{4}{2} = \frac{4!}{2! \cdot 2!} = 6$ ways and one of three physicists can be selected in $\binom{3}{1} = \frac{3!}{1! \cdot 2!} = 3$ ways, Theorem 1.1 shows that the number of committees is $6 \cdot 3 = 18$.

A combination of r objects selected from a set of n distinct objects may be considered a **partition** of the n objects into two subsets containing, respectively, the r objects that are selected and the $n - r$ objects that are left. Often, we are concerned with the more general problem of partitioning a set of n distinct objects into k subsets, which requires that each of the n objects must belong to one and only one of the subsets.† The order of the objects within a subset is of no importance.

EXAMPLE 1.17

In how many ways can a set of four objects be partitioned into three subsets containing, respectively, 2, 1, and 1 of the objects?

Solution

Denoting the four objects by a, b, c, and d, we find by enumeration that there are the twelve possibilities:

$$\begin{array}{cccc} ab|c|d & ab|d|c & ac|b|d & ac|d|b \\ ad|b|c & ad|c|b & bc|a|d & bc|d|a \\ bd|a|c & bd|c|a & cd|a|b & cd|b|a \end{array}$$

The number of partitions for this example is denoted by the symbol

$$\binom{4}{2, 1, 1} = 12$$

where the number at the top represents the total number of objects and the numbers at the bottom represent the number of objects going into each subset.

† Symbolically (see Appendix I at the end of the book), the subsets $A_1, A_2, \ldots,$ and A_k constitute a partition of set A if $A_1 \cup A_2 \cup \cdots \cup A_k = A$ and $A_i \cap A_j = \varnothing$ for all $i \neq j$.

In general, we have the following theorem:

THEOREM 1.8 The number of ways of partitioning a set of n distinct objects into k subsets with n_1 objects in the first subset, n_2 objects in the second subset, . . . , and n_k objects in the kth subset, is

$$\binom{n}{n_1, n_2, \ldots, n_k} = \frac{n!}{n_1! \cdot n_2! \cdot \ldots \cdot n_k!}$$

Proof. First we note that there are $\binom{n}{n_1}$ ways to form the first subset. For each of these there are $\binom{n - n_1}{n_2}$ ways to form the second subset, for each first and second subset there are $\binom{n - n_1 - n_2}{n_3}$ ways to form the third subset, and so forth. Hence, by Theorem 1.2 it follows that

$$\binom{n}{n_1, n_2, \ldots, n_k} = \binom{n}{n_1} \cdot \binom{n - n_1}{n_2} \cdot \ldots \cdot \binom{n - n_1 - n_2 - \cdots - n_{k-1}}{n_k}$$

$$= \frac{n!}{n_1! \cdot (n - n_1)!} \cdot \frac{(n - n_1)!}{n_2! \cdot (n - n_1 - n_2)!} \cdot \ldots \cdot \frac{(n - n_1 - n_2 - \cdots - n_{k-1})!}{n_k! \cdot 0!}$$

$$= \frac{n!}{n_1! \cdot n_2! \cdot \ldots \cdot n_k!}$$

EXAMPLE 1.18

In how many ways can seven scientists be assigned to one triple and two double hotel rooms?

Solution

Substitution of $n = 7$, $n_1 = 3$, $n_2 = 2$, and $n_3 = 2$ into the formula of Theorem 1.8 yields $\binom{7}{3, 2, 2} = \frac{7!}{3! \cdot 2! \cdot 2!} = 210$.

1.3 MATHEMATICAL PRELIMINARY: BINOMIAL COEFFICIENTS

If n is a positive integer and we multiply out $(x + y)^n$ term by term, each term will be the product of x's and y's, with an x or a y coming from each of the n factors $x + y$. For instance, the expansion

$$\begin{aligned}(x + y)^3 &= (x + y)(x + y)(x + y)\\ &= x \cdot x \cdot x + x \cdot x \cdot y + x \cdot y \cdot x + x \cdot y \cdot y\\ &\qquad + y \cdot x \cdot x + y \cdot x \cdot y + y \cdot y \cdot x + y \cdot y \cdot y\\ &= x^3 + 3x^2y + 3xy^2 + y^3\end{aligned}$$

yields terms of the form x^3, x^2y, xy^2, and y^3. Their coefficients are 1, 3, 3, and 1, and the coefficient of xy^2, for example, is $\binom{3}{2} = 3$, the number of ways in which we can choose the two factors providing the y's. Similarly, the coefficient of x^2y is $\binom{3}{1} = 3$, the number of ways in which we can choose the one factor providing the y, and the coefficients of x^3 and y^3 are $\binom{3}{0} = 1$ and $\binom{3}{3} = 1$.

More generally, if n is a positive integer and we multiply out $(x + y)^n$ term by term, the coefficient of $x^{n-r}y^r$ is $\binom{n}{r}$, the number of ways in which we can choose the r factors providing the y's. Accordingly, we refer to $\binom{n}{r}$ as a **binomial coefficient**. We can now state the following theorem:

THEOREM 1.9

$$(x + y)^n = \sum_{r=0}^{n} \binom{n}{r} x^{n-r} y^r \qquad \text{for any positive integer } n$$

(In case the reader is not familiar with the $\sum$ notation, he will find a brief explanation in Appendix II at the end of the book.)

The calculation of binomial coefficients can often be simplified by making use of the three theorems which follow.

THEOREM 1.10 For any positive integers n and $r = 0, 1, 2, \ldots, n$,

$$\binom{n}{r} = \binom{n}{n - r}$$

Proof. We might argue that when we select a subset of r objects from a set of n distinct objects we leave a subset of $n - r$ objects, and, hence, there are as many ways of selecting r objects as there are ways of leaving (or selecting) $n - r$ objects. To prove the theorem algebraically, we write

$$\binom{n}{n-r} = \frac{n!}{(n-r)![n-(n-r)]!} = \frac{n!}{(n-r)!r!} = \frac{n!}{r!(n-r)!} = \binom{n}{r}$$

Theorem 1.10 implies that if we calculate the binomial coefficients for $r = 0, 1, \ldots, \frac{n}{2}$ when n is even and for $r = 0, 1, \ldots, \frac{n-1}{2}$ where n is odd, the remaining binomial coefficients can be obtained by making use of the theorem.

EXAMPLE 1.19

Given $\binom{4}{0} = 1$ and $\binom{4}{1} = 4$, find $\binom{4}{3}$ and $\binom{4}{4}$.

Solution

$$\binom{4}{3} = \binom{4}{4-3} = \binom{4}{1} = 4 \quad \text{and} \quad \binom{4}{4} = \binom{4}{4-4} = \binom{4}{0} = 1.$$

EXAMPLE 1.20

Given $\binom{5}{0} = 1$, $\binom{5}{1} = 5$, and $\binom{5}{2} = 10$, find $\binom{5}{3}$, $\binom{5}{4}$, and $\binom{5}{5}$.

Solution

$$\binom{5}{3} = \binom{5}{5-3} = \binom{5}{2} = 10, \quad \binom{5}{4} = \binom{5}{5-4} = \binom{5}{1} = 5,$$

$$\text{and} \quad \binom{5}{5} = \binom{5}{5-5} = \binom{5}{0} = 1.$$

It is precisely in this fashion that Theorem 1.10 may have to be used in connection with Table VII at the end of the book.

EXAMPLE 1.21

Find $\binom{20}{12}$ and $\binom{17}{10}$.

Solution

To find $\binom{20}{12}$, we make use of the fact that $\binom{20}{12} = \binom{20}{8}$, look up $\binom{20}{8}$, and get $\binom{20}{12} = 125{,}970$; to find $\binom{17}{10}$, we make use of the fact that $\binom{17}{10} = \binom{17}{7}$, look up $\binom{17}{7}$, and get $\binom{17}{10} = 19{,}448$.

THEOREM 1.11 For any positive integer n and $r = 1, 2, \ldots, n - 1$,

$$\binom{n}{r} = \binom{n-1}{r} + \binom{n-1}{r-1}$$

Proof. Substituting $x = 1$ into $(x + y)^n$, let us write

$$\begin{aligned}(1 + y)^n &= (1 + y)(1 + y)^{n-1} \\ &= (1 + y)^{n-1} + y(1 + y)^{n-1}\end{aligned}$$

and equate the coefficient of y^r in $(1 + y)^n$ with that in $(1 + y)^{n-1} + y(1 + y)^{n-1}$. Since the coefficient of y^r in $(1 + y)^n$ is $\binom{n}{r}$ and the coefficient of y^r in $(1 + y)^{n-1} + y(1 + y)^{n-1}$ is the sum of the coefficient of y^r in $(1 + y)^{n-1}$, namely, $\binom{n-1}{r}$, and the coefficient of y^{r-1} in $(1 + y)^{n-1}$, namely, $\binom{n-1}{r-1}$, we obtain

$$\binom{n}{r} = \binom{n-1}{r} + \binom{n-1}{r-1}$$

which completes the proof.

Theorem 1.11 can also be proved by expressing the binomial coefficients on both sides of the equation in terms of factorials and then proceeding algebraically, but we shall leave this to the reader in Exercise 6 on page 19. One important application of Theorem 1.11 is given in Exercise 5 on page 19, where it provides the key for the construction of what is known as **Pascal's triangle**.

To state the third theorem about binomial coefficients, let us make the definition that $\binom{n}{r} = 0$ whenever n is a positive integer and r is a positive integer greater than n. (Clearly, there is no way in which we can select a subset from a set which contains more elements than the set itself.)

THEOREM 1.12

$$\sum_{r=0}^{k} \binom{m}{r}\binom{n}{k-r} = \binom{m+n}{k}$$

Proof. Using the same technique as in the proof of Theorem 1.11, let us prove this theorem by equating the coefficients of y^k in the expressions on both sides of the equation

$$(1+y)^{m+n} = (1+y)^m(1+y)^n$$

The coefficient of y^k in $(1+y)^{m+n}$ is $\binom{m+n}{k}$, and the coefficient of y^k in

$$(1+y)^m(1+y)^n = \left[\binom{m}{0} + \binom{m}{1}y + \cdots + \binom{m}{m}y^m\right] \times \left[\binom{n}{0} + \binom{n}{1}y + \cdots + \binom{n}{n}y^n\right]$$

is the sum of the products which we obtain by multiplying the constant term of the first factor by the coefficient of y^k in the second factor, the coefficient of y in the first factor by the coefficient of y^{k-1} in the second factor, . . . , and the coefficient of y^k in the first factor by the constant term of the second factor. Thus, the coefficient of y^k in $(1+y)^m(1+y)^n$ is

$$\binom{m}{0}\binom{n}{k} + \binom{m}{1}\binom{n}{k-1} + \binom{m}{2}\binom{n}{k-2} + \cdots + \binom{m}{k}\binom{n}{0} = \sum_{r=0}^{k} \binom{m}{r}\binom{n}{k-r}$$

and this completes the proof.

EXAMPLE 1.22

Verify Theorem 1.12 numerically for $m = 2$, $n = 3$, and $k = 4$.

Solution

Substituting these values we get

$$\binom{2}{0}\binom{3}{4} + \binom{2}{1}\binom{3}{3} + \binom{2}{2}\binom{3}{2} + \binom{2}{3}\binom{3}{1} + \binom{2}{4}\binom{3}{0} = \binom{5}{4}$$

and since $\binom{3}{4}$, $\binom{2}{3}$, and $\binom{2}{4}$ equal 0 according to the definition on page 15, the equation reduces to

$$\binom{2}{1}\binom{3}{3} + \binom{2}{2}\binom{3}{2} = \binom{5}{4}$$

which checks, since $2 \cdot 1 + 1 \cdot 3 = 5$.

Using Theorem 1.8, we can extend our discussion to **multinomial coefficients,** namely, to the coefficients that arise in the expansion of $(x_1 + x_2 + \cdots + x_k)^n$. The multinomial coefficient of the term $x_1^{r_1} \cdot x_2^{r_2} \cdot \ldots \cdot x_k^{r_k}$ in the expansion of $(x_1 + x_2 + \cdots + x_k)^n$ is

$$\binom{n}{r_1, r_2, \ldots, r_k} = \frac{n!}{r_1! \cdot r_2! \cdot \ldots \cdot r_k!}$$

EXAMPLE 1.23

What is the coefficient of $x_1^3 x_2 x_3^2$ in the expansion of $(x_1 + x_2 + x_3)^6$?

Solution

Substituting into the above formula, we get

$$\frac{6!}{3! \cdot 1! \cdot 2!} = 60$$

THEORETICAL EXERCISES

1. In a two-team play-off in some sport, the winner is the first team to win m games.

(a) Counting separately the number of play-offs requiring m, $m + 1, \ldots,$ and $2m - 1$ games, show that the total number of different outcomes (sequences of wins and losses) is

$$2\left[\binom{m-1}{m-1} + \binom{m}{m-1} + \cdots + \binom{2m-2}{m-1}\right]$$

(b) How many different outcomes are there in a "2 out of 3" play-off, a "3 out of 5" play-off, and a "4 out of 7" play-off?

2. An operation consists of two steps, of which the first can be made in n_1 ways.

If the first step is made in the ith way, the second step can be made in n_{2i} ways.[†]

(a) What is the total number of ways in which the whole operation can be made?

(b) A student can study 0, 1, 2, or 3 hours for a statistics test on any given day. In how many ways can this student study at most 4 hours for the test on two consecutive days?

3. When n is large, $n!$ can be approximated by means of the expression

$$\sqrt{2\pi n}\left(\frac{n}{e}\right)^n$$

called **Stirling's formula**, where e is the base of natural logarithms. (A derivation of this formula may be found in the book by W. Feller cited among the references at the end of this chapter.)

(a) Use Stirling's formula to obtain approximations for 10! and 12!, and find the percentage errors of these approximations by comparing them with the exact values given in Table VII.

(b) Use Stirling's formula to obtain an approximation for the number of 13-card bridge hands that can be dealt with an ordinary deck of 52 playing cards.

(c) Use Stirling's formula to show that

$$\lim_{n\to\infty} \frac{\binom{2n}{n}\sqrt{\pi n}}{2^{2n}} = 1$$

4. In **occupancy theory** we are concerned with the number of ways in which certain distinguishable or indistinguishable objects can be distributed among a given number of individuals, urns, boxes, or cells.

(a) Find an expression for the number of ways in which r *distinguishable* objects can be distributed among n cells, and use it to find the number of ways in which three new books can be issued to 12 members of a library.

(b) Find an expression for the number of ways in which r *indistinguishable* objects can be distributed among n cells, and use it to find the number of ways in which a baker can sell ten (indistinguishable) loaves of bread to six customers. (*Hint*: For $r = 5$ and $n = 3$, for example, we might argue that 0|000|0 represents the case where one objects goes into the first cell, three into the second cell, and one into the third cell, and we must look for the number of ways in which we can arrange the five 0's and the two vertical bars.)

[†] The use of double subscripts is explained in Appendix II.

(c) Find an expression for the number of ways in which r *indistinguishable* objects can be distributed among n cells, with at least one object in each cell, and rework the numerical part of (b) if each customer must get at least one loaf of bread.

5. When no table is available, it is sometimes convenient to determine binomial coefficients by means of the following arrangement, called **Pascal's triangle**,

$$
\begin{array}{ccccccccccc}
 & & & & & 1 & & & & & \\
 & & & & 1 & & 1 & & & & \\
 & & & 1 & & 2 & & 1 & & & \\
 & & 1 & & 3 & & 3 & & 1 & & \\
 & 1 & & 4 & & 6 & & 4 & & 1 & \\
1 & & 5 & & 10 & & 10 & & 5 & & 1 \\
\cdots & \cdots & \cdots & \cdots & \cdots & \cdots & \cdots & \cdots & \cdots & \cdots & \cdots
\end{array}
$$

where each row begins with a 1, ends with a 1, and each other entry is the sum of the nearest two entries in the row immediately above.

(a) Explain why, with this method of construction, the rth entry of the nth row is the binomial coefficient $\binom{n-1}{r-1}$. (*Hint*: Use mathematical induction and Theorem 1.11.)

(b) Construct the next two (seventh and eighth) rows of the triangle and write the binomial expansions of $(x + y)^6$ and $(x + y)^7$.

6. Prove Theorem 1.11 by expressing all the binomial coefficients in terms of factorials and then simplifying algebraically.

7. Expressing the binomial coefficients in terms of factorials and simplifying algebraically, show that

(a) $\binom{n}{r} = \dfrac{n-r+1}{r} \cdot \binom{n}{r-1}$;

(b) $\binom{n}{r} = \dfrac{n}{n-r} \cdot \binom{n-1}{r}$;

(c) $n\binom{n-1}{r} = (r+1)\binom{n}{r+1}$.

8. Substitute suitable values for x and y into the formula of Theorem 1.9 to show that

(a) $\sum_{r=0}^{n} \binom{n}{r} = 2^n$;

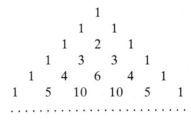

(b) $\sum_{r=0}^{n} (-1)^r \binom{n}{r} = 0;$

(c) $\sum_{r=0}^{n} \binom{n}{r} (a-1)^r = a^n.$

9. Repeatedly apply Theorem 1.11 to show that

$$\binom{n}{r} = \sum_{i=1}^{r+1} \binom{n-i}{r-i+1}$$

10. Use Theorem 1.12 to show that

$$\sum_{r=0}^{n} \binom{n}{r}^2 = \binom{2n}{n}$$

11. Show that $\sum_{r=0}^{n} r\binom{n}{r} = n2^{n-1}$

(a) by setting $x = 1$ in Theorem 1.9, then differentiating the expressions on both sides with respect to y, and finally substituting $y = 1$;
(b) by making use of part (a) of Exercise 8 and part (c) of Exercise 7.

12. If n is not a positive integer or zero, the binomial expansion of $(1 + y)^n$ yields, for $-1 < y < 1$, the infinite series

$$1 + \binom{n}{1}y + \binom{n}{2}y^2 + \binom{n}{3}y^3 + \cdots + \binom{n}{r}y^r + \cdots$$

where $\binom{n}{r} = \dfrac{n(n-1)\cdot\ldots\cdot(n-r+1)}{r!}$ for $r = 1, 2, 3, \ldots$. Use this **generalized definition of binomial coefficients**, which, incidentally, agrees with that on page 13 for positive integral values of n, to evaluate

(a) $\binom{\frac{1}{2}}{4}$ and $\binom{-3}{3}$;
(b) $\sqrt{5}$ by writing $\sqrt{5} = 2(1 + \frac{1}{4})^{\frac{1}{2}}$ and using the first four terms of the binomial expansion of $(1 + \frac{1}{4})^{\frac{1}{2}}$.

Also, show that

(c) $\binom{-1}{r} = (-1)^r;$

(d) $\binom{-n}{r} = (-1)^r \binom{n+r-1}{r}$ for $n > 0$.

13. Find the coefficient of $x^2y^3z^3$ in the expansion of $(x + y + z)^8$.

14. Find the coefficient of $x^3y^2z^3w$ in the expansion of $(2x + 3y - 4z + w)^9$.

15. Show that

$$\binom{n}{n_1, n_2, \ldots, n_k} = \binom{n-1}{n_1 - 1, n_2, \ldots, n_k} + \binom{n-1}{n_1, n_2 - 1, \ldots, n_k} + \cdots + \binom{n-1}{n_1, n_2, \ldots, n_k - 1}$$

by expressing all these multinomial coefficients in terms of factorials and simplifying algebraically.

APPLIED EXERCISES

16. There are four routes, A, B, C, and D, between a person's home and the place where he works, but route B is one-way so that he cannot take it on the way to work, and route C is one-way so that he cannot take it on the way home.

(a) Draw a tree diagram showing the various ways he can go to and from work.

(b) Draw a tree diagram showing the various ways he can go to and from work, without going the same route both ways.

17. A person with \$2 in her pocket bets \$1, even money, on the flip of a coin, and she continues to bet \$1 so long as she has any money. Draw a tree diagram to show the various things that can happen during the first four flips of the coin. In how many of the cases will she be

(a) exactly even;

(b) exactly \$2 ahead?

18. If the NCAA has applications from six universities for hosting its intercollegiate tennis championships in 1984 and 1985, in how many ways can they select the hosts for these championships

(a) if they are not both to be held at the same university;

(b) if they may both be held at the same university?

19. The five finalists in the Miss Universe contest are Miss Argentina, Miss Belgium, Miss U.S.A., Miss Japan, and Miss Norway. In how many ways can the judges choose

(a) the winner and the first runner-up;

(b) the winner, the first runner-up, and the second runner-up?

20. In a primary election, there are four candidates for mayor, five candidates for city treasurer, and two candidates for county attorney.

(a) In how many ways can a voter mark his ballot for all three of these offices?

(b) In how many ways can a person vote if he exercises his option of not voting for a candidate for any or all of these offices?

21. A multiple-choice test consists of 15 questions, each permitting a choice of three alternatives. In how many different ways can a student check off her answers to these questions?

22. The price of a European tour includes four stopovers to be selected from among ten cities. In how many different ways can one plan such a tour
    (a) if the order of the stopovers matters;
    (b) if the order of the stopovers does not matter?

23. In how many ways can a television director schedule a sponsor's six different commercials during the six time slots allocated to commercials during an hour "special"?

24. In how many ways can the television director of Exercise 23 fill the six time slots for commercials
    (a) if the sponsor has three different commercials, each of which is to be shown twice;
    (b) if the sponsor has two different commercials, each of which is to be shown three times?

25. In how many ways can five persons line up to get on a bus? In how many ways can they line up, if two of the persons refuse to follow each other?

26. In how many ways can a family of five sit around a dinner table, if it matters only who sits next to whom?

27. How many distinct permutations are there of the letters in the word "statistics"? How many of these begin and end with the letter s?

28. A college team plays ten football games during a season. In how many ways can it end the season with five wins, four losses, and one tie?

29. In Example 1.4 we showed that a true-false test which consists of 20 questions can be marked in 1,048,576 different ways. In how many ways can they be marked true or false so that
    (a) 7 are right and 13 are wrong;
    (b) 10 are right and 10 are wrong;
    (c) at least 17 are right?

30. Among the seven nominees for two vacancies on a city council are three men and four women. In how many ways can these vacancies be filled
    (a) with any two of the seven nominees;
    (b) with any two of the four women;
    (c) with one of the men and one of the women?

31. A shipment of ten television sets includes three that are defective. In how many ways can a hotel purchase four of these sets and receive at least two of the defective sets?

32. Ms. Jones has four skirts, seven blouses, and three sweaters. In how many ways can she choose two of the skirts, three of the blouses, and one of the sweaters to take along on a trip?

33. How many different bridge hands are possible containing five spades, three diamonds, three clubs, and two hearts?

34. Find the number of ways in which one A, three B's, two C's, and one F can be distributed among seven students taking a course in statistics.

35. If eight persons are having dinner together, in how many different ways can three order chicken, four order steak, and one order lobster?

References

Among the few books on the history of statistics there are

WALKER, H. M., *Studies in the History of Statistical Method.* Baltimore: The Williams & Wilkins Company, 1929,

WESTERGAARD, H., *Contributions to the History of Statistics.* London: P. S. King & Son, 1932,

and the two more recent publications

PEARSON, E. S., and KENDALL, M. G., eds., *Studies in the History of Statistics and Probability.* Darien, Conn.: Hafner Publishing Co., Inc., 1970,

KENDALL, M. G., and PLACKETT, R. L., eds., *Studies in the History of Statistics and Probability*, Vol. II. New York: Macmillan Publishing Co., Inc., 1977.

A wealth of material on combinatorial methods can be found in

EISEN, M., *Elementary Combinatorial Analysis.* New York: Gordon and Breach, Science Publishers, Inc., 1970,

FELLER, W., *An Introduction to Probability Theory and Its Applications*, Vol. I, 3rd ed. New York: John Wiley & Sons, Inc., 1968,

NIVEN, J., *Mathematics of Choice.* New York: Random House, Inc., 1965,

and in

WHITWORTH, W. A., *Choice and Chance*, 5th ed. New York: Hafner Publishing Co., Inc., 1959,

which has become a classic in this field. More advanced treatments may be found in

BECKENBACH, E. F., ed., *Applied Combinatorial Mathematics.* New York: John Wiley & Sons, Inc., 1964,

DAVID, F. N., and BARTON, D. E., *Combinatorial Chance.* New York: Hafner Publishing Co., Inc., 1962,

and

RIORDAN, J., *An Introduction to Combinatorial Analysis.* New York: John Wiley & Sons, Inc., 1958.

probability

2

2.1 INTRODUCTION

Historically, the oldest way of measuring probabilities, the **classical probability concept**, applies when all possible outcomes are equally likely, as is presumably the case in most games of chance. We can then say that *if there are N equally likely possibilities, of which one must occur and n are regarded as favorable, or as a "success," then the probability of a "success" is given by the ratio* $\frac{n}{N}$.

EXAMPLE 2.1

What is the probability of drawing an ace from an ordinary deck of playing cards?

Solution

There are $n = 4$ aces among the $N = 52$ cards, so the probability of drawing an ace is $\frac{4}{52}$.

Although equally likely possibilities are found mostly in games of chance, the classical probability concept applies also in a great variety of situations where gambling devices are used to make random selections—when office space is assigned to teaching assistants by lot, when some of the families in a township are chosen in such a way that each one has the same chance of being included in a sample study, when machine parts are chosen for inspection so that each part produced has the same chance of being selected, and so forth.

A major shortcoming of the classical probability concept is its limited applicability, for there are many situations in which the possibilities that arise cannot all be regarded as equally likely. This would be the case, for instance, if we are concerned with the question whether it will rain on a given day, if we are concerned with the outcome of an election, or if we are concerned with a person's recovery from a disease.

Among the various probability concepts, most widely held is the **frequency interpretation**, according to which *the probability of an event (outcome or happening) is the proportion of the time that events of the same kind will occur in the long run.* If we say that the probability is 0.84 that a jet from Los Angeles to San Francisco will arrive on time, we mean (in accordance with the frequency interpretation) that such flights arrive on time 84 percent of the time. Similarly, if the weather bureau predicts that there is a 30 percent chance for rain (namely, a probability of 0.30), this means that under the same weather conditions it will rain 30 percent of the time. More generally, we say that an event has a probability of, say, 0.90, in the same sense in which we might say that our car will start in cold weather 90 percent of the time. We cannot guarantee what will happen on any particular occasion—the car may start and then it may not—but if we kept records over a long period of time, we should find that the proportion of "successes" is very close to 0.90.

An alternative point of view, which is currently gaining in favor, is to interpret probabilities as **personal** or **subjective evaluations**. Such probabilities express the strength of one's belief with regard to the uncertainties that are involved, and they apply especially when there is little or no direct evidence, so that there is no choice but to consider collateral (indirect) evidence, "educated guesses," and perhaps intuition and other subjective factors.

The approach to probability we shall use in this chapter is the **axiomatic approach**, in which probabilities are defined as "mathematical objects" which behave according to certain well-defined rules. Then, any one of the above probability concepts, or interpretations, can be used in applications, so long as it is consistent with these rules.

2.2 SAMPLE SPACES

Probabilities always pertain to the occurrence or nonoccurrence of events, so let us explain formally what we mean by "event" and by the related terms "experiment" and "sample space."

It is customary in statistics to refer to any process of observation or measurement as an **experiment**. In this sense, an experiment may consist of the simple process of checking whether a switch is turned on or off; it may consist of counting the imperfections in a piece of cloth; or it may consist of the very complicated process of determining the mass of an electron. The results one obtains from an experiment, whether they are instrument readings, counts, "yes" or "no" answers, or values obtained through extensive calculations, are called the **outcomes** of the experiment.

The set of all possible outcomes of an experiment is called the **sample space** and it is usually denoted by the letter S. Each outcome in a sample space is called an **element** of the sample space or simply a **sample point**. If a sample space has a finite number of elements, we may list the elements in the usual set notation; for instance, the sample space for the possible outcomes of one flip of a coin may be written

$$S = \{\mathrm{H}, \mathrm{T}\}$$

where H and T stand for head and tail. Sample spaces with a large or infinite number of elements are best described by a statement or rule; for example, if the possible outcomes of an experiment are the set of automobiles equipped with citizen band radios, the sample space may be written

$$S = \{x | x \text{ is an automobile with a CB radio}\}$$

This is read "S is the set of all x such that x is an automobile with a CB radio." Similarly, if S is the set of odd positive integers, we write

$$S = \{2k + 1 | k = 0, 1, 2, \ldots\}$$

How we formulate the sample space for a given situation will depend on the problem at hand. If an experiment consists of one roll of a die and we are interested in the number on the face turned up, we would use the sample space

$$S_1 = \{1, 2, 3, 4, 5, 6\}$$

If we are interested only in whether the number is even or odd, we would use the sample space

$$S_2 = \{\text{even}, \text{odd}\}$$

This example also illustrates the fact that different sample spaces may be used to describe one and the same experiment. Both S_1 and S_2 represent outcomes for an experiment consisting of one roll of a die. Which one is appropriate depends on the problem at hand, but S_1, clearly, provides more information than S_2. If we know which element in S_1 occurs, we can tell which element in S_2 occurs; however, if we know which element in S_2 occurs, we cannot tell which element in S_1 occurs. *Generally speaking, it is desirable to use a sample space whose elements cannot be "subdivided" into more primitive or more elementary kinds of outcomes; that is, an element of a sample space should not represent two or more outcomes which are distinguishable in some fashion.*

EXAMPLE 2.2

Describe the sample space for an experiment in which we roll a pair of dice, one red and one green.

Solution

The sample space which provides the most information consists of the 36 points given by

$$S_1 = \{(x, y)|x = 1, 2, \ldots, 6; y = 1, 2, \ldots, 6\}$$

where x represents the number of points rolled with the red die and y represents the number of points rolled with the green die. A second sample space, adequate for some purposes though generally less desirable as it provides less information, might be written as

$$S_2 = \{2, 3, 4, \ldots, 12\}$$

where the 11 elements represent the possible totals rolled with the pair of dice.

Sample spaces are usually classified according to the number of elements which they contain. In the preceding example the sample spaces S_1 and S_2 contained a **finite** number of elements, but if a coin is flipped until a head appears for the first time, this could happen on the first flip, the second flip, the third flip, the fourth flip, . . . , and there are infinitely many possibilities. For this experiment we obtain the sample space

$$S = \{\text{H}, \text{TH}, \text{TTH}, \text{TTTH}, \text{TTTTH}, \ldots\}$$

with an unending sequence of elements. But even here the number of elements can be matched one-to-one with the whole numbers, and in this sense the sample space is said to be **countable**. If a sample space contains a finite number of elements, or an infinite though countable number of elements, it is said to be **discrete**.

The outcomes of some experiments are neither finite nor countably infinite. Such is the case, for example, when one conducts an investigation to determine the distance that a certain make of car will travel over a prescribed test course on 5 liters of gasoline. If we assume that distance is a variable that can be measured to any desired degree of accuracy, there is an infinity of possible distances that cannot be matched one-to-one with the whole numbers. Also, if one wants to record the length of time it takes for two chemicals to react, the possible lengths of time making up the sample space are infinite in number and not countable. Thus, sample spaces need not be discrete. If a sample space contains an infinite number of sample points constituting a continuum, such as all the points on a line segment or all the points in a plane, it is said to be **continuous**.

Continuous sample spaces arise in practice whenever the outcomes of experiments are measurements of physical properties such as temperature, speed, pressure, length, . . . , that are measured on continuous scales.

2.3 EVENTS

In many problems we are interested in the occurrence of events that are not given directly by a specific element of a sample space.

EXAMPLE 2.3

With reference to the sample space S_1 on page 26, describe the event A that the number of points rolled with a die is divisible by 3.

Solution

The number of points rolled will be divisible by 3 if the outcome is 3 or 6; namely, if the outcome is an element of the subset $A = \{3, 6\}$ of the sample space.

EXAMPLE 2.4

With reference to the sample space S_1 of Example 2.2, describe the event B that the total number of points rolled with a pair of dice is 7.

Solution

This will occur if the outcome is an element of the subset

$$B = \{(1, 6), (2, 5), (3, 4), (4, 3), (5, 2), (6, 1)\}$$

of the sample space S_1. Note that in Figure 2.1 the event of rolling a total of 7 with a pair of dice is represented by the set of points inside the region bounded by the dotted line.

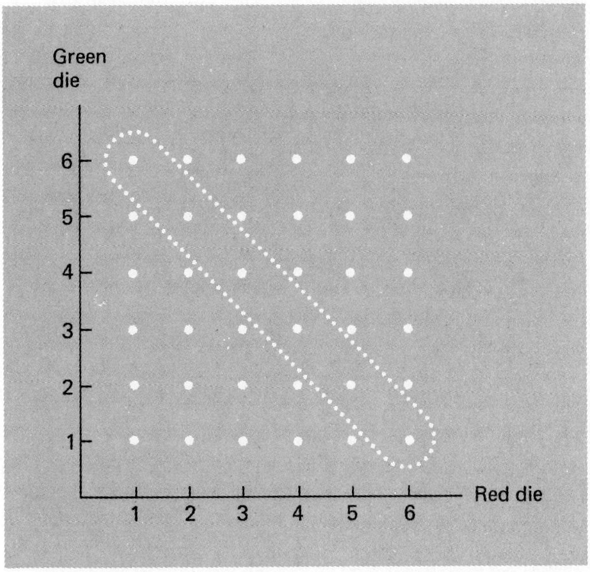

Figure 2.1 The event of rolling a total of seven with a pair of dice.

In the same way, any event can be assigned a collection of sample points, which constitute a subset of an appropriate sample space. This subset represents all the elements for which the event is true and, in probability theory, we identify the subset with the event. Thus, by definition, an **event** is a subset of a sample space.

EXAMPLE 2.5

If someone shoots at a target three times and we are interested only in whether each shot is a hit or a miss, describe the sample space S, the event M that the person will miss the target in each of the three shots, and the event H that the person will hit the target once and miss it twice.

Solution

If we label the outcome of each shot 0 for a miss and 1 for a hit, the eight sample points of S might be displayed as the three-dimensional geometric configuration of Figure 2.2. Then, the subset $M = \{(0, 0, 0)\}$ represents the event of missing the target in each of the three shots, and the subset $H = \{(1, 0, 0), (0, 1, 0), (0, 0, 1)\}$ represents the event of hitting the target once and missing it twice. Of course, the entire sample space represents the event of getting either 0, 1, 2, or 3 hits in three shots, an event that is certain to occur.

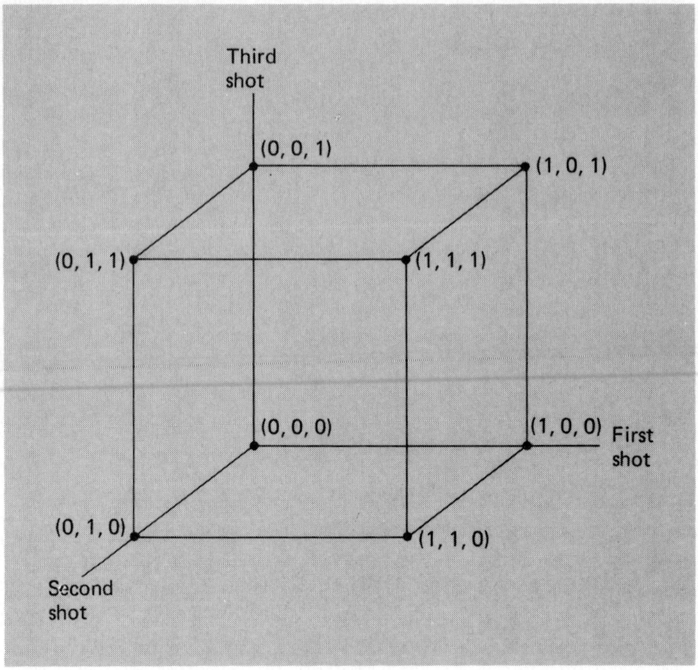

Figure 2.2 Sample space for three shots at a target.

EXAMPLE 2.6

Construct a sample space for the length of the useful life of a certain electronic component and indicate the subset which represents the event F that the component fails before the end of the sixth year.

Solution

If t is the length of the component's useful life in years, the sample space may be written $S = \{t|t \geqslant 0\}$, and the subset $F = \{t|0 \leqslant t < 6\}$ is the event that the component fails before the end of the sixth year.

According to our definition, any event is a subset of an appropriate sample space, but it should be observed that the converse is not necessarily true. For discrete sample spaces all subsets are events, but in the continuous case some rather abstruse point sets must be excluded for mathematical reasons. This is discussed further in some of the more advanced texts listed among the references on page 69, but it is of no consequence so far as the work of this book is concerned.

In many problems of probability we are interested in events which are actually combinations of two or more events, formed by taking **unions**, **inter-**

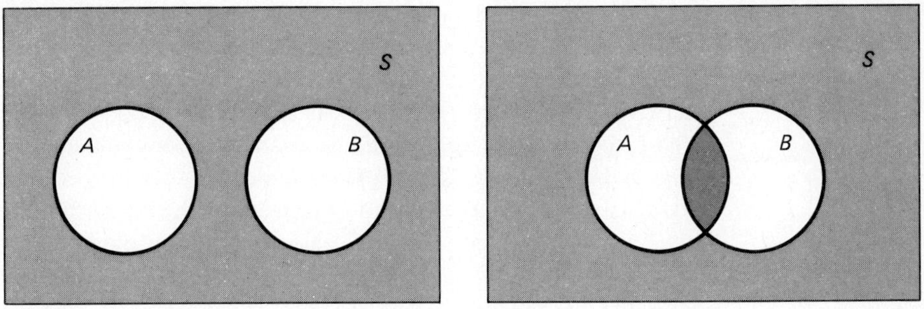

Figure 2.3 Diagrams showing that events A and B are mutually exclusive.

sections, and **complements**. Although the reader must surely be familiar with these terms, let us review briefly that if A and B are any two subsets of a sample space S, their union $A \cup B$ is the subset of S which contains all the elements that are either in A, in B, or in both; their intersection $A \cap B$ is the subset of S which contains all the elements that are in both A and B; and the complement A' of A is the subset of S which contains all the elements of S that are not in A. The various rules which control the formation of unions, intersections, and complements are summarized in Appendix I at the end of the book.

Sample spaces and events, particularly relationships among events, are often depicted by means of **Venn diagrams**, in which the sample space is represented by a rectangle, while events are represented by regions within the rectangle, usually by circles or parts of circles. For instance, the first diagram of Figure 2.3 serves to indicate that events A and B are **mutually exclusive**, namely, that they cannot both occur simultaneously. This same concept is conveyed by the second diagram of Figure 2.3, where the region representing $A \cap B$ is shaded darker to indicate that it is empty.† Symbolically, we write $A \cap B = \varnothing$ when A and B are mutually exclusive; since they have no elements in common, their intersection equals the **empty set** $\varnothing$.

APPLIED EXERCISES

1. If $S = \{1, 2, 3, 4, 5, 6, 7, 8, 9\}$, $A = \{1, 3, 5, 7\}$, $B = \{6, 7, 8, 9\}$, $C = \{2, 4, 8\}$, and $D = \{1, 5, 9\}$, list the elements of the subsets of S corresponding to the following events:

(a) $A' \cap B$; **(b)** $(A' \cap B) \cap C$; **(c)** $B' \cup C$;
(d) $(B' \cup C) \cap D$; **(e)** $A' \cap C$; **(f)** $(A' \cap C) \cap D$.

† It is the custom nowadays to refer to all these diagrams as Venn diagrams, although, strictly speaking, the term was originally meant to apply only when the circles intersect as in the second diagram of Figure 2.3, or as in the diagram of Figure 2.6 on page 45.

2. An electronics firm plans to build a research laboratory in Southern California, and its management has to decide between sites in Los Angeles, San Diego, Long Beach, Pasadena, Santa Barbara, Anaheim, Santa Monica, and Westwood. If A represents the event that they will choose a site in San Diego or Santa Barbara, B represents the event that they will choose a site in San Diego or Long Beach, C represents the event that they will choose a site in Santa Barbara or Anaheim, and D represents the event that they will choose a site in Los Angeles or Santa Barbara, list the elements of each of the following subsets of the sample space, which consists of the eight site selections:

(a) A'; **(b)** D'; **(c)** $C \cap D$;
(d) $B \cap C$; **(e)** $B \cup C$; **(f)** $A \cup B$;
(g) $C \cup D$; **(h)** $(B \cup C)'$; **(i)** $B' \cap C'$.

3. Among the eight cars which a dealer has in his showroom, Car 1 is new, has air-conditioning, power steering, and bucket seats, Car 2 is one year old, has air-conditioning, but neither power steering nor bucket seats, Car 3 is two years old, has air-conditioning and power steering, but no bucket seats, Car 4 is three years old, has air-conditioning, but neither power steering nor bucket seats, Car 5 is new, has no air-conditioning, no power steering, and no bucket seats, Car 6 is one year old, has power steering, but neither air-conditioning nor bucket seats, Car 7 is two years old, has no air-conditioning, no power steering, and no bucket seats, and Car 8 is three years old, has no air-conditioning, but power steering as well as bucket seats. If a customer buys one of these cars, and the event that he chooses a new car, for example, is represented by the set {Car 1, Car 5}, indicate similarly the sets which represent the events that

(a) he chooses a car without air-conditioning;
(b) he chooses a car without power steering;
(c) he chooses a car with bucket seats;
(d) he chooses a car that is either two or three years old.

Also state in words what kind of car he will choose, if his choice is given by

(e) the complement of the set of part (a);
(f) the union of the sets of parts (b) and (c);
(g) the intersection of the sets of parts (c) and (d);
(h) the intersection of the sets of parts (f) and (g).

4. If Ms. Brown buys one of the houses advertised for sale in a Seattle newspaper (on a given Sunday), T is the event that the house has three or more baths, U is the event that it has a fireplace, V is the event that it costs more than $60,000, and W is the event that it is new, describe (in words) each of the following events:

(a) T'; **(b)** U'; **(c)** V';
(d) W'; **(e)** $T \cap U$; **(f)** $T \cap V$;

(g) $U' \cap V$; (h) $V \cup W$; (i) $V' \cup W$;
(j) $T \cup U$; (k) $T \cup V$; (l) $V \cap W$.

5. A resort hotel has two station wagons, which it uses to shuttle its guests to and from the airport. If the larger of the two station wagons can carry 5 passengers and the smaller can carry 4 passengers, the point (0, 3) represents the event that at a given moment the larger station wagon is empty while the smaller one has 3 passengers, the point (4, 2) represents the event that at the given moment the larger station wagon has 4 passengers while the smaller one has 2 passengers, . . . , draw a figure showing the 30 points of the corresponding sample space. Also, if E stands for the event that at least one of the station wagons is empty, F stands for the event that together they carry 2, 4, or 6 passengers, and G stands for the event that each carries the same number of passengers, list the points of the sample space which correspond to each of the following events:

(a) E; (b) F; (c) G;
(d) $E \cup F$; (e) $E \cap F$; (f) $F \cup G$;
(g) $E \cup F'$; (h) $E \cap G'$; (i) $F' \cap E'$.

6. A coin is tossed once. Then, if it comes up heads, a die is thrown once; if it comes up tails, it is tossed twice more. Using the notation in which (H, 2), for example, denotes the event that the coin comes up heads and then the die comes up 2, and (T, T, T) denotes the event that the coin comes up tails three times in a row, list

(a) the ten elements of the sample space S;
(b) the elements of S corresponding to event A that exactly one head occurs;
(c) the elements of S corresponding to event B that at least two tails occur or a number greater than 4 occurs.

7. An electronic game contains three components arranged in the series-parallel circuit shown in Figure 2.4. At any given time, each component may or may not be operative, and the game will operate only if there is a continuous circuit from P to Q. Let A be the event that the game will operate; let B be the event

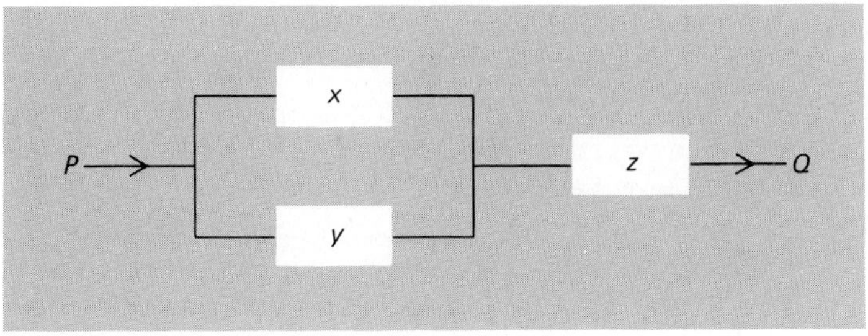

Figure 2.4 Diagram for Exercise 7.

that the game will operate though component x is not operative; and let C be the event that the game will operate though component y is not operative. Using the notation in which (0, 0, 1), for example, denotes that component z is operative but components x and y are not,

(a) list the elements of the sample space S and also the elements of S corresponding to events A, B, and C;
(b) determine which pairs of events are mutually exclusive.

8. An experiment consists of rolling a die until a 3 appears. Describe the sample space and determine

(a) how many elements of the sample space correspond to the event that the 3 appears on the kth roll of the die;
(b) how many elements of the sample space correspond to the event that the 3 appears not later than the kth roll of the die.

9. If $S = \{x|0 < x < 10\}$, $M = \{x|3 < x \leq 8\}$, and $N = \{x|5 < x < 10\}$, find

(a) $M \cup N$; **(b)** $M \cap N$; **(c)** $M \cap N'$; **(d)** $M' \cup N$.

10. Symbolically, describe the sample space S consisting of all points (x, y) on or in the interior of a circle of radius 3 centered at the point $(2, -3)$.

11. In a group of 200 college students 138 are enrolled in a course in psychology, 115 are enrolled in a course in sociology, and 91 are enrolled in both. How many of these students are not enrolled in either course? (*Hint*: Draw a suitable Venn diagram and fill in the numbers associated with the various regions.)

12. A market research organization claims that among 500 shoppers interviewed, 308 regularly buy Product X, 266 regularly buy Product Y, 103 regularly buy both, and 59 buy neither on a regular basis. Using a Venn diagram and filling in the number of shoppers associated with the various regions, check whether the results of this study should be questioned.

13. Among 120 visitors to Disneyland, 74 stayed for at least three hours, 86 spent at least \$8.00, 64 went on the Matterhorn ride, 60 stayed for at least three hours and spent at least \$8.00, 52 stayed for at least three hours and went on the Matterhorn ride, 54 spent at least \$8.00 and went on the Matterhorn ride, and 48 stayed for at least three hours, spent at least \$8.00, and went on the Matterhorn ride. Drawing a Venn diagram with three circles (like that of Figure 2.6 on page 45) and filling in the numbers associated with the various regions, find

(a) how many of the visitors to Disneyland stayed for at least three hours, spent at least \$8.00, but did not go on the Matterhorn ride;
(b) how many of the visitors to Disneyland went on the Matterhorn ride, but stayed less than three hours and spent less than \$8.00;
(c) how many of the visitors to Disneyland stayed less than three hours, spent at least \$8.00, but did not go on the Matterhorn ride.

2.4 THE PROBABILITY OF AN EVENT

To formulate the postulates of probability we shall follow the practice of denoting events by means of capital letters, and we shall write the probability of event A as $P(A)$, the probability of event B as $P(B)$, and so forth. As before, we shall denote the set of all possible outcomes, the sample space, by the letter S.

Probabilities are values of a set function, also called a **probability measure**, for as we shall see, this function assigns real numbers to the various subsets of a sample space S. As we shall formulate them here, the postulates of probability apply only when the sample space S is discrete.

POSTULATE 1 The probability of an event is a non-negative real number; that is, $P(A) \geq 0$ for any subset A of S.

POSTULATE 2 $P(S) = 1$

POSTULATE 3 If $A_1, A_2, A_3, \ldots$, is a finite or infinite sequence of mutually exclusive events of S, then

$$P(A_1 \cup A_2 \cup A_3 \cup \cdots) = P(A_1) + P(A_2) + P(A_3) + \cdots$$

Postulates *per se* require no proof, but if the resulting theory is to be applied, we must show that the postulates are satisfied when we give probabilities a "real" meaning. Let us illustrate this here in connection with the frequency interpretation; the relationship between the postulates and the classical probability concept will be discussed on page 39, while the relationship between the postulates and subjective probabilities is left for the reader to examine in Exercises 12 and 17 on pages 44 and 46.

Since proportions are always positive or zero, the first postulate is in complete agreement with the frequency interpretation. The second postulate states indirectly that certainty is identified with a probability of 1—after all, it is always assumed that one of the possibilities in S must occur, and it is to this certain event that we assign a probability of 1. So far as the frequency interpretation is concerned, a probability of 1 implies that the event in question will occur 100 percent of the time, or in other words, that it is certain to occur.

Taking the third postulate in the simplest case, namely, for two mutually exclusive events A_1 and A_2, it can easily be seen that it is satisfied by the frequency interpretation. If one event occurs, say, 28 percent of the time, another event occurs 39 percent of the time, and the two events cannot both occur at the same time (that is, they are mutually exclusive), then one or the other will occur $28 + 39 = 67$ percent of the time. Thus, the third postulate is satisfied, and the same kind of argument applies when there are more than two mutually exclusive events.

Before we study some of the immediate consequences of the postulates of probability, let us emphasize the point that the three postulates do not tell us how to assign probabilities to events, they merely restrict the ways in which it can be done.

EXAMPLE 2.7

For each of the following, explain why it is not a permissible way of assigning probabilities to the four possible and mutually exclusive outcomes A, B, C, and D of an experiment:
(a) $P(A) = 0.12$, $P(B) = 0.63$, $P(C) = 0.45$, $P(D) = -0.20$;
(b) $P(A) = \frac{9}{120}$, $P(B) = \frac{45}{120}$, $P(C) = \frac{27}{120}$, $P(D) = \frac{46}{120}$.

Solution

In (a) we find that $P(D) = -0.20$ violates Postulate 1, and in (b) we get $P(S) = P(A \cup B \cup C \cup D) = \frac{9}{120} + \frac{45}{120} + \frac{27}{120} + \frac{46}{120} = \frac{127}{120}$, which violates Postulate 2.

Of course, in actual practice probabilities are assigned on the basis of past experience, on the basis of a careful analysis of all underlying conditions, or on the basis of assumptions—sometimes, the assumption that all possible outcomes are equiprobable.

To assign a probability measure to a sample space, it is not necessary to specify the probability of each possible subset. This is fortunate, for a sample space with as few as 20 possible outcomes has already $2^{20} = 1{,}048{,}576$ subsets [the general formula follows directly from part (a) of Exercise 8 on page 19], and the number of subsets grows very rapidly when there are 50 possible outcomes, 100 possible outcomes, or more. Instead of listing the probabilities of all possible subsets, we often list the probabilities of the individual outcomes, or sample points of S, and then make use of the following theorem:

THEOREM 2.1 If A is an event in a discrete sample space S, then $P(A)$ equals the sum of the probabilities of the individual outcomes comprising A.

Proof. Let $O_1, O_2, O_3, \ldots,$ be the finite or infinite sequence of outcomes which comprise the event A. Thus,

$$A = O_1 \cup O_2 \cup O_3 \cdots$$

and since the individual outcomes, the O's, are by definition mutually exclusive, the third postulate of probability yields

$$P(A) = P(O_1) + P(O_2) + P(O_3) + \cdots$$

This completes the proof.

For Theorem 2.1 to be useful, we must be able to assign probabilities to the individual outcomes of experiments. How this may be done in some special situations, is illustrated by the following examples:

EXAMPLE 2.8

If a balanced coin is tossed twice, what is the probability of getting at least one head?

Solution

The sample space for this experiment is

$$S = \{\text{HH, HT, TH, TT}\}$$

Since the coin is balanced, we assume that each of these outcomes is equally likely to occur, and we therefore assign a probability of $\frac{1}{4}$ to each sample point. If A is the event that we will get at least one head, then $A = \{\text{HH, HT, TH}\}$ and

$$\begin{aligned} P(A) &= P(\text{HH}) + P(\text{HT}) + P(\text{TH}) \\ &= \tfrac{1}{4} + \tfrac{1}{4} + \tfrac{1}{4} \\ &= \tfrac{3}{4} \end{aligned}$$

EXAMPLE 2.9

A die is loaded in such a way that each odd number is twice as likely to occur as each even number. If E is the event that a number greater than 3 occurs on a single toss of the die, find $P(E)$.

Solution

The sample space is $S = \{1, 2, 3, 4, 5, 6\}$. If we assign a probability of w to each even number and a probability of $2w$ to each odd number, we find that $2w + w + 2w + w + 2w + w = 9w = 1$ in accordance with Postulate 2. Thus, $w = \frac{1}{9}$ and

$$P(E) = \tfrac{1}{9} + \tfrac{2}{9} + \tfrac{1}{9} = \tfrac{4}{9}$$

If a sample space is discrete but infinite, probabilities will have to be assigned to the individual outcomes by means of some mathematical rule.

EXAMPLE 2.10

If $O_1, O_2, O_3, \ldots,$ represent the infinitely many outcomes of an experiment, verify that a permissible probability measure is given by the rule

$$P(O_i) = \left(\tfrac{1}{2}\right)^i \qquad \text{for } i = 1, 2, 3, \ldots$$

Solution

Since the probabilities are all positive, it remains to be shown that $P(S) = 1$. Getting

$$P(S) = \tfrac{1}{2} + \tfrac{1}{4} + \tfrac{1}{8} + \tfrac{1}{16} + \cdots$$

and making use of the formula for the sum of the terms in an infinite geometric progression, we find that

$$P(S) = \frac{\frac{1}{2}}{1 - \frac{1}{2}} = 1$$

(To be rigorous in a situation like this, the word "sum" in Theorem 2.1 will have to be interpreted so that it includes the value of an infinite series.)

As we shall see in Chapter 5, the probability measure given here would be appropriate if O_i represents the event that a person flipping a balanced coin will get a head for the first time on the ith try. Thus, the probability that the first head will come on the third, fourth, or fifth try is $\left(\frac{1}{2}\right)^3 + \left(\frac{1}{2}\right)^4 + \left(\frac{1}{2}\right)^5 = \frac{7}{32}$, and the probability that the first head will come on an odd-numbered try is

$$\left(\tfrac{1}{2}\right)^1 + \left(\tfrac{1}{2}\right)^3 + \left(\tfrac{1}{2}\right)^5 + \cdots = \frac{\frac{1}{2}}{1 - \frac{1}{4}} = \frac{2}{3}$$

where we again made use of the formula for the sum of the terms in an infinite geometric progression.

If an experiment is such that we can assume equal probabilities for the sample points of S, as was the case in Example 2.8, we can take advantage of the following special case of Theorem 2.1:

THEOREM 2.2 If an experiment can result in any one of N different equally likely outcomes, and if n of these outcomes together constitute event A, then the probability of event A is

$$P(A) = \frac{n}{N}$$

Proof. Let $O_1, O_2, \ldots, O_N$ represent the individual outcomes in S, each with probability $\frac{1}{N}$. If event A is the union of n of these mutually exclusive outcomes, and it does not matter which ones, then

$$\begin{aligned} P(A) &= P(O_1 \cup O_2 \cup \cdots \cup O_n) \\ &= P(O_1) + P(O_2) + \cdots + P(O_n) \\ &= \underbrace{\frac{1}{N} + \frac{1}{N} + \cdots + \frac{1}{N}}_{n \text{ terms}} \\ &= \frac{n}{N} \end{aligned}$$

Observe that the formula $P(A) = \frac{n}{N}$ of Theorem 2.2 is identical with that of the classical probability concept, which we gave on page 24. Indeed, what we have shown here is that the classical probability concept is consistent with the postulates of probability—it follows from them in the special case where the individual outcomes are all equiprobable.

2.5 SOME RULES OF PROBABILITY

By using the three postulates of probability, we can derive many other rules which have important applications. Among the immediate consequences of the postulates, we prove the following theorems:

THEOREM 2.3 If A and A' are complementary events in a sample space S, then

$$P(A') = 1 - P(A)$$

Proof. In the second and third steps we make use of the definition of a complement in Appendix I, according to which A and A' are mutually exclusive and $A \cup A' = S$:

$$\begin{aligned} 1 &= P(S) && \text{(by Postulate 2)} \\ &= P(A \cup A') \\ &= P(A) + P(A') && \text{(by Postulate 3)} \end{aligned}$$

Therefore, $P(A') = 1 - P(A)$.

In connection with the frequency interpretation, this result implies that if an event occurs, say, 37 percent of the time, it does not occur 63 percent of the time.

THEOREM 2.4 $P(\varnothing) = 0$ for any sample space S.

Proof. Since $S \cup \varnothing = S$ and the events S and $\varnothing$ are mutually exclusive (see Appendix I at the end of the book), it follows that

$$\begin{aligned} P(S) &= P(S \cup \varnothing) \\ &= P(S) + P(\varnothing) \qquad \text{(by Postulate 3)} \end{aligned}$$

and, hence, that $P(\varnothing) = 0$.

It is important to note that it does not follow from $P(A) = 0$ that A is necessarily an empty set. In practice, we often assign a probability of 0 to events which, in colloquial terms, would not happen in a million years. For instance, there is the classical example that we assign a probability of 0 to the event that a monkey set loose on a typewriter will type Plato's *Republic* word for word without a mistake. As we shall see in Chapters 3 and 6, the fact that $P(A) = 0$ does not imply $A = \varnothing$ is of relevance, especially, in the continuous case.

THEOREM 2.5 If A and B are events in a sample space S and $A \subset B$, then $P(A) \leq P(B)$.

Proof. Since $A \subset B$, we can write

$$B = A \cup (A' \cap B)$$

as can easily be verified by means of a Venn diagram. Then, since A and $A' \cap B$ are mutually exclusive, we get

$$\begin{aligned} P(B) &= P(A) + P(A' \cap B) \qquad \text{(by Postulate 3)} \\ &\geq P(A) \qquad \text{(by Postulate 1)} \end{aligned}$$

In words, this theorem states that if event A is a subset of event B, then $P(A)$ is no greater than $P(B)$. For instance, the probability of drawing a heart from an ordinary deck of 52 playing cards is no greater than the probability of drawing a red card, namely, $\frac{1}{4}$ compared to $\frac{1}{2}$.

THEOREM 2.6 $0 \leq P(A) \leq 1$ for any event A.

Proof. Using Theorem 2.5 and the fact that $\varnothing \subset A \subset S$ for any event A in S, we have

$$P(\varnothing) \leq P(A) \leq P(S)$$

Then, $P(\varnothing) = 0$ and $P(S) = 1$ leads to the result that

$$0 \leq P(A) \leq 1$$

The third postulate of probability is sometimes referred to as a **special addition rule**; it is special in that the events $A_1, A_2, A_3, \ldots,$ must all be mutually exclusive. For two events A and B there exists the more general **addition rule**:

THEOREM 2.7 If A and B are any two events in a sample space S, then

$$P(A \cup B) = P(A) + P(B) - P(A \cap B)$$

Proof. Assigning the probabilities a, b, and c to the mutually exclusive events $A \cap B$, $A \cap B'$, and $A' \cap B$ as in the Venn diagram of Figure 2.5, we find that

$$\begin{aligned} P(A \cup B) &= a + b + c \\ &= (a + b) + (c + a) - a \\ &= P(A) + P(B) - P(A \cap B) \end{aligned}$$

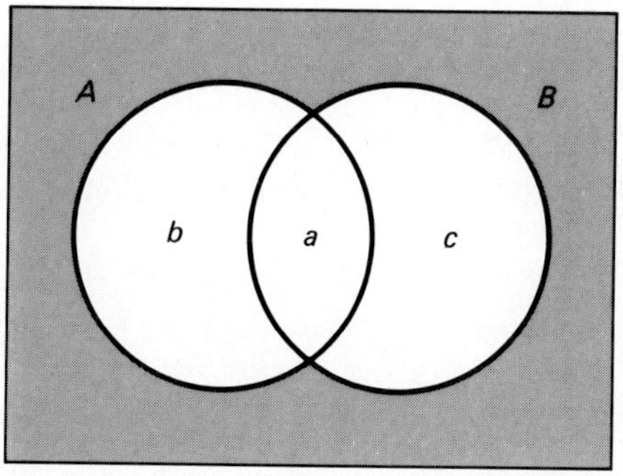

Figure 2.5 Venn diagram for proof of Theorem 2.7.

If the probabilities are, respectively, 0.86, 0.35, and 0.29 that a family (randomly chosen for a sample survey in a large metropolitan area) will own a color television set, a black and white set, or both kinds of sets, what is the probability that such a family will own either kind of set?

Solution

If A is the event that such a family owns a color television set and B is the event that it owns a black and white set, we are given $P(A) = 0.86$, $P(B) = 0.35$, $P(A \cap B) = 0.29$, and substitution into the formula of Theorem 2.7 yields

$$\begin{aligned} P(A \cup B) &= 0.86 + 0.35 - 0.29 \\ &= 0.92 \end{aligned}$$

EXAMPLE 2.12

If the probabilities are, respectively, 0.23, 0.24, and 0.38 that a car stopped at a road block will have faulty brakes, badly worn tires, or faulty brakes and/or badly worn tires, what is the probability that such a car will have both faulty brakes and badly worn tires?

Solution

If B is the event that such a car will have faulty brakes and T is the event that it will have badly worn tires, we are given $P(B) = 0.23$, $P(T) = 0.24$, $P(B \cup T) = 0.38$, and substitution into the formula of Theorem 2.7 yields

$$0.38 = 0.23 + 0.24 - P(B \cap T)$$

Solving this equation for $P(B \cap T)$, we get

$$P(B \cap T) = 0.23 + 0.24 - 0.38 = 0.09$$

Repeatedly applying Theorem 2.7, this addition rule can be generalized so that it applies to any given number of events; for three events we get

THEOREM 2.8 If A, B, and C are any three events in a sample space S, then

$$\begin{aligned} P(A \cup B \cup C) = P(A) + P(B) + P(C) - P(A \cap B) - P(A \cap C) - \\ P(B \cap C) + P(A \cap B \cap C) \end{aligned}$$

Proof. Writing $A \cup B \cup C$ as $A \cup (B \cup C)$ and using Theorem 2.7 twice, once for $P[A \cup (B \cup C)]$ and once for $P(B \cup C)$, we get

$$\begin{aligned} P(A \cup B \cup C) &= P[A \cup (B \cup C)] \\ &= P(A) + P(B \cup C) - P[A \cap (B \cup C)] \\ &= P(A) + P(B) + P(C) - P(B \cap C) - P[A \cap (B \cup C)] \end{aligned}$$

From the first distributive law (see page 501), it follows that

$$\begin{aligned} P[A \cap (B \cup C)] &= P[(A \cap B) \cup (A \cap C)] \\ &= P(A \cap B) + P(A \cap C) - P[(A \cap B) \cap (A \cap C)] \\ &= P(A \cap B) + P(A \cap C) - P(A \cap B \cap C) \end{aligned}$$

and, hence, that

$$P(A \cup B \cup C) = P(A) + P(B) + P(C) - P(A \cap B) - P(A \cap C) - P(B \cap C) + P(A \cap B \cap C)$$

In Exercise 8 on page 44 the reader will be asked to give an alternative proof of Theorem 2.8, based on the method of proof used in the text for the proof of Theorem 2.7.

EXAMPLE 2.13

Suppose that if a person visits his dentist, the probability that he will have his teeth cleaned is 0.44, the probability that he will have a cavity filled is 0.24, the probability that he will have a tooth extracted is 0.21, the probability that he will have his teeth cleaned and a cavity filled is 0.08, the probability that he will have his teeth cleaned and a tooth extracted is 0.11, the probability that he will have a cavity filled and a tooth extracted is 0.07, and the probability that he will have his teeth cleaned, a cavity filled, and a tooth extracted is 0.03. What is the probability that a person visiting his dentist will have at least one of these things done to him?

Solution

If C is the event that the person will have his teeth cleaned, F is the event that he will have a cavity filled, and E is the event that he will have a tooth extracted, we are given $P(C) = 0.44$, $P(F) = 0.24$, $P(E) = 0.21$, $P(C \cap F) = 0.08$, $P(C \cap E) = 0.11$, $P(F \cap E) = 0.07$, $P(C \cap F \cap E) = 0.03$, and substitution into the formula yields

$$\begin{aligned} P(C \cup F \cup E) &= 0.44 + 0.24 + 0.21 - 0.08 - 0.11 - 0.07 + 0.03 \\ &= 0.66 \end{aligned}$$

THEORETICAL EXERCISES

1. Refer to parts (c) and (d) of Exercise 3 on page 502 to show that
(a) $P(A) \geq P(A \cap B)$;
(b) $P(A) \leq P(A \cup B)$.

2. Show that $P(A \cap B') = P(A) - P(A \cap B)$.

3. Show that $P(A' \cap B') = 1 - P(A) - P(B) + P(A \cap B)$.

4. The event that "A or B but not both" will occur can be written $(A \cap B') \cup (A' \cap B)$. Express the probability of this event in terms of $P(A)$, $P(B)$, and $P(A \cap B)$.

5. Use Theorem 2.7 to show that
(a) $P(A \cap B) \leq P(A) + P(B)$;
(b) $P(A \cap B) \geq P(A) + P(B) - 1$.

6. Show that if $P(A) = P(B) = P(C) = 1$, then $P(A \cap B \cap C) = 1$. [*Hint*: Assume that $P(A \cap B \cap C) \neq 1$ and show that this leads to a contradiction.]

7. Give an alternative proof of Theorem 2.7 by making use of the relationships $A \cup B = A \cup (A' \cap B)$ and $B = (A \cap B) \cup (A' \cap B)$.

8. By assigning the probabilities a, b, c, d, e, f, and g as in the Venn diagram of Figure 2.6, duplicate the method by which we proved Theorem 2.7 to prove Theorem 2.8.

9. Duplicate the method of proof of Exercise 8 to show that $P(A \cup B \cup C \cup D) = P(A) + P(B) + P(C) + P(D) - P(A \cap B) - P(A \cap C) - P(A \cap D) - P(B \cap C) - P(B \cap D) - P(C \cap D) + P(A \cap B \cap C) + P(A \cap B \cap D) + P(A \cap C \cap D) + P(B \cap C \cap D) - P(A \cap B \cap C \cap D)$. (*Hint*: Divide each of the eight regions of the Venn diagram of Figure 2.6 into two parts, one inside D and one outside D, and assign the resulting regions the probabilities a, b, c, . . . , n, o, and p.)

10. Prove by induction that

$$P(E_1 \cup E_2 \cup \cdots \cup E_n) \leq \sum_{i=1}^{n} P(E_i)$$

for any finite sequence of events E_1, E_2, . . . , and E_n.

11. The **odds** that an event will occur are given by the ratio of the probability that the event will occur to the probability that it will not occur; they are usually quoted in terms of positive integers having no common factor. If the odds that an event will occur are a to b, show that its probability is $p = \dfrac{a}{a+b}$.

12. Subjective probabilities may be determined by exposing persons to risk-taking situations and finding the odds at which they would consider it fair to bet on the outcome. The odds are then converted into probabilities by means of the formula of the preceding exercise. For instance, if a person feels that 3

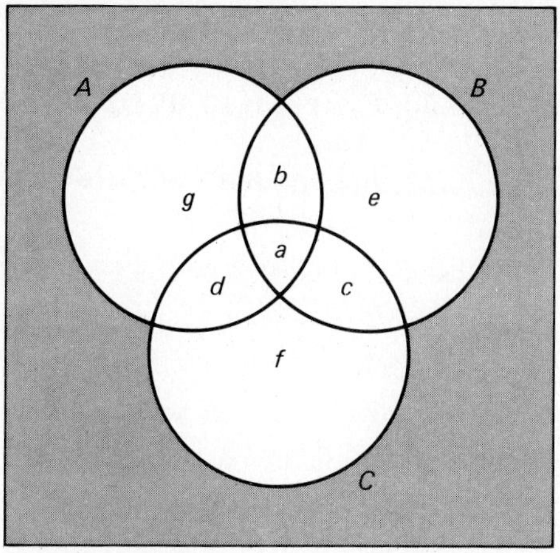

Figure 2.6 Diagram for Exercise 8.

to 2 are fair odds that a business venture will succeed (or that it would be fair to bet \$30 against \$20 that it will succeed), the probability is $\frac{3}{3+2} = 0.6$ that the business venture will succeed.

(a) Show that if subjective probabilities are determined in this way, they satisfy Postulate 1 on page 35.

(b) If a person feels that a to b are fair odds that event A will occur, this implies that the odds are b to a that event A will not occur. Thus, $P(A) = \frac{a}{a+b}$ and $P(A') = \frac{b}{b+a}$, and it should be observed that the sum of these two probabilities is 1. Under what condition can this argument be used to show that subjective probabilities determined in this way satisfy Postulate 2 on page 35?

See also Exercise 17 on page 46.

APPLIED EXERCISES

13. An experiment has five possible outcomes, A, B, C, D, and E. Check for each of the following whether it constitutes a permissible assignment of probability and explain your answers:

(a) $P(A) = 0.20$, $P(B) = 0.20$, $P(C) = 0.20$, $P(D) = 0.20$, and $P(E) = 0.20$;

(b) $P(A) = 0.21$, $P(B) = 0.26$, $P(C) = 0.58$, $P(D) = 0.01$, and $P(E) = 0.06$;

(c) $P(A) = 0.18$, $P(B) = 0.19$, $P(C) = 0.20$, $P(D) = 0.21$, and $P(E) = 0.22$;

(d) $P(A) = 0.10$, $P(B) = 0.30$, $P(C) = 0.10$, $P(D) = 0.60$, and $P(E) = -0.10$;

(e) $P(A) = 0.23$, $P(B) = 0.12$, $P(C) = 0.05$, $P(D) = 0.50$, and $P(E) = 0.08$.

14. If A and B are mutually exclusive events, $P(A) = 0.37$ and $P(B) = 0.44$, find

(a) $P(A')$; **(b)** $P(B')$; **(c)** $P(A \cup B)$;
(d) $P(A \cap B)$; **(e)** $P(A \cap B')$; **(f)** $P(A' \cap B')$.

15. Explain why there must be a mistake in each of the following statements:

(a) The probability that Jean will pass the bar examination is 0.66 and the probability that she will not pass is -0.34.

(b) The probability that the home team will win an upcoming football game is 0.77, the probability that it will tie the game is 0.08, and the probability that it will win or tie the game is 0.95.

(c) The probabilities that a secretary will make 0, 1, 2, 3, 4, or 5 *or more* mistakes in typing a report are, respectively, 0.12, 0.25, 0.36, 0.14, 0.09, and 0.07.

(d) The probabilities that a bank will get 0, 1, 2, or 3 *or more* bad checks on any given day are, respectively, 0.08, 0.21, 0.29, and 0.40.

16. Supposing that each of the 30 points of the sample space of Exercise 5 on page 33 is assigned the probability $\frac{1}{30}$, find the probabilities that at a given moment

(a) at least one of the station wagons is empty;

(b) each of the two station wagons carries the same number of passengers;

(c) the larger station wagon carries more passengers than the smaller station wagon;

(d) together they carry at least six passengers.

17. If subjective probabilities are determined as in Exercise 12, it does not follow that Postulate 3 on page 35 must necessarily be satisfied. However, proponents of the subjective probability concept generally impose this postulate as a **consistency criterion**; in other words, they regard subjective probabilities which do not satisfy Postulate 3 as inconsistent.

(a) The branch manager of a bank feels that the odds are 7 to 5 against her getting a \$1,000 bonus and 11 to 1 against her getting a \$2,000 bonus. Furthermore, she feels that it is an even-money bet (the odds are 1 to 1) that she will get one or the other. Are the corresponding subjective probabilities consistent?

(b) There are two Porsches in a race, and a reporter feels that the odds against their winning are, respectively, 3 to 1 and 4 to 1. To be consistent, what odds should he assign to the event that either car will win?

18. The probabilities that the serviceability of a new X-ray machine will be rated very difficult, difficult, average, easy, or very easy are, respectively, 0.12, 0.17, 0.34, 0.29, and 0.08. Find the probabilities that the serviceability of the machine will be rated

(a) difficult or very difficult;
(b) neither very difficult nor very easy;
(c) average or worse;
(d) average or better.

19. A police department needs new tires for its patrol cars and the probabilities are 0.15, 0.24, 0.03, 0.28, 0.22, and 0.08 that it will buy Uniroyal tires, Goodyear tires, Michelin tires, General tires, Goodrich tires, or Armstrong tires. Find the probabilities that it will buy

(a) Goodyear or Goodrich tires;
(b) Uniroyal, Michelin, or Goodrich tires;
(c) Michelin or Armstrong tires;
(d) Uniroyal, Michelin, General, or Goodrich tires.

20. If each card of an ordinary deck of 52 playing cards has the same probability of being drawn, what is the probability of drawing

(a) a red jack;
(b) a 3, 4, 5, 6, or 8;
(c) a red king or a black ace?

21. A hat contains twenty white slips of paper numbered from 1 through 20, ten red slips of paper numbered from 1 through 10, forty yellow slips of paper numbered from 1 through 40, and ten blue slips of paper numbered from 1 through 10. If these 80 slips of paper are thoroughly shuffled so that each slip has the same probability of being drawn, find the probabilities of drawing a slip of paper which is

(a) blue or white;
(b) numbered 1, 2, 3, 4, or 5;
(c) red or yellow and numbered 1, 2, 3, or 4;
(d) numbered 5, 15, 25, or 35;
(e) white and numbered higher than 12 or yellow and numbered higher than 26.

22. Four candidates are seeking a vacancy on a school board. If A is twice as likely to be elected as B, and B and C are given about the same chance of being elected, while C is twice as likely to be elected as D, what are the probabilities that

(a) C wins;
(b) A does not win?

23. Two cards are randomly selected from a deck of 52 playing cards. Use Theorem 1.7 to find the probability that both cards are greater than 3 and less than 8.

24. In a game of "Yahtzee," where five dice are tossed simultaneously, find the probabilities of getting

(a) two pairs;
(b) three of a kind;
(c) a full house (three of a kind and a pair);
(d) four of a kind.

25. Among the 78 doctors on the staff of a hospital, 64 carry malpractice insurance, 36 are surgeons, and 34 of the surgeons carry malpractice insurance. If one of these doctors is chosen by lot to represent the hospital staff at an A.M.A. convention (that is, each doctor has a probability of $\frac{1}{78}$ of being selected), what is the probability that the one chosen is not a surgeon and does not carry malpractice insurance?

26. Refer to Exercises 1 and 5 to explain why there must be a mistake in each of the following statements:

(a) The probability that it will rain is 0.67 and the probability that it will rain or snow is 0.55.
(b) The probability that a student will get a passing grade in English is 0.82 and the probability that she will get a passing grade in English and French is 0.86.
(c) The probability that a person visiting the San Diego Zoo will see the giraffes is 0.72, the probability that he will see the bears is 0.84, and the probability that he will see both is 0.52.

27. Given $P(A) = 0.59$, $P(B) = 0.30$, and $P(A \cap B) = 0.21$, find

(a) $P(A \cup B)$; **(b)** $P(A \cap B')$;
(c) $P(A' \cup B')$; **(d)** $P(A' \cap B')$.

28. For married couples living in a certain suburb, the probability that the husband will vote in a school board election is 0.21, the probability that his wife will vote in the election is 0.28, and the probability that they will both vote is 0.15. What is the probability that at least one of them will vote?

29. A biology professor has two graduate assistants helping him with his research. The probability that the older of the two assistants will be absent on any given day is 0.08, the probability that the younger of the two will be absent on any given day is 0.05, and the probability that they will both be absent on any given day is 0.02. Find the probabilities that

(a) either or both of the graduate assistants will be absent on any given day;
(b) at least one of the two graduate assistants will not be absent on any given day;
(c) only one of the two graduate assistants will be absent on any given day.

30. At Roanoke College it is known that $\frac{1}{3}$ of the students live off campus. It is also known that $\frac{5}{9}$ of the students are from within the state of Virginia and that $\frac{3}{4}$ of the students are from out-of-state or live in the dormitories. What is the probability that a student selected at random from Roanoke College is from out-of-state and lives on campus?

31. Suppose that if a person visits Disneyland, the probability that he will go on the Jungle Cruise is 0.74, the probability that he will ride the Monorail is 0.70, the probability that he will go on the Matterhorn ride is 0.62, the probability that he will go on the Jungle Cruise and ride the Monorail is 0.52, the probability that he will go on the Jungle Cruise as well as the Matterhorn ride is 0.46, the probability that he will ride the Monorail and go on the Matterhorn ride is 0.44, and the probability that he will go on all three of these rides is 0.34. What is the probability that a person visiting Disneyland will go on at least one of these three rides?

32. Suppose that if a person travels to Europe for the first time, the probability that he will see London is 0.70, the probability that he will see Paris is 0.64, the probability that he will see Rome is 0.58, the probability that he will see Amsterdam is 0.58, the probability that he will see London and Paris is 0.45, the probability that he will see London and Rome is 0.42, the probability that he will see London and Amsterdam is 0.41, the probability that he will see Paris and Rome is 0.35, the probability that he will see Paris and Amsterdam is 0.39, the probability that he will see Rome and Amsterdam is 0.32, the probability that he will see London, Paris, and Rome is 0.23, the probability that he will see London, Paris, and Amsterdam is 0.26, the probability that he will see London, Rome, and Amsterdam is 0.21, the probability that he will see Paris, Rome, and Amsterdam is 0.20, and the probability that he will see all four of these cities is 0.12. What is the probability that a person traveling to Europe for the first time will see at least one of these four cities? (*Hint*: Use the formula of Exercise 9.)

2.6 CONDITIONAL PROBABILITY

Difficulties can easily arise when probabilities are quoted without specification of the sample space. For instance, if we ask for the probability that a lawyer makes more than $50,000 per year, we may well get several different answers, and they may all be correct. One of them might apply to all law school graduates, another might apply to all persons licensed to practice law, a third might apply to all those who are actively engaged in the practice of law, and so forth. Since the choice of the sample space (namely, the set of all possibilities under consideration) is by no means always self-evident, it often helps to use the symbol $P(A|S)$ to denote the **conditional probability** of event A relative to the sample space S, or as we also call it "the probability of A given S." The symbol $P(A|S)$ makes it explicit that we are referring to a particular sample space S, and it is preferable to the abbreviated notation $P(A)$ unless the tacit choice of S is clearly understood. It is also preferable when we want to refer to several sample spaces in the same example. If A is the event that a person makes more than $50,000 per year, G is the event that a person is a law school graduate, L is the event that a person is licensed to practice law, and E is the event that a person is actively engaged in the practice of law, then $P(A|G)$ is the probability that a law school graduate makes more than \$50,000 per year, $P(A|L)$ is the probability that a person licensed to practice law makes more than \$50,000 per year, and $P(A|E)$ is the probability that a person actively engaged in the practice of law makes more than \$50,000 per year.

Some ideas connected with conditional probabilities are illustrated in the following example:

EXAMPLE 2.14

A consumer research organization has studied the services under warranty provided by the 50 new car dealers in a certain city, and its findings are summarized in the following table:

	Good service under warranty	*Poor service under warranty*
In business ten years or more	16	4
In business less than ten years	10	20

If a person randomly selects one of these new car dealers, what is the probability that he gets one who provides good service under warranty? Also, if a person randomly selects one of the dealers who has been in business for ten years or more, what is the probability that he gets one who provides good service under warranty?

Solution

By "randomly" we mean that, in each case, all possible selections are equally likely, and we can therefore use the formula of Theorem 2.2. If we let G denote the selection of a dealer who provides good service under warranty, and if we let $n(G)$ denote the number of elements in G, and $n(S)$ the number of elements in the whole sample space, we get

$$P(G) = \frac{n(G)}{n(S)} = \frac{16 + 10}{50} = 0.52$$

This answers the first question.

For the second question, we limit ourselves to the reduced sample space which consists of the first line of the table, namely, the $16 + 4 = 20$ dealers who have been in business ten years or more. Of these, 16 provide good service under warranty, and we get

$$P(G|T) = \frac{16}{20} = 0.80$$

where T denotes the selection of a dealer who has been in business ten years or more. This answers the second question, and as should have been expected, $P(G|T)$ is considerably higher than $P(G)$.

Since the numerator of $P(G|T)$ is $n(T \cap G) = 16$, the number of dealers who have been in business for ten years or more and provide good service under warranty, and the denominator is $n(T)$, the number of dealers who have been in business ten years or more, we can write symbolically

$$P(G|T) = \frac{n(T \cap G)}{n(T)}$$

Then, if we divide the numerator and the denominator by $n(S)$, the total number of new car dealers in the given city, we get

$$P(G|T) = \frac{\dfrac{n(T \cap G)}{n(S)}}{\dfrac{n(T)}{n(S)}} = \frac{P(T \cap G)}{P(T)}$$

and we have, thus, expressed the conditional probability $P(G|T)$ in terms of two probabilities defined for the whole sample space S.

Generalizing from this example, let us now make the following definition of conditional probability:

DEFINITION 2.1 If A and B are any two events in a sample space S and $P(A) \neq 0$, the **conditional probability** of B given A is

$$P(B|A) = \frac{P(A \cap B)}{P(A)}$$

EXAMPLE 2.15

With reference to Example 2.14, what is the probability that one of the dealers who has been in business less than ten years will provide good service under warranty?

Solution

Since $P(T' \cap G) = \frac{10}{50} = 0.20$ and $P(T') = \frac{10 + 20}{50} = 0.60$, substitution into the formula yields

$$P(G|T') = \frac{P(T' \cap G)}{P(T')} = \frac{0.20}{0.60} = \frac{1}{3}$$

Although we justified the formula of Definition 2.1 with an example in which the possibilities were all equally likely, this is not a requirement for its use.

EXAMPLE 2.16

With reference to the loaded die of Example 2.9, what is the probability that the number of points rolled is a perfect square? Also, what is the probability that it is a perfect square given that it is greater than 3?

Solution

If A is the event that the number of points rolled is greater than 3 and B is the event that it is a perfect square, we have $A = \{4, 5, 6\}$, $B = \{1, 4\}$, and $A \cap B = \{4\}$. Since the probabilities of rolling a 1, 2, 3, 4, 5, or 6 with the die are $\frac{2}{9}$, $\frac{1}{9}$, $\frac{2}{9}$, $\frac{1}{9}$, $\frac{2}{9}$, and $\frac{1}{9}$ (see page 37), we find that the answer to the first question is

$$P(B) = \tfrac{2}{9} + \tfrac{1}{9} = \tfrac{1}{3}$$

To determine $P(B|A)$, we first calculate

$$P(A \cap B) = \tfrac{1}{9} \qquad \text{and} \qquad P(A) = \tfrac{1}{9} + \tfrac{2}{9} + \tfrac{1}{9} = \tfrac{4}{9}$$

Then, substituting into the formula of Definition 2.1, we get

$$P(B|A) = \frac{P(A \cap B)}{P(A)} = \frac{\frac{1}{9}}{\frac{4}{9}} = \frac{1}{4}$$

Originally, we justified Definition 2.1 with an example in which the possibilities were all equally likely. To show that it yields the "right" answer here, where the possibilities are not all equally likely, we need only to assign a probability of v to the two even numbers and a probability of $2v$ to the odd number in the reduced sample space A, such that the sum of the three probabilities is 1. We then have $v + 2v + v = 1$ or $v = \frac{1}{4}$, and hence $P(B|A) = \frac{1}{4}$ as before.

EXAMPLE 2.17

A manufacturer of airplane parts knows from past experience that the probability is 0.80 that an order will be ready for shipment on time, and it is 0.72 that an order will be ready for shipment on time and will also be delivered on time. What is the probability that such an order will be delivered on time given that it was ready for shipment on time?

Solution

If we let R stand for the event that an order is ready for shipment on time and D for the event that it is delivered on time, we have $P(R) = 0.80$ and $P(R \cap D) = 0.72$, and it follows that

$$P(D|R) = \frac{P(R \cap D)}{P(R)} = \frac{0.72}{0.80} = 0.90$$

Thus, 90 percent of the shipment will be delivered on time provided they are shipped on time. Note that $P(R|D)$, the probability that a shipment which is delivered on time was also ready for shipment on time, cannot be determined without further information; for this purpose we would also have to know $P(D)$.

Multiplying the expressions on both sides of the formula of Definition 2.1 by $P(A)$, we obtain the following **multiplication rule**:

THEOREM 2.9 If A and B are any two events in a sample space S and $P(A) \neq 0$, then

$$P(A \cap B) = P(A) \cdot P(B|A)$$

In words, the probability that A and B will both occur is the product of the probability of A and the conditional probability of B given A. Alternatively, it is the product of the probability of B and the conditional probability of A given B; symbolically, $P(A \cap B) = P(B) \cdot P(A|B)$. To derive this alternative form, we interchange A and B in the formula of Theorem 2.9 and make use of the fact that $A \cap B = B \cap A$.

EXAMPLE 2.18

If we randomly pick two television tubes in succession from a shipment of 240 television tubes of which 15 are defective, what is the probability that they will both be defective?

Solution

If we assume equal probabilities for each selection (which is what we mean by "randomly" picking the tubes), the probability that the first tube will be defective is $\frac{15}{240}$, and the probability that the second tube will be defective given that the first tube is defective is $\frac{14}{239}$. Thus, the probability that both tubes will be defective is $\frac{15}{240} \cdot \frac{14}{239} = \frac{7}{1,912}$. This assumes that we are **sampling without replacement**, namely, that the first tube is not replaced before the second tube is picked.

EXAMPLE 2.19

Find the probability of randomly drawing two aces in succession from an ordinary deck of 52 playing cards (a) if we sample without replacement, and (b) if we sample with replacement.

Solution

(a) If the first card is not replaced before the second card is drawn, the probability of getting two aces in succession is

$$\frac{4}{52} \cdot \frac{3}{51} = \frac{1}{221}$$

(b) If the first card is replaced before the second card is drawn, the corresponding probability is

$$\frac{4}{52} \cdot \frac{4}{52} = \frac{1}{169}$$

In the situations described in the two preceding examples there is a definite temporal order between the two events A and B. In general, this need not be the case when we write $P(A|B)$ or $P(B|A)$. For instance, we could ask for the probability that the first card drawn was an ace given that the second card drawn (without replacement) is an ace—the answer would also be $\frac{3}{51}$.

Theorem 2.9 can easily be generalized so that it applies to more than two events; for instance, for three events we have:

THEOREM 2.10 If A, B, and C are any three events in a sample space S, such that $P(A) \neq 0$ and $P(A \cap B) \neq 0$, then

$$P(A \cap B \cap C) = P(A) \cdot P(B|A) \cdot P(C|A \cap B)$$

Proof. Writing $A \cap B \cap C$ as $(A \cap B) \cap C$ and using the formula of Theorem 2.9 twice, we get

$$\begin{aligned} P(A \cap B \cap C) &= P[(A \cap B) \cap C] \\ &= P(A \cap B) \cdot P(C|A \cap B) \\ &= P(A) \cdot P(B|A) \cdot P(C|A \cap B) \end{aligned}$$

Further generalization of Theorems 2.9 and 2.10 to k events is now straightforward, and the resulting formula can be proved by mathematical induction.

EXAMPLE 2.20

A box of fuses contains 20 fuses, of which 5 are defective. If 3 of the fuses are selected at random and removed from the box in succession without replacement, what is the probability that all three fuses are defective?

Solution

If A is the event that the first fuse is defective, B is the event that the second fuse is defective, and C is the event that the third fuse is defective, then $P(A) = \frac{5}{20}$, $P(B|A) = \frac{4}{19}$, $P(C|A \cap B) = \frac{3}{18}$, and substitution into the formula yields

$$\begin{aligned} P(A \cap B \cap C) &= \frac{5}{20} \cdot \frac{4}{19} \cdot \frac{3}{18} \\ &= \frac{1}{114} \end{aligned}$$

2.7 INDEPENDENT EVENTS

Informally speaking, two events A and B are said to be **independent** if the occurrence or nonoccurrence of either of them does not affect the probability of the occurrence of the other. For instance, if in Example 2.20 each fuse is replaced before the next one is randomly drawn, the outcomes of successive selections are all independent—the probability of getting a defective fuse remains $\frac{5}{20}$ in each case.

Symbolically, two events A and B are independent if $P(B|A) = P(B)$ and $P(A|B) = P(A)$, and it can be shown that either of these equalities implies the other when both of the conditional probabilities exist, namely, when neither $P(A)$ nor $P(B)$ equals zero (see Exercise 6 on page 64).

Now, if we substitute $P(B)$ for $P(B|A)$ into the formula of Theorem 2.9, we get

$$\begin{aligned} P(A \cap B) &= P(A) \cdot P(B|A) \\ &= P(A) \cdot P(B) \end{aligned}$$

and we shall use this as our formal definition of independence.

DEFINITION 2.2 Two events A and B are **independent** if and only if

$$P(A \cap B) = P(A) \cdot P(B)$$

If two events are not independent, they are said to be **dependent**. In the derivation of the formula of Definition 2.2 we assumed that $P(B|A)$ exists and, hence, that $P(A) \neq 0$. For mathematical convenience, we shall let the definition apply also when $P(A) = 0$ and/or $P(B) = 0$.

EXAMPLE 2.21

A coin is tossed three times and the eight possible outcomes, HHH, HHT, HTH, THH, HTT, THT, TTH, and TTT, are assumed to be equally likely. If A is the event that a head occurs on each of the first two tosses, B is the event that a tail occurs on the third toss, and C is the event that exactly two tails occur in the three tosses, show that events A and B are independent whereas B and C are dependent.

Solution

Since

$$A = \{\text{HHH, HHT}\}$$

$$B = \{\text{HHT, HTT, THT, TTT}\}$$

$$C = \{\text{HTT, THT, TTH}\}$$

$$A \cap B = \{\text{HHT}\}$$

$$B \cap C = \{\text{HTT, THT}\}$$

the assumption that the eight possible outcomes are all equiprobable yields $P(A) = \frac{1}{4}$, $P(B) = \frac{1}{2}$, $P(C) = \frac{3}{8}$, $P(A \cap B) = \frac{1}{8}$, and $P(B \cap C) = \frac{1}{4}$. Then, since $P(A) \cdot P(B) = \frac{1}{4} \cdot \frac{1}{2} = \frac{1}{8}$ equals $P(A \cap B)$, the events A and B are independent, and since $P(B) \cdot P(C) = \frac{1}{2} \cdot \frac{3}{8} = \frac{3}{16}$ does not equal $P(B \cap C)$, the events B and C are dependent.

With regard to Definition 2.2 it can be shown that either, or both, events can be replaced by their complements. For instance,

THEOREM 2.11 If the two events A and B are independent, then the two events A and B' are also independent.

Proof. Since $A = (A \cap B) \cup (A \cap B')$ and $A \cap B$ and $A \cap B'$ are mutually exclusive, we have

$$\begin{aligned} P(A) &= P[(A \cap B) \cup (A \cap B')] \\ &= P(A \cap B) + P(A \cap B') \end{aligned}$$

or, since A and B are independent,

$$P(A) = P(A) \cdot P(B) + P(A \cap B')$$

It follows that

$$\begin{aligned} P(A \cap B') &= P(A) \cdot [1 - P(B)] \\ &= P(A) \cdot P(B') \end{aligned}$$

and, hence, that A and B' are independent. In Exercise 5 on page 64 the reader will be asked to show that if A and B are independent, then A' and B, and A' and B', are also independent.

To extend the concept of independence to more than two events, let us make the following definition:

DEFINITION 2.3 Events $A_1, A_2, \ldots,$ and A_k are **independent** if and only if the probability of the intersection of any 2, 3, ..., or k of these events equals the product of their respective probabilities.

For three events A, B, and C, for example, independence requires that $P(A \cap B) = P(A) \cdot P(B)$, $P(A \cap C) = P(A) \cdot P(C)$, $P(B \cap C) = P(B) \cdot P(C)$, and $P(A \cap B \cap C) = P(A) \cdot P(B) \cdot P(C)$.

It is of interest to note that three or more events can be **pairwise independent** without being independent.

EXAMPLE 2.22

Consider three events A, B, and C in a sample space S, with probabilities assigned as in the Venn diagram of Figure 2.7. Show that A and B are independent, A and C are independent, B and C are independent, but A, B, and C are not independent.

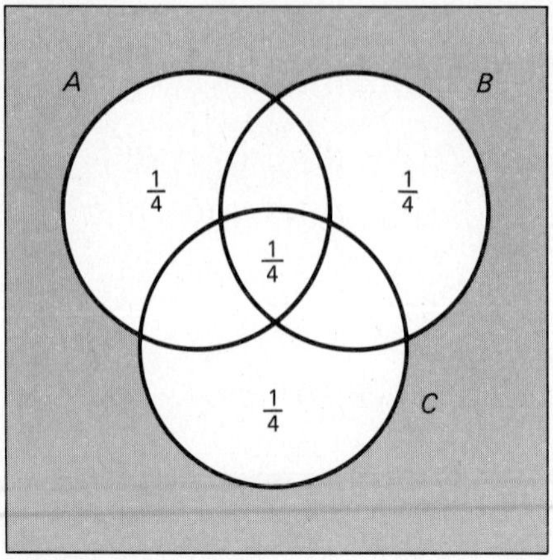

Figure 2.7 Venn diagram for Example 2.22.

Solution

As can be seen from the diagram, $P(A) = P(B) = P(C) = \frac{1}{2}$, $P(A \cap B) = P(A \cap C) = P(B \cap C) = \frac{1}{4}$, and $P(A \cap B \cap C) = \frac{1}{4}$. Thus, $P(A) \cdot P(B) = \frac{1}{4} = P(A \cap B)$, $P(A) \cdot P(C) = \frac{1}{4} = P(A \cap C)$, $P(B) \cdot P(C) = \frac{1}{4} = P(B \cap C)$, but $P(A) \cdot P(B) \cdot P(C) = \frac{1}{8} \neq P(A \cap B \cap C)$, and this proves what we set out to show.

Incidentally, this example can be given a "real" interpretation by considering a large room which has three separate switches controlling the ceiling lights. These lights will be on when all three switches are "up," and hence also when one of the switches is "up" and the other two are "down." If A is the event that the first switch is "up," B is the event that the second switch is "up," and C is the event that the third switch is "up," the Venn diagram of Figure 2.7 shows a possible set of probabilities associated with the switches being "up" or "down" when the ceiling lights are on.

It can also happen that $P(A \cap B \cap C) = P(A) \cdot P(B) \cdot P(C)$ without A, B, and C being pairwise independent—this the reader will be asked to verify in part (a) of Exercise 7 on page 64.

Of course, if certain events are given as independent, the probability that they will all occur is simply the product of their respective probabilities.

EXAMPLE 2.23

Find the probability of getting three heads in three (independent) tosses of a balanced coin, and also the probability of first rolling four fives and then another number in five (independent) rolls of a fair die.

Solution

Multiplying the respective probabilities, we get

$$\frac{1}{2} \cdot \frac{1}{2} \cdot \frac{1}{2} = \frac{1}{8}$$

for the probability of getting three heads, and

$$\frac{1}{6} \cdot \frac{1}{6} \cdot \frac{1}{6} \cdot \frac{1}{6} \cdot \frac{5}{6} = \frac{5}{7776}$$

for the probability of first rolling four fives and then another number.

2.8 BAYES' THEOREM

There are many problems in which the ultimate outcome of an experiment depends on what happens in various intermediate stages.

EXAMPLE 2.24

In the simplest case there is one intermediate stage consisting of two alternatives. Suppose, for instance, that we are concerned with the completion of a highway construction job, which may be delayed because of a strike. Suppose, furthermore, that the probabilities are 0.60 that there will be a strike, 0.85 that the job will be completed on time if there is no strike, and 0.35 that the job will be completed on time if there is a strike. What is the probability that the job will be completed on time?

Solution

If A is the event that the job will be completed on time and B is the event that there will be a strike, the given information can be written as $P(B) = 0.60$, $P(A|B) = 0.35$, and $P(A|B') = 0.85$. Since A is the union of the two mutually exclusive events $A \cap B$ and $A \cap B'$ (see Figure 2.8), we can write

$$\begin{aligned} P(A) &= P[(A \cap B) \cup (A \cap B')] \\ &= P(A \cap B) + P(A \cap B') \\ &= P(B) \cdot P(A|B) + P(B') \cdot P(A|B') \end{aligned}$$

Then, substitution of the given numerical values yields

$$\begin{aligned} P(A) &= (0.60)(0.35) + (1 - 0.60)(0.85) \\ &= 0.55 \end{aligned}$$

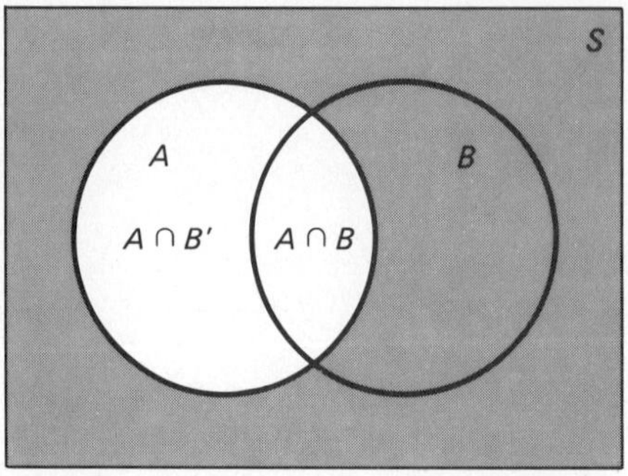

Figure 2.8 Venn diagram for Example 2.24.

An immediate generalization of this kind of problem to the case where the intermediate stage permits k alternatives (whose occurrence is denoted B_1, $B_2, \ldots, B_k$) is taken care of by the following theorem, sometimes called the **rule of elimination**:

THEOREM 2.12 If the events $B_1, B_2, \ldots,$ and B_k constitute a partition of the sample space S and $P(B_i) \neq 0$ for $i = 1, 2, \ldots, k$, then for any event A in S

$$P(A) = \sum_{i=1}^{k} P(B_i) \cdot P(A|B_i)$$

As was defined in the footnote to page 11, the B's constitute a partition of the sample space if they are pairwise mutually exclusive and if their union equals S. A formal proof of Theorem 2.12 consists, essentially, of the same steps which we used in Example 2.24, and it will be left to the reader in Exercise 11 on page 64.

EXAMPLE 2.25

The members of a consulting firm rent cars from three rental agencies: 60 percent from agency 1, 30 percent from agency 2, and 10 percent from agency 3. If 9 percent of the cars from agency 1 need a tune-up, 20 percent of the cars from agency 2 need a tune-up, and 6 percent of the cars from agency 3 need a tune-up, what is the probability that a rental car delivered to the firm will need a tune-up?

Solution

If A is the event that the car needs a tune-up, and B_1, B_2, and B_3 are, respectively, the events that the car comes from rental agencies 1, 2, or 3, we have $P(B_1) = 0.60$, $P(B_2) = 0.30$, $P(B_3) = 0.10$, $P(A|B_1) = 0.09$, $P(A|B_2) = 0.20$, and

$P(A|B_3) = 0.06$. Substitution of these values into the formula of Theorem 2.12 yields

$$P(A) = (0.60)(0.09) + (0.30)(0.20) + (0.10)(0.06)$$
$$= 0.12$$

Thus, 12 percent of all the rental cars delivered to this firm need a tune-up.

With reference to the preceding example, suppose that we are interested in the following question: If a rental car delivered to the consulting firm needs a tune-up, what is the probability that it came from rental agency 2? To answer questions of this kind, we need the following theorem, called **Bayes' theorem**:

THEOREM 2.13 If the events $B_1, B_2, \ldots,$ and B_k constitute a partition of the sample space S and $P(B_i) \neq 0$ for $i = 1, 2, \ldots, k$, then for any event A in S such that $P(A) \neq 0$

$$P(B_r|A) = \frac{P(B_r) \cdot P(A|B_r)}{\sum_{i=1}^{k} P(B_i) \cdot P(A|B_i)}$$

for $r = 1, 2, \ldots, k$.

In words, the probability that event A was reached via the rth branch of the tree diagram of Figure 2.9, given that it was reached via one of its k branches, is the *ratio* of the probability associated with the rth branch to the sum of the probabilities associated with all k branches of the tree.

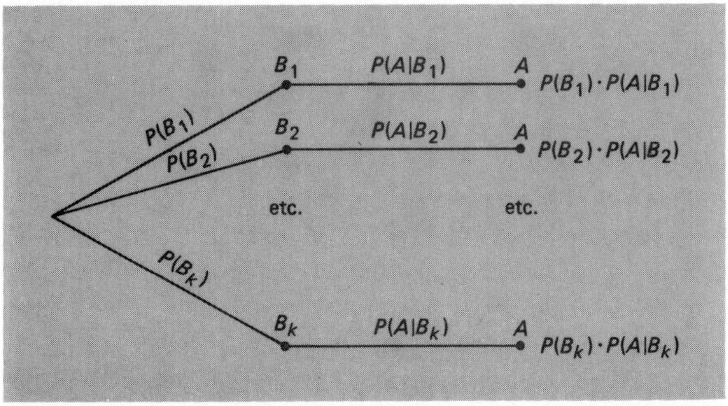

Figure 2.9 Tree diagram for Bayes' theorem.

Proof. Writing $P(B_r|A) = \dfrac{P(A \cap B_r)}{P(A)}$ in accordance with the definition of conditional probability, we have only to substitute $P(B_r) \cdot P(A|B_r)$ for $P(A \cap B_r)$ and the expression in the formula of Theorem 2.12 for $P(A)$.

EXAMPLE 2.26

With reference to Example 2.25, if a rental car delivered to the consulting firm needs a tune-up, what is the probability that it came from rental agency 2?

Solution

Substituting the probabilities on page 60 into the formula of Theorem 2.13, we get

$$\begin{aligned} P(B_2|A) &= \frac{(0.30)(0.20)}{(0.60)(0.09) + (0.30)(0.20) + (0.10)(0.06)} \\ &= \frac{0.060}{0.120} \\ &= 0.5 \end{aligned}$$

Observe that although only 30 percent of the cars delivered to the firm come from agency 2, 50 percent of those requiring a tune-up come from that agency.

EXAMPLE 2.27

In a certain state, 25 percent of all cars emit excessive amounts of pollutants. If the probability is 0.99 that a car emitting excessive amounts of pollutants will fail the state's vehicular emission test, and the probability is 0.17 that a car not emitting excessive amounts of pollutants will nevertheless fail the test, what is the probability that a car which fails the test actually emits excessive amounts of pollutants?

Solution

Picturing this situation as in Figure 2.10, we find that the probabilities associated with the two branches of the tree diagram are $(0.25)(0.99) = 0.2475$ and $(1 - 0.25)(0.17) = 0.1275$. Thus, the probability that a car which fails the test actually emits excessive amounts of pollutants is

$$\frac{0.2475}{0.2475 + 0.1275} = 0.66$$

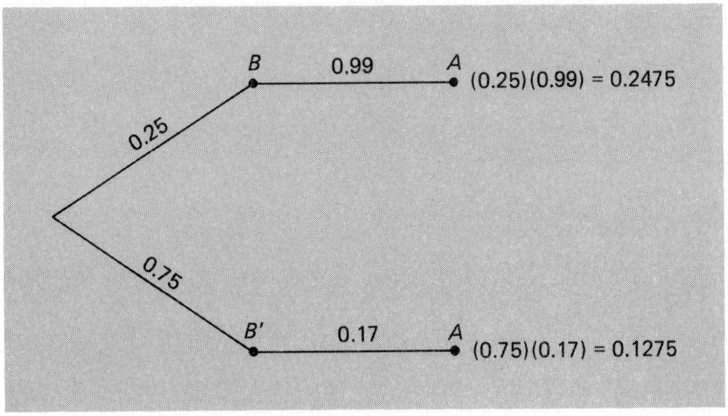

Figure 2.10 Tree diagram for Example 2.27.

Of course, this result could also have been obtained without the diagram, by substituting directly into the formula of Bayes' theorem.

Although Bayes' theorem follows from the postulates of probability and the definition of conditional probability, it has been the subject of extensive controversy. There can be no question about the validity of Bayes' theorem, but considerable arguments have been raised about the interpretation of the **prior probabilities** $P(B_i)$. Also, a good deal of mysticism surrounding Bayes' theorem is due to the fact that it entails a "backward" or "inverse" sort of reasoning, namely, reasoning "from effect to cause," as in Example 2.27.

THEORETICAL EXERCISES

1. Show that the three postulates of probability are satisfied by conditional probabilities; that is, show that with $P(B) > 0$,

(a) $P(A|B) \geqslant 0$;
(b) $P(B|B) = 1$;
(c) $P(A_1 \cup A_2 \cup \cdots |B) = P(A_1|B) + P(A_2|B) + \cdots$ for any sequence of mutually exclusive events $A_1, A_2, \ldots$.

2. Show by means of numerical examples that $P(B|A) + P(B|A')$

(a) may equal 1;
(b) need not equal 1.

3. Duplicating the method of proof of Theorem 2.10, show that $P(A \cap B \cap C \cap D) = P(A) \cdot P(B|A) \cdot P(C|A \cap B) \cdot P(D|A \cap B \cap C)$ provided $P(A \cap B \cap C) \neq 0$.

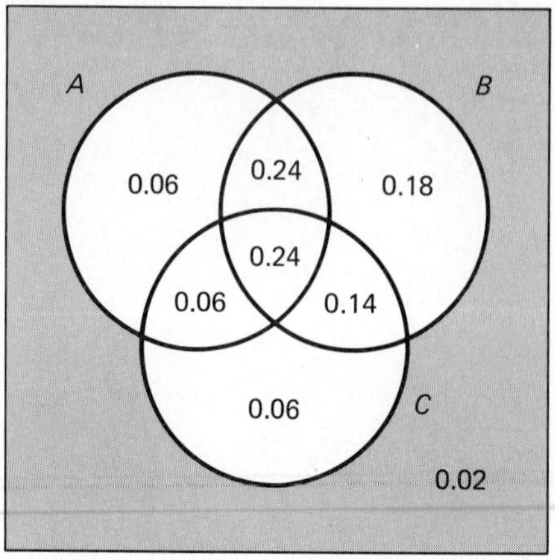

Figure 2.11 Diagram for Exercise 7.

4. Given three events A, B, and C such that $P(A \cap B \cap C) \neq 0$ and $P(C|A \cap B) = P(C|B)$, show that $P(A|B \cap C) = P(A|B)$.

5. Show that if the two events A and B are independent, then

(a) the two events A' and B are also independent;

(b) the two events A' and B' are also independent.

6. Show that if $P(B|A) = P(B)$ and $P(B) \neq 0$, then $P(A|B) = P(A)$.

7. Refer to Figure 2.11 to show that

(a) $P(A \cap B \cap C) = P(A) \cdot P(B) \cdot P(C)$ does not necessarily imply that the events A, B, and C are all pairwise independent;

(b) if A is independent of B and A is independent of C, then B is not necessarily independent of C;

(c) if A is independent of B and A is independent of C, then A is not necessarily independent of $B \cup C$.

8. If the three events A, B, and C are independent, show that

(a) A and $B \cap C$ are independent;

(b) A and $B \cup C$ are independent;

(c) A' and $B \cap C'$ are independent.

9. Show that $2^k - k - 1$ conditions must be satisfied for k events to be independent.

10. For any event A, show that A and $\varnothing$ are independent.

11. Prove Theorem 2.12, making use of the following generalization of the first distributive law given in Appendix I on page 501:

$$A \cap (B_1 \cup B_2 \cup \cdots \cup B_k) = (A \cap B_1) \cup (A \cap B_2) \cup \cdots \cup (A \cap B_k)$$

APPLIED EXERCISES

12. There are 90 applicants for a job with the news department of a television station. Some of them are college graduates and some are not, some of them have at least three years experience and some have not, with the exact breakdown being

	College graduates	*Not college graduates*
At least three years experience	18	9
Less than three years experience	36	27

If the order in which the applicants are interviewed by the station manager is random, G is the event that the first applicant interviewed is a college graduate, and T is the event that the first applicant interviewed has at least three years experience, determine each of the following probabilities directly from the entries and the row and column totals of the table:

(a) $P(G)$; **(b)** $P(T')$; **(c)** $P(G \cap T)$;
(d) $P(G' \cap T')$; **(e)** $P(T|G)$; **(f)** $P(G'|T')$.

Use these results to verify that

(g) $P(T|G) = \dfrac{P(G \cap T)}{P(G)}$; **(h)** $P(G'|T') = \dfrac{P(G' \cap T')}{P(T')}$.

13. With reference to Exercise 25 on page 48, what is the probability that the doctor chosen to represent the hospital staff at the convention carries malpractice insurance given that he or she is a surgeon?

14. With reference to Exercise 28 on page 48, what is the probability that a husband will vote in the given school board election given that his wife will vote?

15. With reference to Exercise 30 on page 48, what is the probability that one of the students lives in a dormitory given that he or she is from out-of-state?

16. Basketball teams from Universities A, B, C, and D are quoted as having probabilities of 0.2, 0.4, 0.3, and 0.1, respectively, of winning the playoff for the national championship. If University B is placed on probation and declared ineligible to participate (and it is not replaced by another university), what is the probability that University A will win the national championship?

17. With reference to Exercise 31 on page 49, find the probabilities that a person who visits Disneyland will

(a) ride the Monorail given that he will go on the Jungle Cruise;
(b) go on the Matterhorn ride given that he will go on the Jungle Cruise and ride the Monorail;

(c) not go on the Jungle Cruise given that he will ride the Monorail and/or go on the Matterhorn ride;

(d) go on the Matterhorn ride and the Jungle Cruise given that he will not ride the Monorail.

(*Hint*: Draw a Venn diagram and fill in the various probabilities.)

18. The probability of surviving a certain transplant operation is 0.5. If a patient survives the operation, the probability that his or her body will reject the transplant within a month is 0.2. What is the probability of surviving both of these critical stages?

19. Crates of eggs are inspected for blood clots by randomly removing three eggs in succession and examining their contents. If all three eggs are good, the crate is shipped; otherwise it is rejected. What is the probability that a crate containing 120 eggs of which 10 have blood clots will be shipped?

20. Suppose that in Vancouver, B.C., the probability that a rainy fall day is followed by a rainy day is 0.80 and the probability that a sunny fall day is followed by a rainy day is 0.60. Find the probabilities that a rainy day is followed by

(a) a rainy day, a sunny day, and another rainy day;

(b) two sunny days and then a rainy day;

(c) two rainy days and then two sunny days;

(d) rain two days later.

[*Hint*: In part (c) use the formula of Exercise 3.]

21. Use the formula of Exercise 3 to find the probability of randomly choosing (without replacement) four healthy guinea pigs from a cage containing 20 guinea pigs of which 15 are healthy and 5 are diseased.

22. A balanced die is tossed twice. If A is the event that an even number comes up on the first toss, B is the event that an even number comes up on the second toss, and C is the event that both tosses result in the same number, are the events A, B, and C independent?

23. If three persons, selected at random, are stopped on a street, what are the probabilities that

(a) all were born on a Friday;

(b) two were born on a Friday and the other on a Tuesday;

(c) none was born on a Monday?

24. A marksman hits a target with probability $\frac{3}{4}$. Assuming independence for successive firings, find the probabilities of getting

(a) one hit followed by two misses;

(b) two hits in three firings.

25. A certain coin is loaded so that heads is four times as likely as tails. If this coin is tossed three times, find the probabilities of getting

(a) all heads;

(b) two tails and a head (not necessarily in that order).

26. An urn contains four black balls and three white balls. If four balls are drawn in succession, each ball being replaced in the urn before the next one is drawn, what are the probabilities that

(a) three of the four balls are black and the other is white;

(b) the first and last balls drawn are both white?

27. Medical records show that one out of ten individuals in a certain town has a low thyroid condition. If 20 persons in this town are randomly chosen and tested, what is the probability that at least one of them will have a low thyroid condition?

28. With reference to Figure 2.12 verify that events A, B, C, and D are independent. Note that the region which represents event A consists of two circles, and so do the regions representing events B and C.

29. At an electronics plant, it is known from past experience that the probability is 0.84 that a new worker who has attended the company's training program will meet the production quota, and that the corresponding probability is 0.49 for a new worker who has not attended the company's training program. If 70 percent of all new workers attend the training program, what is the probability that a new worker will meet the production quota?

30. In a T-maze, a rat is given food if it turns left and an electric shock if it turns right. On the first trial there is a fifty-fifty chance that a rat will turn either way; then, if it receives food on the first trial the probability is 0.68 that it will turn left on the next trial, and if it receives a shock on the first trial the

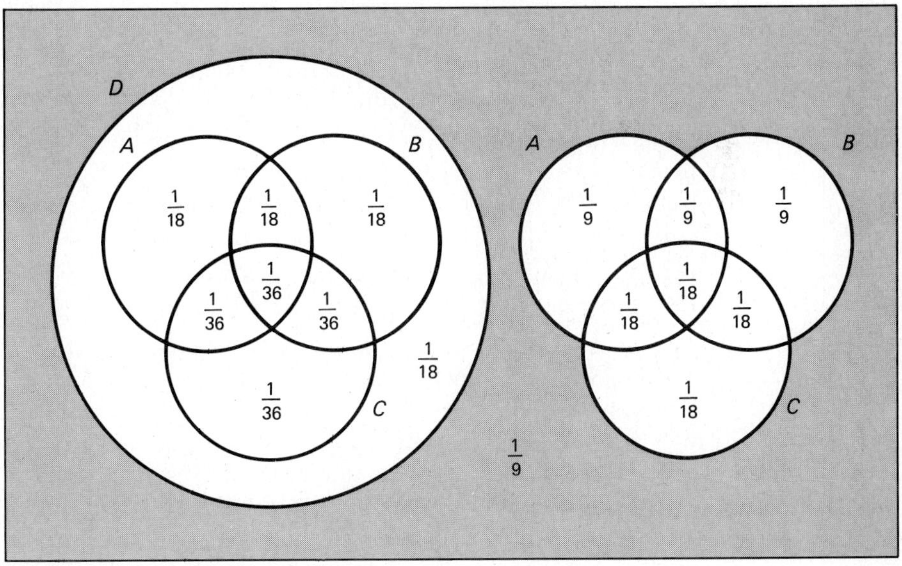

Figure 2.12 Diagram for Exercise 28.

probability is 0.84 that it will turn left on the next trial. What is the probability that a rat will turn left on the second trial?

31. A mail-order house employs three stock clerks, U, V, and W, who pull items from shelves and assemble them for subsequent verification and packaging. U makes a mistake in an order (gets a wrong item or the wrong quantity) one time in a hundred, V makes a mistake in an order five times in a hundred, and W makes a mistake in an order three times in a hundred. If U, V, and W fill, respectively, 30, 40, and 30 percent of all orders, what is the probability of a mistake in an order?

32. In a certain community, 8 percent of all adults over 50 have diabetes. If a health service in this community correctly diagnoses 95 percent of all persons with diabetes as having the disease and incorrectly diagnoses 2 percent of all persons without diabetes as having the disease, find the probabilities that

(a) the community health service will diagnose an adult over 50 as having diabetes;

(b) a person over 50 diagnosed by the health service as having diabetes actually has the disease.

33. With reference to Exercise 29, what is the probability that a new worker who meets the production quota attended the company's training program?

34. With reference to Exercise 31, if a mistake is found in an order, what is the probability that it was filled by clerk V?

35. With reference to Example 2.24 on page 59, if we discover later that the job was completed on time, what is the probability that, nevertheless, there had been a strike?

36. An explosion at a construction site could have occurred as the result of static electricity, malfunctioning of equipment, carelessness, or sabotage. Interviews with construction engineers analyzing the risks involved led to the estimates that such an explosion would occur with probability 0.25 as a result of static electricity, 0.20 as a result of malfunctioning of equipment, 0.40 as a result of carelessness, and 0.75 as a result of sabotage. It is also felt that the prior probabilities of the four causes of the explosion are, respectively, 0.20, 0.40, 0.25, and 0.15. Based on all this information, what is

(a) the most likely cause of the explosion;

(b) the least likely cause of the explosion?

37. The manager of a restaurant knows that the odds are 2 to 1 that a customer will not have a cocktail before dinner. If he has a cocktail, the odds are 3 to 2 that he will not order steak, and if he does not have a cocktail, the odds are 3 to 1 that he will not order steak. If a customer has a cocktail and a steak, the odds are 9 to 1 that he will not have dessert; if he has a cocktail but does not order steak, the odds are 7 to 1 that he will not have dessert; if he does not have a cocktail but orders steak, the odds are 3 to 1 that he will not have dessert; and if he does not have a cocktail and does not order steak, the odds are 2 to 1 that he will not have dessert.

(a) What is the probability that any one customer of this restaurant will order dessert?

(b) What is the probability that a customer who orders dessert had a cocktail before dinner?

(c) What is the probability that a customer who orders steak and dessert also had a cocktail before dinner?

38. An art dealer receives a shipment of five old paintings from abroad, and, on the basis of past experience, she feels that the probabilities are, respectively, 0.76, 0.09, 0.02, 0.01, 0.02, and 0.10 that 0, 1, 2, 3, 4, or all 5 of them are forgeries. Since the cost of authentication is fairly high, she decides to select one of the five paintings at random and send it away for authentication. If it turns out that this painting is a forgery, what probability should she now assign to the possibility that all the other paintings are also forgeries?

References

Among the numerous textbooks on probability published in recent years, one of the most popular is

FELLER, W., *An Introduction to Probability Theory and Its Applications*, Vol. I, 3rd ed. New York: John Wiley & Sons, Inc., 1968.

More elementary treatments may be found in

BURFORD, R. L., *Introduction to Finite Probability*. Columbus, Ohio: Charles E. Merrill Publishing Company, 1967,

FREUND, J. E., *Introduction to Probability*. Encino, Calif.: Dickenson Publishing Co., Inc., 1973,

GOLDBERG, S., *Probability—An Introduction*. Englewood Cliffs, N.J.: Prentice-Hall, Inc., 1960,

HODGES, J. L., and LEHMANN, E. L., *Elements of Finite Probability*. San Francisco: Holden-Day, Inc., 1965,

NOSAL, M., *Basic Probability and Applications*. Philadelphia: W. B. Saunders Company, 1977,

SCHAEFFER, R. L., and MENDENHALL, *Introduction to Probability: Theory and Applications*. North Scituate, Mass.: Duxbury Press, 1975.

More advanced treatments are given in many texts, for instance, in

HOEL, P., PORT, S. C., and STONE, C. J., *Introduction to Probability Theory*. Boston: Houghton Mifflin Company, 1971,

KHAZANIE, R., *Basic Probability Theory and Applications*. Pacific Palisades, Calif.: Goodyear Publishing Company, Inc., 1976,

MCCORD, J. R., and MORONEY, R. M., *Introduction to Probability Theory*. New York: Macmillan Publishing Co., Inc., 1964,

PARZEN, E., *Modern Probability Theory and Its Applications*. New York: John Wiley & Sons, Inc., 1960,

ROSS, S., *A First Course in Probability*. New York: Macmillan Publishing Company, Inc., 1976.

probability distributions

3

3.1 RANDOM VARIABLES

In most applications of probability theory we are interested only in a particular aspect (or in two or a few particular aspects) of the outcomes of experiments. For instance, when we roll a pair of dice we are usually interested only in the total, and not in the outcome for each die; when we interview a randomly chosen married couple we may be interested in the size of their family and in their joint income, but not in the number of years they have been married or their total assets; and when we sample mass-produced light bulbs we may be interested in their durability or their brightness, but not in their price.

In each of these examples we are interested in numbers which are associated with the outcomes of chance experiments, namely, in the values which are taken on by so-called **random variables**. In the language of probability and statistics, the total we roll with a pair of dice is a random variable, the size of the family of a randomly chosen married couple and their joint income are random variables, and so are the durability and the brightness of a light bulb randomly picked for inspection.

To be more explicit, consider Figure 3.1, which (like Figure 2.1 on page 29) pictures the sample space for an experiment in which we roll a pair of dice, and let us assume that each of the 36 possible outcomes has the probability $\frac{1}{36}$. Note, however, that in Figure 3.1 we have attached a number to each point: for instance, we attached the number 2 to the point (1, 1), the number 6 to the point (1, 5), the

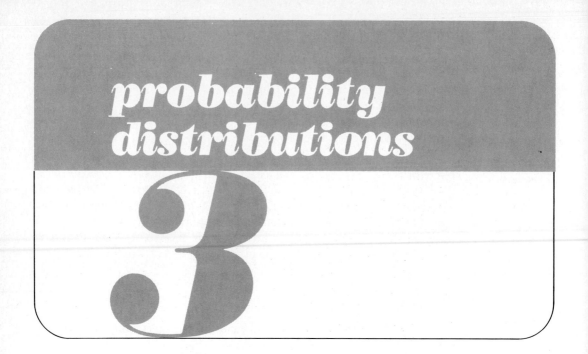

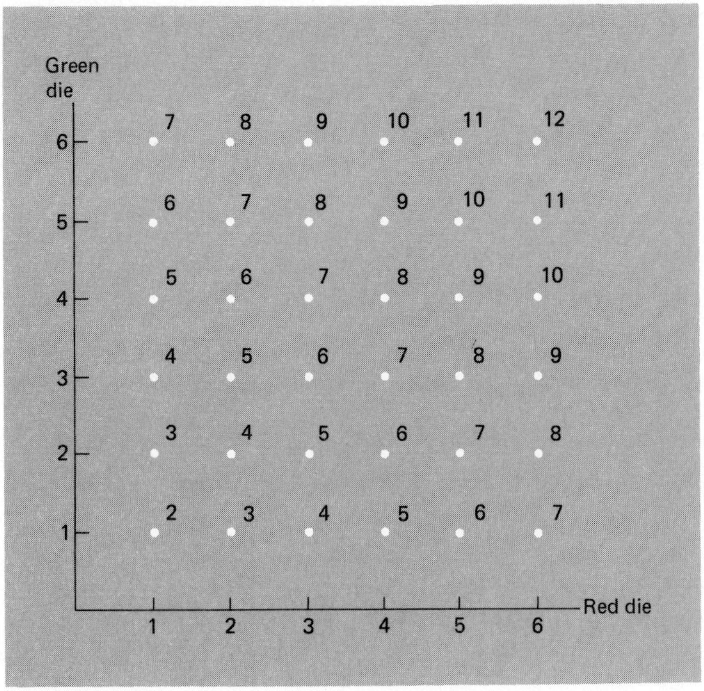

Figure 3.1 The total number of points rolled with a pair of dice.

number 8 to the point (6, 2), the number 11 to the point (5, 6), and so forth. Evidently, we associated with each point the value of a random variable, namely, the corresponding total rolled with the pair of dice.

Since "associating a number with each point (element) of a sample space" is merely another way of saying that we are "defining a function over the points of a sample space," let us now make the following definition:

DEFINITION 3.1 If S is a sample space with a probability measure and $\mathbf{x}$ is a real-valued function defined over the elements of S, then $\mathbf{x}$ is called a **random variable**.†

In this book we shall always write random variables in boldface type and their values in the corresponding lightface type; for instance, we shall write x to denote a value of the random variable $\mathbf{x}$. This practice is not universally accepted (some

† Instead of "random variable," the terms "chance variable," "stochastic variable," and "variate" are also used in some books.

authors use capital letters to denote random variables), but it is consistent with the notation of modern mathematics. When actually writing on note paper or on a blackboard, it is convenient to indicate random variables by underlining the respective symbols, perhaps with wavy lines, the usual typesetting notation for boldface type.

With reference to the above example, observe that the random variable $\mathbf{x}$ takes on the value 9, and we write $\mathbf{x} = 9$, for the subset

$$\{(6, 3), (5, 4), (4, 5), (3, 6)\}$$

of the sample space S. Thus, $\mathbf{x} = 9$ is to be interpreted as the set of elements of S for which the total is 9, and more generally, $\mathbf{x} = x$ is to be interpreted as the set of elements of the sample space for which the random variables $\mathbf{x}$ takes on the value x.

EXAMPLE 3.1

Two socks are selected at random and removed in succession from a drawer containing five brown socks and three green socks. List the elements of the sample space, the corresponding probabilities, and the corresponding values w of the random variable $\mathbf{w}$, where $\mathbf{w}$ is the number of brown socks selected.

Solution

If B and G stand for brown and green, the probabilities for BB, BG, GB, and GG are, respectively, $\frac{5}{8} \cdot \frac{4}{7} = \frac{5}{14}$, $\frac{5}{8} \cdot \frac{3}{7} = \frac{15}{56}$, $\frac{3}{8} \cdot \frac{5}{7} = \frac{15}{56}$, and $\frac{3}{8} \cdot \frac{2}{7} = \frac{3}{28}$, and the results are shown in the following table:

Element of sample space	*Probability*	w
BB	$\frac{5}{14}$	2
BG	$\frac{15}{56}$	1
GB	$\frac{15}{56}$	1
GG	$\frac{3}{28}$	0

Also, we can write $P(\mathbf{w} = 2) = \frac{5}{14}$, for example, for the probability of the event that the random variable $\mathbf{w}$ will take on the value 2.

EXAMPLE 3.2

A balanced coin is tossed four times. List the elements of the sample space which are presumably all equally likely, and the corresponding values x of the random variable $\mathbf{x}$, the total number of heads.

Solution

If H and T stand for heads and tails, the results are as shown in the following table:

Element of sample space	*Probability*	x
HHHH	$\frac{1}{16}$	4
HHHT	$\frac{1}{16}$	3
HHTH	$\frac{1}{16}$	3
HTHH	$\frac{1}{16}$	3
THHH	$\frac{1}{16}$	3
HHTT	$\frac{1}{16}$	2
HTHT	$\frac{1}{16}$	2
HTTH	$\frac{1}{16}$	2
THHT	$\frac{1}{16}$	2
THTH	$\frac{1}{16}$	2
TTHH	$\frac{1}{16}$	2
HTTT	$\frac{1}{16}$	1
THTT	$\frac{1}{16}$	1
TTHT	$\frac{1}{16}$	1
TTTH	$\frac{1}{16}$	1
TTTT	$\frac{1}{16}$	0

Also, we can write $P(\mathbf{x} = 3) = \frac{4}{16}$, for example, for the probability of the event that the random variable $\mathbf{x}$ will take on the value 3.

The fact that the definition on page 71 is limited to real-valued functions does not impose any undue restrictions. If the numbers we want to assign to the outcomes of an experiment are complex numbers, we can always look upon the real and the imaginary parts separately as values taken on by two random variables. Also, if we want to describe the outcomes of an experiment quantitatively, say, by giving the color of a person's hair, we can arbitrarily make the descriptions real-valued by coding the various colors; perhaps, by representing them with the numbers 1, 2, 3, etc.

In all of the examples of this section we have limited our discussion to discrete sample spaces, and hence to **discrete random variables**, namely, random variables whose range is finite or countably infinite. Continuous random variables defined over continuous sample spaces will be taken up in Section 3.3.

3.2 DISCRETE PROBABILITY DISTRIBUTIONS

As we already saw in Examples 3.1 and 3.2, the probability measure defined over a discrete sample space automatically provides the probabilities that a random variable will take on any given value within its range.

For instance, having assigned the probability $\frac{1}{36}$ to each element of the sample space of Figure 3.1, we immediately find that the random variable $\mathbf{x}$, the total rolled with the pair of dice, takes on the value 9 with probability $\frac{4}{36}$, since $\mathbf{x} = 9$ (described on page 72) contains 4 of the 36 equally likely sample points. The probabilities associated with all possible values of $\mathbf{x}$ are shown in the following table:

x	$P(\mathbf{x} = x)$
2	$\frac{1}{36}$
3	$\frac{2}{36}$
4	$\frac{3}{36}$
5	$\frac{4}{36}$
6	$\frac{5}{36}$
7	$\frac{6}{36}$
8	$\frac{5}{36}$
9	$\frac{4}{36}$
10	$\frac{3}{36}$
11	$\frac{2}{36}$
12	$\frac{1}{36}$

Instead of displaying the probabilities associated with the values of a random variable in a table, as we did in the preceding illustration, it is usually preferable to give a formula, that is, to express the probabilities by means of a function such that its values, $f(x)$, equal $P(\mathbf{x} = x)$ for each x within the range of the random variable $\mathbf{x}$. For instance, for the total rolled with a pair of dice we can write

$$f(x) = \frac{6 - |x - 7|}{36} \qquad \text{for } x = 2, 3, \ldots, 12$$

as can easily be verified by substitution. Clearly,

$$f(2) = \frac{6 - |2 - 7|}{36} = \frac{6 - 5}{36} = \frac{1}{36}$$

$$f(3) = \frac{6 - |3 - 7|}{36} = \frac{6 - 4}{36} = \frac{2}{36}$$

$$\cdots\cdots\cdots\cdots\cdots\cdots$$

$$f(12) = \frac{6 - |12 - 7|}{36} = \frac{6 - 5}{36} = \frac{1}{36}$$

and all these values agree with the ones shown in the table above.

DEFINITION 3.2 If $\mathbf{x}$ is a discrete random variable, the function given by $f(x) = P(\mathbf{x} = x)$ for each x within the range of $\mathbf{x}$ is called the **probability function**, or **probability distribution**, of $\mathbf{x}$.

Based on the postulates of probability, it immediately follows that

THEOREM 3.1 A function can serve as the probability distribution of a discrete random variable $\mathbf{x}$ if and only if its values, $f(x)$, satisfy the conditions

1. $f(x) \geqslant 0$ for each value within its domain;
2. $\sum_x f(x) = 1$, where the summation extends over all the values within its domain.

EXAMPLE 3.3

Find a formula for the probability distribution of the total number of heads obtained in four tosses of a balanced coin.

Solution

Based on the probabilities in the table on page 73, we find that $P(\mathbf{x} = 0) = \frac{1}{16}$, $P(\mathbf{x} = 1) = \frac{4}{16}$, $P(\mathbf{x} = 2) = \frac{6}{16}$, $P(\mathbf{x} = 3) = \frac{4}{16}$, and $P(\mathbf{x} = 4) = \frac{1}{16}$. Observing that the numerators of these five fractions, 1, 4, 6, 4, and 1, are the binomial coefficients $\binom{4}{0}$, $\binom{4}{1}$, $\binom{4}{2}$, $\binom{4}{3}$, and $\binom{4}{4}$, we find that we can write the formula for the probability distribution as

$$f(x) = \frac{\binom{4}{x}}{16} \qquad \text{for } x = 0, 1, 2, 3, 4$$

A theoretical justification for this formula, and a more general treatment for n tosses of a balanced coin, will be given in Section 5.4.

EXAMPLE 3.4

Check whether the function given by

$$f(x) = \frac{x + 2}{25} \qquad \text{for } x = 1, 2, 3, 4, 5$$

can serve as the probability function of a random variable.

Solution

Substituting the different values of x, we get $f(1) = \frac{3}{25}$, $f(2) = \frac{4}{25}$, $f(3) = \frac{5}{25}$, $f(4) = \frac{6}{25}$, and $f(5) = \frac{7}{25}$. Since these values are all non-negative, the first condition of Theorem 3.1 is satisfied, and since

$$\begin{aligned} f(1) + f(2) + f(3) + f(4) + f(5) &= \tfrac{3}{25} + \tfrac{4}{25} + \tfrac{5}{25} + \tfrac{6}{25} + \tfrac{7}{25} \\ &= 1 \end{aligned}$$

the second condition of Theorem 3.1 is satisfied. Thus, the given function can serve as the probability distribution of a random variable having the range $\{1, 2, 3, 4, 5\}$. Of course, whether any given random variable actually has this probability distribution is an entirely different matter.

In some problems it is desirable to present probability distributions graphically, and two kinds of graphical presentations used for this purpose are shown in Figures 3.2 and 3.3. The one shown in Figure 3.2, called a **probability histogram**,

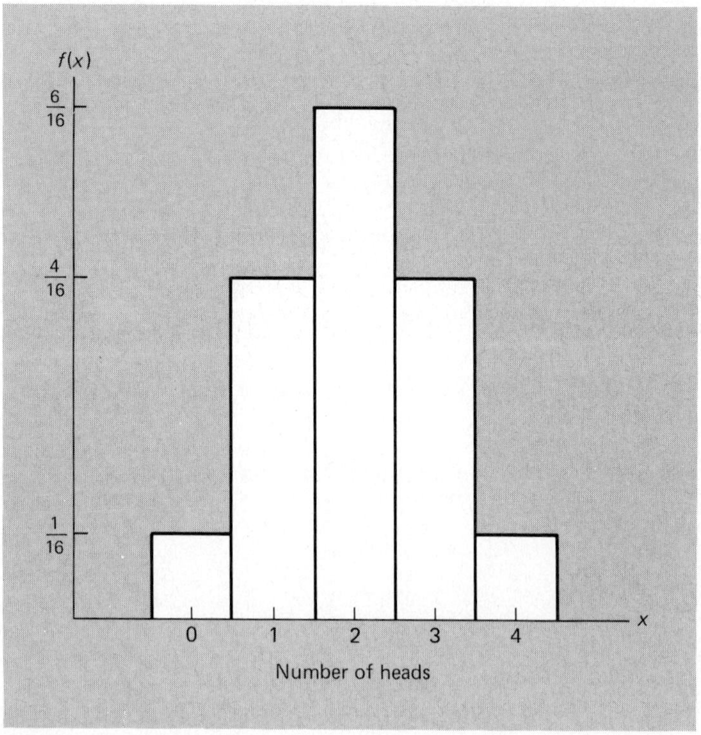

Figure 3.2 Probability histogram.

represents the probability distribution of Example 3.3. The height of each rectangle equals the probability that **x** takes on the value which corresponds to the midpoint of its base. By representing 0 with the interval from −0.5 to 0.5, 1 with the interval from 0.5 to 1.5, . . . , and 4 with the interval from 3.5 to 4.5, we are so to speak "spreading" the values of the given discrete random variable over a continuous scale.

Since each rectangle of the histogram of Figure 3.2 has unit width, we could have said that the *areas* of the rectangles, rather than their heights, equal the corresponding probabilities. There are certain advantages to identifying the areas of the rectangles with the probabilities; for instance, when we wish to approximate the graph of a discrete probability distribution with a continuous curve. This can be done even when the rectangles of a histogram do not all have unit width, by adjusting the heights of the rectangles or by modifying the vertical scale.

The graph of Figure 3.3 is called a **bar chart**. As in Figure 3.2, the height of each rectangle, or bar, equals the probability of the corresponding value of the

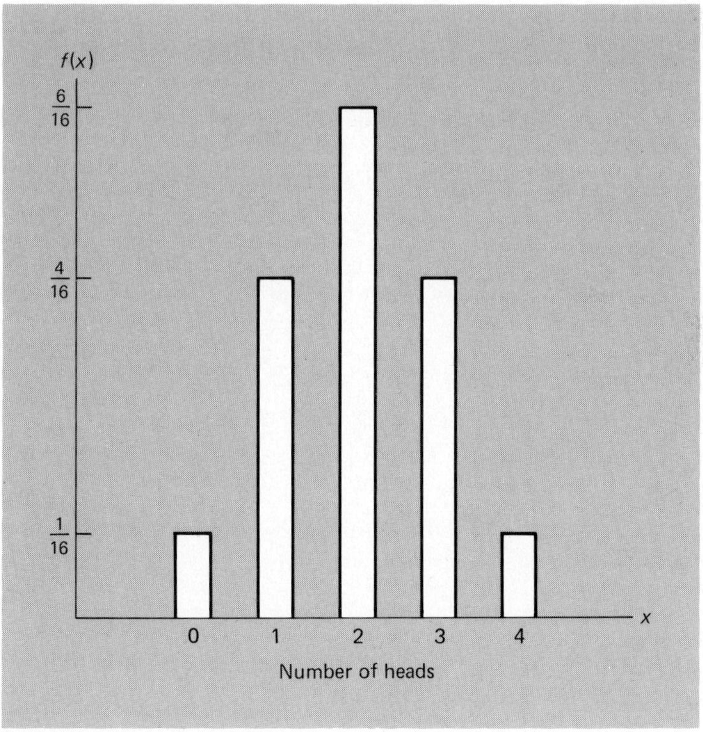

Figure 3.3 Bar chart.

random variable, but there is no pretense of having a continuous horizontal scale. Although there are several occasions where we shall use such charts in this text, histograms and bar charts are used mainly in descriptive statistics to convey visually the information provided by a probability distribution or a distribution of actual data.

There are many problems in which it is of interest to know the probability that the value of a random variable is less than or equal to some real number x. Thus, let us write the probability that $\mathbf{x}$ takes on a value less than or equal to x as $F(x) = P(\mathbf{x} \leq x)$, and refer to this function defined for all real numbers x as the **distribution function**, or **cumulative distribution**, of the random variable $\mathbf{x}$.

DEFINITION 3.3 If $\mathbf{x}$ is a discrete random variable, the function given by

$$F(x) = P(\mathbf{x} \leq x) = \sum_{t \leq x} f(t) \qquad \text{for } -\infty < x < \infty$$

where $f(t)$ is the value of the probability distribution of $\mathbf{x}$ at t, is called the **distribution function**, or **cumulative distribution**, of $\mathbf{x}$.

Based on the postulates of probability and some of its immediate consequences, it follows that

THEOREM 3.2 The values, $F(x)$, of the distribution function of a discrete random variable $\mathbf{x}$ satisfy the conditions

1. $F(-\infty) = 0$;
2. $F(\infty) = 1$;
3. if $a < b$, then $F(a) \leq F(b)$ for any real numbers a and b.

If we are given a discrete probability distribution, the corresponding distribution function is generally easy to find.

EXAMPLE 3.5

Find the distribution function of the total number of heads obtained in four tosses of a balanced coin.

Solution

Given $f(0) = \frac{1}{16}$, $f(1) = \frac{4}{16}$, $f(2) = \frac{6}{16}$, $f(3) = \frac{4}{16}$, and $f(4) = \frac{1}{16}$ from Example 3.3, it follows that

$$F(0) = f(0) = \tfrac{1}{16}$$
$$F(1) = f(0) + f(1) = \tfrac{5}{16}$$
$$F(2) = f(0) + f(1) + f(2) = \tfrac{11}{16}$$
$$F(3) = f(0) + f(1) + f(2) + f(3) = \tfrac{15}{16}$$
$$F(4) = f(0) + f(1) + f(2) + f(3) + f(4) = 1$$

Hence, the distribution function is given by

$$F(x) = \begin{cases} 0 & \text{for } x < 0 \\ \frac{1}{16} & \text{for } 0 \leqslant x < 1 \\ \frac{5}{16} & \text{for } 1 \leqslant x < 2 \\ \frac{11}{16} & \text{for } 2 \leqslant x < 3 \\ \frac{15}{16} & \text{for } 3 \leqslant x < 4 \\ 1 & \text{for } x \geqslant 4 \end{cases}$$

Observe that this distribution function is defined not only for the values taken on by the given random variable, but for all real numbers. For instance, we can write $F(1.7) = \frac{5}{16}$ and $F(100) = 1$, although the probabilities of getting "at most 1.7 heads" or "at most 100 heads" in four tosses of a balanced coin may not be of particular significance.

EXAMPLE 3.6

Find the distribution function of the random variable **w** of Example 3.1 and plot its graph.

Solution

Based on the probabilities given in the table on page 72, we can write $f(0) = \frac{3}{28}$, $f(1) = \frac{15}{56} + \frac{15}{56} = \frac{15}{28}$, and $f(2) = \frac{5}{14}$, so that

$$F(0) = f(0) = \tfrac{3}{28}$$
$$F(1) = f(0) + f(1) = \tfrac{9}{14}$$
$$F(2) = f(0) + f(1) + f(2) = 1$$

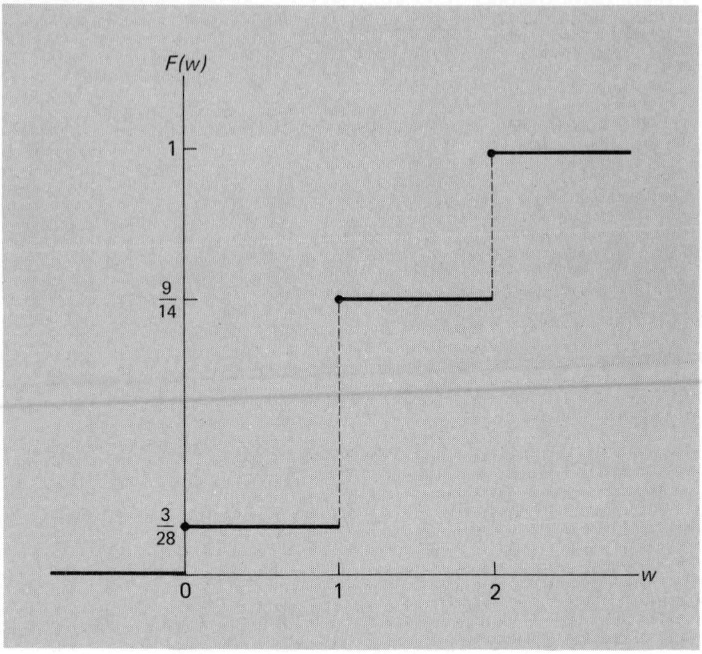

Figure 3.4 Graph of distribution function of Example 3.6.

Hence, the distribution function of **w** is given by

$$F(w) = \begin{cases} 0 & \text{for } w < 0 \\ \frac{3}{28} & \text{for } 0 \leq w < 1 \\ \frac{9}{14} & \text{for } 1 \leq w < 2 \\ 1 & \text{for } w \geq 2 \end{cases}$$

The graph of this distribution function, shown in Figure 3.4, was obtained by first plotting the points $(w, F(w))$ for $w = 0$, 1, and 2. Then, the step function is completed as indicated, and it should be observed that at all points of discontinuity it takes on the greater of the two values.

We can also reverse the process illustrated in the two preceding examples, namely, obtain values of the probability distribution of a random variable from its distribution function. To this end, we use the following result:

THEOREM 3.3 If the range of a random variable **x** consists of the values $x_1 < x_2 < x_3 < \cdots < x_n$, then $f(x_1) = F(x_1)$ and

$$f(x_i) = F(x_i) - F(x_{i-1}) \qquad \text{for } i = 2, 3, \ldots, n$$

EXAMPLE 3.7

Given the distribution function

$$F(x) = \begin{cases} 0 & \text{for } x < 2 \\ \frac{1}{36} & \text{for } 2 \leq x < 3 \\ \frac{3}{36} & \text{for } 3 \leq x < 4 \\ \frac{6}{36} & \text{for } 4 \leq x < 5 \\ \frac{10}{36} & \text{for } 5 \leq x < 6 \\ \frac{15}{36} & \text{for } 6 \leq x < 7 \\ \frac{21}{36} & \text{for } 7 \leq x < 8 \\ \frac{26}{36} & \text{for } 8 \leq x < 9 \\ \frac{30}{36} & \text{for } 9 \leq x < 10 \\ \frac{33}{36} & \text{for } 10 \leq x < 11 \\ \frac{35}{36} & \text{for } 11 \leq x < 12 \\ 1 & \text{for } x \geq 12 \end{cases}$$

find the values of the probability distribution of this random variable.

Solution

Making use of Theorem 3.3, we get $f(2) = \frac{1}{36}$, $f(3) = \frac{3}{36} - \frac{1}{36} = \frac{2}{36}$, $f(4) = \frac{6}{36} - \frac{3}{36} = \frac{3}{36}$, $f(5) = \frac{10}{36} - \frac{6}{36} = \frac{4}{36}, \ldots, f(12) = 1 - \frac{35}{36} = \frac{1}{36}$, and comparison with the probabilities in the table on page 74 reveals that the random variable we are concerned with here is the total number of points rolled with a pair of dice.

In the remainder of this chapter we will be concerned with continuous random variables and their distributions, and with problems relating to the simultaneous occurrence of the values of two or more random variables. In Chapter 5 we shall return to discrete probability distributions; in fact, all of that chapter will be devoted to discrete probability distributions which provide especially important models for applications.

THEORETICAL EXERCISES

1. Verify that $f(x) = \dfrac{2x}{k(k+1)}$ for $x = 1, 2, 3, \ldots, k$ can serve as the probability distribution of a random variable.

2. For each of the following, determine c so that the function can serve as the probability distribution of a random variable:

(a) $f(x) = cx$ for $x = 1, 2, 3, 4, 5$;

(b) $f(x) = c\binom{5}{x}$ for $x = 0, 1, 2, 3, 4, 5$;
(c) $f(x) = cx^2$ for $x = 1, 2, 3, \ldots, k$;
(d) $f(x) = c2^x$ for $x = 1, 2, 3, \ldots, k$;
(e) $f(x) = c(\frac{1}{4})^x$ for $x = 1, 2, 3, \ldots$.

[*Hint*: For part (c) refer to Appendix II at the end of the book.]

3. For what values of k can

$$f(x) = (1 - k)k^x \qquad \text{for } x = 0, 1, 2, \ldots$$

serve as the probability distribution of a random variable?

4. Show that there are no values of c such that the following can serve as probability distributions:

(a) $f(x) = \frac{c}{x}$ for $x = 1, 2, 3, \ldots$;
(b) $f(x) = c2^x$ for $x = 1, 2, 3, \ldots$.

5. Construct a probability histogram for each of the following discrete probability distributions:

(a) $f(x) = \dfrac{\binom{2}{x}\binom{4}{3-x}}{\binom{6}{3}}$ for $x = 0, 1, 2$;

(b) $f(x) = \binom{5}{x}\left(\frac{1}{5}\right)^x\left(\frac{4}{5}\right)^{5-x}$ for $x = 0, 1, 2, 3, 4, 5$.

6. Prove Theorem 3.2.

7. Find the distribution function which corresponds to the probability distribution of part (a) of Exercise 5 and plot its graph.

8. Given that $\mathbf{x}$ has the distribution function

$$F(x) = \begin{cases} 0 & \text{for } x < -1 \\ \frac{1}{4} & \text{for } -1 \leqslant x < 1 \\ \frac{1}{2} & \text{for } 1 \leqslant x < 3 \\ \frac{3}{4} & \text{for } 3 \leqslant x < 5 \\ 1 & \text{for } x \geqslant 5 \end{cases}$$

find

(a) $P(\mathbf{x} \leqslant 3)$; (b) $P(\mathbf{x} = 3)$; (c) $P(\mathbf{x} < 3)$;
(d) $P(\mathbf{x} \geqslant 1)$; (e) $P(-0.4 < \mathbf{x} < 4)$; (f) $P(\mathbf{x} = 5)$.

9. With reference to Example 3.4, verify that for $x = 1, 2, 3, 4$, and 5, the values of the distribution function are given by

$$F(x) = \frac{x^2 + 5x}{50}$$

10. With reference to Theorem 3.3, verify that

(a) $P(\mathbf{x} > x_i) = 1 - F(x_i)$ for $i = 1, 2, 3, \ldots, n$;
(b) $P(\mathbf{x} \geqslant x_i) = 1 - F(x_{i-1})$ for $i = 2, 3, \ldots, n$, and $P(\mathbf{x} \geqslant x_1) = 1$.

APPLIED EXERCISES

11. With reference to Example 3.3, find the probability distribution of **y**, the difference between the number of heads and the number of tails obtained in four tosses of a balanced coin.

12. An urn contains four balls numbered 1, 2, 3, and 4, respectively. If two balls are drawn from the urn without replacement and **x** is the sum of the numbers on the two balls drawn, find

(a) the probability distribution of **x** and represent it by means of a histogram;
(b) the distribution function of **x** and draw its graph.

13. A tape recorder contains six transistors, of which two are defective. If two of these transistors are selected at random, removed from the tape recorder, and inspected, and if **x** is the number of defectives observed, find

(a) the probability distribution of **x**;
(b) the distribution function of **x**.

Also draw a histogram of the probability distribution and a graph of the distribution function.

14. A coin is biased so that heads is three times as likely as tails. For three independent tosses of the coin, find

(a) the probability distribution of **x**, the total number of heads;
(b) the probability of getting at most two heads.

15. With reference to Exercise 14, find the distribution function of the random variable **x** and plot its graph. Using this distribution function, find

(a) $P(1 < \mathbf{x} \leqslant 3)$;
(b) $P(\mathbf{x} > 2)$.

16. The probability distribution of **x**, the weekly number of accidents at a certain intersection, is given by $f(0) = 0.40$, $f(1) = 0.30$, $f(2) = 0.20$, and $f(3) = 0.10$. Construct the distribution function of this random variable and draw its graph.

17. The probabilities that a person shopping at a certain department store will not make a purchase, make at most one purchase, at most two purchases, at most three purchases, or at most four purchases, are, respectively, 0.22, 0.54, 0.87, 0.91, and 1.00. Find the probabilities that a person shopping at this department store will make

(a) two purchases; **(b)** more than two purchases;
(c) three purchases; **(d)** at least one purchase.

3.3 CONTINUOUS RANDOM VARIABLES

In Section 3.1 we introduced the concept of a random variable as a real-valued function defined over the points of a sample space with a probability measure, and in Figure 3.1 we illustrated this by assigning the total rolled with a pair of dice to each of the 36 equally likely points of the sample space. In the continuous case, where random variables can take on values on a continuous scale, the procedure is very much the same. The outcomes of experiments are represented by the points on line segments or lines, and the values of random variables are numbers appropriately assigned to the points by means of rules or equations. When the value of a random variable is given directly by a measurement or observation, we generally do not bother to distinguish between the value of the random variable (the measurement which we obtain) and the outcome of the experiment (the corresponding point on the real axis). Thus, if an experiment consists of determining the actual content of a 230-gram jar of instant coffee, the result itself, say, 225.3 grams, is the value of the random variable with which we are concerned, and there is no real need to add that the sample space consists of a certain continuous interval of points on the positive real axis.

The problem of defining probabilities in connection with continuous sample spaces and continuous random variables involves some complications. To illustrate, let us consider the following situation:

EXAMPLE 3.8

Suppose that we are concerned with the possibility that an accident will occur on a freeway which is 200 kilometers long, and that we are interested in the probability that it will occur at a given location, or perhaps on a given stretch of the road. The sample space of this "experiment" consists of a continuum of points, those on the interval from 0 to 200, and we shall assume, for the sake of argument, that the probability that an accident will occur on any interval of length D is $\frac{D}{200}$, with D measured in kilometers. Note that this assignment of probabilities is consistent with Postulates 1 and 2 on page 35, since the probabilities $\frac{D}{200}$ are all non-negative and $P(S) = \frac{200}{200} = 1$. So far, this assignment of probabilities applies only to intervals on the line segment from 0 to 200, but if we use Postulate 3, we can also obtain probabilities for the union of any finite or countably infinite sequence of non-overlapping intervals. For instance, the probability that an accident will occur on either of two non-overlapping intervals of length D_1 and D_2 is

$$\frac{D_1 + D_2}{200}$$

and the probability that it will occur on one of a countably infinite sequence of non-overlapping intervals of length $D_1, D_2, D_3, \ldots$, is

$$\frac{D_1 + D_2 + D_3 + \cdots}{200}$$

Then, if we apply Theorem 2.7, we can extend the probability assignment to the union of intervals that overlap, and since the intersection of two intervals is an interval and the complement of an interval is either an interval or the union of two intervals, we can extend the probability assignment to any subset of the sample space which can be obtained by forming unions or intersections of finitely many or countably many intervals, or by forming complements.

Thus, in extending the concept of probability to the continuous case, we have again used Postulates 1, 2, and 3, but to do this in general we must exclude from our definition of "event" all subsets of the sample space which cannot be obtained by forming unions or intersections of finitely many or countably many intervals, or by forming complements. Practically speaking, this is of no consequence, for we simply do not assign probabilities to such abstruse kinds of sets.

With reference to Example 3.8, observe also that the probability of the accident occurring on a very short interval, say, an interval of 1 centimeter, is only 0.00000005, which is very small. As the length of the interval approaches zero, the probability that an accident will occur on it also approaches zero; indeed, in the continuous case we always assign zero probability to individual points. This does not mean that the corresponding events cannot occur—after all, when an accident occurs on the 200-kilometer stretch of road, it has to occur at some point even though each point has zero probability.

3.4 PROBABILITY DENSITY FUNCTIONS

The way in which we assigned probabilities in Example 3.8 is very special, and it is similar in nature to the way in which we assign equal probabilities to the six faces of a die, heads and tails, the 52 playing cards in a standard deck, and so forth. To treat the problem of associating probabilities with continuous random variables more generally, suppose that a bottler of soft drinks is concerned about the actual amount of a soft drink which his bottling machine puts into 16-ounce bottles. Evidently, the amount will vary somewhat from bottle to bottle and hence is a continuous random variable. However, if he rounds the amounts to the nearest tenth of an ounce, he will be dealing with a discrete random variable which has a probability distribution, and this probability distribution may be pictured as a histogram in which the probabilities are given by the areas of rectangles, say, as in the diagram at the top of Figure 3.5. If he rounds the amounts to the nearest hundredth of an ounce, he will again be dealing with a discrete random variable (a different one) which has a probability distribution, and this probability distribution may be pictured as a histogram in which the probabilities are given by the areas of rectangles, say, as in the diagram in the middle of Figure 3.5.

It should be apparent that if he rounded the amounts to the nearest thousandth of an ounce, or to the nearest ten-thousandth of an ounce, the histograms of the probability distributions of the corresponding discrete random variables will approach the continuous curve shown in the diagram at the bottom of Figure 3.5, and the sum of the areas of the rectangles which represent the probability that the amount falls within any specified interval approaches the corresponding area under the curve.

Indeed, the definition of probability in the continuous case presumes for each random variable the existence of a function, called a **probability density function**, so that areas under the curve give the probabilities associated with the corresponding intervals along the horizontal axis. In other words, a probability density function, integrated from a to b (with $a \leq b$), gives the probability that the corresponding random variable will take on a value on the interval from a to b.

DEFINITION 3.4 A function with values $f(x)$, defined over the set of all real numbers, is called a **probability density function** of the continuous random variable $\mathbf{x}$ if and only if

$$P(a \leq \mathbf{x} \leq b) = \int_a^b f(x)\, dx$$

for any real constants a and b with $a \leq b$.

Probability density functions are also referred to, more briefly, as **probability densities**, **density functions**, **densities**, and **p.d.f.**'s.

Note that $f(c)$, the value of the probability density function of $\mathbf{x}$ at c does not give $P(\mathbf{x} = c)$ as in the discrete case. In connection with continuous random variables, probabilities are always given by integrals evaluated over intervals,

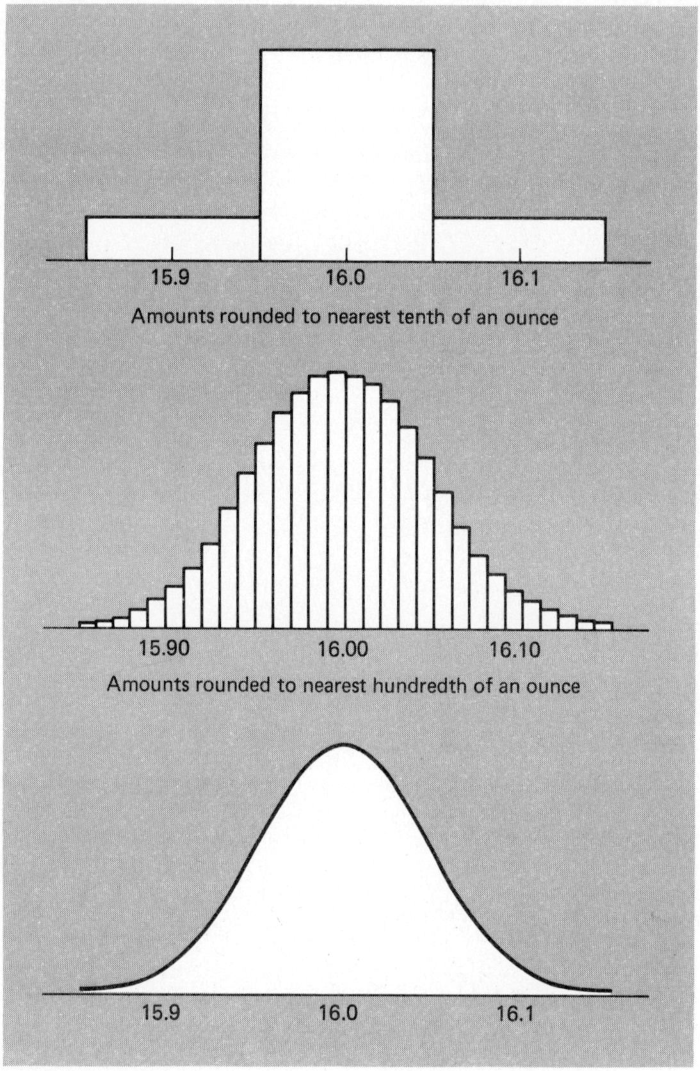

Figure 3.5 Definition of probability in the continuous case.

whereas $P(\mathbf{x} = c) = 0$ for any real constant c. This agrees with our discussion on page 85 and it also follows directly from Definition 3.4 with $a = b = c$.

Because of this property, the value of a probability density function can be changed for some of the values of a random variable without changing any of the probabilities, and this is why we said in Definition 3.4 that $f(x)$ is the value of *a* probability density, not *the* probability density, of the random $\mathbf{x}$ at x. Also because of this property, it does not matter whether we include the endpoints of the interval from a to b; symbolically,

THEOREM 3.4 If $\mathbf{x}$ is a continuous random variable and a and b are two real constants with $a \leqslant b$, then

$$P(a \leqslant \mathbf{x} \leqslant b) = P(a \leqslant \mathbf{x} < b) = P(a < \mathbf{x} \leqslant b) = P(a < \mathbf{x} < b)$$

Analogous to Theorem 3.1, let us now state the following properties of probability density functions, which again follow directly from the postulates of probability:

THEOREM 3.5 A function can serve as a probability density function of a continuous random variable $\mathbf{x}$ if its values, $f(x)$, satisfy the conditions†

1. $f(x) \geqslant 0$ for $-\infty < x < \infty$;

2. $\int_{-\infty}^{\infty} f(x)\, dx = 1.$

EXAMPLE 3.9

The probability density function of the random variable $\mathbf{x}$ is given by

$$f(x) = \begin{cases} k \cdot e^{-3x} & \text{for } x > 0 \\ 0 & \text{elsewhere} \end{cases}$$

Find k and $P(0.5 \leqslant \mathbf{x} \leqslant 1)$.

Solution

To satisfy the second condition of Theorem 3.5, we must have

$$\int_{-\infty}^{\infty} f(x)\, dx = \int_{0}^{\infty} k \cdot e^{-3x}\, dx = k \cdot \lim_{t \to \infty} \frac{e^{-3x}}{-3}\bigg|_0^t = \frac{k}{3} = 1$$

† The conditions are not "if and only if" as in Theorem 3.1 because $f(x)$ could be negative for some values of the random variable without affecting any of the probabilities. However, both conditions of Theorem 3.5 will be satisfied by all the probability density functions we shall study in this text.

and it follows that $k = 3$. For the probability we get

$$P(0.5 \le \mathbf{x} \le 1) = \int_{0.5}^{1} 3e^{-3x}\, dx = -e^{-3x}\Big|_{0.5}^{1} = -e^{-3} + e^{-1.5} = 0.173$$

Although the random variable of Example 3.9 cannot take on negative values, we artificially extended the domain of its probability density to include all the real numbers. This is a practice we shall follow throughout this text.

As in the discrete case, there are many problems in which it is of interest to know the probability that the value of a continuous random variable is less than or equal to some real number x. Thus, let us make the following definition analogous to Definition 3.3:

DEFINITION 3.5 If $\mathbf{x}$ is a continuous random variable, the function given by

$$F(x) = P(\mathbf{x} \le x) = \int_{-\infty}^{x} f(t)\, dt \qquad \text{for } -\infty < x < \infty$$

where $f(t)$ is the value of the probability density function of $\mathbf{x}$ at t, is called the **distribution function**, or **cumulative distribution**, of $\mathbf{x}$.

The properties of distribution functions given in Theorem 3.2 hold also for the continuous case; that is, $F(-\infty) = 0$, $F(\infty) = 1$, and $F(a) \le F(b)$ when $a < b$. Furthermore, it immediately follows from Definition 3.5 that

THEOREM 3.6 If $f(x)$ and $F(x)$ are, respectively, values of the probability distribution and the distribution function of $\mathbf{x}$ at x, then

$$P(a \le \mathbf{x} \le b) = F(b) - F(a)$$

for any real constants a and b with $a \le b$, and

$$f(x) = \frac{dF(x)}{dx}$$

where the derivative exists.

EXAMPLE 3.10

Find the distribution function which corresponds to the probability density function of Example 3.9. Also, use this distribution function to reevaluate $P(0.5 \le \mathbf{x} \le 1)$.

Solution

For $x > 0$,

$$F(x) = \int_{-\infty}^{x} f(t)\,dt = \int_{0}^{x} 3e^{-3t}\,dt = -e^{-3t}\Big|_{0}^{x} = 1 - e^{-3x}$$

and we can write

$$F(x) = \begin{cases} 0 & \text{for } x \leqslant 0 \\ 1 - e^{-3x} & \text{for } x > 0 \end{cases}$$

To find the probability $P(0.5 \leqslant \mathbf{x} \leqslant 1)$ we make use of the formula of the first part of Theorem 3.6, getting

$$\begin{aligned} P(0.5 \leqslant \mathbf{x} \leqslant 1) &= F(1) - F(0.5) \\ &= (1 - e^{-3}) - (1 - e^{-1.5}) \\ &= 0.173 \end{aligned}$$

This agrees with the result obtained by using the probability density function in Example 3.9.

EXAMPLE 3.11

Find a probability density function for the random variable whose distribution function is given by

$$F(x) = \begin{cases} 0 & \text{for } x \leqslant 0 \\ x & \text{for } 0 < x < 1 \\ 1 & \text{for } x \geqslant 1 \end{cases}$$

Solution

The graph of this distribution function is shown in Figure 3.6, and it can be seen that it is continuous and differentiable everywhere except at $x = 0$ and $x = 1$. Differentiating the distribution function for $x < 0$, for $0 < x < 1$, and for $x > 1$, we get

$$f(x) = \begin{cases} 0 & \text{for } x < 0 \\ 1 & \text{for } 0 < x < 1 \\ 0 & \text{for } x > 1 \end{cases}$$

and to fill the two gaps we let $f(0)$ and $f(1)$ both equal zero. Actually, it does not

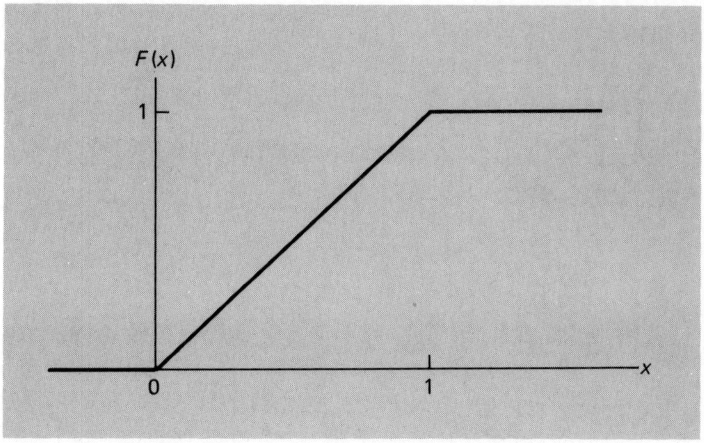

Figure 3.6 Graph of the distribution function of Example 3.11.

matter how the probability density function is defined at these two points, but there are certain advantages (which will be explained on page 232) for choosing the values in such a way that the probability density function is non-zero over an open interval. Thus, we can write the probability density function as

$$f(x) = \begin{cases} 1 & \text{for } 0 < x < 1 \\ 0 & \text{elsewhere} \end{cases}$$

and the graph of this function is shown in Figure 3.7.

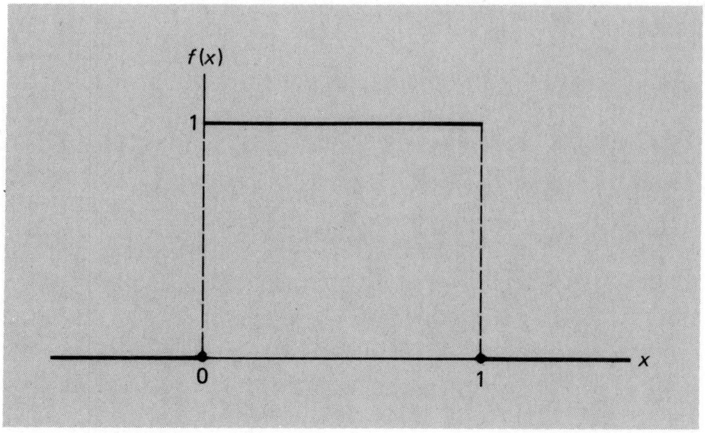

Figure 3.7 Graph of the probability density function of Example 3.11.

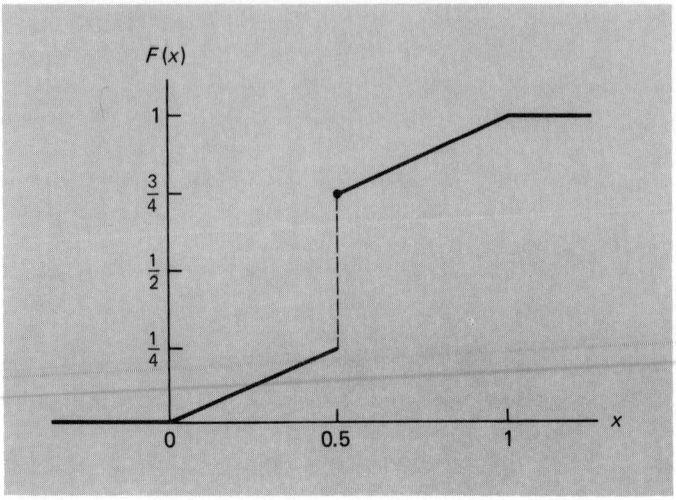

Figure 3.8 Graph of the distribution function of a mixed random variable.

In most practical applications we encounter random variables that are either discrete or continuous, so that the corresponding distribution functions have either a step-like appearance as in Figure 3.4 or they are continuous curves as in Figure 3.6. Discontinuous distribution functions like that of Figure 3.8 arise when random variables are **mixed**. The distribution function of such a random variable will be discontinuous at each point having a non-zero probability and continuous elsewhere. As in the discrete case, the height of the step at a point of discontinuity gives the corresponding probability that the random variable will take on that value. With reference to Figure 3.8, $P(\mathbf{x} = 0.5) = \frac{3}{4} - \frac{1}{4} = \frac{1}{2}$, but otherwise the random variable is like a continuous random variable. With the exception of Exercise 13 on page 96, we shall limit ourselves in this text to random variables which are either discrete or continuous, with the latter having distribution functions which are differentiable for all but a finite set of values of the random variables.

THEORETICAL EXERCISES

1. The probability density function of the continuous random variable $\mathbf{x}$ is given by

$$f(x) = \begin{cases} \frac{1}{5} & \text{for } 2 < x < 7 \\ 0 & \text{elsewhere} \end{cases}$$

(a) Show that the area under the curve (above the x-axis) is equal to 1.
(b) Find $P(3 < \mathbf{x} < 5)$.

2. If the probability density of the continuous random variable $\mathbf{y}$ is given by

$$f(y) = \begin{cases} \frac{1}{8}(y + 1) & \text{for } 2 < y < 4 \\ 0 & \text{elsewhere} \end{cases}$$

find

(a) $P(\mathbf{y} < 3.2)$;
(b) $P(2.9 < \mathbf{y} < 3.2)$.

3. If the p.d.f. of the random variable $\mathbf{x}$ is given by

$$f(x) = \begin{cases} \dfrac{c}{\sqrt{x}} & \text{for } 0 < x < 4 \\ 0 & \text{elsewhere} \end{cases}$$

find

(a) the value of c;
(b) the distribution function of this random variable;
(c) $P(\mathbf{x} > 1)$.

4. If the probability density function of the random variable $\mathbf{z}$ is given by

$$f(z) = \begin{cases} kz\, e^{-z^2} & \text{for } z > 0 \\ 0 & \text{for } z \leqslant 0 \end{cases}$$

find

(a) the value of k;
(b) the distribution function of this random variable.

Also sketch the graphs of the probability density and distribution functions.

5. If the density function of the random variable $\mathbf{x}$ is given by

$$g(x) = \begin{cases} 6x(1 - x) & \text{for } 0 < x < 1 \\ 0 & \text{elsewhere} \end{cases}$$

find

(a) $P(\mathbf{x} > \frac{1}{2})$;
(b) the distribution function of this random variable;
(c) the value of m such that $G(m) = 0.5$, namely, such that m is the **median** of the distribution of $\mathbf{x}$.

6. If the probability density function of the random variable **w** is given by

$$f(w) = \begin{cases} cw^2 + w & \text{for } 0 < w < 1 \\ 0 & \text{elsewhere} \end{cases}$$

find

(a) the value of c;
(b) the distribution function of this random variable and plot its graph;
(c) $P(0 \leqslant \mathbf{w} \leqslant \frac{1}{2})$.

7. Find the distribution function of the random variable **x** whose probability density is given by

$$f(x) = \begin{cases} \frac{1}{3} & \text{for } 0 < x < 1 \\ \frac{1}{3} & \text{for } 2 < x < 4 \\ 0 & \text{elsewhere} \end{cases}$$

Sketch the graphs of the probability density and distribution functions.

8. If the density function of the random variable **z** is given by

$$h(z) = \begin{cases} -kz & \text{for } -1 < z < 0 \\ kz & \text{for } 0 \leqslant z < 1 \\ 0 & \text{elsewhere} \end{cases}$$

find

(a) the value of k;
(b) the distribution function of this random variable and sketch its graph;
(c) $P(-\frac{1}{2} < \mathbf{z} < \frac{1}{2})$.

9. Find the distribution function of the random variable **x** whose density function is given by

$$f(x) = \begin{cases} \dfrac{x}{2} & \text{for } 0 < x \leqslant 1 \\ \dfrac{1}{2} & \text{for } 1 < x \leqslant 2 \\ \dfrac{3 - x}{2} & \text{for } 2 < x < 3 \\ 0 & \text{elsewhere} \end{cases}$$

Also sketch the graphs of the density and distribution functions.

10. If the distribution function of the random variable $\mathbf{x}$ is given by

$$F(x) = \begin{cases} 0 & \text{for } x < -1 \\ \dfrac{x+1}{2} & \text{for } -1 \leqslant x < 1 \\ 1 & \text{for } x \geqslant 1 \end{cases}$$

find

(a) $P(-\frac{1}{2} < \mathbf{x} < \frac{1}{2})$;
(b) $P(2 < \mathbf{x} < 3)$;
(c) the probability density of this random variable and use it to recalculate the probability of part (a).

11. If the distribution function of the random variable $\mathbf{y}$ is given by

$$F(y) = \begin{cases} 1 - \dfrac{9}{y^2} & \text{for } y > 3 \\ 0 & \text{elsewhere} \end{cases}$$

find

(a) $P(\mathbf{y} \leqslant 5)$;
(b) $P(\mathbf{y} > 8)$;
(c) the probability density of $\mathbf{y}$, letting its value equal zero wherever it is undefined.

Also sketch the graphs of the two functions.

12. If the distribution function of the random variable $\mathbf{x}$ is given by

$$F(x) = \begin{cases} 1 - (1 + x)\,e^{-x} & \text{for } x > 0 \\ 0 & \text{for } x \leqslant 0 \end{cases}$$

find

(a) $P(\mathbf{x} \leqslant 2)$;
(b) $P(1 < \mathbf{x} < 3)$;
(c) $P(\mathbf{x} > 4)$;
(d) the probability density function of $\mathbf{x}$. Are there any points at which it is undefined?

Also sketch the graphs of the distribution and density functions of $\mathbf{x}$.

13. If the distribution function of the random variable **z** is given by

$$F(z) = \begin{cases} 0 & \text{for } z < -2 \\ \dfrac{z+4}{8} & \text{for } -2 \leqslant z < 2 \\ 1 & \text{for } z \geqslant 2 \end{cases}$$

find

(a) $P(\mathbf{z} = -2)$; **(b)** $P(\mathbf{z} = 2)$;
(c) $P(-2 < \mathbf{z} < 1)$; **(d)** $P(0 \leqslant \mathbf{z} \leqslant 2)$.

APPLIED EXERCISES

14. The actual amount of coffee (in grams) in a 230-gram jar filled by a certain machine is a random variable whose probability density function is given by

$$f(x) = \begin{cases} 0 & \text{for } x \leqslant 227.5 \\ \frac{1}{5} & \text{for } 227.5 < x < 232.5 \\ 0 & \text{for } x \geqslant 232.5 \end{cases}$$

Find the probabilities that a 230-gram jar filled by this machine will contain
(a) at most 228.65 grams of coffee;
(b) anywhere from 229.34 to 231.66 grams of coffee;
(c) at least 229.85 grams of coffee.

15. The number of minutes that a flight from Phoenix to Tucson is early or late is a random variable whose probability density is given by

$$f(x) = \begin{cases} \frac{1}{288}(36 - x^2) & \text{for } -6 < x < 6 \\ 0 & \text{elsewhere} \end{cases}$$

where negative values are indicative of the flight's being early and positive values are indicative of its being late. Find the probabilities that one of these flights will be
(a) at least 2 minutes early;
(b) at least 1 minute late;
(c) anywhere from 1 to 3 minutes early;
(d) exactly 5 minutes late.

16. The shelf life (in hours) of a certain perishable packaged food is a random variable whose probability density function is given by

$$f(x) = \begin{cases} \dfrac{20{,}000}{(x+100)^3} & \text{for } x > 0 \\ 0 & \text{elsewhere} \end{cases}$$

Find the probabilities that one of these packages will have a shelf life of

(a) at least 200 hours;
(b) at most 100 hours;
(c) anywhere from 80 to 120 hours.

17. The tread wear (in thousands of kilometers) which car owners get with a certain kind of tire is a random variable whose probability density function is given by

$$f(x) = \begin{cases} \frac{1}{30} e^{-\frac{x}{30}} & \text{for } x > 0 \\ 0 & \text{for } x \leq 0 \end{cases}$$

Find the probabilities that one of these tires will last

(a) at most 19,000 kilometers;
(b) anywhere from 29,000 to 38,000 kilometers;
(c) at least 48,000 kilometers.

18. In a certain city the daily consumption of water (in millions of liters) is a random variable whose probability density is given by

$$f(x) = \begin{cases} \frac{1}{9} x e^{-\frac{x}{3}} & \text{for } x > 0 \\ 0 & \text{elsewhere} \end{cases}$$

What are the probabilities that on a given day

(a) the water consumption in this city is no more than 6 million liters;
(b) the water supply is inadequate if the daily capacity of this city is 9 million liters?

19. The total lifetime (in years) of five-year-old dogs of a certain breed is a random variable whose distribution function is given by

$$F(x) = \begin{cases} 0 & \text{for } x \leq 5 \\ 1 - \dfrac{25}{x^2} & \text{for } x > 5 \end{cases}$$

Find the probabilities that such a five-year-old dog will live

(a) beyond 10 years;
(b) less than 8 years;
(c) anywhere from 12 to 15 years.

3.5 MULTIVARIATE DISTRIBUTIONS

In Section 3.1 we defined a random variable as a real-valued function defined over a sample space with a probability measure, and it stands to reason that many different random variables can be defined over one and the same sample space. With reference to the sample space of Figure 3.1, for example, we considered only the random variable whose values were the totals rolled with the pair of dice, but we could also have considered the random variable whose values are the products of the numbers rolled with the two dice, the random variable whose values are the differences between the numbers rolled with the red die and the green die, the random variable whose values are 0, 1, or 2 depending on the number of dice which come up 2, and so forth. Closer to life, an experiment may consist of randomly choosing some of the 345 students attending an elementary school, and the principal may be interested in their I.Q.'s, the school nurse in their weights, their teachers in the number of days they have been absent, and so forth.

In this section we shall be concerned first with the **bivariate case**, that is, with situations where we are interested at the same time in a pair of random variables defined over a joint sample space. Later, we shall extend this discussion to the **multivariate case**, covering any finite number of random variables.

If $\mathbf{x}$ and $\mathbf{y}$ are two discrete random variables, we write the probability that $\mathbf{x}$ will take on the value x and $\mathbf{y}$ will take on the value y as $P(\mathbf{x} = x, \mathbf{y} = y)$; thus, $P(\mathbf{x} = x, \mathbf{y} = y)$ is the probability of the intersection of the events $\mathbf{x} = x$ and $\mathbf{y} = y$. As in the **univariate case** (see page 74), the probabilities associated with all possible pairs of values (x, y) can be displayed by means of a table.

EXAMPLE 3.12

Two tablets are selected at random from a bottle containing 3 aspirin, 2 sedative, and 4 laxative tablets. If $\mathbf{x}$ and $\mathbf{y}$ are, respectively, the number of aspirin tablets and the number of sedative tablets included among the two tablets drawn from the bottle, find the probabilities associated with all possible pairs of values (x, y).

Solution

The possible pairs are (0, 0), (0, 1), (1, 0), (1, 1), (0, 2), and (2, 0). To find the probability associated with (1, 0), for example, observe that we are concerned with the event of getting 1 aspirin, none of the sedatives, and hence 1 laxative tablet. The number of ways of selecting 1 of the 3 aspirin tablets and 1 of the 4 laxative tablets is $\binom{3}{1}\binom{4}{1} = 12$, and the total number of equally likely ways of selecting 2 of the 9 tablets is $\binom{9}{2} = 36$. By Theorem 2.2 it follows that the probability associated with (1, 0) is $\frac{12}{36} = \frac{1}{3}$. Similarly, the probability associated with (1, 1) is $\dfrac{\binom{3}{1}\binom{2}{1}}{36} = \dfrac{1}{6}$, and, continuing this way, we obtain the values shown in the following table:

		x = 0	x = 1	x = 2
y	0	$\frac{1}{6}$	$\frac{1}{3}$	$\frac{1}{12}$
	1	$\frac{2}{9}$	$\frac{1}{6}$	
	2	$\frac{1}{36}$		

As in the univariate case, it is usually preferable to represent probabilities like these by means of a formula; that is, express the probabilities by means of a function with the values $f(x, y) = P(\mathbf{x} = x, \mathbf{y} = y)$ for any pair of values (x, y) within the range of the random variables **x** and **y**. For instance, we shall see in Chapter 5 that we can write

$$f(x, y) = \frac{\binom{3}{x}\binom{2}{y}\binom{4}{2-x-y}}{\binom{9}{2}} \qquad \text{for } x = 0, 1, 2; \quad y = 0, 1, 2; \quad 0 \leqslant x + y \leqslant 2$$

for the pair of random variables of Example 3.12.

DEFINITION 3.6 If **x** and **y** are discrete random variables, the function given by $f(x, y) = P(\mathbf{x} = x, \mathbf{y} = y)$ for each pair of values (x, y) within the range of **x** and **y** is called the **joint probability function**, or **joint probability distribution**, of **x** and **y**.

Analogous to Theorem 3.1, it immediately follows from the postulates of probability that

THEOREM 3.7 A bivariate function can serve as the joint probability distribution of a pair of discrete random variables **x** and **y** if and only if its values, $f(x, y)$, satisfy the conditions

1. $f(x, y) \geqslant 0$ for each pair of values (x, y) within its domain;
2. $\sum_x \sum_y f(x, y) = 1$, where the double summation extends over all possible pairs (x, y) within its domain.

EXAMPLE 3.13

Determine the value of k so that the function given by

$$f(x, y) = kxy \qquad \text{for } x = 1, 2, 3; \quad y = 1, 2, 3$$

can serve as a joint probability distribution.

Solution

Substituting the various values of x and y, we get $f(1, 1) = k$, $f(1, 2) = 2k$, $f(1, 3) = 3k$, $f(2, 1) = 2k$, $f(2, 2) = 4k$, $f(2, 3) = 6k$, $f(3, 1) = 3k$, $f(3, 2) = 6k$, and $f(3, 3) = 9k$. To satisfy the first condition of Theorem 3.7, the constant k must be non-negative, and to satisfy the second condition,

$$k + 2k + 3k + 2k + 4k + 6k + 3k + 6k + 9k = 1$$

so that $36k = 1$ and $k = \frac{1}{36}$.

As in the univariate case, there are many problems in which it is of interest to know the probability that the values of two random variables are less than or equal to some real numbers x and y.

DEFINITION 3.7 If $\mathbf{x}$ and $\mathbf{y}$ are discrete random variables, the function given by

$$F(x, y) = P(\mathbf{x} \leq x, \mathbf{y} \leq y) = \sum_{s \leq x} \sum_{t \leq y} f(s, t) \qquad \text{for } -\infty < x < \infty, \; -\infty < y < \infty$$

where $f(s, t)$ is the value of the joint probability distribution of $\mathbf{x}$ and $\mathbf{y}$ at (s, t), is called the **joint distribution function**, or **joint cumulative distribution**, of $\mathbf{x}$ and $\mathbf{y}$.

In Exercise 6 on page 109, the reader will be asked to prove properties of joint distribution functions which are analogous to those of Theorem 3.2.

EXAMPLE 3.14

With reference to Example 3.12, find $F(1, 1)$.

Solution

$$\begin{aligned} F(1,1) &= P(\mathbf{x} \leqslant 1, \mathbf{y} \leqslant 1) \\ &= f(0,0) + f(0,1) + f(1,0) + f(1,1) \\ &= \tfrac{1}{6} + \tfrac{2}{9} + \tfrac{1}{3} + \tfrac{1}{6} \\ &= \tfrac{8}{9} \end{aligned}$$

As in the univariate case, the joint distribution function of two random variables is defined for all real numbers; for instance, $F(-2, 1) = P(\mathbf{x} \leqslant -2, \mathbf{y} \leqslant 1) = 0$ and $F(3.7, 4.5) = P(\mathbf{x} \leqslant 3.7, \mathbf{y} \leqslant 4.5) = 1$.

Let us now extend the concepts introduced so far in this section to the continuous case.

DEFINITION 3.8 A bivariate function with values $f(x, y)$, defined over the xy-plane, is called a **joint probability density function** of the continuous random variables $\mathbf{x}$ and $\mathbf{y}$ if and only if

$$P[(\mathbf{x}, \mathbf{y}) \in A] = \iint_A f(x, y)\, dx\, dy$$

for any region A in the xy-plane.

Analogous to Theorem 3.5, it immediately follows from the postulates of probability that

THEOREM 3.8 A bivariate function can serve as a joint probability density function of a pair of continuous random variables $\mathbf{x}$ and $\mathbf{y}$ if its values, $f(x, y)$, satisfy the conditions

1. $f(x, y) \geqslant 0 \quad$ for $-\infty < x < \infty, -\infty < y < \infty$;
2. $\displaystyle\int_{-\infty}^{\infty}\int_{-\infty}^{\infty} f(x, y)\, dx\, dy = 1$.

EXAMPLE 3.15

If the joint probability density function of **x** and **y** is given by

$$f(x, y) = \begin{cases} \frac{3}{5}x(y + x) & \text{for } 0 < x < 1, 0 < y < 2 \\ 0 & \text{elsewhere} \end{cases}$$

find $P[(\mathbf{x}, \mathbf{y}) \in A]$, where A is the region $\{(x, y)|0 < x < \frac{1}{2}, 1 < y < 2\}$.

Solution

$$\begin{aligned} P[(\mathbf{x}, \mathbf{y}) \in A] &= P(0 < \mathbf{x} < \tfrac{1}{2}, 1 < \mathbf{y} < 2) \\ &= \int_1^2 \int_0^{\frac{1}{2}} \tfrac{3}{5}x(y + x)\, dx\, dy \\ &= \int_1^2 \frac{3x^2y}{10} + \frac{3x^3}{15}\bigg|_{x=0}^{x=\frac{1}{2}} dy \\ &= \int_1^2 \left(\frac{3y}{40} + \frac{1}{40}\right) dy = \frac{3y^2}{80} + \frac{y}{40}\bigg|_1^2 \\ &= \frac{11}{80} \end{aligned}$$

Analogous to Definition 3.7, we have the following definition of the **joint distribution function** of two continuous random variables:

DEFINITION 3.9 If **x** and **y** are continuous random variables, the function given by

$$F(x, y) = P(\mathbf{x} \leq x, \mathbf{y} \leq y) = \int_{-\infty}^{y} \int_{-\infty}^{x} f(s, t)\, ds\, dt \qquad \text{for } -\infty < x < \infty, \; -\infty < y < \infty$$

where $f(s, t)$ is the value of the joint probability density of x and y at (s, t), is called the **joint distribution function** of **x** and **y**.

The properties of joint distribution functions which the reader will be asked to prove in Exercise 6 on page 109 for the discrete case hold also for the continuous case.

As in Section 3.4, we shall limit our discussion to random variables whose joint distribution function is continuous everywhere and partially differentiable with respect to each variable for all but a finite set of values of each of the two random variables.

Analogous to the relationship $f(x) = \frac{dF(x)}{dx}$ of Theorem 3.6, partial differentiation in Definition 3.9 leads to

$$f(x, y) = \frac{\partial^2}{\partial x\, \partial y} F(x, y)$$

wherever these partial derivatives exist. As in Section 3.4, the joint distribution function of two continuous random variables determines their **joint density** (short for joint probability density function) at all points (x, y) where the joint density is continuous. Also as in Section 3.4, we generally let the values of joint probability densities equal zero wherever they are not defined by the above relationship.

EXAMPLE 3.16

If the joint probability density of **x** and **y** is given by

$$f(x, y) = \begin{cases} x + y & \text{for } 0 < x < 1, 0 < y < 1 \\ 0 & \text{elsewhere} \end{cases}$$

find the corresponding joint distribution function.

Solution

If either $x < 0$ or $y < 0$, it follows immediately that $F(x, y) = 0$. For $0 < x < 1$ and $0 < y < 1$ (Region I of Figure 3.9) we get

$$F(x, y) = \int_0^y \int_0^x (s + t)\, ds\, dt = \tfrac{1}{2}xy(x + y)$$

for $x > 1$ and $0 < y < 1$ (Region II of Figure 3.9) we get

$$F(x, y) = \int_0^y \int_0^1 (s + t)\, ds\, dt = \tfrac{1}{2}y(y + 1)$$

for $0 < x < 1$ and $y > 1$ (Region III of Figure 3.9) we get

$$F(x, y) = \int_0^1 \int_0^x (s + t)\, ds\, dt = \tfrac{1}{2}x(x + 1)$$

and for $x > 1$ and $y > 1$ (Region IV of Figure 3.9) we get

$$F(x, y) = \int_0^1 \int_0^1 (s + t)\, ds\, dt = 1$$

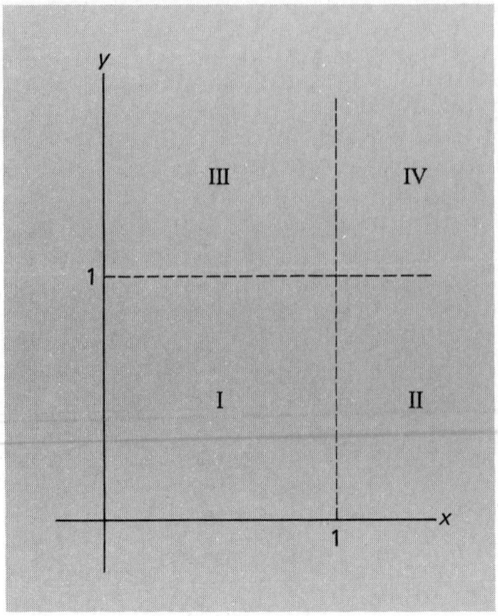

Figure 3.9 Diagram for Example 3.16.

Since the joint distribution function is everywhere continuous, the boundaries between any two of these regions can be included in either one, and we can write

$$F(x, y) = \begin{cases} 0 & \text{for } x \leqslant 0 \text{ or } y \leqslant 0 \\ \frac{1}{2}xy(x + y) & \text{for } 0 < x < 1, 0 < y < 1 \\ \frac{1}{2}y(y + 1) & \text{for } x \geqslant 1, 0 < y < 1 \\ \frac{1}{2}x(x + 1) & \text{for } 0 < x < 1, y \geqslant 1 \\ 1 & \text{for } x \geqslant 1, y \geqslant 1 \end{cases}$$

EXAMPLE 3.17

If the joint distribution function of **x** and **y** is given by

$$F(x, y) = \begin{cases} (1 - e^{-x})(1 - e^{-y}) & \text{for } x > 0 \text{ and } y > 0 \\ 0 & \text{elsewhere} \end{cases}$$

find $P(1 < \mathbf{x} < 3, 1 < \mathbf{y} < 2)$.

Solution

This probability may be obtained directly from the joint distribution function (see Exercise 13 on page 110), but let us find it here by first getting the joint probability density function and then integrating it over the appropriate region. Partial differentiation yields

$$\frac{\partial^2}{\partial x\, \partial y} F(x, y) = e^{-(x+y)}$$

for $x > 0$ and $y > 0$, and 0 elsewhere, so that the joint probability density of the two random variables is given by

$$f(x, y) = \begin{cases} e^{-(x+y)} & \text{for } x > 0 \text{ and } y > 0 \\ 0 & \text{elsewhere} \end{cases}$$

Now, the probability that $\mathbf{x}$ will take on a value on the interval from 1 to 3 and $\mathbf{y}$ will take on a value on the interval from 1 to 2 is given by the double integral

$$\begin{aligned} \int_1^2 \int_1^3 e^{-(x+y)}\, dx\, dy &= (e^{-1} - e^{-3})(e^{-1} - e^{-2}) \\ &= e^{-2} - e^{-3} - e^{-4} + e^{-5} \\ &= 0.074 \end{aligned}$$

For two random variables, the joint probability density is, geometrically speaking, a surface, and the probability which we have calculated in Example 3.17 is given by the volume under this surface shown in Figure 3.10.

All of the preceding definitions concerning two random variables can be generalized to the multivariate case, where there are n random variables. The values of the joint probability distribution of the discrete random variables $\mathbf{x}_1, \mathbf{x}_2, \ldots,$ and $\mathbf{x}_n$, defined over the same sample space S, are given by

$$f(x_1, x_2, \ldots, x_n) = P(\mathbf{x}_1 = x_1, \mathbf{x}_2 = x_2, \ldots, \mathbf{x}_n = x_n)$$

for each n-tuple $(x_1, x_2, \ldots, x_n)$ within the range of the random variables. Also, the values of their joint distribution function are given by

$$F(x_1, x_2, \ldots, x_n) = P(\mathbf{x}_1 \leqslant x_1, \mathbf{x}_2 \leqslant x_2, \ldots, \mathbf{x}_n \leqslant x_n)$$

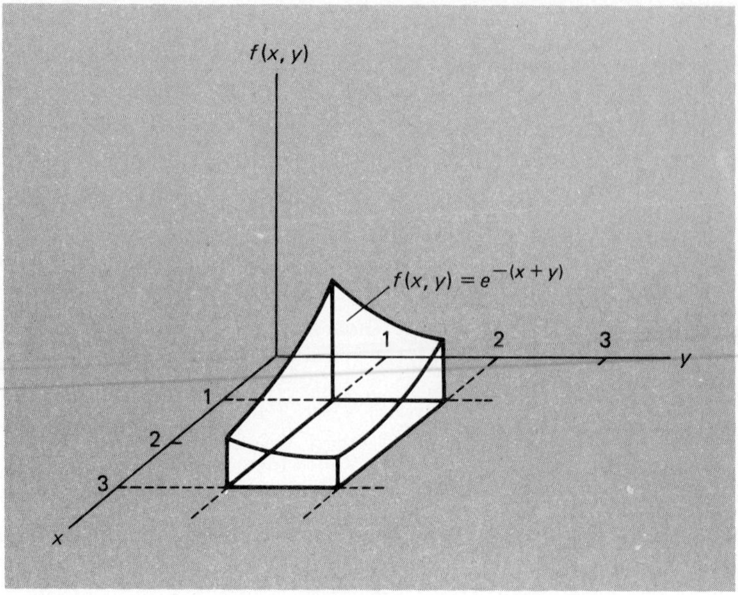

Figure 3.10 Diagram for Example 3.17.

EXAMPLE 3.18

Considering n flips of a balanced coin, let $\mathbf{x}_1$ be the number of heads (0 or 1) obtained on the first flip, $\mathbf{x}_2$ the number of heads obtained on the second flip, . . . , and $\mathbf{x}_n$ the number of heads obtained on the nth flip. Find the joint probability distribution of these n random variables.

Solution

$$f(x_1, x_2, \ldots, x_n) = \tfrac{1}{2} \cdot \tfrac{1}{2} \cdot \tfrac{1}{2} \cdot \ldots \cdot \tfrac{1}{2} = (\tfrac{1}{2})^n$$

where each x can take on the value 0 or 1.

In the continuous case, probabilities are again obtained by integrating the joint probability density, and the joint distribution function is given by

$$F(x_1, x_2, \ldots, x_n) = \int_{-\infty}^{x_n} \cdots \int_{-\infty}^{x_2} \int_{-\infty}^{x_1} f(t_1, t_2, \ldots, t_n)\, dt_1\, dt_2 \cdots dt_n$$

for $-\infty < x_1 < \infty, -\infty < x_2 < \infty, \ldots, -\infty < x_n < \infty$. Also, partial differentiation yields the formula

$$f(x_1, x_2, \ldots, x_n) = \frac{\partial^n}{\partial x_1\, \partial x_2 \cdots \partial x_n} F(x_1, x_2, \ldots, x_n)$$

wherever these partial derivatives exist.

EXAMPLE 3.19

If the **trivariate** probability density of $\mathbf{x}_1$, $\mathbf{x}_2$, and $\mathbf{x}_3$ is given by

$$f(x_1, x_2, x_3) = \begin{cases} (x_1 + x_2)\, e^{-x_3} & \text{for } 0 < x_1 < 1, 0 < x_2 < 1, x_3 > 0 \\ 0 & \text{elsewhere} \end{cases}$$

find
(a) $P[(\mathbf{x}_1, \mathbf{x}_2, \mathbf{x}_3) \in A]$, where A is the region

$$\{(x_1, x_2, x_3) | 0 < x_1 < \tfrac{1}{2}, \tfrac{1}{2} < x_2 < 1, x_3 < 1\}$$

(b) the joint distribution function of the three random variables $\mathbf{x}_1$, $\mathbf{x}_2$, and $\mathbf{x}_3$.

Solution

(a)
$$\begin{aligned} P[(\mathbf{x}_1, \mathbf{x}_2, \mathbf{x}_3) \in A] &= P(0 < \mathbf{x}_1 < \tfrac{1}{2}, \tfrac{1}{2} < \mathbf{x}_2 < 1, \mathbf{x}_3 < 1) \\ &= \int_0^1 \int_{\frac{1}{2}}^1 \int_0^{\frac{1}{2}} (x_1 + x_2)\, e^{-x_3}\, dx_1\, dx_2\, dx_3 \\ &= \int_0^1 \int_{\frac{1}{2}}^1 \left(\frac{1}{8} + \frac{x_2}{2}\right) e^{-x_3}\, dx_2\, dx_3 \\ &= \int_0^1 \tfrac{1}{4} e^{-x_3}\, dx_3 \\ &= \tfrac{1}{4}(1 - e^{-1}) = 0.158 \end{aligned}$$

(b) It will be left to the reader in Exercise 16 on page 110 to use the results of Example 3.16 to show that the distribution function of the three random variables is given by

$$F(x_1, x_2, x_3)$$
$$= \begin{cases} 0 & \text{for } x_1 \leqslant 0, x_2 \leqslant 0, \text{or } x_3 \leqslant 0 \\ \tfrac{1}{2}x_1 x_2(x_1 + x_2)(1 - e^{-x_3}) & \text{for } 0 < x_1 < 1, 0 < x_2 < 1, x_3 > 0 \\ \tfrac{1}{2}x_2(x_2 + 1)(1 - e^{-x_3}) & \text{for } x_1 \geqslant 1, 0 < x_2 < 1, x_3 > 0 \\ \tfrac{1}{2}x_1(x_1 + 1)(1 - e^{-x_3}) & \text{for } 0 < x_1 < 1, x_2 \geqslant 1, x_3 > 0 \\ 1 - e^{-x_3} & \text{for } x_1 \geqslant 1, x_2 \geqslant 1, x_3 > 0 \end{cases}$$

THEORETICAL EXERCISES

1. If **x** and **y** have the joint probabilities shown in the following table

		x = 0	1	2
y	0	$\frac{1}{12}$	$\frac{1}{6}$	$\frac{1}{24}$
	1	$\frac{1}{4}$	$\frac{1}{4}$	$\frac{1}{40}$
	2	$\frac{1}{8}$	$\frac{1}{20}$	
	3	$\frac{1}{120}$		

find

(a) $P(\mathbf{x} = 1, \mathbf{y} = 2)$; **(b)** $P(\mathbf{x} = 0, 1 \leq \mathbf{y} < 3)$;
(c) $P(\mathbf{x} + \mathbf{y} \leq 1)$; **(d)** $P(\mathbf{x} > \mathbf{y})$.

2. If the joint probability distribution of **x** and **y** is given by

$$f(x, y) = c(x^2 + y^2) \quad \text{for } x = -1, 0, 1, 3;\ y = -1, 2, 3$$

find

(a) the value of c; **(b)** $P(\mathbf{x} = 0, \mathbf{y} \leq 2)$;
(c) $P(\mathbf{x} \leq 1, \mathbf{y} > 2)$; **(d)** $P(\mathbf{x} > 2 - \mathbf{y})$.

3. Show that

$$f(x, y) = ky(2y - x) \quad \text{for } x = 0, 3;\ y = 0, 1, 2$$

cannot serve as the joint probability distribution of two random variables for any value of k.

4. With reference to Exercise 1, find the following values of the joint distribution function of the two random variables:

(a) $F(1.2, 0.9)$; **(b)** $F(2, 0)$;
(c) $F(-3, 1.5)$; **(d)** $F(4, 2.7)$.

5. If the joint probability distribution of **x** and **y** is given by

$$f(x, y) = \tfrac{1}{30}(x + y) \quad \text{for } x = 0, 1, 2, 3;\ y = 0, 1, 2$$

construct a table showing the values of the joint distribution function of the two random variables at the twelve points $(0, 0), (0, 1), \ldots, (3, 2)$.

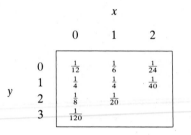

6. If $F(x, y)$ is the value of the joint distribution function of two discrete random variables $\mathbf{x}$ and $\mathbf{y}$ at (x, y), show that

(a) $F(-\infty, -\infty) = 0$;
(b) $F(\infty, \infty) = 1$;
(c) if $a < b$ and $c < d$, then $F(a, c) \leq F(b, d)$.

7. Determine k so that

$$f(x, y) = \begin{cases} kx(x - y) & \text{for } 0 < x < 1, -x < y < x \\ 0 & \text{elsewhere} \end{cases}$$

can serve as a probability density function.

8. If the joint probability density function of $\mathbf{x}$ and $\mathbf{y}$ is given by

$$f(x, y) = \begin{cases} 2 & \text{for } x > 0, y > 0, x + y < 1 \\ 0 & \text{elsewhere} \end{cases}$$

find

(a) $P(\mathbf{x} \leq \frac{1}{2}, \mathbf{y} \leq \frac{1}{2})$; **(b)** $P(\mathbf{x} + \mathbf{y} > \frac{2}{3})$;
(c) $P(\mathbf{x} > 2\mathbf{y})$.

9. If the joint probability density of $\mathbf{x}$ and $\mathbf{y}$ is given by

$$f(x, y) = \begin{cases} \dfrac{1}{y} & \text{for } 0 < x < y, 0 < y < 1 \\ 0 & \text{elsewhere} \end{cases}$$

find the probability that the sum of the values taken on by the two random variables will exceed $\frac{1}{2}$.

10. Find expressions for the values of the joint distribution function of the random variables $\mathbf{x}$ and $\mathbf{y}$ of Exercise 8 which apply when

(a) $x \leq 0$ or $y \leq 0$;
(b) $x \geq 1, y \geq 1$;
(c) $x > 0, y > 0, x + y < 1$, and use it to verify the result of part (a) of Exercise 8.

For how many other parts of the xy-plane will the joint distribution function of these random variables have to be given separately?

11. If the joint distribution function of two random variables $\mathbf{x}$ and $\mathbf{y}$ is given by

$$F(x, y) = \begin{cases} (1 - e^{-x^2})(1 - e^{-y^2}) & \text{for } x > 0, y > 0 \\ 0 & \text{elsewhere} \end{cases}$$

find $P(\mathbf{x} \leq 1, \mathbf{y} \leq 2)$.

12. With reference to Exercise 11, find

(a) the joint probability density of the two random variables;
(b) $P(1 < \mathbf{x} \leq 2, 1 < \mathbf{y} \leq 2)$.

13. If $F(x, y)$ is the value of the joint distribution function of two continuous random variables **x** and **y** at (x, y), express $P(a < \mathbf{x} \leq b, c < \mathbf{y} \leq d)$ in terms of $F(a, c)$, $F(a, d)$, $F(b, c)$, and $F(b, d)$. Observe that the result holds also for discrete random variables.

14. Use the formula asked for in Exercise 13 to rework

(a) Example 3.17 on page 104;
(b) part (b) of Exercise 12.

15. If the joint probability distribution of **x**, **y**, and **z** is given by

$$f(x, y, z) = kxyz \qquad \text{for } x = 1, 2;\ y = 1, 2, 3;\ z = 1, 2$$

find

(a) the value of k;
(b) $P(\mathbf{x} = 1, \mathbf{y} \leq 2, \mathbf{z} = 1)$;
(c) $P(\mathbf{x} = 2, \mathbf{y} + \mathbf{z} = 4)$.

If $F(x, y, z)$ is the value of the joint distribution function of these three random variables at (x, y, z), also find

(d) $F(2, 1, 2)$;
(e) $F(1, 0, 1)$;
(f) $F(4, 4, 4)$.

16. Verify the result of part (b) of Example 3.19.

17. If the joint probability density of the three random variables **x**, **y**, and **z** is given by

$$f(x, y, z) = \begin{cases} \frac{1}{3}(2x + 3y + z) & \text{for } 0 < x < 1, 0 < y < 1, 0 < z < 1 \\ 0 & \text{elsewhere} \end{cases}$$

find

(a) $P(\mathbf{x} = \frac{1}{2}, \mathbf{y} = \frac{1}{2}, \mathbf{z} = \frac{1}{2})$;
(b) $P(\mathbf{x} < \frac{1}{2}, \mathbf{y} < \frac{1}{2}, \mathbf{z} < \frac{1}{2})$.

APPLIED EXERCISES

18. Suppose that we roll a pair of balanced dice, **x** is the number of dice that come up 1, and **y** is the number of dice that come up 4, 5, or 6.

(a) Draw a diagram like that of Figure 3.1, showing the values of **x** and **y** associated with each of the 36 equally likely points of the sample space.
(b) Construct a table showing the values of the joint probability distribution of **x** and **y**.

19. Two textbooks are selected at random from a shelf that contains 3 statistics texts, 2 mathematics texts, and 3 physics texts. If **x** is the number of statistics texts and **y** is the number of mathematics texts selected, construct a table showing the values of the joint probability distribution of **x** and **y**.

20. Let **x** denote the number of heads and **y** the number of heads minus the number of tails obtained in three tosses of a balanced coin. Find the values of the joint probability distribution of **x** and **y**.

21. A marksman is aiming at a circular target of radius 1. If we draw a rectangular system of coordinates with its origin at the center of the target, the coordinates of the point of impact, (**x**, **y**), are random variables having the joint probability density

$$f(x, y) = \begin{cases} \dfrac{1}{\pi} & \text{for } 0 < x^2 + y^2 < 1 \\ 0 & \text{elsewhere} \end{cases}$$

Find

(a) $P[(\mathbf{x}, \mathbf{y}) \in A]$, where A is the sector of the circle in the first quadrant between the radii along the lines $y = 0$ and $y = x$;

(b) $P[(\mathbf{x}, \mathbf{y}) \in B]$, where $B = \{(x, y) | 0 < x^2 + y^2 < \frac{1}{2}\}$.

22. A certain college gives aptitude tests in the sciences and the humanities to all entering freshmen. If **x** and **y** are, respectively, the proportions of correct answers a student gets on the tests, the distribution of these random variables can be approximated with the joint probability density

$$f(x, y) = \begin{cases} \frac{2}{5}(2x + 3y) & \text{for } 0 < x < 1, 0 < y < 1 \\ 0 & \text{elsewhere} \end{cases}$$

What proportions of the students will get

(a) less than 0.4 on both tests;

(b) more than 0.8 on the science test and less than 0.5 on the humanities test?

23. If **p**, the price of a certain commodity (in dollars), and **s**, total sales (in 10,000 units), are random variables whose distribution can be approximated with the joint probability density

$$f(p, s) = \begin{cases} 5pe^{-ps} & \text{for } 0.20 < p < 0.40, s > 0 \\ 0 & \text{elsewhere} \end{cases}$$

find the probabilities that

(a) the price will be less than 30 cents and sales will exceed 20,000 units;

(b) the price will be between 25 cents and 30 cents and sales will be less than 10,000 units.

3.6 MARGINAL DISTRIBUTIONS

To introduce the concept of a marginal distribution, let us consider the following example:

EXAMPLE 3.20

In Example 3.12 on page 98 we derived the joint distribution of the random variables **x** and **y**, the number of aspirin tablets and the number of sedative tablets included among two tablets drawn at random from a bottle containing three aspirin, two sedative, and four laxative tablets. Find the probability distribution of **x** alone and that of **y** alone.

Solution

The results of Example 3.12 are shown in the following table, together with the **marginal totals**, that is, the totals of the respective rows and columns:

		$x = 0$	$x = 1$	$x = 2$	
y	0	$\frac{1}{6}$	$\frac{1}{3}$	$\frac{1}{12}$	$\frac{7}{12}$
	1	$\frac{2}{9}$	$\frac{1}{6}$		$\frac{7}{18}$
	2	$\frac{1}{36}$			$\frac{1}{36}$
		$\frac{5}{12}$	$\frac{1}{2}$	$\frac{1}{12}$	

Observe that the numbers in the bottom margin are, in fact, the probabilities that the random variable **x** will take on the values 0, 1, and 2. In other words, the column totals give the values $g(x)$ of the probability distribution of **x**; that is,

$$g(0) = f(0, 0) + f(0, 1) + f(0, 2) = \sum_{y=0}^{2} f(0, y)$$

$$g(1) = f(1, 0) + f(1, 1) + f(1, 2) = \sum_{y=0}^{2} f(1, y)$$

$$g(2) = f(2, 0) + f(2, 1) + f(2, 2) = \sum_{y=0}^{2} f(2, y)$$

More compactly,

$$g(x) = \sum_{y=0}^{2} f(x, y) \qquad \text{for } x = 0, 1, 2$$

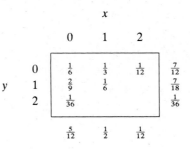

and by the same token, the values $h(y)$ of the probability distribution of **y** are given by the row totals

$$h(y) = \sum_{x=0}^{2} f(x, y) \qquad \text{for } y = 0, 1, 2$$

We are thus led to the following general definition:

DEFINITION 3.10 If **x** and **y** are discrete random variables and $f(x, y)$ is the value of their joint probability distribution at (x, y), the function given by

$$g(x) = \sum_{y} f(x, y)$$

for each x within the range of **x**, is called the **marginal distribution** of **x**. Correspondingly, the function given by

$$h(y) = \sum_{x} f(x, y)$$

for each y within the range of **y**, is called the **marginal distribution** of **y**.

When **x** and **y** are continuous random variables, the probability distributions are replaced by probability densities, the summations are replaced by integrals, and we get

DEFINITION 3.11 If **x** and **y** are continuous random variables and $f(x, y)$ is the value of their joint probability density at (x, y), the function given by

$$g(x) = \int_{-\infty}^{\infty} f(x, y)\, dy \qquad \text{for } -\infty < x < \infty$$

is called the **marginal density** of **x**. Correspondingly, the function given by

$$h(y) = \int_{-\infty}^{\infty} f(x, y)\, dx \qquad \text{for } -\infty < y < \infty$$

is called the **marginal density** of **y**.

EXAMPLE 3.21

Given the joint density

$$f(x, y) = \begin{cases} \frac{2}{3}(x + 2y) & \text{for } 0 < x < 1, 0 < y < 1 \\ 0 & \text{elsewhere} \end{cases}$$

find the marginal densities of $\mathbf{x}$ and $\mathbf{y}$.

Solution

Performing the necessary integrations, we get

$$g(x) = \int_{-\infty}^{\infty} f(x, y)\, dy = \int_0^1 \frac{2}{3}(x + 2y)\, dy = \frac{2}{3}(x + 1)$$

for $0 < x < 1$, and $g(x) = 0$ elsewhere. Likewise,

$$h(y) = \int_{-\infty}^{\infty} f(x, y)\, dx = \int_0^1 \frac{2}{3}(x + 2y)\, dx = \frac{1}{3}(1 + 4y)$$

for $0 < y < 1$, and $h(y) = 0$ elsewhere.

When we are dealing with more than two random variables, we can speak not only of the marginal distributions of the individual random variables, but also of the **joint marginal distributions** of several of the random variables. If the joint probability distribution of the discrete random variables $\mathbf{x}_1, \mathbf{x}_2, \ldots,$ and $\mathbf{x}_n$ has the values $f(x_1, x_2, \ldots, x_n)$, the marginal distribution of $\mathbf{x}_1$ alone is given by

$$g(x_1) = \sum_{x_2} \cdots \sum_{x_n} f(x_1, x_2, \ldots, x_n)$$

for all values within the range of $\mathbf{x}_1$, the joint marginal distribution of $\mathbf{x}_1$, $\mathbf{x}_2$, and $\mathbf{x}_3$ is given by

$$m(x_1, x_2, x_3) = \sum_{x_4} \cdots \sum_{x_n} f(x_1, x_2, \ldots, x_n)$$

for all values within the range of $\mathbf{x}_1$, $\mathbf{x}_2$, and $\mathbf{x}_3$, and other marginal distributions can be defined in the same way. For the continuous case, probability distributions are replaced by probability densities, summations are replaced by integrals, and if the joint probability density of the continuous random variables $\mathbf{x}_1, \mathbf{x}_2, \ldots,$ and $\mathbf{x}_n$

has the values $f(x_1, x_2, \ldots, x_n)$, the marginal density of $\mathbf{x}_2$ alone is given by

$$h(x_2) = \int_{-\infty}^{\infty} \cdots \int_{-\infty}^{\infty} f(x_1, x_2, \ldots, x_n)\, dx_1\, dx_3 \cdots dx_n$$

for $-\infty < x_2 < \infty$, the joint marginal density of $\mathbf{x}_1$ and $\mathbf{x}_n$ is given by

$$\varphi(x_1, x_n) = \int_{-\infty}^{\infty} \cdots \int_{-\infty}^{\infty} f(x_1, x_2, \ldots, x_n)\, dx_2\, dx_3 \cdots dx_{n-1}$$

for $-\infty < x_1 < \infty$ and $-\infty < x_n < \infty$, and so forth.

EXAMPLE 3.22

Considering again the trivariate density

$$f(x_1, x_2, x_3) = \begin{cases} (x_1 + x_2)\, e^{-x_3} & \text{for } 0 < x_1 < 1, 0 < x_2 < 1, x_3 > 0 \\ 0 & \text{elsewhere} \end{cases}$$

of Example 3.19, find the joint marginal density of $\mathbf{x}_1$ and $\mathbf{x}_3$, and the marginal density of $\mathbf{x}_1$.

Solution

Performing the necessary integration, we find that the joint marginal density of $\mathbf{x}_1$ and $\mathbf{x}_3$ is given by

$$m(x_1, x_3) = \int_0^1 (x_1 + x_2)e^{-x_3}\, dx_2 = (x_1 + \tfrac{1}{2})e^{-x_3}$$

for $0 < x_1 < 1$ and $x_3 > 0$, and $m(x_1, x_3) = 0$ elsewhere. Using this result, we find that the marginal density of x_1 alone is given by

$$g(x_1) = \int_0^{\infty} \int_0^1 f(x_1, x_2, x_3)\, dx_2\, dx_3 = \int_0^{\infty} m(x_1, x_3)\, dx_3$$

$$= \int_0^{\infty} (x_1 + \tfrac{1}{2})\, e^{-x_3}\, dx_3 = x_1 + \tfrac{1}{2}$$

for $0 < x_1 < 1$, and $g(x_1) = 0$ elsewhere.

Corresponding to the various marginal and joint marginal distributions and densities we have introduced in this section, we can also define **marginal** and **joint marginal distribution functions**. Some problems relating to such distribution functions will be left to the reader in Exercises 5, 10, and 11 on pages 122 and 123.

3.7 CONDITIONAL DISTRIBUTIONS

In Chapter 2 we defined the conditional probability of event A given event B as

$$P(A|B) = \frac{P(A \cap B)}{P(B)}$$

provided $P(B) \neq 0$. Suppose now that A and B are the events $\mathbf{x} = x$ and $\mathbf{y} = y$, so that we can write

$$P(\mathbf{x} = x|\mathbf{y} = y) = \frac{P(\mathbf{x} = x, \mathbf{y} = y)}{P(\mathbf{y} = y)}$$

$$= \frac{f(x, y)}{h(y)}$$

provided $P(\mathbf{y} = y) = h(y) \neq 0$, where $f(x, y)$ is the value of the joint probability distribution of $\mathbf{x}$ and $\mathbf{y}$ at (x, y) and $h(y)$ is the value of the marginal distribution of $\mathbf{y}$ at y. Denoting the probability by $f(x|y)$ to indicate that x is a variable and y is fixed, let us now make the following definition:

DEFINITION 3.12 If $f(x, y)$ is the value of the joint probability distribution of the discrete random variables $\mathbf{x}$ and $\mathbf{y}$ at (x, y) and $h(y)$ is the value of the marginal distribution of $\mathbf{y}$ at y, the function given by

$$f(x|y) = \frac{f(x, y)}{h(y)} \qquad h(y) \neq 0$$

for each x within the range of $\mathbf{x}$, is called the **conditional distribution** of $\mathbf{x}$ given $\mathbf{y} = y$. Correspondingly, if $g(x)$ is the value of the marginal distribution of $\mathbf{x}$ at x, the function given by

$$w(y|x) = \frac{f(x, y)}{g(x)} \qquad g(x) \neq 0$$

for each y within the range of $\mathbf{y}$, is called the **conditional distribution** of $\mathbf{y}$ given $\mathbf{x} = x$.

EXAMPLE 3.23

Referring to Example 3.20, find the conditional distribution of $\mathbf{x}$ given $\mathbf{y} = 1$.

Solution

Substituting the appropriate values from the table on page 112, we get

$$f(0|1) = \frac{\frac{2}{9}}{\frac{7}{18}} = \frac{4}{7}$$

$$f(1|1) = \frac{\frac{1}{6}}{\frac{7}{18}} = \frac{3}{7}$$

$$f(2|1) = \frac{0}{\frac{7}{18}} = 0$$

When **x** and **y** are continuous random variables, the probability distributions are replaced by probability densities, and we get

DEFINITION 3.13 If $f(x, y)$ is the value of the joint density of the continuous random variables **x** and **y** at (x, y) and $h(y)$ is the value of the marginal density of **y** at y, the function given by

$$f(x|y) = \frac{f(x, y)}{h(y)} \qquad h(y) \neq 0$$

for $-\infty < x < \infty$, is called the **conditional density** of **x** given $\mathbf{y} = y$. Correspondingly, if $g(x)$ is the value of the marginal density of **x** at x, the function given by

$$w(y|x) = \frac{f(x, y)}{g(x)} \qquad g(x) \neq 0$$

for $-\infty < y < \infty$, is called the **conditional density** of **y** given $\mathbf{x} = x$.

EXAMPLE 3.24

Referring to Example 3.21, find the conditional density of **x** given $\mathbf{y} = y$, and use it to evaluate the probability $P(\mathbf{x} \leqslant \frac{1}{2}|\mathbf{y} = \frac{1}{2})$.

Solution

Using the results obtained on page 114, we have

$$f(x|y) = \frac{f(x, y)}{h(y)} = \frac{\frac{2}{3}(x + 2y)}{\frac{1}{3}(1 + 4y)}$$

$$= \frac{2x + 4y}{1 + 4y}$$

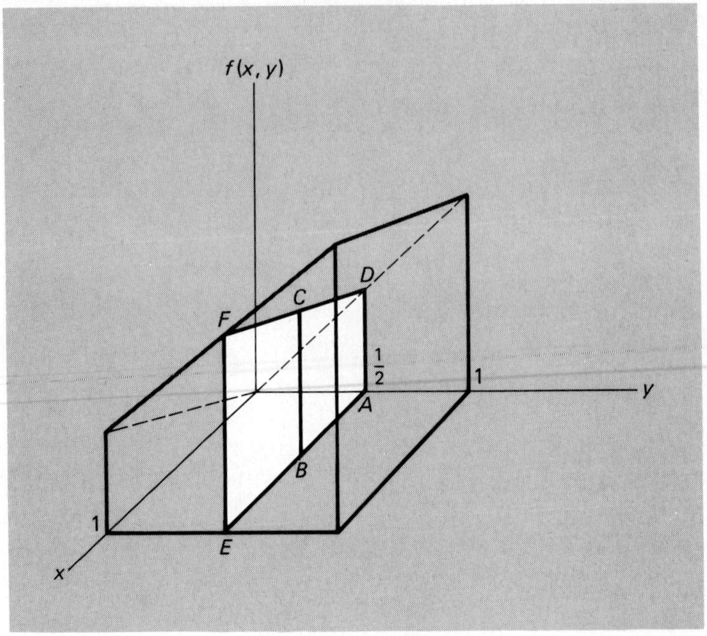

Figure 3.11 Diagram for Example 3.24.

for $0 < x < 1$, and $f(x|y) = 0$ elsewhere. Now, $f(x|\frac{1}{2}) = \dfrac{2x + 4 \cdot \frac{1}{2}}{1 + 4 \cdot \frac{1}{2}} = \dfrac{2x + 2}{3}$ and we can write

$$P(\mathbf{x} \leq \tfrac{1}{2} | \mathbf{y} = \tfrac{1}{2}) = \int_0^{\frac{1}{2}} \frac{2x + 2}{3}\, dx = \tfrac{5}{12}$$

It is of interest to note that in Figure 3.11 this probability is given by the ratio of the area of trapezoid *ABCD* to the area of the trapezoid *AEFD*.

EXAMPLE 3.25

Given the joint density function

$$f(x, y) = \begin{cases} 4xy & \text{for } 0 < x < 1, 0 < y < 1 \\ 0 & \text{elsewhere} \end{cases}$$

find the marginal densities of **x** and **y**, and the conditional density of **x** given $\mathbf{y} = y$.

Solution

Performing the necessary integrations, we get

$$g(x) = \int_{-\infty}^{\infty} f(x, y)\, dy = \int_0^1 4xy\, dy$$

$$= 2xy^2 \Big|_{y=0}^{y=1} = 2x$$

for $0 < x < 1$, and $g(x) = 0$ elsewhere; also

$$h(y) = \int_{-\infty}^{\infty} f(x, y)\, dx = \int_0^1 4xy\, dx$$

$$= 2x^2y \Big|_{x=0}^{x=1} = 2y$$

for $0 < y < 1$, and $h(y) = 0$ elsewhere. Then, substituting into the formula for a conditional density, we get

$$f(x|y) = \frac{f(x, y)}{h(y)} = \frac{4xy}{2y} = 2x$$

for $0 < x < 1$, and $f(x|y) = 0$ elsewhere.

When we are dealing with more than two random variables, whether continuous or discrete, we can consider various different kinds of conditional distributions or densities. For instance, if $f(x_1, x_2, x_3, x_4)$ is the value of the joint distribution of the discrete random variables $\mathbf{x}_1$, $\mathbf{x}_2$, $\mathbf{x}_3$, and $\mathbf{x}_4$ at (x_1, x_2, x_3, x_4), we can write

$$p(x_3|x_1, x_2, x_4) = \frac{f(x_1, x_2, x_3, x_4)}{g(x_1, x_2, x_4)} \qquad g(x_1, x_2, x_4) \neq 0$$

for the value of the conditional distribution of $\mathbf{x}_3$ given $\mathbf{x}_1 = x_1$, $\mathbf{x}_2 = x_2$, and $\mathbf{x}_4 = x_4$, at x_3, where $g(x_1, x_2, x_4)$ is the value of the joint marginal distribution of $\mathbf{x}_1$, $\mathbf{x}_2$, and $\mathbf{x}_4$ at (x_1, x_2, x_4). We can also write

$$q(x_2, x_4|x_1, x_3) = \frac{f(x_1, x_2, x_3, x_4)}{m(x_1, x_3)} \qquad m(x_1, x_3) \neq 0$$

for the value of the **joint conditional distribution** of $\mathbf{x}_2$ and $\mathbf{x}_4$ given $\mathbf{x}_1 = x_1$ and $\mathbf{x}_3 = x_3$, at (x_2, x_4), or

$$r(x_2, x_3, x_4|x_1) = \frac{f(x_1, x_2, x_3, x_4)}{b(x_1)} \qquad b(x_1) \neq 0$$

for the value of the joint conditional distribution of $\mathbf{x}_2$, $\mathbf{x}_3$, and $\mathbf{x}_4$ given $\mathbf{x}_1 = x_1$, at (x_2, x_3, x_4).

When we are dealing with two or more random variables, questions of **independence** are usually of great importance. In Example 3.25 we see that $f(x|y) = 2x$ does not depend on the given value $\mathbf{y} = y$, but this is clearly not the case in Example 3.24 where $f(x|y) = \frac{2x + 4y}{1 + 4y}$. Whenever the values of the conditional distribution of $\mathbf{x}$ given $\mathbf{y} = y$ do not depend on y, it follows that $f(x|y) = g(x)$, and hence the formulas of Definitions 3.12 and 3.13 yield

$$f(x, y) = f(x|y) \cdot h(y) = g(x) \cdot h(y)$$

That is, the values of the joint distribution are given by the products of the corresponding values of the two marginal distributions. Generalizing from this observation, let us now make the following definition:

DEFINITION 3.14 If $f(x_1, x_2, \ldots, x_n)$ is the value of the joint probability distribution of the n discrete random variables $\mathbf{x}_1, \mathbf{x}_2, \ldots, \mathbf{x}_n$ at $(x_1, x_2, \ldots, x_n)$, and $f_i(x_i)$ is the value of the marginal distribution of $\mathbf{x}_i$ at x_i for $i = 1, 2, \ldots, n$, these random variables are **independent** if and only if

$$f(x_1, x_2, \ldots, x_n) = f_1(x_1) \cdot f_2(x_2) \cdot \ldots \cdot f_n(x_n)$$

for all $(x_1, x_2, \ldots, x_n)$ within their range.

To give a corresponding definition for continuous random variables, we simply substitute the word "density" for the word "distribution."

With this definition of independence, it can easily be verified that the three random variables of Example 3.22 are not independent, but that the two random variables $\mathbf{x}_1$ and $\mathbf{x}_3$, and also the two random variables $\mathbf{x}_2$ and $\mathbf{x}_3$, are pairwise independent (see Exercise 12 on page 123).

The following example serves to illustrate the use of Definition 3.14 in finding probabilities relating to several independent random variables:

EXAMPLE 3.26

Given the independent random variables $\mathbf{x}_1$, $\mathbf{x}_2$, and $\mathbf{x}_3$ with the probability densities

$$f_1(x_1) = \begin{cases} e^{-x_1} & \text{for } x_1 > 0 \\ 0 & \text{elsewhere} \end{cases}$$

$$f_2(x_2) = \begin{cases} 2e^{-2x_2} & \text{for } x_2 > 0 \\ 0 & \text{elsewhere} \end{cases}$$

$$f_3(x_3) = \begin{cases} 3e^{-3x_3} & \text{for } x_3 > 0 \\ 0 & \text{elsewhere} \end{cases}$$

find their joint probability density, and use it to evaluate the probability $P(\mathbf{x}_1 + \mathbf{x}_2 \leqslant 1, \mathbf{x}_3 > 1)$.

Solution

According to Definition 3.14, the values of the joint probability density are given by

$$\begin{aligned} f(x_1, x_2, x_3) &= f_1(x_1) \cdot f_2(x_2) \cdot f_3(x_3) \\ &= e^{-x_1} \cdot 2e^{-2x_2} \cdot 3e^{-3x_3} \\ &= 6e^{-x_1-2x_2-3x_3} \end{aligned}$$

for $x_1 > 0$, $x_2 > 0$, $x_3 > 0$, and $f(x_1, x_2, x_3) = 0$ elsewhere. Thus,

$$\begin{aligned} P(\mathbf{x}_1 + \mathbf{x}_2 \leqslant 1, \mathbf{x}_3 > 1) &= \int_1^{\infty} \int_0^1 \int_0^{1-x_2} 6e^{-x_1-2x_2-3x_3}\, dx_1\, dx_2\, dx_3 \\ &= (1 - 2e^{-1} + e^{-2})e^{-3} \\ &= 0.020 \end{aligned}$$

THEORETICAL EXERCISES

1. If the values of the joint probability distribution of $\mathbf{x}$ and $\mathbf{y}$ are as shown in the following table

		$x = -1$	$x = 1$
y	-1	$\frac{1}{8}$	$\frac{1}{2}$
	0	0	$\frac{1}{4}$
	1	$\frac{1}{8}$	0

find

(a) the marginal distribution of $\mathbf{x}$;
(b) the marginal distribution of $\mathbf{y}$;
(c) the conditional distribution of $\mathbf{x}$ given $\mathbf{y} = -1$.

2. With reference to Exercise 1 on page 108, find

(a) the marginal distribution of $\mathbf{x}$;
(b) the marginal distribution of $\mathbf{y}$;
(c) the conditional distribution of $\mathbf{x}$ given $\mathbf{y} = 1$;
(d) the conditional distribution of $\mathbf{y}$ given $\mathbf{x} = 0$.

3. Given the joint probability distribution

$$f(x, y, z) = \frac{xyz}{108} \qquad \text{for } x = 1, 2, 3; \quad y = 1, 2, 3; \quad z = 1, 2$$

find

(a) the joint marginal distribution of $\mathbf{x}$ and $\mathbf{y}$;
(b) the joint marginal distribution of $\mathbf{x}$ and $\mathbf{z}$;
(c) the marginal distribution of $\mathbf{x}$;
(d) the conditional distribution of $\mathbf{z}$ given $\mathbf{x} = 1$ and $\mathbf{y} = 2$;
(e) the joint conditional distribution of $\mathbf{y}$ and $\mathbf{z}$ given $\mathbf{x} = 3$.

4. Check whether the random variables $\mathbf{x}$ and $\mathbf{y}$ are independent, if their joint probability distribution is given by

(a) $f(x, y) = \frac{1}{4}$ for $x = -1$ and $y = -1$, $x = -1$ and $y = 1$, $x = 1$ and $y = -1$, and $x = 1$ and $y = 1$;
(b) $f(x, y) = \frac{1}{3}$ for $x = 0$ and $y = 0$, $x = 0$ and $y = 1$, and $x = 1$ and $y = 1$.

5. With reference to Example 3.20 on page 112, find

(a) the **marginal distribution function** of $\mathbf{x}$, namely, the function given by $G(x) = P(\mathbf{x} \leq x)$ for $-\infty < x < \infty$.
(b) the **conditional distribution function** of $\mathbf{x}$ given $\mathbf{y} = 1$, namely the function given by $F(x|1) = P(\mathbf{x} \leq x|\mathbf{y} = 1)$ for $-\infty < x < \infty$.

6. If the joint density function of $\mathbf{x}$ and $\mathbf{y}$ is given by

$$f(x, y) = \begin{cases} \frac{1}{4}(2x + y) & \text{for } 0 < x < 1, 0 < y < 2 \\ 0 & \text{elsewhere} \end{cases}$$

find

(a) the marginal density of $\mathbf{x}$;
(b) the marginal density of $\mathbf{y}$;
(c) the conditional density of $\mathbf{x}$ given $\mathbf{y} = 1$;
(d) the conditional density of $\mathbf{y}$ given $\mathbf{x} = \frac{1}{4}$.

7. If the random variables $\mathbf{x}$ and $\mathbf{y}$ have the joint density function given by

$$f(x, y) = \begin{cases} 24y(1 - x - y) & \text{for } x > 0, y > 0, x + y < 1 \\ 0 & \text{elsewhere} \end{cases}$$

find

(a) the marginal density of $\mathbf{x}$;
(b) the marginal density of $\mathbf{y}$;

Also determine whether the two random variables are independent.

8. With reference to the joint density of Exercise 9 on page 109, find

(a) the marginal density of $\mathbf{x}$;
(b) the marginal density of $\mathbf{y}$.

Also determine whether the two random variables are independent.

9. With reference to Example 3.22 on page 115, find

(a) the conditional density of $\mathbf{x}_2$ given $\mathbf{x}_1 = \frac{1}{3}$ and $\mathbf{x}_3 = 2$;
(b) the joint conditional density of $\mathbf{x}_2$ and $\mathbf{x}_3$ given $\mathbf{x}_1 = \frac{1}{2}$.

10. If $F(x, y)$ is the value of the joint distribution function of the random variables $\mathbf{x}$ and $\mathbf{y}$ at (x, y), show that the **marginal distribution function** of $\mathbf{x}$ is given by

$$G(x) = F(x, \infty) \qquad \text{for } -\infty < x < \infty$$

Use this result to find the marginal distribution function of $\mathbf{x}$ for two random variables $\mathbf{x}$ and $\mathbf{y}$ having the joint distribution function of Exercise 11 on page 109.

11. If $F(x_1, x_2, x_3)$ is the value of the joint distribution function of the random variables $\mathbf{x}_1$, $\mathbf{x}_2$, and $\mathbf{x}_3$ at (x_1, x_2, x_3), show that the **joint marginal distribution function** of $\mathbf{x}_1$ and $\mathbf{x}_3$ is given by

$$M(x_1, x_3) = F(x_1, \infty, x_3) \qquad \text{for } -\infty < x_1 < \infty, -\infty < x_3 < \infty$$

and that the **marginal distribution function** of $\mathbf{x}_1$ is given by

$$G(x_1) = F(x_1, \infty, \infty) \qquad \text{for } -\infty < x_1 < \infty$$

Use these results to find the joint marginal distribution function of $\mathbf{x}_1$ and $\mathbf{x}_3$, and the marginal distribution function of $\mathbf{x}_1$, for three random variables $\mathbf{x}_1$, $\mathbf{x}_2$, and $\mathbf{x}_3$ having the joint distribution function obtained in Example 3.19 on page 107.

12. With reference to Example 3.22 on page 115 verify that the three random variables are not independent, but that the two random variables $\mathbf{x}_1$ and $\mathbf{x}_3$, and also the two random variables $\mathbf{x}_2$ and $\mathbf{x}_3$, are pairwise independent.

13. If $\mathbf{x}$ and $\mathbf{y}$ have the joint density function

$$f(x, y) = \begin{cases} \frac{1}{8}(6 - x - y) & \text{for } 0 < x < 2, 2 < y < 4 \\ 0 & \text{elsewhere} \end{cases}$$

determine the value of $P(\mathbf{x} < 1|\mathbf{y} < 3)$.

14. Given the independent random variables $\mathbf{x}$ and $\mathbf{y}$ with the probability densities

$$f(x) = \begin{cases} \frac{1}{2} & \text{for } 0 < x < 2 \\ 0 & \text{elsewhere} \end{cases}$$

$$\pi(y) = \begin{cases} \frac{1}{3} & \text{for } 0 < y < 3 \\ 0 & \text{elsewhere} \end{cases}$$

find

(a) the joint probability density of $\mathbf{x}$ and $\mathbf{y}$;
(b) the probability $P(\mathbf{x}^2 + \mathbf{y}^2 > 1)$.

APPLIED EXERCISES

15. Two cards are drawn without replacement from an ordinary deck of 52 playing cards. If $\mathbf{z}$ is the number of aces obtained in the first draw and $\mathbf{w}$ is the total number of aces obtained in both draws, find

(a) the joint probability distribution of $\mathbf{z}$ and $\mathbf{w}$;
(b) the marginal distribution of $\mathbf{z}$;
(c) the marginal distribution of $\mathbf{w}$;
(d) the conditional distribution of $\mathbf{w}$ given $\mathbf{z} = 1$.

16. With reference to Exercise 19 on page 111, find

(a) the marginal distribution of $\mathbf{x}$;
(b) the marginal distribution of $\mathbf{y}$;
(c) the conditional distribution of $\mathbf{x}$ given $\mathbf{y} = 1$;
(d) the conditional distribution of $\mathbf{y}$ given $\mathbf{x} = 0$.

17. If $\mathbf{x}$ is the proportion of persons who will respond to one kind of mail-order solicitation, $\mathbf{y}$ is the proportion of persons who will respond to another kind of mail-order solicitation, and the joint probability density function of $\mathbf{x}$ and $\mathbf{y}$ is given by

$$f(x, y) = \begin{cases} \frac{2}{5}(x + 4y) & \text{for } 0 < x < 1, 0 < y < 1 \\ 0 & \text{elsewhere} \end{cases}$$

find

(a) the marginal density of $\mathbf{x}$;
(b) the probability that there will be at least a 30 percent response to the first kind of mail-order solicitation;
(c) the conditional density of $\mathbf{y}$ given $\mathbf{x} = x$;
(d) the probability that there will be at most a 50 percent response to the second kind of mail-order solicitation given that there has been only a 20 percent response to the first kind of mail-order solicitation.

18. With reference to Exercise 23 on page 111, find

(a) the marginal density function of $\mathbf{p}$;
(b) the probability that the price per unit will be less than 28 cents;
(c) the conditional density function of $\mathbf{s}$ given $\mathbf{p} = p$;
(d) the probability that sales will be less than 30,000 units when $p = 25$ cents.

19. If $\mathbf{x}$ is the amount (in dollars) a salesperson spends on gasoline during a day and $\mathbf{y}$ is the amount (in dollars) for which the salesperson is reimbursed, and the joint density of these random variables is given by

$$f(x, y) = \begin{cases} \dfrac{1}{25}\left(\dfrac{20 - x}{x}\right) & \text{for } 10 < x < 20, \dfrac{x}{2} < y < x \\ 0 & \text{elsewhere} \end{cases}$$

find

(a) the marginal densities of $\mathbf{x}$ and $\mathbf{y}$;
(b) the conditional density of $\mathbf{y}$ given $\mathbf{x} = 12$;
(c) the probability that the salesperson will be reimbursed at least \$8 when spending \$12.

20. Show that the two random variables of Exercise 22 on page 111 are not independent.

21. The useful life (in hours) of a certain kind of vacuum tube is a random variable having the probability density

$$f(x) = \begin{cases} \dfrac{20{,}000}{(x + 100)^3} & \text{for } x > 0 \\ 0 & \text{elsewhere} \end{cases}$$

If three of these tubes operate independently, find

(a) the joint probability density function of $\mathbf{x}_1$, $\mathbf{x}_2$, and $\mathbf{x}_3$, representing the lengths of their useful lives;
(b) the probability $P(\mathbf{x}_1 < 100, \mathbf{x}_2 < 100, \mathbf{x}_3 \geqslant 200)$.

References

More advanced, or more detailed treatments of the material in this chapter may be found in

BRUNK, H. D., *An Introduction to Mathematical Statistics*, 3rd ed. Lexington, Mass.: Xerox College Publishing, 1975,

DEGROOT, M. H., *Probability and Statistics*. Reading, Mass.: Addison-Wesley Publishing Company, Inc., 1975,

FRASER, D. A. S., *Probability and Statistics: Theory and Applications*. North Scituate, Mass.: Duxbury Press, 1976,

HOGG, R. V., and CRAIG, A. T., *Introduction to Mathematical Statistics*, 4th ed. New York: Macmillan Publishing Co., Inc., 1978.

KENDALL, M. G., and STUART, A., *The Advanced Theory of Statistics, Vol. 1*, 3rd ed. New York: Hafner Publishing Company, 1969.

mathematical expectation

4

4.1 INTRODUCTION

Originally, the concept of a **mathematical expectation** arose in connection with games of chance, and in its simplest form it is the product of the amount a player stands to win and the probability that he or she will win. For instance, if we hold one of 10,000 tickets in a raffle for which the grand prize is a car worth \$4,800, our mathematical expectation is $4{,}800 \cdot \frac{1}{10{,}000} = \$0.48$. This figure will have to be interpreted in the sense of an average—altogether the 10,000 tickets pay \$4,800, or on the average $\frac{\$4{,}800}{10{,}000} = \$0.48$ per ticket.

If there is also a second prize worth \$1,200 and a third prize worth \$400, we can argue that altogether the 10,000 tickets pay $\$4{,}800 + \$1{,}200 + \$400 = \$6{,}400$, or on the average $\frac{\$6{,}400}{10{,}000} = \$0.64$ per ticket. Looking at this in a different way, we could argue that if the raffle is repeated many times, we would lose 99.97 percent of the time (or with probability 0.9997) and win each of the prizes 0.01 percent of the time (or with probability 0.0001). On the average we would thus win

$$0(0.9997) + 4{,}800(0.0001) + 1{,}200(0.0001) + 400(0.0001) = \$0.64$$

which is the sum of the products obtained by multiplying each amount by the corresponding probability.

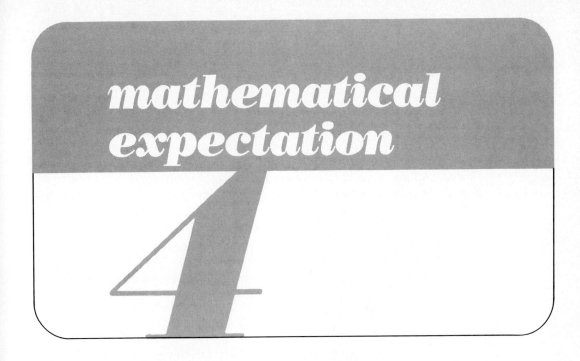

4.2 THE EXPECTED VALUE OF A RANDOM VARIABLE

In the illustration of the preceding section, the amount we stood to win was a random variable, and the mathematical expectation of this random variable was the sum of the products obtained by multiplying each value of the random variable by the corresponding probability. Referring to the mathematical expectation of a random variable simply as its **expected value**, and extending the definition to the continuous case by replacing the operation of summation by integration, we thus have

DEFINITION 4.1 If $\mathbf{x}$ is a discrete random variable and $f(x)$ is the value of its probability distribution at x, the **expected value** of this random variable is

$$E(\mathbf{x}) = \sum_{x} x \cdot f(x)$$

Correspondingly, if $\mathbf{x}$ is a continuous random variable and $f(x)$ is the value of its probability density at x, the **expected value** of this random variable is

$$E(\mathbf{x}) = \int_{-\infty}^{\infty} x \cdot f(x)\, dx$$

In this definition it is assumed, of course, that the sum or integral exists; otherwise, the mathematical expectation does not exist.

EXAMPLE 4.1

A lot of twelve television sets includes two that are defective. If three of the sets are chosen at random for shipment to a hotel, how many defective sets can they expect?

Solution

We can select x of the 2 defective sets and $3 - x$ of the 10 good sets in $\binom{2}{x}\binom{10}{3-x}$ ways, and we can select 3 of the 12 sets in $\binom{12}{3}$ ways. Assuming that the $\binom{12}{3}$ possibilities are all equally likely, we find that the probability distribution of **x**, the number of defective sets shipped to the hotel, is given by

$$f(x) = \frac{\binom{2}{x}\binom{10}{3-x}}{\binom{12}{3}} \qquad \text{for } x = 0, 1, 2$$

or, in tabular form,

x	0	1	2
$f(x)$	$\frac{6}{11}$	$\frac{9}{22}$	$\frac{1}{22}$

Now,

$$E(\mathbf{x}) = 0 \cdot \tfrac{6}{11} + 1 \cdot \tfrac{9}{22} + 2 \cdot \tfrac{1}{22} = \tfrac{1}{2}$$

and since they cannot possibly get half a defective set, it should be clear that the term "expect" is not used in its colloquial sense. Indeed, it must be interpreted as an average pertaining to repeated shipments made under the given conditions.

EXAMPLE 4.2

Certain coded measurements of the pitch diameter of threads of a fitting have the probability density

$$f(x) = \begin{cases} \dfrac{4}{\pi(1 + x^2)} & \text{for } 0 < x < 1 \\ 0 & \text{elsewhere} \end{cases}$$

Find the expected value of this random variable.

Solution

Using Definition 4.1, we have

$$\begin{aligned} E(\mathbf{x}) &= \int_0^1 x \cdot \frac{4}{\pi(1 + x^2)}\, dx \\ &= \frac{4}{\pi} \int_0^1 \frac{x}{1 + x^2}\, dx \\ &= \frac{\ln 4}{\pi} = 0.4413 \end{aligned}$$

In many problems of statistics we are interested not only in the expected value of a random variable $\mathbf{x}$, but also in the expected values of random variables related to $\mathbf{x}$. Thus, we might be interested in the random variable $\mathbf{y}$, whose values are related to those of $\mathbf{x}$ by means of the equation $y = g(x)$. To simplify our notation we denote this random variable by $g(\mathbf{x})$; for instance, $g(\mathbf{x})$ might be $\mathbf{x}^3$, so that when $\mathbf{x}$ takes on the value 2, $g(\mathbf{x})$ takes on the value 8. If we want to find the expected value of such a random variable $g(\mathbf{x})$, we could first find its probability distribution or probability density (by methods to be discussed in Chapter 7) and then use Definition 4.1, but it is usually easier and more straightforward to use the following theorem:

THEOREM 4.1 If $\mathbf{x}$ is a discrete random variable and $f(x)$ is the value of its probability distribution at x, the expected value of the random variable $g(\mathbf{x})$ is given by

$$E[g(\mathbf{x})] = \sum_x g(x) \cdot f(x)$$

Correspondingly, if $\mathbf{x}$ is a continuous random variable and $f(x)$ is the value of its probability density at x, the expected value of the random variable $g(\mathbf{x})$ is given by

$$E[g(\mathbf{x})] = \int_{-\infty}^{\infty} g(x) \cdot f(x)\, dx$$

Proof. Since a more general proof is beyond the scope of this text, we shall prove this theorem here only for the case where $\mathbf{x}$ is discrete and takes on a finite set of values. Since $y = g(x)$ does not necessarily define a one-to-one correspondence, suppose that $g(x)$ takes on the value g_i when x takes on the values $x_{i1}, x_{i2}, \ldots, x_{in_i}$. Then the probability that $g(\mathbf{x})$ will take on the value g_i is

$$P\Big[g(\mathbf{x}) = g_i\Big] = \sum_{j=1}^{n_i} f(x_{ij})$$

and if $g(x)$ takes on the values $g_1, g_2, \ldots, g_m$, it follows that

$$\begin{aligned} E[g(\mathbf{x})] &= \sum_{i=1}^{m} g_i \cdot P[g(\mathbf{x}) = g_i] \\ &= \sum_{i=1}^{m} g_i \cdot \sum_{j=1}^{n_i} f(x_{ij}) \\ &= \sum_{i=1}^{m} \sum_{j=1}^{n_i} g_i \cdot f(x_{ij}) \\ &= \sum_x g(x) \cdot f(x) \end{aligned}$$

where the summation extends over all values of $\mathbf{x}$.

EXAMPLE 4.3

If $\mathbf{x}$ is the number of points rolled with a balanced die, find the expected value of the random variable $g(\mathbf{x}) = 2\mathbf{x}^2 + 1$.

Solution

Since each possible outcome has the probability $\frac{1}{6}$, we get

$$\begin{aligned} E[g(\mathbf{x})] &= \sum_{x=1}^{6} (2x^2 + 1) \cdot \tfrac{1}{6} \\ &= (2 \cdot 1^2 + 1) \cdot \tfrac{1}{6} + \cdots + (2 \cdot 6^2 + 1) \cdot \tfrac{1}{6} \\ &= \tfrac{94}{3} \end{aligned}$$

EXAMPLE 4.4

If $\mathbf{x}$ has the probability density

$$f(x) = \begin{cases} e^{-x} & \text{for } x > 0 \\ 0 & \text{elsewhere} \end{cases}$$

find the expected value of the random variable $g(\mathbf{x}) = e^{3\mathbf{x}/4}$.

Solution

According to Theorem 4.1, we have

$$\begin{aligned} E[e^{3\mathbf{x}/4}] &= \int_0^{\infty} e^{3x/4} \cdot e^{-x}\, dx \\ &= \int_0^{\infty} e^{-x/4}\, dx \\ &= 4 \end{aligned}$$

The determination of mathematical expectations can often be simplified by using the following theorems, which enable us to calculate expected values from other known or easily computed expectations. Since the steps are essentially the same, some proofs will be given either for the discrete case or the continuous case; others are left for the reader as exercises.

THEOREM 4.2 If a and b are constants, then

$$E(a\mathbf{x} + b) = aE(\mathbf{x}) + b$$

Proof. Using Theorem 4.1 with $g(\mathbf{x}) = a\mathbf{x} + b$, we get

$$E(a\mathbf{x} + b) = \int_{-\infty}^{\infty} (ax + b) \cdot f(x)\,dx$$

$$= a\int_{-\infty}^{\infty} x \cdot f(x)\,dx + b\int_{-\infty}^{\infty} f(x)\,dx$$

$$= aE(\mathbf{x}) + b$$

If we set, respectively, $b = 0$ and $a = 0$, it follows from Theorem 4.2 that

COROLLARY 1 If a is a constant, then

$$E(a\mathbf{x}) = aE(\mathbf{x})$$

COROLLARY 2 If b is a constant, then

$$E(b) = b$$

Observe that if we write $E(b)$, the constant b may be looked upon as a random variable which always takes on the value b.

THEOREM 4.3 If $c_1, c_2, \ldots,$ and c_n are constants, then

$$E\left[\sum_{i=1}^{n} c_i g_i(\mathbf{x})\right] = \sum_{i=1}^{n} c_i E[g_i(\mathbf{x})]$$

Proof. According to Theorem 4.1 with $g(\mathbf{x}) = \sum_{i=1}^{n} c_i g_i(\mathbf{x})$,

$$E\left[\sum_{i=1}^{n} c_i g_i(\mathbf{x})\right] = \sum_{x}\left[\sum_{i=1}^{n} c_i g_i(x)\right] f(x)$$

$$= \sum_{i=1}^{n}\sum_{x} c_i g_i(x) f(x)$$

$$= \sum_{i=1}^{n} c_i \sum_{x} g_i(x) f(x)$$

$$= \sum_{i=1}^{n} c_i E[g_i(\mathbf{x})]$$

EXAMPLE 4.5

Making use of the fact that $E(\mathbf{x}^2) = (1^2 + 2^2 + 3^2 + 4^2 + 5^2 + 6^2) \cdot \frac{1}{6} = \frac{91}{6}$, rework Example 4.3.

Solution

$$E(2\mathbf{x}^2 + 1) = 2E(\mathbf{x}^2) + 1 = 2 \cdot \tfrac{91}{6} + 1 = \tfrac{94}{3}$$

EXAMPLE 4.6

If the probability density of $\mathbf{x}$ is given by

$$f(x) = \begin{cases} 2(1 - x) & \text{for } 0 < x < 1 \\ 0 & \text{elsewhere} \end{cases}$$

show that $E(\mathbf{x}^r) = \dfrac{2}{(r + 1)(r + 2)}$, and use this result to evaluate $E[(2\mathbf{x} + 1)^2]$.

Solution

$E(\mathbf{x}^r) = \displaystyle\int_0^1 x^r \cdot 2(1 - x)\, dx = 2\int_0^1 (x^r - x^{r+1})\, dx = 2\left(\frac{1}{r + 1} - \frac{1}{r + 2}\right)$
$= \dfrac{2}{(r + 1)(r + 2)}$. Since

$$E[(2\mathbf{x} + 1)^2] = 4E(\mathbf{x}^2) + 4E(\mathbf{x}) + 1$$

and $E(\mathbf{x}) = \dfrac{2}{2 \cdot 3} = \dfrac{1}{3}$ and $E(\mathbf{x}^2) = \dfrac{2}{3 \cdot 4} = \dfrac{1}{6}$, we get

$$E[(2\mathbf{x} + 1)^2] = 4 \cdot \tfrac{1}{6} + 4 \cdot \tfrac{1}{3} + 1 = 3$$

EXAMPLE 4.7

Show that

$$E[(a\mathbf{x} + b)^n] = \sum_{i=0}^{n} \binom{n}{i} a^{n-i} b^i E(\mathbf{x}^{n-i})$$

Solution

According to Theorem 1.9, we can write

$$(ax + b)^n = \sum_{i=0}^{n} \binom{n}{i} (ax)^{n-i} b^i$$

and it follows that

$$E[(a\mathbf{x} + b)^n] = E\left[\sum_{i=0}^{n} \binom{n}{i} a^{n-i} b^i \mathbf{x}^{n-i}\right]$$

$$= \sum_{i=0}^{n} \binom{n}{i} a^{n-i} b^i E(\mathbf{x}^{n-i})$$

The concept of mathematical expectation can easily be extended to situations involving more than one random variable. For instance, if **z** is the random variable whose values are related to those of the two random variables **x** and **y** by means of the equation $z = g(x, y)$, it can be shown that

THEOREM 4.4 If **x** and **y** are discrete random variables and $f(x, y)$ is the value of their joint probability distribution at (x, y), the expected value of the random variable $g(\mathbf{x}, \mathbf{y})$ is given by

$$E[g(\mathbf{x}, \mathbf{y})] = \sum_x \sum_y g(x, y) \cdot f(x, y)$$

Correspondingly, if **x** and **y** are continuous random variables and $f(x, y)$ is the value of their joint density at (x, y), the expected value of the random variable $g(\mathbf{x}, \mathbf{y})$ is given by

$$E[g(\mathbf{x}, \mathbf{y})] = \int_{-\infty}^{\infty} \int_{-\infty}^{\infty} g(x, y) f(x, y)\, dx\, dy$$

Generalization of this theorem for functions of any finite number of random variables is straightforward.

EXAMPLE 4.8

With reference to Example 3.12 on page 98, find the expected value of $g(\mathbf{x}, \mathbf{y}) = \mathbf{x} + \mathbf{y}$.

Solution

$$E(\mathbf{x} + \mathbf{y}) = \sum_{x=0}^{2} \sum_{y=0}^{2} (x + y) \cdot f(x, y)$$

$$= (0 + 0) \cdot \tfrac{1}{6} + (0 + 1) \cdot \tfrac{2}{9} + (0 + 2) \cdot \tfrac{1}{36} + (1 + 0) \cdot \tfrac{1}{3} + (1 + 1) \cdot \tfrac{1}{6} + (2 + 0) \cdot \tfrac{1}{12}$$

$$= \tfrac{10}{9}$$

EXAMPLE 4.9

If the joint density function of **x** and **y** is given by

$$f(x, y) = \begin{cases} \frac{2}{7}(x + 2y) & \text{for } 0 < x < 1, 1 < y < 2 \\ 0 & \text{elsewhere} \end{cases}$$

find the expected value of $g(\mathbf{x}, \mathbf{y}) = \mathbf{x}/\mathbf{y}^3$.

Solution

$$E(\mathbf{x}/\mathbf{y}^3) = \int_1^2 \int_0^1 \frac{2x(x + 2y)}{7y^3}\, dx\, dy$$

$$= \frac{2}{7} \int_1^2 \left(\frac{1}{3y^3} + \frac{1}{y^2}\right) dy$$

$$= \frac{15}{84}$$

The following is another theorem which finds useful applications in subsequent work. It is a generalization of Theorem 4.3, and its proof parallels the one of that theorem.

THEOREM 4.5 If $c_1, c_2, \ldots,$ and c_n are constants, then

$$E\left[\sum_{i=1}^{n} c_i g_i(\mathbf{x}_1, \mathbf{x}_2, \ldots, \mathbf{x}_k)\right] = \sum_{i=1}^{n} c_i E[g_i(\mathbf{x}_1, \mathbf{x}_2, \ldots, \mathbf{x}_k)]$$

THEORETICAL EXERCISES

1. To illustrate the proof of Theorem 4.1 with an example, consider the random variable **x** which takes on the values $-2, -1, 0, 1, 2$, and 3 with the respective probabilities $f(-2)$, $f(-1)$, $f(0)$, $f(1)$, $f(2)$, and $f(3)$. If $g(\mathbf{x}) = \mathbf{x}^2$, find

(a) the four possible values g_1, g_2, g_3, and g_4 of $g(x)$;

(b) the probabilities $P[g(\mathbf{x}) = g_i]$ for $i = 1, 2, 3, 4$;

(c) $E[g(\mathbf{x})] = \sum_{i=1}^{4} g_i \cdot P[g(\mathbf{x}) = g_i]$, and show that it equals $\sum_x g(x) \cdot f(x)$.

2. Prove Theorem 4.2 for the discrete case.

3. Prove Theorem 4.3 for the continuous case.

4. Prove Theorem 4.5 for the discrete case.

5. Given the two continuous random variables **x** and **y**, use Theorem 4.4 to express $E(\mathbf{x})$ in terms of
   (a) the joint density of **x** and **y**;
   (b) the marginal density of **x**.

6. Find the expected value of the random variable **x** having the probability distribution $f(x) = \dfrac{|x - 2|}{7}$ for $x = -1, 0, 1, 3$.

7. Find the expected value of the random variable **y** whose probability density is given by

$$f(y) = \begin{cases} \frac{1}{8}(y + 1) & \text{for } 2 < y < 4 \\ 0 & \text{elsewhere} \end{cases}$$

8. The random variable **x** takes on the values 0, 1, 2, and 3 with respective probabilities of $\frac{1}{125}$, $\frac{12}{125}$, $\frac{48}{125}$, and $\frac{64}{125}$.
   (a) Find $E(\mathbf{x})$ and $E(\mathbf{x}^2)$.
   (b) Use the results of part (a) to find $E[(3\mathbf{x} + 2)^2]$.

9. The density function of the continuous random variable **x** is given by

$$f(x) = \begin{cases} \dfrac{1}{x(\ln 3)} & \text{for } 1 < x < 3 \\ 0 & \text{elsewhere} \end{cases}$$

   (a) Find $E(\mathbf{x})$, $E(\mathbf{x}^2)$, and $E(\mathbf{x}^3)$.
   (b) Use the results of part (a) to determine the value of $E(\mathbf{x}^3 + 2\mathbf{x}^2 - 3\mathbf{x} + 1)$.

10. With reference to Exercise 5 on page 108, find $E(2\mathbf{x} - \mathbf{y})$.

11. With reference to Exercise 9 on page 109, find $E(\mathbf{x}/\mathbf{y})$.

12. If **x**, **y**, and **z** have the joint probability distribution of Exercise 15 on page 110, find the expected value of the random variable $\mathbf{u} = \mathbf{x} + \mathbf{y} + \mathbf{z}$.

13. If **x**, **y**, and **z** have the joint probability density of Exercise 17 on page 110, find the expected value of the random variable $\mathbf{w} = \mathbf{x}^2 - \mathbf{yz}$.

14. If **x** has the probability distribution $f(x) = (\frac{1}{2})^x$ for $x = 1, 2, 3, \ldots$, show that $E(2^{\mathbf{x}})$ does not exist. This is the famous **Petersburg paradox**, according to which a player's expectation is infinite (does not exist) if he is to receive 2^x dollars when, in a series of flips of a balanced coin, the first head appears on the xth flip.

APPLIED EXERCISES

15. The probability that Ms. Brown will sell a piece of property at a profit of \$3,000 is $\frac{3}{20}$, the probability that she will sell it at a profit of \$1,500 is $\frac{7}{20}$, the probability that she will break even is $\frac{7}{20}$, and the probability that she will lose \$1,500 is $\frac{3}{20}$. What is her expected profit?

16. A game of chance is considered **fair**, or **equitable**, if each player's expectation is equal to zero. If someone pays us \$10 each time we roll a 3 or a 4 with a balanced die, how much should we pay him when we roll a 1, 2, 5, or 6 to make the game equitable?

17. The manager of a bakery knows that the number of chocolate cakes he can sell on any given day is a random variable having the probability distribution $f(x) = \frac{1}{6}$ for $x = 0, 1, 2, 3, 4$, and 5. He also knows that there is a profit of \$1.00 for each cake which he sells and a loss (due to spoilage) of \$0.40 for each cake he does not sell. Assuming that each cake can be sold only on the day it is made, find the baker's expected profit for a day on which he bakes

(a) 3 of the cakes;
(b) 4 of the cakes;
(c) 5 of the cakes.

18. If a contractor's profit on a construction job can be looked upon as a continuous random variable having the probability density

$$f(x) = \begin{cases} \frac{1}{18}(x+1) & \text{for } -1 < x < 5 \\ 0 & \text{elsewhere} \end{cases}$$

where the units are in thousand dollars, what is his expected profit?

19. With reference to Exercise 17 on page 97, what tread wear can a car owner expect to get with one of the tires?

20. With reference to Exercise 18 on page 97, what is the city's expected water consumption for any given day?

21. With reference to Exercise 23 on page 111, find $E(\mathbf{ps})$, the expected receipts for the commodity.

22. Mr. Adams and Ms. Smith are betting on repeated flips of a coin. At the start of the game Mr. Adams has a dollars, Ms. Smith has b dollars, at each flip the loser pays the winner one dollar, and the game continues until either player is "ruined." Making use of the fact that in an equitable game each player's mathematical expectation is zero, find the probability that Mr. Adams will win Ms. Smith's b dollars before he loses his a dollars.

4.3 MOMENTS

Among the mathematical expectations that are of special importance in statistics, there are the **moments of the distribution of a random variable**, or simply the **moments of a random variable**.

DEFINITION 4.2 The rth **moment about the origin** of the random variable $\mathbf{x}$, denoted by μ'_r, is the expected value of $\mathbf{x}^r$; symbolically,

$$\mu'_r = E(\mathbf{x}^r) = \sum_x x^r \cdot f(x)$$

for $r = 0, 1, 2, 3, \ldots$, when $\mathbf{x}$ is discrete, and

$$\mu'_r = E(\mathbf{x}^r) = \int_{-\infty}^{\infty} x^r \cdot f(x)\,dx$$

when $\mathbf{x}$ is continuous.

It is of interest to note that the term "moment" comes from the field of physics—if the quantities $f(x)$ in the discrete case were point masses acting perpendicularly to the x-axis at distances x from the origin, μ'_1 would be the x-coordinate of the center of gravity, namely, the first moment divided by $\sum f(x) = 1$, and μ'_2 would be the moment of inertia. This also explains why the moments μ'_r are called moments about the origin—in the analogy to physics, the length of the lever arm is in each case the distance from the origin. The analogy applies also in the continuous case, where μ'_1 and μ'_2 might be the x-coordinate of the center of gravity and the moment of inertia of a rod of variable density.

When $r = 0$, we have $\mu'_0 = E(\mathbf{x}^0) = E(1) = 1$, by Corollary 2 of Theorem 4.2, and this is as it should be in accordance with Theorems 3.1 and 3.5. When $r = 1$, we have $\mu'_1 = E(\mathbf{x})$, which is just the expected value of the random variable $\mathbf{x}$ itself; in view of its importance in statistics, we give it a special symbol and a special name.

DEFINITION 4.3 μ'_1 is called the **mean of the distribution** of $\mathbf{x}$, or simply the **mean** of $\mathbf{x}$, and it is denoted by μ.

The special moments we shall define next are of importance in statistics because they serve to describe the shape of the distribution of a random variable, namely, the shape of the graph of its probability distribution or probability density.

DEFINITION 4.4 The rth **moment about the mean** of the random variable $\mathbf{x}$, denoted by μ_r, is the expected value of $(\mathbf{x} - \mu)^r$; symbolically,

$$\mu_r = E[(\mathbf{x} - \mu)^r] = \sum_x (x - \mu)^r \cdot f(x)$$

for $r = 0, 1, 2, 3, \ldots$, when $\mathbf{x}$ is discrete, and

$$\mu_r = E[(\mathbf{x} - \mu)^r] = \int_{-\infty}^{\infty} (x - \mu)^r \cdot f(x)\,dx$$

when $\mathbf{x}$ is continuous.

Note that $\mu_0 = 1$ and $\mu_1 = 0$ for any random variable for which μ exists (see Exercise 1 on page 147).

The second moment about the mean is of special importance in statistics because it is indicative of the spread or dispersion of the distribution of a random variable; thus, it is given a special symbol and a special name.

DEFINITION 4.5 μ_2 is called the **variance of the distribution** of $\mathbf{x}$, or simply the **variance** of $\mathbf{x}$, and it is denoted by σ^2, $\text{var}(\mathbf{x})$, or $V(\mathbf{x})$; σ, the positive square root of the variance, is called the **standard deviation**.

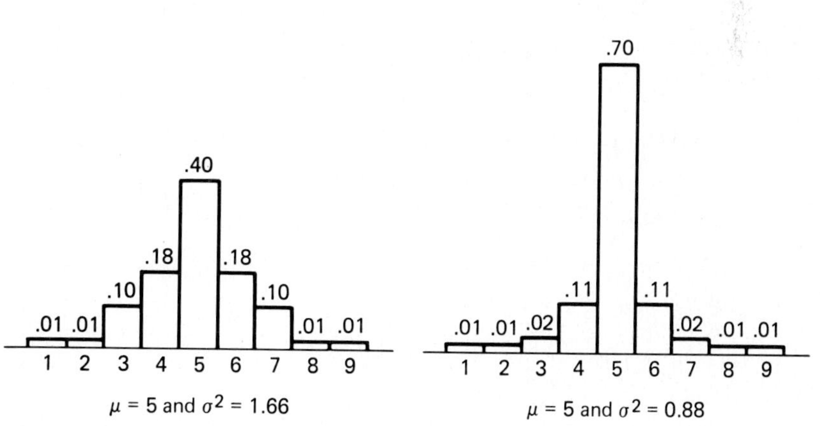

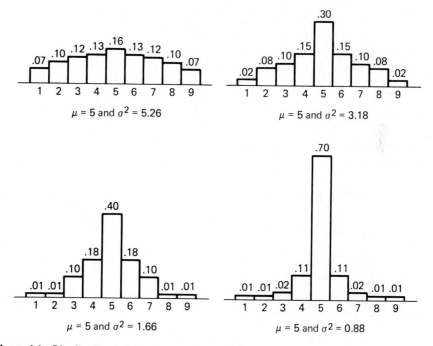

Figure 4.1 Distributions with different dispersions.

Figure 4.1 shows how the variance reflects the spread or dispersion of the distribution of a random variable. Here we show the histograms of the probability distributions of four random variables with the same mean $\mu = 5$, but variances equalling 5.26, 3.18, 1.66, and 0.88. As can be seen, a small value of σ^2 suggests that we are likely to get a value close to the mean, and a large value of σ^2 suggests that there is a greater probability of getting a value that is not close to the mean. This will be discussed further in Section 4.4. A brief discussion of how μ_3, the third moment about the mean, describes the **symmetry** or **skewness** (lack of symmetry) of a distribution is given in Exercise 8 on page 148.

In many instances moments about the mean are obtained by first calculating moments about the origin and then expressing the μ_r in terms of the μ'_r. To serve this purpose, the reader will be asked to verify a general formula in Exercise 7 on page 148. Here, let us merely derive the following computing formula for σ^2:

THEOREM 4.6

$$\sigma^2 = \mu'_2 - \mu^2$$

Proof.

$$\begin{aligned}
\sigma^2 &= E[(\mathbf{x} - \mu)^2] \\
&= E(\mathbf{x}^2 - 2\mu\mathbf{x} + \mu^2) \\
&= E(\mathbf{x}^2) - 2\mu E(\mathbf{x}) + E(\mu^2) \\
&= E(\mathbf{x}^2) - 2\mu \cdot \mu + \mu^2 \\
&= \mu'_2 - \mu^2
\end{aligned}$$

EXAMPLE 4.10

Find the variance of the random variable $\mathbf{x}$ of Example 4.6.

Solution

Having shown on page 133 that $E(\mathbf{x}) = \frac{1}{3}$ and $E(\mathbf{x}^2) = \frac{1}{6}$, it follows that $\sigma^2 = \frac{1}{6} - (\frac{1}{3})^2 = \frac{1}{18}$.

EXAMPLE 4.11

Find the standard deviation of the random variable $\mathbf{x}$ of Example 4.2.

Solution

On page 129 we showed that $\mu = E(\mathbf{x}) = 0.4413$. Now

$$\mu_2' = E(\mathbf{x}^2) = \frac{4}{\pi}\int_0^1 \frac{x^2}{1+x^2}\,dx$$

$$= \frac{4}{\pi}\int_0^1 \left(1 - \frac{1}{1+x^2}\right)dx$$

$$= \frac{4}{\pi} - 1$$

$$= 0.2732$$

and it follows that

$$\sigma^2 = 0.2732 - (0.4413)^2 = 0.0785$$

Hence, $\sigma = \sqrt{0.0785} = 0.2802$.

The following is another theorem that is of importance in work connected with standard deviations or variances:

THEOREM 4.7 If $\mathbf{x}$ has the variance σ^2, then

$$\text{var}(a\mathbf{x} + b) = a^2\sigma^2$$

The proof of this theorem will be left to the reader, but let us point out the following corollaries: For $a = 1$ we find that the addition of a constant to the values of a random variable, resulting in a shift of all the values of $\mathbf{x}$ to the left or to the right, in no way affects the spread of its distribution; for $b = 0$ we find that if the values of a random variable are multiplied by a constant, the variance is multiplied by the square of that constant, resulting in a corresponding change in the spread of the distribution.

4.4 CHEBYSHEV'S THEOREM

To demonstrate how σ or σ^2 is indicative of the spread or dispersion of the distribution of a random variable, let us now prove the following theorem, called **Chebyshev's theorem** after the nineteenth-century Russian mathematician

P. L. Chebyshev. We shall prove it here only for the continuous case, leaving the discrete case as an exercise.

THEOREM 4.8 (Chebyshev's Theorem) If μ and σ are, respectively, the mean and the standard deviation of the random variable $\mathbf{x}$, then for any positive constant k the probability is *at least* $1 - \frac{1}{k^2}$ that $\mathbf{x}$ will take on a value within k standard deviations of the mean; symbolically,

$$P(|\mathbf{x} - \mu| < k\sigma) \geq 1 - \frac{1}{k^2}$$

Proof. From Definitions 4.4 and 4.5 we write

$$\sigma^2 = E[(\mathbf{x} - \mu)^2] = \int_{-\infty}^{\infty} (x - \mu)^2 \cdot f(x)\,dx$$

and, dividing the integral into three parts as shown in Figure 4.2, we get

$$\sigma^2 = \int_{-\infty}^{\mu-k\sigma} (x - \mu)^2 \cdot f(x)\,dx + \int_{\mu-k\sigma}^{\mu+k\sigma} (x - \mu)^2 \cdot f(x)\,dx + \int_{\mu+k\sigma}^{\infty} (x - \mu)^2 \cdot f(x)\,dx$$

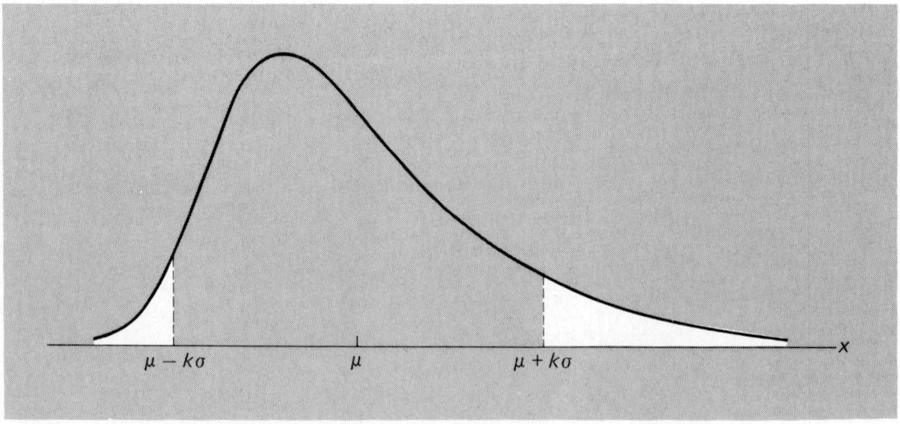

Figure 4.2 Diagram for proof of Chebyshev's theorem.

Since the integrand $(x - \mu)^2 \cdot f(x)$ is non-negative, we can form the inequality

$$\sigma^2 \geqslant \int_{-\infty}^{\mu-k\sigma} (x - \mu)^2 \cdot f(x)\,dx + \int_{\mu+k\sigma}^{\infty} (x - \mu)^2 \cdot f(x)\,dx$$

by deleting the second integral. Now, since $(x - \mu)^2 \geqslant k^2\sigma^2$ for $x \leqslant \mu - k\sigma$ or $x \geqslant \mu + k\sigma$, it follows that

$$\sigma^2 \geqslant \int_{-\infty}^{\mu-k\sigma} k^2\sigma^2 \cdot f(x)\,dx + \int_{\mu+k\sigma}^{\infty} k^2\sigma^2 \cdot f(x)\,dx$$

and, hence, that

$$\frac{1}{k^2} \geqslant \int_{-\infty}^{\mu-k\sigma} f(x)\,dx + \int_{\mu+k\sigma}^{\infty} f(x)\,dx$$

provided $\sigma^2 \neq 0$. Since the sum of the two integrals in this inequality represents the probability that **x** will take on a value less than or equal to $\mu - k\sigma$ or greater than or equal to $\mu + k\sigma$, we have, thus, shown that

$$P(|\mathbf{x} - \mu| \geqslant k\sigma) \leqslant \frac{1}{k^2}$$

and it follows that

$$P(|\mathbf{x} - \mu| < k\sigma) \geqslant 1 - \frac{1}{k^2}$$

For instance, the probability is at least $\frac{3}{4}$ that **x** will take on a value within two standard deviations of the mean, the probability is at least $\frac{8}{9}$ that **x** will take on a value within three standard deviations of the mean, and the probability is at least $\frac{24}{25}$ that **x** will take on a value within five standard deviations of the mean. It is in this sense that σ controls the spread or dispersion of the distribution of a random variable. Clearly, the probability given by Chebyshev's theorem is only a lower bound; whether the probability that a given random variable will take on a value within k standard deviations of the mean is actually greater than $1 - \frac{1}{k^2}$, and if so by how much, we cannot say, but Chebyshev's theorem assures us that this probability cannot be less than $1 - \frac{1}{k^2}$. Only when the distribution of a random variable is known can we calculate the exact probability.

EXAMPLE 4.12

If the probability density of the random variable $\mathbf{x}$ is given by

$$f(x) = \begin{cases} 630x^4(1-x)^4 & \text{for } 0 < x < 1 \\ 0 & \text{elsewhere} \end{cases}$$

find the probability that $\mathbf{x}$ will take on a value within two standard deviations of the mean and compare it with the lower bound provided by Chebyshev's theorem.

Solution

Straightforward integration shows that $\mu = \frac{1}{2}$ and $\sigma^2 = \frac{1}{44}$, so that $\sigma = \sqrt{\frac{1}{44}} =$ 0.15 (approximately). Thus, the probability that $\mathbf{x}$ will take on a value within two standard deviations of the mean is the probability that it will take on a value between 0.20 and 0.80, namely,

$$\begin{aligned} P(0.20 < \mathbf{x} < 0.80) &= \int_{0.20}^{0.80} 630x^4(1-x)^4\,dx \\ &= 0.96 \end{aligned}$$

Observe that the statement "the probability is 0.96" is a much stronger statement than "the probability is at least 0.75," which is provided by Chebyshev's theorem.

4.5 MOMENT-GENERATING FUNCTIONS

Although the moments of most distributions can be determined directly by evaluating the necessary integrals or sums, there exists an alternative procedure which sometimes provides considerable simplifications. This technique utilizes **moment-generating functions.**

DEFINITION 4.6 The **moment-generating function** of the random variable $\mathbf{x}$, where it exists, is given by

$$M_{\mathbf{x}}(t) = E(e^{t\mathbf{x}}) = \sum_x e^{tx} \cdot f(x)$$

when $\mathbf{x}$ is discrete, and

$$M_{\mathbf{x}}(t) = E(e^{t\mathbf{x}}) = \int_{-\infty}^{\infty} e^{tx} \cdot f(x)\,dx$$

when $\mathbf{x}$ is continuous.

The independent variable is t, and we are usually interested in values of t in the neighborhood of 0.

To explain why we refer to this function as a "moment-generating" function, let us substitute for e^{tx} its Maclaurin's series expansion, namely,

$$e^{tx} = 1 + tx + \frac{t^2x^2}{2!} + \frac{t^3x^3}{3!} + \cdots + \frac{t^rx^r}{r!} + \cdots$$

For the discrete case, we thus get

$$\begin{aligned} M_{\mathbf{x}}(t) &= \sum_x \left[1 + tx + \frac{t^2x^2}{2!} + \cdots + \frac{t^rx^r}{r!} + \cdots\right] f(x) \\ &= \sum_x f(x) + t \cdot \sum_x xf(x) + \frac{t^2}{2!} \cdot \sum_x x^2 f(x) + \cdots + \frac{t^r}{r!} \cdot \sum_x x^r f(x) + \cdots \\ &= 1 + \mu t + \mu_2' \cdot \frac{t^2}{2!} + \cdots + \mu_r' \cdot \frac{t^r}{r!} + \cdots \end{aligned}$$

and it can be seen that in the Maclaurin's series of the moment-generating function of $\mathbf{x}$ the coefficient of $\frac{t^r}{r!}$ is μ_r', the rth moment about the origin of the random variable $\mathbf{x}$. In the continuous case, the argument is the same.

EXAMPLE 4.13

Find the moment-generating function of the random variable $\mathbf{x}$, whose probability density is given by

$$f(x) = \begin{cases} e^{-x} & \text{for } x > 0 \\ 0 & \text{elsewhere} \end{cases}$$

and use it to find an expression for μ_r'.

Solution

By definition

$$\begin{aligned} M_{\mathbf{x}}(t) = E(e^{t\mathbf{x}}) &= \int_0^\infty e^{tx} \cdot e^{-x}\, dx \\ &= \int_0^\infty e^{-x(1-t)}\, dx \\ &= \frac{1}{1-t} \qquad \text{for } |t| < 1 \end{aligned}$$

As is well known, the Maclaurin's series for this moment-generating function is

$$\begin{aligned} M_{\mathbf{x}}(t) &= 1 + t + t^2 + t^3 + \cdots + t^r + \cdots \\ &= 1 + 1! \cdot \frac{t}{1!} + 2! \cdot \frac{t^2}{2!} + 3! \cdot \frac{t^3}{3!} + \cdots + r! \cdot \frac{t^r}{r!} + \cdots \end{aligned}$$

and, hence, $\mu_r' = r!$ for $r = 0, 1, 2, \ldots$.

The main difficulty in using the Maclaurin's series of a moment-generating function to determine the moments of a random variable is usually *not* that of finding the moment-generating function, but that of expanding it into a Maclaurin's series. If we are interested only in the first few moments of a random variable, say, μ_1' and μ_2', their determination can usually be simplified by using the following theorem:

THEOREM 4.9

$$\left.\frac{d^r M_{\mathbf{x}}(t)}{dt^r}\right|_{t=0} = \mu_r'$$

This follows from the fact that if a function is expanded as a power series in t, the coefficient of $\frac{t^r}{r!}$ is the rth derivative of the function with respect to t at $t = 0$.

EXAMPLE 4.14

Given that $\mathbf{x}$ has the probability distribution $f(x) = \frac{1}{8}\binom{3}{x}$ for $x = 0, 1, 2,$ and 3, find the moment-generating function of this random variable and use it to determine μ_1' and μ_2'.

Solution

Substituting in accordance with Definition 4.6, we get

$$\begin{aligned} M_{\mathbf{x}}(t) = E(e^{t\mathbf{x}}) &= \tfrac{1}{8} \cdot \sum_{x=0}^{3} e^{tx}\binom{3}{x} \\ &= \tfrac{1}{8}(1 + 3e^t + 3e^{2t} + e^{3t}) \\ &= \tfrac{1}{8}(1 + e^t)^3 \end{aligned}$$

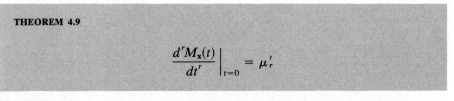

Then, by Theorem 4.9,

$$\mu_1' = M_{\mathbf{x}}'(0) = \tfrac{3}{8}(1 + e^t)^2 e^t \Big|_{t=0} = \tfrac{3}{2}$$

and

$$\mu_2' = M_{\mathbf{x}}''(0) = \tfrac{3}{4}(1 + e^t)\, e^{2t} + \tfrac{3}{8}(1 + e^t)^2 e^t \Big|_{t=0} = 3$$

Often the work involved in using moment-generating functions can be simplified by making use of the following theorem:

THEOREM 4.10 If a and b are constants, then

1. $M_{\mathbf{x}+a}(t) = E[e^{(\mathbf{x}+a)t}] = e^{at} \cdot M_{\mathbf{x}}(t)$
2. $M_{b\mathbf{x}}(t) = E(e^{b\mathbf{x}t}) = M_{\mathbf{x}}(bt)$
3. $M_{\frac{\mathbf{x}+a}{b}}(t) = E\left[e^{\left(\frac{\mathbf{x}+a}{b}\right)t}\right] = e^{\frac{a}{b}t} \cdot M_{\mathbf{x}}\left(\frac{t}{b}\right)$

The proof of this theorem will be left to the reader in Exercise 20 on page 150. As we shall see later, the first part of the theorem is of special importance when $a = -\mu$, and the third part is of special importance when $a = -\mu$ and $b = \sigma$, in which case

$$M_{\frac{\mathbf{x}-\mu}{\sigma}}(t) = e^{-\frac{\mu t}{\sigma}} \cdot M_{\mathbf{x}}\left(\frac{t}{\sigma}\right)$$

THEORETICAL EXERCISES

1. Show that $\mu_0 = 1$ and $\mu_1 = 0$ for any random variable for which $E(\mathbf{x})$ exists.
2. Find μ, μ_2', and σ^2 for a random variable which has the probability distribution $f(x) = \frac{1}{2}$ for $x = -2$ and $x = 2$.
3. Find μ, μ_2', and σ for a random variable which has the probability density

$$f(x) = \begin{cases} \dfrac{x}{2} & \text{for } 0 < x < 2 \\ 0 & \text{elsewhere} \end{cases}$$

4. Prove Theorem 4.7.

5. If the random variable **x** has the mean μ and the variance σ^2, show that the random variable **z**, whose values are related to those of **x** by means of the equation $z = \frac{x - \mu}{\sigma}$, has $E(\mathbf{z}) = 0$ and $\text{var}(\mathbf{z}) = 1$. A distribution which has the mean 0 and the variance 1 is said to be in **standard form**, and when we perform the above change of variable, $z = \frac{x - \mu}{\sigma}$, we are said to be **standardizing** the distribution of **x**.

6. If the probability density of the random variable **x** is given by

$$f(x) = \begin{cases} 2x^{-3} & \text{for } x > 1 \\ 0 & \text{elsewhere} \end{cases}$$

check whether its mean and its variance exist.

7. Show that

$$\mu_r = \mu'_r - \binom{r}{1}\mu'_{r-1} \cdot \mu + \cdots + (-1)^i\binom{r}{i}\mu'_{r-i} \cdot \mu^i + \cdots + (-1)^{r-1}(r-1) \cdot \mu^r$$

for $r = 1, 2, 3, \ldots$, and use this formula to determine expressions for μ_3 and μ_4.

8. The **symmetry** or **skewness** (lack of symmetry) of a distribution is often measured by means of the quantity

$$\alpha_3 = \frac{\mu_3}{\sigma^3}$$

Use the formula for μ_3 obtained in Exercise 7 to find α_3 for each of the following distributions, which (as can easily be verified) have equal means and standard deviations:

(a) $f(1) = 0.05$, $f(2) = 0.15$, $f(3) = 0.30$, $f(4) = 0.30$, $f(5) = 0.15$, and $f(6) = 0.05$;

(b) $f(1) = 0.05$, $f(2) = 0.20$, $f(3) = 0.15$, $f(4) = 0.45$, $f(5) = 0.10$, and $f(6) = 0.05$.

Also draw histograms of the two distributions and note that whereas the first is symmetrical, the second has a "tail" on the left-hand side and is said to be **negatively skewed**.

9. The extent to which a distribution is peaked or flat, also called the **kurtosis** of the distribution, is often measured by means of the quantity

$$\alpha_4 = \frac{\mu_4}{\sigma^4}$$

Use the formula for μ_4 obtained in Exercise 7 to find α_4 for each of the following symmetrical distributions, of which the first is more peaked (narrow humped) than the second:

(a) $f(-3) = 0.06$, $f(-2) = 0.09$, $f(-1) = 0.10$, $f(0) = 0.50$, $f(1) = 0.10$, $f(2) = 0.09$, and $f(3) = 0.06$;

(b) $f(-3) = 0.04$, $f(-2) = 0.11$, $f(-1) = 0.20$, $f(0) = 0.30$, $f(1) = 0.20$, $f(2) = 0.11$, and $f(3) = 0.04$.

10. Duplicating the steps used in the text, prove Chebyshev's theorem for a discrete random variable **x**.

11. Show that if **x** is a random variable with non-negative values (that is, $f(x) = 0$ for $x < 0$) and the mean μ, then for any positive constant a,

$$P(\mathbf{x} \geqslant a) \leqslant \frac{\mu}{a}$$

This inequality is called **Markov's inequality**, and we have given it here mainly because it leads to a relatively easy alternative proof of Chebyshev's theorem.

12. Use the inequality of Exercise 11 to prove Chebyshev's theorem. [*Hint:* Substitute $(\mathbf{x} - \mu)^2$ for **x**.]

13. What is the least value of k in Chebyshev's theorem for which the probability that a random variable takes on a value between $\mu - k\sigma$ and $\mu + k\sigma$ is

(a) at least 0.95;

(b) at least 0.99?

14. If we let $k\sigma = c$ in Chebyshev's theorem, what does this theorem assert about the probability that a random variable will take on a value between $\mu - c$ and $\mu + c$?

15. Find the moment-generating function of the discrete random variable **x** which has the probability distribution $f(x) = 2(\frac{1}{3})^x$ for $x = 1, 2, 3, \ldots$, and use it to determine the values of μ'_1 and μ'_2.

16. Find the moment-generating function of the continuous random variable **x** whose probability density is given by

$$f(x) = \begin{cases} 1 & \text{for } 0 < x < 1 \\ 0 & \text{elsewhere} \end{cases}$$

and use it to find μ'_1, μ'_2, and σ^2.

17. If we let $R_\mathbf{x}(t) = \ln M_\mathbf{x}(t)$, show that $R'_\mathbf{x}(0) = \mu$ and $R''_\mathbf{x}(0) = \sigma^2$. Also, use these results to find the mean and the variance of a random variable **x** having the moment-generating function

$$M_\mathbf{x}(t) = e^{4(e^t - 1)}$$

18. Explain why there can be no random variable for which $M_\mathbf{x}(t) = \dfrac{t}{1 - t}$.

19. Show that if a random variable has the probability density $f(x) = \frac{1}{2}e^{-|x|}$ for $-\infty < x < \infty$, its moment-generating function is given by

$$M_{\mathbf{x}}(t) = \frac{1}{1 - t^2}$$

Also find the variance of the distribution of this random variable

(a) by expanding the moment-generating function as an infinite series and reading off the necessary coefficients;

(b) by using Theorem 4.9.

20. Prove all three parts of Theorem 4.10.

21. Given the moment-generating function $M_{\mathbf{x}}(t) = e^{3t+8t^2}$ of the random variable $\mathbf{x}$, find the moment generating function of the random variable $\mathbf{z} = \frac{1}{4}(\mathbf{x} - 3)$, and use it to find the mean and the variance of $\mathbf{z}$.

APPLIED EXERCISES

22. With reference to Example 4.1 on page 128, find the variance of the distribution of the number of defective sets.

23. The length of time for one individual to be served at a cafeteria is a random variable with the probability density

$$f(x) = \begin{cases} \frac{1}{4}e^{-\frac{x}{4}} & \text{for } x > 0 \\ 0 & \text{elsewhere} \end{cases}$$

Find the mean and the variance of this distribution.

24. With reference to Exercise 15 on page 96, find the mean and the variance of the distribution of the random variable.

25. With reference to Exercise 16 on page 83, find the mean and the variance of the weekly number of accidents at the intersection.

26. The following are some applications of the Markov inequality of Exercise 11:

(a) The scores which high school juniors get on the verbal part of the PSAT/NMSQT test may be looked upon as a random variable with the mean $\mu = 41$. Find an upper bound to the probability that one of the students will get a score of 65 or more.

(b) The weight of certain animals may be looked upon as a random variable with a mean of 212 grams. If none of the animals weighs less than 165 grams, find an upper bound to the probability that such an animal will weigh at least 250 grams.

27. The number of marriage licenses issued in a certain city during the month of June may be looked upon as a random variable with $\mu = 124$ and $\sigma = 7.5$. According to Chebyshev's theorem, with what probability can we assert that between 64 and 184 marriage licenses will be issued there during a month of June?

28. A study of the nutritional value of a certain kind of bread shows that the amount of thiamine (vitamin B_1) in a slice may be looked upon as a random variable with $\mu = 0\ 260$ milligram and $\sigma = 0.005$ milligram. According to Chebyshev's theorem, between what values must be the thiamine content of

(a) at least $\frac{35}{36}$ of all slices of this bread;
(b) at least $\frac{143}{144}$ of all slices of this bread?

29. With reference to Exercise 23, what can we assert about the length of time it takes a person to be served at the cafeteria, if we use Chebyshev's theorem with $k = 1.5$? What is the corresponding exact probability?

4.6 PRODUCT MOMENTS

To continue the discussion of Section 4.3, let us now present the various **product moments** of two random variables.

DEFINITION 4.7 The ***r*th and *s*th product moment about the origin** of the random variables $\mathbf{x}$ and $\mathbf{y}$, denoted by $\mu'_{r,s}$, is the expected value of $\mathbf{x}^r\mathbf{y}^s$; symbolically,

$$\mu'_{r,s} = E(\mathbf{x}^r\mathbf{y}^s) = \sum_x \sum_y x^r y^s \cdot f(x, y)$$

for $r = 0, 1, 2, \ldots,$ and $s = 0, 1, 2, \ldots,$ when $\mathbf{x}$ and $\mathbf{y}$ are discrete, and

$$\mu'_{r,s} = E(\mathbf{x}^r\mathbf{y}^s) = \int_{-\infty}^{\infty} \int_{-\infty}^{\infty} x^r y^s \cdot f(x, y)\, dx\, dy$$

when $\mathbf{x}$ and $\mathbf{y}$ are continuous.

In the discrete case, the double summation extends over the entire joint range of the two random variables. Note that $\mu'_{1,0} = E(\mathbf{x})$, which will be denoted here by $\mu_{\mathbf{x}}$, and that $\mu'_{0,1} = E(\mathbf{y})$, which will be denoted here by $\mu_{\mathbf{y}}$.

Analogous to Definition 4.4, let us now make the following definition of product moments about the respective means:

DEFINITION 4.8 **The *r*th and *s*th product moment about the respective means** of the random variables $\mathbf{x}$ and $\mathbf{y}$, denoted by $\mu_{r,s}$, is the expected value of $(\mathbf{x} - \mu_\mathbf{x})^r(\mathbf{y} - \mu_\mathbf{y})^s$; symbolically,

$$\mu_{r,s} = E[(\mathbf{x} - \mu_\mathbf{x})^r(\mathbf{y} - \mu_\mathbf{y})^s] = \sum_x \sum_y (x - \mu_x)^r(y - \mu_y)^s \cdot f(x, y)$$

for $r = 0, 1, 2, \ldots$, and $s = 0, 1, 2, \ldots$, when $\mathbf{x}$ and $\mathbf{y}$ are discrete, and

$$\mu_{r,s} = E[(\mathbf{x} - \mu_\mathbf{x})^r(y - \mu_\mathbf{y})^s]$$
$$= \int_{-\infty}^{\infty}\int_{-\infty}^{\infty} (x - \mu_x)^r(y - \mu_y)^s \cdot f(x, y)\, dx\, dy$$

when $\mathbf{x}$ and $\mathbf{y}$ are continuous.

In statistics, $\mu_{1,1}$ is of special importance because it is indicative of the relationship, if any, between the values of $\mathbf{x}$ and $\mathbf{y}$; thus, it is given a special symbol and a special name.

DEFINITION 4.9 $\mu_{1,1}$ is called the **covariance** of $\mathbf{x}$ and $\mathbf{y}$, and it is denoted by $\sigma_{\mathbf{xy}}$, $\text{cov}(\mathbf{x}, \mathbf{y})$, or $C(\mathbf{x}, \mathbf{y})$.

Observe that if there is a high probability that large values of $\mathbf{x}$ will go with large values of $\mathbf{y}$ and small values of $\mathbf{x}$ with small values of $\mathbf{y}$, the covariance will be positive; if there is a high probability that large values of $\mathbf{x}$ will go with small values of $\mathbf{y}$ and vice versa, the covariance will be negative. It is in this sense that the covariance measures the relationship, or association, between the values of $\mathbf{x}$ and $\mathbf{y}$.

Analogous to Theorem 4.6, let us now prove the following result, which is useful in actually determining the values of covariances:

THEOREM 4.11

$$\sigma_{\mathbf{xy}} = \mu'_{1,1} - \mu_\mathbf{x}\mu_\mathbf{y}$$

Proof. Using the various theorems about expected values, we can write

$$\begin{aligned}\sigma_{\mathbf{xy}} &= E[(\mathbf{x}-\mu_{\mathbf{x}})(\mathbf{y}-\mu_{\mathbf{y}})] \\ &= E(\mathbf{xy}-\mathbf{x}\mu_{\mathbf{y}}-\mathbf{y}\mu_{\mathbf{x}}+\mu_{\mathbf{x}}\mu_{\mathbf{y}}) \\ &= E(\mathbf{xy})-\mu_{\mathbf{y}}E(\mathbf{x})-\mu_{\mathbf{x}}E(\mathbf{y})+\mu_{\mathbf{x}}\mu_{\mathbf{y}} \\ &= E(\mathbf{xy})-\mu_{\mathbf{y}}\mu_{\mathbf{x}}-\mu_{\mathbf{x}}\mu_{\mathbf{y}}+\mu_{\mathbf{x}}\mu_{\mathbf{y}} \\ &= \mu'_{1,1}-\mu_{\mathbf{x}}\mu_{\mathbf{y}}\end{aligned}$$

EXAMPLE 4.15

With reference to Example 3.12 on page 98, find the covariance of the two random variables **x** and **y**, representing, respectively, the number of aspirin tablets and the number of sedative tablets among the two tablets drawn from the bottle.

Solution

Referring to the table on page 99, we get

$$\begin{aligned}E(\mathbf{xy}) &= 0\cdot 0\cdot \tfrac{1}{6}+0\cdot 1\cdot \tfrac{2}{9}+0\cdot 2\cdot \tfrac{1}{36}+1\cdot 0\cdot \tfrac{1}{3}+1\cdot 1\cdot \tfrac{1}{6}+2\cdot 0\cdot \tfrac{1}{12} \\ &= \tfrac{1}{6}\end{aligned}$$

and using the marginal distributions shown in the table of Example 3.20 on page 112, we get

$$E(\mathbf{x}) = 0\cdot \tfrac{5}{12}+1\cdot \tfrac{1}{2}+2\cdot \tfrac{1}{12} = \tfrac{2}{3}$$

and

$$E(\mathbf{y}) = 0\cdot \tfrac{7}{12}+1\cdot \tfrac{7}{18}+2\cdot \tfrac{1}{36} = \tfrac{4}{9}$$

It follows that

$$\sigma_{\mathbf{xy}} = \tfrac{1}{6}-\tfrac{2}{3}\cdot\tfrac{4}{9} = -\tfrac{7}{54}$$

The negative result suggests that the more aspirin tablets we get the fewer sedative tablets we will get, and vice versa, and this, of course, makes sense.

EXAMPLE 4.16

Find the covariance of the two random variables whose joint density is given by

$$f(x, y) = \begin{cases} 2 & \text{for } x > 0,\ y > 0,\ x+y < 1 \\ 0 & \text{elsewhere} \end{cases}$$

Solution

Evaluating the necessary integrals, we get

$$\mu_{\mathbf{x}} = \int_0^1 \int_0^{1-x} 2x\, dy\, dx = \tfrac{1}{3}$$

$$\mu_{\mathbf{y}} = \int_0^1 \int_0^{1-x} 2y\, dy\, dx = \tfrac{1}{3}$$

and

$$\mu'_{1,1} = \int_0^1 \int_0^{1-x} 2xy\, dy\, dx = \tfrac{1}{12}$$

It follows that

$$\sigma_{\mathbf{xy}} = \tfrac{1}{12} - \tfrac{1}{3} \cdot \tfrac{1}{3} = -\tfrac{1}{36}$$

So far as the relationship between **x** and **y** is concerned, observe that if **x** and **y** are independent, their covariance is zero; symbolically,

> THEOREM 4.12 If **x** and **y** are independent, then $E(\mathbf{xy}) = E(\mathbf{x}) \cdot E(\mathbf{y})$ and $\sigma_{\mathbf{xy}} = 0$.

Proof. For the discrete case, we have by definition

$$E(\mathbf{xy}) = \sum_x \sum_y xy \cdot f(x, y)$$

Since **x** and **y** are independent, we can write $f(x, y) = g(x)h(y)$, where $g(x)$ and $h(y)$ are the values of the respective marginal distributions of **x** and **y**, and we get

$$\begin{aligned} E(\mathbf{xy}) &= \sum_x \sum_y xy \cdot g(x)h(y) \\ &= \left[\sum_x x \cdot g(x)\right]\left[\sum_y y \cdot h(y)\right] \\ &= E(\mathbf{x}) \cdot E(\mathbf{y}) \end{aligned}$$

Hence,

$$\begin{aligned}\sigma_{\mathbf{xy}} &= \mu'_{1,1} - \mu_x\mu_y \\ &= E(\mathbf{x}) \cdot E(\mathbf{y}) - E(\mathbf{x}) \cdot E(\mathbf{y}) \\ &= 0\end{aligned}$$

It is of interest to note that the independence of two random variables implies a zero covariance, but a zero covariance does not necessarily imply their independence. This is illustrated by the following example.

EXAMPLE 4.17

Given two discrete random variables **x** and **y** with the joint probability distribution

		x = −1	x = 0	x = 1	
	−1	$\frac{1}{6}$	$\frac{1}{3}$	$\frac{1}{6}$	$\frac{2}{3}$
y	0	0	0	0	0
	1	$\frac{1}{6}$	0	$\frac{1}{6}$	$\frac{1}{3}$
		$\frac{1}{3}$	$\frac{1}{3}$	$\frac{1}{3}$	

show that the covariance is zero even though the two random variables are not independent.

Solution

Using the values in the table as well as the marginal totals, we get

$$\mu_{\mathbf{x}} = (-1) \cdot \tfrac{1}{3} + 0 \cdot \tfrac{1}{3} + 1 \cdot \tfrac{1}{3} = 0$$

$$\mu_{\mathbf{y}} = (-1) \cdot \tfrac{2}{3} + 0 \cdot 0 + 1 \cdot \tfrac{1}{3} = -\tfrac{1}{3}$$

and

$$\begin{aligned}\mu'_{1,1} &= (-1)(-1) \cdot \tfrac{1}{6} + 0(-1) \cdot \tfrac{1}{3} + 1(-1) \cdot \tfrac{1}{6} + (-1)1 \cdot \tfrac{1}{6} + 1 \cdot 1 \cdot \tfrac{1}{6} \\ &= 0\end{aligned}$$

Thus, $\sigma_{\mathbf{xy}} = 0 - 0(-\frac{1}{3}) = 0$, but the two random variables are not independent since $f(x, y) \neq g(x) \cdot h(y)$, for example, for $x = -1$ and $y = -1$.

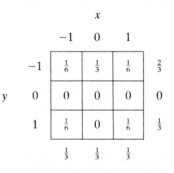

Product moments can also be defined for the case where there are more than two random variables. Here, let us merely state the important result that

THEOREM 4.13 If $\mathbf{x}_1, \mathbf{x}_2, \ldots,$ and $\mathbf{x}_n$ are independent, then

$$E(\mathbf{x}_1\mathbf{x}_2 \cdot \ldots \cdot \mathbf{x}_n) = E(\mathbf{x}_1) \cdot E(\mathbf{x}_2) \cdot \ldots \cdot E(\mathbf{x}_n)$$

This is a generalization of the first part of Theorem 4.12; in fact, the proof of this theorem, based on Definition 3.14, is essentially like that of the first part of Theorem 4.12.

4.7 MOMENTS OF LINEAR COMBINATIONS OF RANDOM VARIABLES

In this section we shall derive expressions for the mean and the variance of a linear combination of n random variables and the covariance of two linear combinations of n random variables. Applications of these results will be treated later in our discussion of sampling theory and problems of statistical inference.

THEOREM 4.14 If $\mathbf{x}_1, \mathbf{x}_2, \ldots, \mathbf{x}_n$ are random variables and

$$\mathbf{y} = \sum_{i=1}^{n} a_i \mathbf{x}_i$$

where $a_1, a_2, \ldots, a_n$ are constants, then

$$E(\mathbf{y}) = \sum_{i=1}^{n} a_i E(\mathbf{x}_i)$$

and

$$\text{var}(\mathbf{y}) = \sum_{i=1}^{n} a_i^2 \cdot \text{var}(\mathbf{x}_i) + 2 \sum_{i<j}\sum a_i a_j \cdot \text{cov}(\mathbf{x}_i, \mathbf{x}_j)$$

where the double sum extends over all the values of i and j, from 1 to n, for which $i < j$.

Proof. From Theorem 4.5 with $g_i(\mathbf{x}_1, \mathbf{x}_2, \ldots, \mathbf{x}_k) = \mathbf{x}_i$ for $i = 0, 1, 2, \ldots, n$, it follows immediately that

$$E(\mathbf{y}) = E\left(\sum_{i=1}^{n} a_i\mathbf{x}_i\right) = \sum_{i=1}^{n} a_i E(\mathbf{x}_i)$$

and this proves the first part of the theorem. To obtain the expression for the variance of $\mathbf{y}$, let us write μ_i for $E(\mathbf{x}_i)$, so that we get

$$\text{var}(\mathbf{y}) = E([\mathbf{y} - E(\mathbf{y})]^2) = E\left(\left[\sum_{i=1}^{n} a_i\mathbf{x}_i - \sum_{i=1}^{n} a_i E(\mathbf{x}_i)\right]^2\right)$$

$$= E\left(\left[\sum_{i=1}^{n} a_i(\mathbf{x}_i - \mu_i)\right]^2\right)$$

Then expanding by means of the multinomial theorem according to which $(a + b + c + d)^2$, for example, equals $a^2 + b^2 + c^2 + d^2 + 2ab + 2ac + 2ad + 2bc + 2bd + 2cd$, and again referring to Theorem 4.5, we get

$$\text{var}(\mathbf{y}) = \sum_{i=1}^{n} a_i^2 E[(\mathbf{x}_i - \mu_i)^2] + 2 \cdot \sum_{i<j}\sum a_i a_j E[(\mathbf{x}_i - \mu_i)(\mathbf{x}_j - \mu_j)]$$

$$= \sum_{i=1}^{n} a_i^2 \cdot \text{var}(\mathbf{x}_i) + 2 \cdot \sum_{i<j}\sum a_i a_j \cdot \text{cov}(\mathbf{x}_i, \mathbf{x}_j)$$

Note that we have tacitly made use of the fact that $\text{cov}(\mathbf{x}_i, \mathbf{x}_j) = \text{cov}(\mathbf{x}_j, \mathbf{x}_i)$.

Since $\text{cov}(\mathbf{x}_i, \mathbf{x}_j) = 0$ when $\mathbf{x}_i$ and $\mathbf{x}_j$ are independent, it follows immediately that

COROLLARY If the random variables $\mathbf{x}_1, \mathbf{x}_2, \ldots, \mathbf{x}_n$ are independent and $\mathbf{y} = \sum_{i=1}^{n} a_i\mathbf{x}_i$, then

$$\text{var}(\mathbf{y}) = \sum_{i=1}^{n} a_i^2 \cdot \text{var}(\mathbf{x}_i)$$

EXAMPLE 4.18

If the random variables $\mathbf{x}$, $\mathbf{y}$, and $\mathbf{z}$ have the means $\mu_\mathbf{x} = 2$, $\mu_\mathbf{y} = -3$, $\mu_\mathbf{z} = 4$, the variances $\sigma_\mathbf{x}^2 = 1$, $\sigma_\mathbf{y}^2 = 5$, $\sigma_\mathbf{z}^2 = 2$, and the covariances $\text{cov}(\mathbf{x}, \mathbf{y}) = -2$,

$\text{cov}(\mathbf{x}, \mathbf{z}) = -1$, $\text{cov}(\mathbf{y}, \mathbf{z}) = 1$, find the mean and the variance of

$$\mathbf{w} = 3\mathbf{x} - \mathbf{y} + 2\mathbf{z}$$

Solution

By Theorem 4.14,

$$\begin{aligned} E(\mathbf{w}) &= E(3\mathbf{x} - \mathbf{y} + 2\mathbf{z}) \\ &= 3E(\mathbf{x}) - E(\mathbf{y}) + 2E(\mathbf{z}) \\ &= 3 \cdot 2 - (-3) + 2 \cdot 4 \\ &= 17 \end{aligned}$$

and

$$\begin{aligned} \text{var}(\mathbf{w}) &= 9\,\text{var}(\mathbf{x}) + \text{var}(\mathbf{y}) + 4\,\text{var}(\mathbf{z}) - 6\,\text{cov}(\mathbf{x}, \mathbf{y}) + 12\,\text{cov}(\mathbf{x}, \mathbf{z}) \\ &\quad - 4\,\text{cov}(\mathbf{y}, \mathbf{z}) \\ &= 9 \cdot 1 + 5 + 4 \cdot 2 - 6(-2) + 12(-1) - 4 \cdot 1 \\ &= 18 \end{aligned}$$

The following is another important theorem about linear combinations of random variables; it concerns the covariance of two linear combinations of n random variables:

THEOREM 4.15 If $\mathbf{x}_1, \mathbf{x}_2, \ldots, \mathbf{x}_n$ are random variables and

$$\mathbf{y}_1 = \sum_{i=1}^{n} a_i \mathbf{x}_i \quad \text{and} \quad \mathbf{y}_2 = \sum_{i=1}^{n} b_i \mathbf{x}_i$$

where $a_1, a_2, \ldots, a_n, b_1, b_2, \ldots, b_n$ are constants, then

$$\text{cov}(\mathbf{y}_1, \mathbf{y}_2) = \sum_{i=1}^{n} a_i b_i \cdot \text{var}(\mathbf{x}_i) + \sum_{i<j}\sum (a_i b_j + a_j b_i) \cdot \text{cov}(\mathbf{x}_i, \mathbf{x}_j)$$

The proof of this theorem, which is very similar to that of Theorem 4.14, will be left to the reader in Exercise 10 on page 162.

Since $\text{cov}(\mathbf{x}_i, \mathbf{x}_j) = 0$ when $\mathbf{x}_i$ and $\mathbf{x}_j$ are independent, it follows immediately that

COROLLARY If the random variables $\mathbf{x}_1, \mathbf{x}_2, \ldots, \mathbf{x}_n$ are independent, $\mathbf{y}_1 = \sum_{i=1}^{n} a_i\mathbf{x}_i$, and $\mathbf{y}_2 = \sum_{i=1}^{n} b_i\mathbf{x}_i$, then

$$\text{cov}(\mathbf{y}_1, \mathbf{y}_2) = \sum_{i=1}^{n} a_i b_i \cdot \text{var}(\mathbf{x}_i)$$

EXAMPLE 4.19

If the random variables $\mathbf{x}$, $\mathbf{y}$, and $\mathbf{z}$ have the means $\mu_\mathbf{x} = 3$, $\mu_\mathbf{y} = 5$, $\mu_\mathbf{z} = 2$, the variances $\sigma_\mathbf{x}^2 = 8$, $\sigma_\mathbf{y}^2 = 12$, $\sigma_\mathbf{z}^2 = 18$, and the covariances $\text{cov}(\mathbf{x}, \mathbf{y}) = 1$, $\text{cov}(\mathbf{x}, \mathbf{z}) = -3$, $\text{cov}(\mathbf{y}, \mathbf{z}) = 2$, find $\text{cov}(\mathbf{u}, \mathbf{v})$, where $\mathbf{u} = \mathbf{x} + 4\mathbf{y} + 2\mathbf{z}$ and $\mathbf{v} = 3\mathbf{x} - \mathbf{y} - \mathbf{z}$.

Solution

By Theorem 4.15,

$$\begin{aligned}
\text{cov}(\mathbf{u}, \mathbf{v}) &= \text{cov}(\mathbf{x} + 4\mathbf{y} + 2\mathbf{z}, 3\mathbf{x} - \mathbf{y} - \mathbf{z}) \\
&= 3\,\text{var}(\mathbf{x}) - 4\,\text{var}(\mathbf{y}) - 2\,\text{var}(\mathbf{z}) + 11\,\text{cov}(\mathbf{x}, \mathbf{y}) + 5\,\text{cov}(\mathbf{x}, \mathbf{z}) - \\
&\qquad 6\,\text{cov}(\mathbf{y}, \mathbf{z}) \\
&= 3 \cdot 8 - 4 \cdot 12 - 2 \cdot 18 + 11 \cdot 1 + 5(-3) - 6 \cdot 2 \\
&= -76
\end{aligned}$$

4.8 CONDITIONAL EXPECTATIONS

In Section 3.7 we obtained conditional probabilities by adding the values of conditional probability distributions, or integrating the values of conditional density functions. **Conditional expectations** of random variables are likewise defined in terms of their conditional distributions.

DEFINITION 4.10 If $\mathbf{x}$ is a discrete random variable and $f(x \mid y)$ is the value of the conditional probability distribution of $\mathbf{x}$ given $\mathbf{y} = y$ at x, the **conditional expectation** of $u(\mathbf{x})$ given $\mathbf{y} = y$ is

$$E[u(\mathbf{x})|y] = \sum_{x} u(x) \cdot f(x|y)$$

Correspondingly, if $\mathbf{x}$ is a continuous random variable and $f(x|y)$ is the value of the conditional probability density of $\mathbf{x}$ given $\mathbf{y} = y$ at x, the **conditional expectation** of $u(\mathbf{x})$ given $\mathbf{y} = y$ is

$$E[u(\mathbf{x})|y] = \int_{-\infty}^{\infty} u(x) \cdot f(x|y)\, dx$$

Similar expressions based on the conditional probability distribution or density of $\mathbf{y}$ given $\mathbf{x} = x$ define the conditional expectation of $v(\mathbf{y})$ given $\mathbf{x} = x$.

If we let $u(\mathbf{x}) = \mathbf{x}$ in Definition 4.10, we obtain the **conditional mean** of the random variable $\mathbf{x}$ given $\mathbf{y} = y$, which we denote by

$$\mu_{\mathbf{x}|y} = E(\mathbf{x}|y)$$

The conditional variance of $\mathbf{x}$ given $\mathbf{y} = y$ is

$$\begin{aligned} \sigma^2_{\mathbf{x}|y} &= E[(\mathbf{x} - \mu_{\mathbf{x}|y})^2|y] \\ &= E(\mathbf{x}^2|y) - \mu^2_{\mathbf{x}|y} \end{aligned}$$

where $E(\mathbf{x}^2|y)$ is given by Definition 4.10 with $u(\mathbf{x}) = \mathbf{x}^2$. The reader should not experience any difficulty in generalizing Definition 4.10 for conditional expectations involving more than two random variables.

EXAMPLE 4.20

If the joint density function of two random variables $\mathbf{x}$ and $\mathbf{y}$ is given by

$$f(x, y) = \begin{cases} \frac{2}{3}(x + 2y) & \text{for } 0 < x < 1, 0 < y < 1 \\ 0 & \text{elsewhere} \end{cases}$$

find the conditional mean and the conditional variance of $\mathbf{x}$ given $\mathbf{y} = \frac{1}{2}$.

Solution

In Example 3.24 on page 117 we showed that the conditional density of $\mathbf{x}$ given $\mathbf{y} = y$ is

$$f(x|y) = \begin{cases} \dfrac{2x + 4y}{1 + 4y} & \text{for } 0 < x < 1 \\ 0 & \text{elsewhere} \end{cases}$$

so that

$$f(x|\tfrac{1}{2}) = \begin{cases} \frac{2}{3}(x+1) & \text{for } 0 < x < 1 \\ 0 & \text{elsewhere} \end{cases}$$

Thus, $\mu_{\mathbf{x}|\frac{1}{2}}$ is given by

$$E(\mathbf{x}|\tfrac{1}{2}) = \int_0^1 \tfrac{2}{3}x(x+1)\,dx$$
$$= \tfrac{5}{9}$$

Next we find

$$E(\mathbf{x}^2|\tfrac{1}{2}) = \int_0^1 \tfrac{2}{3}x^2(x+1)\,dx$$
$$= \tfrac{7}{18}$$

and it follows that

$$\sigma^2_{\mathbf{x}|\frac{1}{2}} = \tfrac{7}{18} - (\tfrac{5}{9})^2 = \tfrac{13}{162}$$

THEORETICAL EXERCISES

1. If $\mathbf{x}$ and $\mathbf{y}$ have the joint probability distribution $f(x, y) = \frac{1}{4}$ for $x = -3$ and $y = -5$, $x = -1$ and $y = -1$, $x = 1$ and $y = 1$, and $x = 3$ and $y = 5$, find $\text{cov}(\mathbf{x}, \mathbf{y})$.
2. With reference to Exercise 1 on page 108, find $\text{cov}(\mathbf{x}, \mathbf{y})$.
3. With reference to Example 3.22 on page 115, find $\text{cov}(\mathbf{x}_1, \mathbf{x}_3)$.
4. With reference to Exercise 6 on page 122, find $\text{cov}(\mathbf{x}, \mathbf{y})$.
5. If the joint distribution of $\mathbf{x}$ and $\mathbf{y}$ has the values $f(-1, 0) = 0$, $f(-1, 1) = \frac{1}{4}$, $f(0, 0) = \frac{1}{6}$, $f(0, 1) = 0$, $f(1, 0) = \frac{1}{12}$, and $f(1, 1) = \frac{1}{2}$, show that $\mathbf{x}$ and $\mathbf{y}$ are dependent, and find their covariance.
6. For k random variables $\mathbf{x}_1, \mathbf{x}_2, \ldots, \mathbf{x}_k$, the values of their **joint moment-generating function** are given by

$$E(e^{t_1\mathbf{x}_1 + t_2\mathbf{x}_2 + \cdots + t_k\mathbf{x}_k})$$

   (a) Show for either the discrete case or the continuous case that the partial derivative of the joint moment-generating function with respect to t_i at $t_1 = t_2 = \cdots = t_k = 0$ is $E(\mathbf{x}_i)$.
   (b) Show for either the discrete case or the continuous case that the second partial derivative of the joint moment-generating function with respect to t_i and t_j, $i \neq j$, at $t_1 = t_2 = \cdots = t_k = 0$ is $E(\mathbf{x}_i\mathbf{x}_j)$.

(c) If two random variables have the joint density given by

$$f(x, y) = \begin{cases} e^{-x-y} & \text{for } x > 0, y > 0 \\ 0 & \text{elsewhere} \end{cases}$$

find their joint moment-generating function and use it to determine the values of $E(\mathbf{xy})$, $E(\mathbf{x})$, $E(\mathbf{y})$, and hence $\text{cov}(\mathbf{x}, \mathbf{y})$.

7. If the independent random variables $\mathbf{x}_1$, $\mathbf{x}_2$, and $\mathbf{x}_3$ have the means 4, 9, 3, and the variances 3, 7, 5, find the mean and the variance of

(a) $\mathbf{y} = 2\mathbf{x}_1 - 3\mathbf{x}_2 + 4\mathbf{x}_3$;

(b) $\mathbf{z} = \mathbf{x}_1 + 2\mathbf{x}_2 - \mathbf{x}_3$.

8. Repeat both parts of Exercise 7, dropping the assumption of independence and adding the information that $\text{cov}(\mathbf{x}_1, \mathbf{x}_2) = 1$, $\text{cov}(\mathbf{x}_2, \mathbf{x}_3) = -2$, and $\text{cov}(\mathbf{x}_1, \mathbf{x}_3) = -3$.

9. If the joint density of $\mathbf{x}$ and $\mathbf{y}$ is given by

$$f(x, y) = \begin{cases} \frac{1}{3}(x + y) & \text{for } 0 < x < 1, 0 < y < 2 \\ 0 & \text{elsewhere} \end{cases}$$

find the variance of $\mathbf{w} = 3\mathbf{x} + 4\mathbf{y} - 5$.

10. Prove Theorem 4.15.

11. Express $\text{var}(\mathbf{x} + \mathbf{y})$, $\text{var}(\mathbf{x} - \mathbf{y})$, and $\text{cov}(\mathbf{x} + \mathbf{y}, \mathbf{x} - \mathbf{y})$ in terms of the variances and covariance of $\mathbf{x}$ and $\mathbf{y}$.

12. If $\mathbf{x}_1$, $\mathbf{x}_2$, and $\mathbf{x}_3$ have the variances 5, 4, 7, and $\text{cov}(\mathbf{x}_1, \mathbf{x}_2) = 3$, $\text{cov}(\mathbf{x}_1, \mathbf{x}_3) = -2$, while $\mathbf{x}_2$ and $\mathbf{x}_3$ are independent, find the covariance of $\mathbf{y}_1 = \mathbf{x}_1 - 2\mathbf{x}_2 + 3\mathbf{x}_3$ and $\mathbf{y}_2 = -2\mathbf{x}_1 + 3\mathbf{x}_2 + 4\mathbf{x}_3$.

13. With reference to Exercise 7, find $\text{cov}(\mathbf{y}, \mathbf{z})$.

14. With reference to Exercise 1 on page 121, find the conditional mean and the conditional variance of $\mathbf{x}$ given $\mathbf{y} = -1$.

15. With reference to Exercise 3 on page 122, find the conditional expectation of the random variable $u(\mathbf{z}) = \mathbf{z}^2$ given $\mathbf{x} = 1$ and $\mathbf{y} = 2$.

16. With reference to Exercise 6 on page 122, find the conditional mean and the conditional variance of $\mathbf{y}$ given $\mathbf{x} = \frac{1}{4}$.

17. With reference to Example 3.22 on page 115 and part (b) of Exercise 9 on page 123, find the expected value of $\mathbf{x}_2^2\mathbf{x}_3$ given $\mathbf{x}_1 = \frac{1}{2}$.

APPLIED EXERCISES

18. A penny, which is unbalanced so that the probability of heads is 0.40, is tossed twice. What is the covariance of $\mathbf{z}$, the number of heads obtained on the first toss, and $\mathbf{w}$, the total number of heads obtained in the two tosses of the coin?

19. The inside diameter of a cylindrical tube is a random variable with a mean of 3 inches and a standard deviation of 0.02 inch, the thickness of the tube is a random variable with a mean of 0.3 inch and a standard deviation of 0.005 inch, and the two random variables are independent. Find the mean and the standard deviation of the outside diameter of the tube.

20. The length of certain bricks is a random variable with a mean of 8 inches and a standard deviation of 0.1 inch, and the thickness of the mortar between two bricks is a random variable with a mean of 0.5 inch and a standard deviation of 0.03 inch. What is the mean and the standard deviation of the length of a wall made of 50 of these bricks laid side by side, if we can assume that all the random variables involved are independent?

21. If heads is a success when we flip a coin, getting a six is a success when we roll a die, and getting an ace is a success when we draw a card from an ordinary deck of 52 playing cards, find the mean and the standard deviation of the total number of successes when we

(a) flip a balanced coin, roll a balanced die, and then draw a card from a well-shuffled deck;

(b) flip a balanced coin three times, roll a balanced die twice, and then draw a card from a well-shuffled deck.

22. If we alternately flip a balanced coin and a coin which is loaded so that the probability of getting heads is 0.45, what are the mean and the variance of the number of heads which we obtain in ten flips of these coins?

23. With reference to Exercise 19 on page 111 and part (c) of Exercise 16 on page 124, find the expected number of statistics texts given that one mathematics text is selected.

24. With reference to Exercise 19 on page 125, by how much can a salesperson who spends $12 on gasoline expect to be reimbursed?

References

Further information about the material in this chapter may be found in the mathematical statistics texts listed on page 126.

special probability distributions

5

5.1 INTRODUCTION

In this chapter we shall study some of the probability distributions which figure most prominently in statistical theory and in applications. We shall also study their **parameters**, that is, the quantities which are constants for particular distributions, but which can take on different values for different members of families of distributions of the same kind. The most common parameters are the lower moments, mainly μ and σ^2, and as we saw in the preceding chapter, there are essentially two ways in which they can be obtained: We can evaluate the necessary sums directly or we can work with moment-generating functions. Although it would seem logical to use in each case whichever method is simplest, we shall sometimes use both. In some instances this will be done because the results are needed later; in others it will merely serve to provide the reader with experience in the application of the respective mathematical techniques. Also, to keep the size of this chapter within bounds, many of the details will be left as exercises.

5.2 THE DISCRETE UNIFORM DISTRIBUTION

If a random variable can take on k different values with equal probabilities, we say that it has a **discrete uniform distribution**; symbolically,

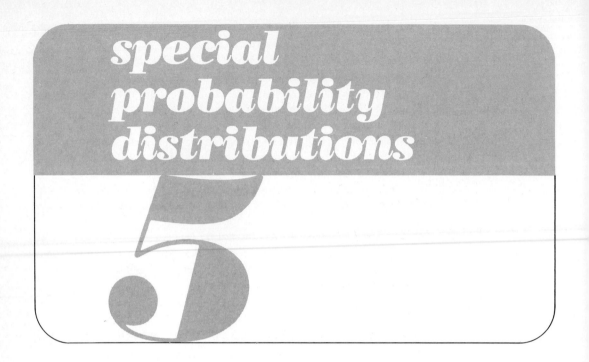

DEFINITION 5.1 A random variable **x** has a **discrete uniform distribution**, and it is referred to as a discrete uniform random variable, if and only if its probability distribution is given by

$$f(x) = \frac{1}{k} \qquad \text{for } x = x_1, x_2, \ldots, x_k$$

where $x_i \neq x_j$ when $i \neq j$.

In accordance with Definitions 4.2 and 4.4, the mean and the variance of this distribution are $\mu = \sum_{i=1}^{k} x_i \cdot \frac{1}{k}$ and $\sigma^2 = \sum_{i=1}^{k} (x_i - \mu)^2 \cdot \frac{1}{k}$.

In the special case where $x_i = i$, the discrete uniform distribution becomes $f(x) = \frac{1}{k}$ for $x = 1, 2, \ldots, k$, and in this form it applies, for example, to the number of points we roll with a balanced die. The mean and the variance of this discrete uniform distribution, and its moment-generating function, are treated in Exercises 1 and 2 on page 171.

5.3 THE BERNOULLI DISTRIBUTION

If an experiment has two possible outcomes, "success" and "failure," and their probabilities are, respectively, θ and $1 - \theta$, then the number of successes, 0 or 1, has a **Bernoulli distribution**; symbolically,

DEFINITION 5.2 A random variable **x** has a **Bernoulli distribution**, and it is referred to as a Bernoulli random variable, if and only if its probability distribution is given by

$$f(x; \theta) = \theta^x (1 - \theta)^{1-x} \qquad \text{for } x = 0, 1$$

Thus, $f(0; \theta) = 1 - \theta$ and $f(1; \theta) = \theta$ are combined into a single formula. Observe that we used the notation $f(x; \theta)$ to indicate explicitly that the Bernoulli distribution has the one parameter θ. Since the Bernoulli distribution is a special case of the distribution of Section 5.4, we shall not discuss it here in any detail.

In connection with the Bernoulli distribution, a success may be getting heads with a balanced coin, it may be catching pneumonia, it may be passing (or failing) an examination, and it may be losing a race. This inconsistency is a carryover from the days when probability theory was applied only to games of chance (and one player's failure was the other's success). Also for this reason, we refer to an experiment to which the Bernoulli distribution applies as a **Bernoulli trial**, or simply a **trial**, and to sequences of such experiments as **repeated trials**.

5.4 THE BINOMIAL DISTRIBUTION

Repeated trials play a very important role in probability and statistics, especially when the number of trials is fixed, the parameter θ (the probability of a success) is the same for each trial, and the trials are all independent. As we shall see, there are several random variables that arise in connection with repeated trials. The one we shall study here concerns the total number of successes; others will be given in Section 5.5.

The theory which we shall discuss in this section has many applications; for instance, it applies if we want to know the probability of getting 5 heads in 12 flips of a coin, the probability that 7 of 10 persons will recover from a tropical disease, or the probability that 35 of 80 persons will respond to a mail-order solicitation. However, this is the case only if each of the 10 persons has the same chance of recovering from the disease and their recoveries are independent (say, they are treated by different doctors in different hospitals), and if the probability of getting a reply to the mail-order solicitation is the same for each of the 80 persons and there is independence (say, no two of them belong to the same household).

To derive a formula for the probability of getting "x successes in n trials" under the stated conditions, observe that the probability of getting x successes and $n - x$ failures *in a specific order* is $\theta^x(1-\theta)^{n-x}$. There is one factor θ for each success, one factor $1 - \theta$ for each failure, and the x factors θ and $n - x$ factors $1 - \theta$ are all multiplied together by virtue of the assumption of independence. Since this probability applies to any sequence of n trials in which there are x successes and $n - x$ failures, we have only to count how many sequences of this kind there are, and then multiply $\theta^x(1-\theta)^{n-x}$ by that number. Clearly, the number of ways in which we can select the x trials on which there is to be a success is $\binom{n}{x}$, and it follows that the desired probability for "x successes in n trials" is $\binom{n}{x}\theta^x(1-\theta)^{n-x}$.

DEFINITION 5.3 A random variable **x** has a **binomial distribution**, and it is referred to as a binomial random variable, if and only if its probability distribution is given by

$$b(x; n, \theta) = \binom{n}{x}\theta^x(1-\theta)^{n-x} \qquad \text{for } x = 0, 1, 2, \ldots, n$$

Thus, the number of successes in n trials is a random variable having a binomial distribution with the parameters n and θ. The name "binomial distribution" derives from the fact that the values of $b(x; n, \theta)$ for $x = 0, 1, 2, \ldots$, and n are the successive terms of the binomial expansion of $[(1-\theta)+\theta]^n$; this shows also that the sum of the probabilities equals 1, as it should.

EXAMPLE 5.1

Find the probability of getting 5 heads and 7 tails in 12 flips of a balanced coin.

Solution

Substituting $x = 5$, $n = 12$, and $\theta = \frac{1}{2}$ into the formula for the binomial distribution, we get

$$b(5; 12, \tfrac{1}{2}) = \binom{12}{5}(\tfrac{1}{2})^5(1-\tfrac{1}{2})^{12-5}$$

and, looking up the value of $\binom{12}{5}$ in Table VII, we find that the result is $792(\frac{1}{2})^{12}$ or approximately 0.19.

EXAMPLE 5.2

Find the probability that 7 of 10 persons will recover from a tropical disease, given that the probability is 0.80 that any one of them will recover from the disease.

Solution

Substituting $x = 7$, $n = 10$, and $\theta = 0.80$ into the formula for the binomial distribution, we get

$$b(7; 10, 0.80) = \binom{10}{7}(0.80)^7(1-0.80)^{10-7}$$

and, looking up the value of $\binom{10}{7}$ in Table VII, we find that the result is $120(0.80)^7(0.20)^3$ or approximately 0.20.

If we tried to calculate the third probability asked for on page 166, the one concerning the responses to the mail-order solicitation, by substituting $x = 35$, $n = 80$, and, say, $\theta = 0.15$, into the formula for the binomial distribution, we would find that this requires a prohibitive amount of work. In actual practice, binomial probabilities are rarely calculated directly, for they are tabulated extensively for various values of θ and n—in the National Bureau of Standards table for $n = 2$ to $n = 49$ and in the book by H. G. Romig for $n = 50$ to $n = 100$.[†] At the end of this book, Table I gives the values of $b(x; n, \theta)$, to four decimals, for $n = 1$ to $n = 20$ and $\theta = 0.05, 0.10, 0.15, \ldots, 0.45$, and 0.50. To use this table when θ is greater than 0.50, we employ the identity

THEOREM 5.1

$$b(x; n, \theta) = b(n - x; n, 1 - \theta)$$

which the reader will be asked to prove in Exercise 4 on page 172. For instance, to find $b(11; 18, 0.70)$, we look up $b(7; 18, 0.30)$, getting 0.1376.

There are also several ways in which binomial probabilities can be approximated when n is large. One of these will be mentioned in Section 5.7, and another in Section 6.6.

Let us now find formulas for the mean and the variance of the binomial distribution.

THEOREM 5.2 The mean and the variance of the binomial distribution are

$$\mu = n\theta \quad \text{and} \quad \sigma^2 = n\theta(1 - \theta)$$

Proof. To determine the mean, let us directly evaluate the sum

$$\begin{aligned}\mu &= \sum_{x=0}^{n} x \cdot \binom{n}{x} \theta^x (1 - \theta)^{n-x} \\ &= \sum_{x=1}^{n} \frac{n!}{(x - 1)!(n - x)!} \theta^x (1 - \theta)^{n-x}\end{aligned}$$

where we omitted the term corresponding to $x = 0$, which is 0, and canceled the x against the first factor of $x! = x(x - 1)!$ in the denominator of $\binom{n}{x}$.

† These books are listed among the references on page 193.

Then, factoring out the factor n in $n! = n(n - 1)!$ and one factor θ, we get

$$\mu = n\theta \cdot \sum_{x=1}^{n} \binom{n-1}{x-1} \theta^{x-1}(1 - \theta)^{n-x}$$

and, letting $y = x - 1$ and $m = n - 1$, this becomes

$$\mu = n\theta \cdot \sum_{y=0}^{m} \binom{m}{y} \theta^{y}(1 - \theta)^{m-y} = n\theta$$

since the last summation is the sum of all the values of a binomial distribution with the parameters m and θ, and hence equal to 1.

To find expressions for μ_2' and then σ^2, let us make use of the fact that $E(\mathbf{x}^2) = E[\mathbf{x}(\mathbf{x} - 1)] + E(\mathbf{x})$, and evaluate $E[\mathbf{x}(\mathbf{x} - 1)]$, duplicating for all practical purposes the steps used above. We thus get

$$\begin{aligned} E[\mathbf{x}(\mathbf{x} - 1)] &= \sum_{x=0}^{n} x(x - 1)\binom{n}{x}\theta^{x}(1 - \theta)^{n-x} \\ &= \sum_{x=2}^{n} \frac{n!}{(x-2)!(n-x)!}\theta^{x}(1 - \theta)^{n-x} \\ &= n(n - 1)\theta^2 \cdot \sum_{x=2}^{n} \binom{n-2}{x-2}\theta^{x-2}(1 - \theta)^{n-x} \end{aligned}$$

and, letting $y = x - 2$ and $m = n - 2$, this becomes

$$\begin{aligned} E[\mathbf{x}(\mathbf{x} - 1)] &= n(n - 1)\theta^2 \cdot \sum_{y=0}^{m} \binom{m}{y}\theta^{y}(1 - \theta)^{m-y} \\ &= n(n - 1)\theta^2 \end{aligned}$$

Therefore,

$$\mu_2' = E[\mathbf{x}(\mathbf{x} - 1)] + E(\mathbf{x}) = n(n - 1)\theta^2 + n\theta$$

and, finally,

$$\begin{aligned} \sigma^2 &= \mu_2' - \mu^2 \\ &= n(n - 1)\theta^2 + n\theta - n^2\theta^2 \\ &= n\theta(1 - \theta) \end{aligned}$$

It should not have come as a surprise that the mean of the binomial distribution is given by the product $n\theta$. After all, if a balanced coin is flipped 200 times, we expect (in the sense of a mathematical expectation) $200 \cdot \frac{1}{2} = 100$ heads and 100 tails; similarly, if a balanced die is rolled 240 times we expect $240 \cdot \frac{1}{6} = 40$ sixes, and if the probability is 0.80 that a person shopping at a department store will make a purchase, we would expect $400(0.80) = 320$ of 400 persons shopping at the department store to make a purchase.

The formula for the variance of the binomial distribution, being a measure of variation, has many important applications, but to emphasize its significance let us consider the random variable $\mathbf{y} = \frac{\mathbf{x}}{n}$, where $\mathbf{x}$ is a random variable having a binomial distribution with the parameters n and θ. This random variable is the proportion of successes in n trials, and in Exercise 6 on page 172 the reader will be asked to prove the following result:

THEOREM 5.3 If $\mathbf{x}$ has a binomial distribution with the parameters n and θ and $\mathbf{y} = \frac{\mathbf{x}}{n}$, then

$$E(\mathbf{y}) = \theta \quad \text{and} \quad \sigma_{\mathbf{y}}^2 = \frac{\theta(1-\theta)}{n}$$

Now, if we apply Chebyshev's theorem with $k\sigma = c$ (see Exercise 14 on page 149), we can assert that *for any positive constant c, the probability is at least*

$$1 - \frac{\theta(1-\theta)}{nc^2}$$

that the proportion of successes in n trials falls between $\theta - c$ and $\theta + c$. Hence, *when $n \to \infty$, the probability approaches 1 that the proportion of successes will differ from θ by less than any arbitrary constant c.* This result is called a **law of large numbers**, and it should be observed that it applies to the proportion of successes, not to their actual number. It is a fallacy to suppose that when n is large, the number of successes must necessarily be close to $n\theta$.

Since the moment-generating function of the binomial distribution is easy to obtain, let us find it and use it to verify the results of Theorem 5.2.

THEOREM 5.4 The moment-generating function of the binomial distribution is given by

$$M_{\mathbf{x}}(t) = [1 + \theta(e^t - 1)]^n$$

Proof. By Definitions 4.6 and 5.3, we get

$$M_{\mathbf{x}}(t) = \sum_{x=0}^{n} e^{xt}\binom{n}{x}\theta^x(1-\theta)^{n-x}$$

$$= \sum_{x=0}^{n} \binom{n}{x}(\theta e^t)^x(1-\theta)^{n-x}$$

and this summation is easily recognizable as the binomial expansion of $[\theta e^t + (1-\theta)]^n = [1 + \theta(e^t - 1)]^n$.

If we differentiate $M_{\mathbf{x}}(t)$ twice with respect to t, we obtain

$$M'_{\mathbf{x}}(t) = n\theta e^t[1 + \theta(e^t - 1)]^{n-1}$$

$$M''_{\mathbf{x}}(t) = n\theta e^t[1 + \theta(e^t - 1)]^{n-1} + n(n-1)\theta^2 e^{2t}[1 + \theta(e^t - 1)]^{n-2}$$

$$= n\theta e^t(1 - \theta + n\theta e^t)[1 + \theta(e^t - 1)]^{n-2}$$

and, upon substituting $t = 0$, we get $\mu'_1 = n\theta$ and $\mu'_2 = n\theta(1 - \theta + n\theta)$. Thus, $\mu = n\theta$ and $\sigma^2 = \mu'_2 - \mu^2 = n\theta(1 - \theta + n\theta) - (n\theta)^2 = n\theta(1 - \theta)$, which agrees with the formulas given in Theorem 5.2.

From the work of this section it may seem easier to find the moments of the binomial distribution with the moment-generating function than to evaluate them directly, but it should be apparent that the differentiation becomes fairly involved if we want to determine, say, μ'_3 or μ'_4. Actually, there exists yet an easier way of determining the moments of the binomial distribution; it is based on its **factorial moment-generating function**, which is explained in Exercise 11 on page 173.

THEORETICAL EXERCISES

1. If $\mathbf{x}$ has the discrete uniform distribution $f(x) = \dfrac{1}{k}$ for $x = 1, 2, \ldots, k$, show that

(a) its mean is $\mu = \dfrac{k+1}{2}$;

(b) its variance is $\sigma^2 = \dfrac{k^2 - 1}{12}$.

(*Hint*: Refer to Appendix II.)

2. If $\mathbf{x}$ has the discrete uniform distribution $f(x) = \dfrac{1}{k}$ for $x = 1, 2, \ldots, k$ show that its moment-generating function is given by

$$M_{\mathbf{x}}(t) = \frac{e^t(1 - e^{kt})}{k(1 - e^t)}$$

Also find the mean of this distribution by evaluating $\lim_{t \to 0} M'_{\mathbf{x}}(t)$, and compare the result with that obtained in part (a) of Exercise 1.

3. The Bernoulli distribution was not studied in any detail in Section 5.3 because it can be looked upon as a binomial distribution with $n = 1$. Show that for the Bernoulli distribution $\mu'_r = \theta$ for $r = 1, 2, 3, \ldots,$

(a) by evaluating the sum $\sum_{x=0}^{1} x^r \cdot f(x; \theta)$;

(b) by letting $n = 1$ in the moment-generating function of the binomial distribution and examining its Maclaurin's series.

Also show that

(c) $\alpha_3 = \dfrac{1 - 2\theta}{\sqrt{\theta(1 - \theta)}}$, where α_3 is the measure of skewness defined in Exercise 8 on page 148;

(d) $\alpha_4 = \dfrac{1 - 3\theta(1 - \theta)}{\theta(1 - \theta)}$, where α_4 is the measure of peakedness defined in Exercise 9 on page 148.

4. Prove Theorem 5.1.

5. If $B(x; n, \theta) = \sum_{k=0}^{x} b(k; n, \theta)$ for $x = 0, 1, 2, \ldots, n$, show that

(a) $b(x; n, \theta) = B(x; n, \theta) - B(x - 1; n, \theta)$;
(b) $b(x; n, \theta) = B(n - x; n, 1 - \theta) - B(n - x - 1; n, 1 - \theta)$;
(c) $B(x; n, \theta) = 1 - B(n - x - 1; n, 1 - \theta)$.

6. Prove Theorem 5.3.

7. When calculating all the values of a binomial distribution, the work can usually be simplified by first calculating $b(0; n, \theta)$ and then using the recursion formula

$$b(x + 1; n, \theta) = \frac{\theta(n - x)}{(x + 1)(1 - \theta)} \cdot b(x; n, \theta)$$

Verify this formula and use it to calculate the values of the binomial distribution with $n = 7$ and $\theta = 0.25$.

8. Use the recursion formula of Exercise 7 to show that for $\theta = \frac{1}{2}$ the binomial distribution has a maximum at $x = \dfrac{n}{2}$ when n is even, and maxima at $x = \dfrac{n - 1}{2}$ and $x = \dfrac{n + 1}{2}$ when n is odd.

9. If $\mathbf{x}$ is a binomial random variable, for what value of θ is the probability $b(x; n, \theta)$ a maximum?

10. In the proof of Theorem 5.2 we determined the quantity $E[\mathbf{x}(\mathbf{x} - 1)]$, called the second **factorial moment**; in general, the rth factorial moment of $\mathbf{x}$ is given by

$$\mu'_{(r)} = E[\mathbf{x}(\mathbf{x} - 1)(\mathbf{x} - 2) \cdot \ldots \cdot (\mathbf{x} - r + 1)]$$

Express μ'_2, μ'_3, and μ'_4 in terms of factorial moments.

11. The **factorial moment-generating function** of a discrete random variable $\mathbf{x}$ is given by

$$F_{\mathbf{x}}(t) = E(t^{\mathbf{x}}) = \sum_x t^x \cdot f(x)$$

(a) Show that the rth derivative of $F_{\mathbf{x}}(t)$ with respect to t at $t = 1$ is $\mu'_{(r)}$, the rth factorial moment defined in Exercise 10.

(b) Show that for the Bernoulli distribution the factorial moment-generating function is given by $F_{\mathbf{x}}(t) = 1 - \theta + \theta t$, and hence that $\mu'_{(1)} = \theta$ and $\mu'_{(r)} = 0$ for $r > 1$.

(c) Find the factorial moment-generating function of the binomial distribution and use it to find μ and σ^2.

12. If we let $a = -\mu$ in the first part of Theorem 4.10, we get

$$M_{\mathbf{y}}(t) = M_{\mathbf{x}-\mu}(t) = e^{-\mu t} \cdot M_{\mathbf{x}}(t)$$

(a) Show that the rth derivative of $M_{\mathbf{x}-\mu}(t)$ with respect to t at $t = 0$ gives the rth moment about the mean of $\mathbf{x}$.

(b) Find such a generating function for moments about the mean of the binomial distribution, and verify that the second derivative at $t = 0$ is $n\theta(1 - \theta)$.

(c) Use the result of part (b) to show that for the binomial distribution

$$\alpha_3 = \frac{1 - 2\theta}{\sqrt{n\theta(1 - \theta)}}$$

where α_3 is the measure of skewness defined in Exercise 8 on page 148. What can we conclude about the skewness of the binomial distribution when (i) $\theta = \frac{1}{2}$, and (ii) n is large?

APPLIED EXERCISES

13. A multiple-choice test consists of eight questions and three answers to each question (of which only one is correct). If a student answers each question by rolling a balanced die and checking the first answer if he gets a 1 or 2, the second answer if he gets a 3 or 4, and the third answer if he gets a 5 or 6, what is the probability that he will get exactly four correct answers?

14. An automobile safety engineer claims that 1 in 10 automobile accidents is due to driver fatigue. Using the formula for the binomial distribution and rounding to four decimals, what is the probability that at least 3 of 5 automobile accidents are due to driver fatigue?

15. If 40 percent of the mice used in an experiment will become very aggressive within one minute after having been administered an experimental drug, find the probability that exactly six of fifteen mice which have been administered the drug will become very aggressive within one minute, using

(a) the formula for the binomial distribution;
(b) Table I.

16. In a certain city, incompatibility is given as the legal reason in 70 percent of all divorce cases. Find the probability that five of the next six divorce cases filed in this city will claim incompatibility as the reason, using

(a) the formula for the binomial distribution;
(b) Table I.

17. A social scientist claims that only 50 percent of all high school seniors capable of doing college work actually go to college. Assuming that this claim is true, use Table I to find the probabilities that among 18 high school seniors capable of doing college work

(a) exactly 10 will go to college;
(b) at least 10 will go to college;
(c) at most eight will go to college.

18. A quality control engineer wants to check whether (in accordance with specifications) 95 percent of the electronic components shipped by her company are without flaws. To this end, she randomly selects 20 from each large lot ready to be shipped and passes the lot if they are all without flaws; otherwise each component in the lot is checked. Assuming that the lots are so large that we can use the binomial distribution as an approximation, use Table I to find the probabilities that she will commit the error of

(a) holding a lot for a complete check even though 95 percent of the components are without flaws;
(b) passing a lot even though only 90 percent of the components are without flaws;
(c) passing a lot even though only 80 percent of the components are without flaws;
(d) passing a lot even though only 70 percent of the components are without flaws.

19. In planning the operation of a new school, one school board member claims that 4 out of 5 newly hired teachers will stay with the school for more than a year, while another school board member claims that it would be correct to say 3 out of 5. In the past, the two board members have been about equally reliable in their predictions, so that in the absence of any other information we would assign their judgments equal weight. If one or the other has to be right, what probabilities would we assign to their claims if it were found that 11 of 12 newly hired teachers stayed with the school for more than a year?

20. Use Chebyshev's theorem and Theorem 5.3 to verify that the probability is at least $\frac{35}{36}$ that

(a) in 900 flips of a balanced coin the proportion of heads will be between 0.40 and 0.60;

(b) in 10,000 flips of a balanced coin the proportion of heads will be between 0.47 and 0.53;

(c) in 1,000,000 flips of a balanced coin the proportion of heads will be between 0.497 and 0.503.

Note that this serves to illustrate the law of large numbers.

5.5 THE NEGATIVE BINOMIAL AND GEOMETRIC DISTRIBUTIONS

In connection with repeated Bernoulli trials, we are sometimes interested in the number of the trial on which the kth success occurs. For instance, we may be interested in the probability that the tenth child exposed to a contagious disease will be the third to catch it, the probability that the fifth person to hear a rumor will be the first one to believe it, or the probability that a burglar will be caught for the second time on his eighth job.

If the kth success is to occur on the xth trial, there must be $k - 1$ successes on the first $x - 1$ trials, and the probability for this is

$$b(k-1; x-1, \theta) = \binom{x-1}{k-1}\theta^{k-1}(1-\theta)^{x-k}$$

The probability of a success on the kth trial is θ, and the probability that the kth success occurs on the xth trial is, therefore,

$$\theta \cdot b(k-1; x-1, \theta) = \binom{x-1}{k-1}\theta^{k}(1-\theta)^{x-k}$$

DEFINITION 5.4 A random variable **x** has a **negative binomial distribution**, and it is referred to as a negative binomial random variable, if and only if its probability distribution is given by

$$b^*(x; k, \theta) = \binom{x-1}{k-1}\theta^{k}(1-\theta)^{x-k}$$

for $x = k, k+1, k+2, \ldots$.

Thus, the number of the trial on which the kth success occurs is a random variable having a negative binomial distribution with the parameters k and θ. The name "negative binomial distribution" derives from the fact that the values of $b^*(x; k, \theta)$ for $x = k, k + 1, k + 2, \ldots$, are the successive terms of the binomial expansion of $\left(\frac{1}{\theta} - \frac{1 - \theta}{\theta}\right)^{-k}$.† In the literature of statistics, negative binomial distributions are also referred to as **binomial waiting-time distributions** or as **Pascal distributions**.

EXAMPLE 5.3

If the probability is 0.40 that a child exposed to a certain contagious disease will catch it, what is the probability that the tenth child exposed to the disease will be the third to catch it.

Solution

Substituting $x = 10$, $k = 3$, and $\theta = 0.40$ into the formula for the negative binomial distribution, we get

$$
\begin{aligned}
b^*(10; 3, 0.40) &= \binom{9}{2}(0.40)^3(0.60)^7 \\
&= 0.0645
\end{aligned}
$$

When a table of binomial probabilities is available, the determination of negative binomial probabilities can generally be simplified by making use of the identity

THEOREM 5.5

$$
b^*(x; k, \theta) = \frac{k}{x} \cdot b(k; x, \theta)
$$

which the reader will be asked to verify in Exercise 2 on page 185.

† Binomial expansions with negative exponents are explained in the book by W. Feller listed among the references on page 69.

EXAMPLE 5.4

Use Theorem 5.5 and Table I to rework Example 5.3.

Solution

Substituting $x = 10$, $k = 3$, and $\theta = 0.40$ into the formula of Theorem 5.5, we get

$$\begin{aligned} b^*(10; 3, 0.40) &= \tfrac{3}{10} \cdot b(3; 10, 0.40) \\ &= \tfrac{3}{10}(0.2150) \\ &= 0.0645 \end{aligned}$$

Moments of the negative binomial distribution may be obtained by proceeding as in the proof of Theorem 5.2. For the mean and the variance we get

THEOREM 5.6 The mean and the variance of the negative binomial distribution are

$$\mu = \frac{k}{\theta} \quad \text{and} \quad \sigma^2 = \frac{k}{\theta}\left(\frac{1}{\theta} - 1\right)$$

as the reader will be asked to verify in Exercise 3 on page 185.

Since the negative binomial distribution with $k = 1$ has many important applications, it is given a special name; it is called the **geometric distribution**.

DEFINITION 5.5 A random variable **x** has a **geometric distribution**, and it is referred to as a geometric random variable, if and only if its probability distribution is given by

$$g(x; \theta) = \theta(1 - \theta)^{x-1} \qquad \text{for } x = 1, 2, 3, \ldots$$

EXAMPLE 5.5

If the probability is 0.75 that an applicant for a driver's licence will pass the road test on any given try, what is the probability that an applicant will finally pass the test on the fourth try.

Solution

Substituting $x = 4$ and $\theta = 0.75$ into the formula for the geometric distribution, we get

$$\begin{aligned} g(4; 0.75) &= 0.75(1 - 0.75)^{4-1} \\ &= 0.75(0.25)^3 \\ &= 0.0117 \end{aligned}$$

Of course, this result is based on the assumption that the trials are all independent, and there may be some question here about its validity.

5.6 THE HYPERGEOMETRIC DISTRIBUTION

In Chapter 2 we used sampling with and without replacement to illustrate the multiplication rules for independent and dependent events. To obtain a formula analogous to that of the binomial distribution which applies to sampling without replacement, in which case the trials are not independent, let us consider a set of N elements of which k are looked upon as successes and the other $N - k$ as failures. As in connection with the binomial distribution, we are interested in the probability of getting x successes in n trials, but now we are choosing, without replacement, n of the N elements contained in the set.

There are $\binom{k}{x}$ ways of choosing x of the k successes and $\binom{N-k}{n-x}$ ways of choosing $n - x$ of the $N - k$ failures, and, hence, $\binom{k}{x}\binom{N-k}{n-x}$ ways of choosing x successes and $n - x$ failures. Since there are $\binom{N}{n}$ ways of choosing n of the N elements in the set and we shall assume that they are all equally likely (which is what we mean when we say that the selection is random), it follows from Theorem 2.2 on page 38 that the probability of "x successes in n trials" is $\binom{k}{x}\binom{N-k}{n-x}\Big/\binom{N}{n}$.

DEFINITION 5.6 A random variable **x** has a **hypergeometric distribution**, and it is referred to as a hypergeometric random variable, if and only if its probability distribution is given by

$$h(x; n, N, k) = \frac{\binom{k}{x}\binom{N-k}{n-x}}{\binom{N}{n}} \qquad \text{for } x = 0, 1, 2, \ldots, n,$$
$$x \leq k \text{ and } n - x \leq N - k$$

Thus, for sampling without replacement, the number of successes in n trials is a random variable having a hypergeometric distribution with the parameters n, N, and k.

EXAMPLE 5.6

As part of an air pollution survey, an inspector decides to examine the exhaust of 6 of a company's 24 trucks. If 4 of the company's trucks emit excessive amounts of pollutants, what is the probability that none of them will be included in the inspector's sample?

Solution

Substituting $x = 0$, $n = 6$, $N = 24$, and $k = 4$ into the formula for the hypergeometric distribution, we get

$$h(0; 6, 24, 4) = \frac{\binom{4}{0}\binom{20}{6}}{\binom{24}{6}}$$

$$= 0.2880$$

The method by which we find the mean and the variance of the hypergeometric distribution is very similar to that employed in the proof of Theorem 5.2.

THEOREM 5.7 The mean and the variance of the hypergeometric distribution are

$$\mu = \frac{nk}{N} \quad \text{and} \quad \sigma^2 = \frac{nk(N-k)(N-n)}{N^2(N-1)}$$

Proof. To determine the mean, let us directly evaluate the sum

$$\mu = \sum_{x=0}^{n} x \cdot \frac{\binom{k}{x}\binom{N-k}{n-x}}{\binom{N}{n}}$$

$$= \sum_{x=1}^{n} \frac{k!}{(x-1)!(k-x)!} \cdot \frac{\binom{N-k}{n-x}}{\binom{N}{n}}$$

where we omitted the term corresponding to $x = 0$, which is 0, and cancelled the x against the first factor of $x! = x(x-1)!$ in the denominator of $\binom{k}{x}$. Then, factoring out $k \Big/ \binom{N}{n}$, we get

$$\mu = \frac{k}{\binom{N}{n}} \cdot \sum_{x=1}^{n} \binom{k-1}{x-1}\binom{N-k}{n-x}$$

and, letting $y = x - 1$ and $m = n - 1$, this becomes

$$\mu = \frac{k}{\binom{N}{n}} \cdot \sum_{y=0}^{m} \binom{k-1}{y}\binom{N-k}{m-y}$$

Finally, using Theorem 1.12 on page 16, we get

$$\mu = \frac{k}{\binom{N}{n}} \cdot \binom{N-1}{m} = \frac{k}{\binom{N}{n}} \cdot \binom{N-1}{n-1} = \frac{nk}{N}$$

To obtain the formula for σ^2, we proceed as on page 169, namely, by first evaluating $E[\mathbf{x}(\mathbf{x}-1)]$ and then making use of the fact that $E(\mathbf{x}^2) = E[\mathbf{x}(\mathbf{x}-1)] + E(\mathbf{x})$ and, hence, $\sigma^2 = E[\mathbf{x}(\mathbf{x}-1)] + E(\mathbf{x}) - \mu^2$. Leaving it to the reader in Exercise 10 on page 186 to show that

$$E[\mathbf{x}(\mathbf{x}-1)] = \frac{k(k-1)n(n-1)}{N(N-1)}$$

we thus obtain

$$\sigma^2 = \frac{k(k-1)n(n-1)}{N(N-1)} + \frac{nk}{N} - \left(\frac{nk}{N}\right)^2$$

$$= \frac{nk(N-k)(N-n)}{N^2(N-1)}$$

Since the moment-generating function of the hypergeometric distribution is fairly complicated, it will not be treated in this book. It may be found, however, in the book by M. G. Kendall and A. Stuart listed among the references on page 126.

When N is large and n is relatively small compared to N (the usual rule of thumb is that n should not exceed 5 percent of N), there is not much difference between sampling with replacement and sampling without replacement, and the formula for the binomial distribution with the parameters n and $\theta = \frac{k}{N}$ may be used to approximate hypergeometric probabilities.

EXAMPLE 5.7

Among the 120 applicants for a job only 80 are actually qualified. If 5 of these applicants are randomly selected for an "in-depth" interview, find the probability that only 2 of the 5 will be qualified for the job by using (a) the hypergeometric distribution, and (b) the binomial distribution as an approximation.

Solution

(a) Substituting $x = 2$, $n = 5$, $N = 120$, and $k = 80$ into the formula for the hypergeometric distribution, we get

$$h(2; 5, 120, 80) = \frac{\binom{80}{2}\binom{40}{3}}{\binom{120}{5}}$$

$$= 0.164$$

rounded to three decimals; (b) substituting $x = 2$, $n = 5$, and $\theta = \frac{80}{120} = \frac{2}{3}$ into the formula for the binomial distribution, we get

$$b\left(2; 5, \frac{2}{3}\right) = \binom{5}{2}\left(\frac{2}{3}\right)^2\left(1 - \frac{2}{3}\right)^3$$

$$= 0.165$$

rounded to three decimals. As can be seen from these results, the approximation is very close.

5.7 THE POISSON DISTRIBUTION

When n is large, the calculation of binomial probabilities with the use of the formula of Definition 5.3 will usually involve a prohibitive amount of work. For instance, to calculate the probability that 18 of 3,000 persons watching a parade on a very hot summer day will suffer from heat exhaustion, we first have to determine $\binom{3{,}000}{18}$, and if the probability is 0.005 that any one of the 3,000 persons watching the parade will suffer from heat exhaustion, we also have to calculate the value of $(0.005)^{18}(0.995)^{2{,}982}$.

In this section we shall present a probability distribution which can be used to approximate binomial probabilities of this kind. Specifically, we shall investigate the limiting form of the binomial distribution when $n \to \infty$, $\theta \to 0$, while $n\theta$ remains constant. Letting this constant be λ, that is, $n\theta = \lambda$ and, hence, $\theta = \frac{\lambda}{n}$, we can write

$$\begin{aligned} b(x; n, \theta) &= \binom{n}{x}\left(\frac{\lambda}{n}\right)^x\left(1 - \frac{\lambda}{n}\right)^{n-x} \\ &= \frac{n(n-1)(n-2)\cdot \ldots \cdot(n-x+1)}{x!}\left(\frac{\lambda}{n}\right)^x\left(1-\frac{\lambda}{n}\right)^{n-x} \end{aligned}$$

Then, if we divide one of the x factors n in $\left(\frac{\lambda}{n}\right)^x$ into each factor of the product $n(n-1)(n-2)\cdot \ldots \cdot(n-x+1)$ and write

$$\left(1-\frac{\lambda}{n}\right)^{n-x} \quad \text{as} \quad \left[\left(1-\frac{\lambda}{n}\right)^{-n/\lambda}\right]^{-\lambda}\left(1-\frac{\lambda}{n}\right)^{-x}$$

we obtain

$$\frac{1\left(1-\frac{1}{n}\right)\left(1-\frac{2}{n}\right)\cdots\left(1-\frac{x-1}{n}\right)}{x!}(\lambda)^x\left[\left(1-\frac{\lambda}{n}\right)^{-n/\lambda}\right]^{-\lambda}\left(1-\frac{\lambda}{n}\right)^{-x}$$

Finally, if we let $n \to \infty$ while x and λ remain fixed, we find that

$$1\left(1-\frac{1}{n}\right)\left(1-\frac{2}{n}\right)\cdots\left(1-\frac{x-1}{n}\right) \to 1$$

$$\left(1-\frac{\lambda}{n}\right)^{-x} \to 1$$

$$\left(1-\frac{\lambda}{n}\right)^{-n/\lambda} \to e$$

and, hence, that the limiting distribution becomes

$$p(x; \lambda) = \frac{\lambda^x e^{-\lambda}}{x!} \qquad \text{for } x = 0, 1, 2, \ldots$$

DEFINITION 5.7 A random variable **x** has a **Poisson distribution**, and it is referred to as a Poisson random variable, if and only if its probability distribution is given by

$$p(x; \lambda) = \frac{\lambda^x e^{-\lambda}}{x!} \qquad \text{for } x = 0, 1, 2, \ldots$$

Thus, in the limit when $n \to \infty$, $\theta \to 0$, and $n\theta = \lambda$ remains constant, the number of successes is a random variable having a Poisson distribution with the parameter λ. This distribution is named after the French mathematician Simeon Poisson (1781–1840).

EXAMPLE 5.8

If the probability is 0.005 that any one person attending a parade on a very hot day will suffer from heat exhaustion, what is the probability that 18 of the 3,000 persons attending the parade will suffer from heat exhaustion?

Solution

Substituting $x = 18$ and $\lambda = 3{,}000(0.005) = 15$ into the formula for the Poisson distribution, we get

$$p(18; 15) = \frac{15^{18} \cdot e^{-15}}{18!}$$

Since this is tedious to evaluate, we refer instead to Table II and find that the answer is $p(18; 15) = 0.0706$.

In general, the Poisson distribution will provide a good approximation to binomial probabilities when $n \geqslant 20$ and $\theta \leqslant 0.05$; when $n \geqslant 100$ and $n\theta \leqslant 10$, the approximation will generally be excellent. Observe that, as is explained in Exercise 14 on page 186, the Poisson distribution has many applications which have no direct connection with binomial distributions.

Having derived the Poisson distribution as a limiting form of the binomial distribution, we can obtain formulas for its mean and its variance by applying the same limiting conditions ($n \to \infty$, $\theta \to 0$, and $n\theta = \lambda$ remains constant) to the mean and the variance of the binomial distribution. For the mean we get $\mu = n\theta = \lambda$ and for the variance we get $\sigma^2 = n\theta(1 - \theta) = \lambda(1 - \theta)$, which approaches λ when $\theta \to 0$.

THEOREM 5.8 The mean and the variance of the Poisson distribution are given by

$$\mu = \lambda \quad \text{and} \quad \sigma^2 = \lambda$$

These results can also be obtained by evaluating the necessary sums (see Exercise 16 on page 187) or by working with the moment-generating function.

THEOREM 5.9 The moment-generating function of the Poisson distribution is given by

$$M_{\mathbf{x}}(t) = e^{\lambda(e^t - 1)}$$

Proof. By Definitions 4.6 and 5.7,

$$M_{\mathbf{x}}(t) = \sum_{x=0}^{\infty} e^{xt} \cdot \frac{\lambda^x e^{-\lambda}}{x!} = e^{-\lambda} \cdot \sum_{x=0}^{\infty} \frac{(\lambda e^t)^x}{x!}$$

where $\sum_{x=0}^{\infty} \frac{(\lambda e^t)^x}{x!}$ can be recognized as the Maclaurin's series of e^z with $z = \lambda e^t$. Thus,

$$M_{\mathbf{x}}(t) = e^{-\lambda} \cdot e^{\lambda e^t} = e^{\lambda(e^t - 1)}$$

Then, differentiating $M_{\mathbf{x}}(t)$ twice with respect to t, we get

$$M'_{\mathbf{x}}(t) = \lambda e^t e^{\lambda(e^t - 1)}$$

$$M''_{\mathbf{x}}(t) = \lambda e^t e^{\lambda(e^t - 1)} + \lambda^2 e^{2t} e^{\lambda(e^t - 1)}$$

so that $\mu'_1 = M'_{\mathbf{x}}(0) = \lambda$ and $\mu'_2 = M''_{\mathbf{x}}(0) = \lambda + \lambda^2$. Thus, $\mu = \lambda$ and $\sigma^2 = \mu'_2 - \mu^2 = (\lambda + \lambda^2) - \lambda^2 = \lambda$, which agrees with Theorem 5.8.

THEORETICAL EXERCISES

1. The negative binomial distribution is sometimes defined differently than in this text as the distribution of the number of failures that precede the kth success. If the kth success occurs on the xth trial, it must be preceded by $x - k$ failures. Thus, find the distribution of $\mathbf{y} = \mathbf{x} - k$, where $\mathbf{x}$ has the distribution of Definition 5.4. Also, use Theorem 5.6 to find expressions for $\mu_{\mathbf{y}}$ and $\sigma^2_{\mathbf{y}}$.

2. Prove Theorem 5.5.

3. Derive the formulas for the mean and the variance of the negative binomial distribution by first determining $E(\mathbf{x})$ and $E[\mathbf{x}(\mathbf{x}+1)]$.

4. Show that the moment-generating function of the geometric distribution is given by

$$M_{\mathbf{x}}(t) = \frac{\theta e^t}{1 - e^t(1-\theta)}$$

and use it to verify that $\mu = \dfrac{1}{\theta}$ and $\sigma^2 = \dfrac{1-\theta}{\theta^2}$.

5. Differentiating with respect to θ the expressions on both sides of the equation

$$\sum_{x=1}^{\infty} \theta(1-\theta)^{x-1} = 1$$

show that the mean of the geometric distribution is given by $\mu = \dfrac{1}{\theta}$. Then, differentiating again with respect to θ show that $\mu_2' = \dfrac{2-\theta}{\theta^2}$, and, hence, $\sigma^2 = \dfrac{1-\theta}{\theta^2}$.

6. If $\mathbf{x}$ is a geometric random variable, show that

$$P(\mathbf{x} = x + n | \mathbf{x} > n) = P(\mathbf{x} = x)$$

7. If the probability is $f(x)$ that a product fails the xth time it is being used, that is, on the xth trial, then its **failure rate** at the xth trial is the probability that it will fail on the xth trial given that it has not failed on the first $x - 1$ trials; symbolically, it is given by

$$Z(x) = \frac{f(x)}{1 - F(x-1)}$$

where $F(x)$ is the value of the corresponding distribution function at x. Show that if $\mathbf{x}$ is a geometric random variable, its failure rate is constant, and equal to θ.

8. Hypergeometric sampling (sampling without replacement) is one variation of binomial sampling (sampling with replacement); another variation arises when the trials are independent but the probability of a success on the ith trial is θ_i, and the θ's are not all equal. If $\mathbf{x}$ is the number of successes obtained with this kind of sampling in n trials, show that

(a) $\mu_{\mathbf{x}} = n\theta$, where $\theta = \frac{1}{n} \cdot \sum_{i=1}^{n} \theta_i$;

(b) $\sigma_{\mathbf{x}}^2 = n\theta(1 - \theta) - n\sigma_{\theta}^2$, where θ is as defined in part (a) and $\sigma_{\theta}^2 = \frac{1}{n} \cdot \sum_{i=1}^{n} (\theta_i - \theta)^2$.

9. When calculating all the values of a hypergeometric distribution, the work can often be simplified by first calculating $h(0; n, N, k)$ and then using the recursion formula

$$h(x + 1; n, N, k) = \frac{(n - x)(k - x)}{(x + 1)(N - k - n + x + 1)} \cdot h(x; n, N, k)$$

Verify this formula and use it to calculate the values of the hypergeometric distribution with $n = 4$, $N = 9$, and $k = 5$.

10. Verify the expression given for $E[\mathbf{x}(\mathbf{x} - 1)]$ in the proof of Theorem 5.7.

11. Show that if we let $\theta = \frac{k}{N}$ in Theorem 5.7, the mean and the variance of the hypergeometric distribution can be written as $\mu = n\theta$ and $\sigma^2 = n\theta(1 - \theta) \cdot \frac{N - n}{N - 1}$. How do these results tie in with the discussion on page 181?

12. Find a recursion formula for the Poisson distribution which expresses $p(x + 1; \lambda)$ in terms of $p(x; \lambda)$, and use it to verify the values given in Table II for $\lambda = 2$. (Use $e^{-2} = 0.1353$.)

13. Approximate the binomial probability $b(3; 100, 0.10)$ by looking up the Poisson probability $p(3; 10)$ in Table II, and compare the result with the exact probability calculated with the use of logarithms.

14. The Poisson distribution has many important applications which have no direct connection with the binomial distribution. Suppose, for instance, that $f(x, t)$ is the probability of getting x successes during a time interval of length t when (i) the probability of a success during a very small time interval from t to $t + \Delta t$ is $\alpha \cdot \Delta t$, (ii) the probability of more than one success during such a time interval is negligible, and (iii) the probability of a success during such a time interval does not depend on what happened prior to time t.

(a) Show that under these conditions

$$f(x, t + \Delta t) = f(x, t)[1 - \alpha \cdot \Delta t] + f(x - 1, t)\alpha \cdot \Delta t$$

and, hence, that

$$\frac{d[f(x, t)]}{dt} = \alpha[f(x - 1, t) - f(x, t)]$$

(b) Show by direct substitution that a solution of this infinite system of differential equations (there is one for each value of x) is given by the Poisson distribution with $\lambda = \alpha t$.

Problems where these assumptions are met may be found among the applied exercises that follow.

15. Use integration by parts to show that

$$\sum_{y=0}^{x} \frac{\lambda^y e^{-\lambda}}{y!} = \frac{1}{x!} \cdot \int_{\lambda}^{\infty} t^x e^{-t}\, dt$$

This result is important because values of the distribution function of a Poisson random variable may, thus, be obtained by referring to a table of incomplete gamma functions.

16. Derive the formulas for the mean and the variance of the Poisson distribution by first evaluating $E(\mathbf{x})$ and $E[\mathbf{x}(\mathbf{x} - 1)]$.

17. Show that if the limiting conditions $n \to \infty$, $\theta \to 0$, while $n\theta$ remains constant, are applied to the moment-generating function of the binomial distribution, we get the moment-generating function of the Poisson distribution. [*Hint*: Make use of the fact that $\lim_{n\to\infty}\left(1 + \frac{z}{n}\right)^n = e^z$.]

18. Use Theorem 5.9 to show that for the Poisson distribution $\alpha_3 = \frac{1}{\sqrt{\lambda}}$, where α_3 is the measure of skewness defined in Exercise 8 on page 148.

19. Differentiating with respect to λ the expressions on both sides of the equation

$$\mu_r = \sum_{x=0}^{\infty} (x - \lambda)^r \cdot \frac{\lambda^x e^{-\lambda}}{x!}$$

derive the following recursion formula for the moments about the mean of the Poisson distribution:

$$\mu_{r+1} = \lambda\left[r\mu_{r-1} + \frac{d\mu_r}{d\lambda}\right]$$

for $r = 1, 2, 3, \ldots$. Also, use this recursion formula and the fact that $\mu_0 = 1$ and $\mu_1 = 0$ to find μ_2, μ_3, μ_4, and verify the formula given for α_3 in Exercise 18.

20. Use Theorem 5.9 to find the moment-generating function of $\mathbf{y} = \mathbf{x} - \lambda$, where $\mathbf{x}$ is a random variable having a Poisson distribution with the parameter λ, and use it to verify that $\sigma_{\mathbf{x}}^2 = \lambda$.

APPLIED EXERCISES

21. If the probability is 0.75 that a person will believe a rumor about the transgressions of a certain politician, find the probabilities that

(a) the eighth person to hear the rumor will be the fifth to believe it;
(b) the fifteenth person to hear the rumor will be the tenth to believe it.

22. If the probabilities of having a male or female child are both 0.50, find the probabilities that

(a) a family's fourth child is their first son;
(b) a family's seventh child is their second daughter;
(c) a family's tenth child is their fourth or fifth son.

23. Use Theorem 5.5 and Table I to check the result of Example 5.5 on page 177.

24. When taping a television commercial, the probability is 0.30 that a certain actor will get his lines straight on any one take. What is the probability that he will get his lines straight for the first time on the sixth take?

25. In a "torture test" a light switch is turned on and off until it fails. If the probability is 0.001 that the switch will fail any time it is turned on or off, what is the probability that the switch will not fail during the first 800 times it is turned on or off? Assume that the conditions underlying the geometric distribution are met and use logarithms.

26. A quality control engineer inspects a random sample of two hand-held calculators from each incoming lot of size 18, and accepts the lot if they are both in good working condition; otherwise, the entire lot is inspected with the cost charged to the vendor. What are the probabilities that such a lot will be accepted without further inspection if it contains

(a) 4 calculators that are not in good working condition;
(b) 8 calculators that are not in good working condition;
(c) 12 calculators that are not in good working condition?

27. Among the 16 applicants for a job, ten have college degrees. If three of the applicants are randomly chosen for interviews, what are the probabilities that

(a) none has a college degree;
(b) one has a college degree;
(c) two have college degrees;
(d) all three have college degrees?

28. What is the probability that an I.R.S. auditor will catch only two income tax returns with illegitimate deductions, if she randomly selects five returns from among 15 returns of which 9 contain illegitimate deductions?

29. A shipment of 80 burglar alarms contains 4 that are defective. If 3 of these are randomly selected and shipped to a customer, find the probability that the customer will get exactly one bad unit using

(a) the formula of the hypergeometric distribution;
(b) the binomial distribution as an approximation.

30. A large shipment of books contains 3 percent with imperfect bindings. Use the Poisson approximation to determine the probabilities that among 400 books randomly selected from the shipment

(a) exactly 10 will have imperfect bindings;
(b) at least 10 will have imperfect bindings.

31. It is known from experience that 1.4 percent of the calls received by a switchboard are wrong numbers. Use the Poisson approximation to the binomial distribution to determine the probability that among 150 calls received by the switchboard two are wrong numbers.

32. Records show that the probability is 0.0004 that a car will break down while driving through a certain tunnel. Use the Poisson approximation to the binomial distribution to find the probability that among 2,000 cars driving through the tunnel at most one will break down.

33. In a certain desert region the number of persons who become seriously ill each year from eating a given kind of poisonous plant is a random variable having a Poisson distribution with $\lambda = 1.6$. What are the probabilities of 0, 1, and 2 such illnesses in a given year?

34. In the inspection of a fabric produced in continuous rolls, the number of imperfections per ten yards is a random variable having a Poisson distribution with $\lambda = 2.8$. Find the probabilities that ten yards of the fabric will have

(a) three imperfections;
(b) at most three imperfections.

5.8 THE MULTINOMIAL DISTRIBUTION

An immediate generalization of the binomial distribution arises when each trial has more than two possible outcomes, the probabilities of the respective outcomes are the same for each trial, and the trials are all independent. This would be the case, for instance, when persons interviewed by an opinion poll are asked whether they are for a candidate, against her, or undecided, or when samples of manufactured products are rated excellent, above average, average, or inferior.

To treat this kind of problem in general, let us consider the case where there are n independent trials permitting k mutually exclusive outcomes whose respective probabilities are $\theta_1, \theta_2, \ldots, \theta_k$ $\left(\text{with } \sum_{i=1}^{k} \theta_i = 1\right)$. Referring to the outcomes as being of the first kind, the second kind, ..., and the kth kind, we shall be interested in the probability of getting x_1 outcomes of the first kind, x_2 outcomes of the second kind, ..., and x_k outcomes of the kth kind $\left(\text{with } \sum_{i=1}^{k} x_i = n\right)$.

Proceeding as in the derivation of the formula for the binomial distribution, we first find that the probability of getting x_1 outcomes of the first kind, x_2 outcomes of the second kind, ..., and x_k outcomes of the kth kind *in a specific order* is $\theta_1^{x_1} \cdot \theta_2^{x_2} \cdot \ldots \cdot \theta_k^{x_k}$. To get the corresponding probability for that many outcomes of each kind *in any order*, we shall have to multiply the probability for any specific order by

$$\binom{n}{x_1, x_2, \ldots, x_k} = \frac{n!}{x_1! \cdot x_2! \cdot \ldots \cdot x_k!}$$

according to Theorem 1.8 on page 12.

DEFINITION 5.8 The random variables $\mathbf{x}_1, \mathbf{x}_2, \ldots$, and $\mathbf{x}_k$, have a **multinomial distribution**, and they are referred to as multinomial random variables, if and only if their joint probability distribution is given by

$$f(x_1, x_2, \ldots, x_k; n, \theta_1, \theta_2, \ldots, \theta_k) = \binom{n}{x_1, x_2, \ldots, x_k} \cdot \theta_1^{x_1} \cdot \theta_2^{x_2} \cdot \ldots \cdot \theta_k^{x_k}$$

for $x_i = 0, 1, \ldots, n$ for each i, where $\sum_{i=1}^{k} x_i = n$ and $\sum_{i=1}^{k} \theta_i = 1$.

Thus, the numbers of outcomes of the different kinds are random variables having the multinomial distribution with the parameters $n, \theta_1, \theta_2, \ldots$, and θ_k. The name "multinomial" derives from the fact that for various values of the x_i, the probabilities equal corresponding terms of the multinomial expansion of $(\theta_1 + \theta_2 + \cdots + \theta_k)^n$.

EXAMPLE 5.9

In a given city on a Saturday night, Channel 12 has 50 percent of the viewing audience, Channel 10 has 30 percent of the viewing audience, and Channel 3 has 20 percent of the viewing audience. Find the probability that among eight television viewers in that city, randomly chosen on a Saturday night, five will be watching Channel 12, two will be watching Channel 10, and one will be watching Channel 3.

Solution

Substituting $x_1 = 5, x_2 = 2, x_3 = 1, \theta_1 = 0.50, \theta_2 = 0.30, \theta_3 = 0.20$, and $n = 8$ into the formula of Definition 5.8, we get

$$f(5, 2, 1; 8, 0.50, 0.30, 0.20) = \frac{8!}{5! \cdot 2! \cdot 1!}(0.50)^5(0.30)^2(0.20)$$
$$= 0.0945$$

5.9 THE MULTIVARIATE HYPERGEOMETRIC DISTRIBUTION

Just as the hypergeometric distribution takes the place of the binomial distribution for sampling without replacement, there also exists a multivariate distribution analogous to the multinomial distribution which applies to sampling without replacement. To derive its formula, let us consider a set of N elements, of which a_1 are elements of the first kind, a_2 are elements of the second kind, ..., and a_k are elements of the kth kind, such that $\sum_{i=1}^{k} a_i = N$. As in connection with the multinomial distribution, we are interested in the probability of getting x_1 elements (outcomes) of the first kind, x_2 elements of the second kind, ..., and x_k elements of the kth kind, but now we are choosing, without replacement, n of the N elements of the set.

There are $\binom{a_1}{x_1}$ ways of choosing x_1 of the a_1 elements of the first kind, $\binom{a_2}{x_2}$ ways of choosing x_2 of the a_2 elements of the second kind, ..., and $\binom{a_k}{x_k}$ ways of choosing x_k of the a_k elements of the kth kind, and, hence, $\binom{a_1}{x_1}\binom{a_2}{x_2}\cdot \ldots \cdot \binom{a_k}{x_k}$ ways of choosing the required $\sum_{i=1}^{k} x_i = n$ elements. Since there are $\binom{N}{n}$ ways of choosing n of the N elements in the set and we assume that they are all equally likely (which is what we mean when we say that the selection is random), it follows that the desired probability is given by $\binom{a_1}{x_1}\binom{a_2}{x_2}\cdot \ldots \cdot \binom{a_k}{x_k} \Big/ \binom{N}{n}$.

DEFINITION 5.9 The random variables $\mathbf{x}_1, \mathbf{x}_2, \ldots,$ and $\mathbf{x}_k$, have a **multivariate hypergeometric distribution**, and they are referred to as multivariate hypergeometric random variables, if and only if their joint probability distribution is given by

$$f(x_1, x_2, \ldots, x_k; n, a_1, a_2, \ldots, a_k) = \frac{\binom{a_1}{x_1}\binom{a_2}{x_2}\cdot \ldots \cdot \binom{a_k}{x_k}}{\binom{N}{n}}$$

for $x_i = 0, 1, \ldots, n$ and $x_i \leq a_i$ for each i, where $\sum_{i=1}^{k} x_i = n$ and $\sum_{i=1}^{k} a_i = N$.

Thus, the joint distribution of the random variables under consideration, namely, the distribution of the numbers of outcomes of the different kinds, is a multivariate hypergeometric distribution with the parameters n, $a_1, a_2, \ldots,$ and a_k.

EXAMPLE 5.10

A panel of prospective jurors includes six married men, three single men, seven married women, and four single women. If the selection is random, what is the probability that the jury will consist of four married men, one single man, five married women, and two single women?

Solution

Substituting $x_1 = 4$, $x_2 = 1$, $x_3 = 5$, $x_4 = 2$, $a_1 = 6$, $a_2 = 3$, $a_3 = 7$, $a_4 = 4$, $N = 20$, and $n = 12$ into the formula of Definition 5.9, we get

$$f(4, 1, 5, 2; 12, 6, 3, 7, 4) = \frac{\binom{6}{4}\binom{3}{1}\binom{7}{5}\binom{4}{2}}{\binom{20}{12}}$$

$$= 0.0450$$

APPLIED EXERCISES

1. The probabilities are 0.40, 0.50, and 0.10 that, in city driving, a certain kind of compact car will average less than 22 miles per gallon, from 22 to 26 miles per gallon, or more than 26 miles per gallon. Find the probability that among ten such cars tested, three will average less than 22 miles per gallon, six will average from 22 to 26 miles per gallon, and one will average more than 26 miles per gallon.
2. Suppose that the probabilities are 0.60, 0.20, 0.10, and 0.10 that a state income tax return will be filled out correctly, that it will contain only errors favoring the taxpayer, that it will contain only errors favoring the state, or that it will contain both kinds of errors. What is the probability that among 12 such income tax returns randomly chosen for audit, five will be filled out correctly, four will contain only errors favoring the taxpayer, two will contain only errors favoring the state, and one will contain both kinds of errors?
3. According to the Mendelian theory of heredity, if plants with round yellow seeds are crossbred with plants with wrinkled green seeds, the probabilities of getting a plant that produces round yellow seeds, wrinkled yellow seeds, round green seeds, or wrinkled green seeds are, respectively, $\frac{9}{16}$, $\frac{3}{16}$, $\frac{3}{16}$, and $\frac{1}{16}$. What is the probability that among nine plants thus obtained there will be four that produce round yellow seeds, two that produce wrinkled yellow

seeds, three that produce round green seeds, and none that produce wrinkled green seeds?

4. If 18 defective glass bricks include 10 that have cracks but no discoloration, five that have discoloration but no cracks, and three that have cracks and discoloration, what is the probability that among six of the bricks (chosen at random for further checks) three will have cracks but no discoloration, one will have discoloration but no cracks, and two will have cracks and discoloration?

5. Among 25 silver dollars struck in 1903 there are 15 from the Philadelphia mint, seven from the New Orleans mint, and three from the San Francisco mint. If five of these silver dollars are picked at random, find the probabilities of getting

(a) four from the Philadelphia mint and one from the New Orleans mint;

(b) three from the Philadelphia mint and one from each of the other two mints.

References

Useful information about various special probability distributions, in outline form, may be found in

HASTINGS, N. A. J., and PEACOCK, J. B., *Statistical Distributions.* London: Butterworth & Co. Ltd., 1975,

and

JOHNSON, N. L., and KOTZ, S., *Discrete Distributions.* Boston: Houghton Mifflin Co., 1969.

Binomial probabilities for $n = 2$ to $n = 49$ may be found in

Tables of the Binomial Probability Distribution, National Bureau of Standards Applied Mathematics Series No. 6, Washington, D.C.: U.S. Government Printing Office, 1950,

and for $n = 50$ to $n = 100$ in

ROMIG, H. G., *50–100 Binomial Tables.* New York: John Wiley & Sons, Inc., 1953.

The most widely used table of Poisson probabilities is

MOLINA, E. C., *Poisson's Exponential Binomial Limit.* New York: D. Van Nostrand Co., Inc., 1943.

special probability densities

6

6.1 INTRODUCTION

In this chapter we shall study some of the probability densities which figure most prominently in statistical theory and in applications. In addition to the ones given in the text, several others are introduced in the exercises following Section 6.4, and three probability densities that are of basic importance in the theory of sampling will be taken up in Chapter 8. As in Chapter 5, we shall derive parameters and moment-generating functions, again leaving as exercises some of the details.

6.2 THE UNIFORM DENSITY

The probabilities of Examples 3.8 and 3.11 are special cases of the **uniform density**, whose graph may be pictured as in Figure 3.7 on page 91.

DEFINITION 6.1 A random variable **x** has a **uniform density**, and it is referred to as a continuous uniform random variable, if and only if its probability density is given by

$$f(x) = \begin{cases} \dfrac{1}{\beta - \alpha} & \text{for } \alpha < x < \beta \\ 0 & \text{elsewhere} \end{cases}$$

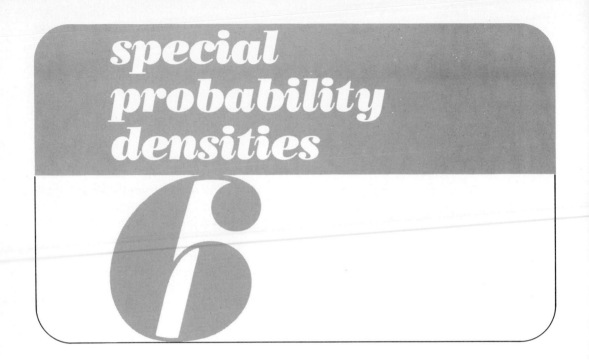

The parameters α and β of this probability density are real constants with $\alpha < \beta$.

Although the uniform density has some direct applications, one of which will be discussed in Section 7.2, its main value is that, due to its simplicity, it lends itself readily to the task of illustrating various aspects of statistical theory. The reader will be asked to verify in Exercise 2 on page 202 that

THEOREM 6.1 The mean and the variance of the uniform density are given by

$$\mu = \frac{\alpha + \beta}{2} \quad \text{and} \quad \sigma^2 = \tfrac{1}{12}(\beta - \alpha)^2$$

6.3 THE GAMMA, EXPONENTIAL, AND CHI-SQUARE DISTRIBUTIONS

Some of the examples and exercises of Chapters 3 and 4 dealt with random variables having densities of the form

$$f(x) = \begin{cases} kx^{\alpha-1}e^{-x/\beta} & \text{for } x > 0 \\ 0 & \text{elsewhere} \end{cases}$$

where $\alpha > 0$, $\beta > 0$, and k must be such that the total area under the curve is equal to 1. To evaluate k, we first make the substitution $y = \frac{x}{\beta}$, which yields

$$\int_0^\infty kx^{\alpha-1}e^{-x/\beta}\,dx = k\beta^\alpha \int_0^\infty y^{\alpha-1}e^{-y}\,dy$$

The integral thus obtained depends on α alone, and it defines the well-known **gamma function**

$$\Gamma(\alpha) = \int_0^\infty y^{\alpha-1}e^{-y}\,dy \qquad \text{for } \alpha > 0$$

which is treated in detail in most advanced calculus texts. Integrating by parts, which will be left to the reader in Exercise 4 on page 202, we find that the gamma function satisfies the recursion formula

$$\Gamma(\alpha) = (\alpha - 1)\cdot\Gamma(\alpha - 1)$$

and, hence, that $\Gamma(\alpha) = (\alpha - 1)!$ when α is a positive integer. Also, an important special value is $\Gamma(\frac{1}{2}) = \sqrt{\pi}$, as the reader will be asked to verify in Exercise 6 on page 202.

Returning now to the problem of evaluating k, we equate the integral we obtained to 1, getting

$$\int_0^\infty kx^{\alpha-1} e^{-x/\beta}\, dx = k\beta^\alpha \Gamma(\alpha) = 1$$

and, hence,

$$k = \frac{1}{\beta^\alpha \Gamma(\alpha)}$$

This leads to the following definition of the **gamma distribution**:

DEFINITION 6.2 A random variable **x** has a **gamma distribution**, and it is referred to as a gamma random variable, if and only if its probability density is given by

$$f(x) = \begin{cases} \dfrac{1}{\beta^\alpha \Gamma(\alpha)} x^{\alpha-1} e^{-x/\beta} & \text{for } x > 0 \\ 0 & \text{elsewhere} \end{cases}$$

where $\alpha > 0$ and $\beta > 0$.

When α is not a positive integer, the value of $\Gamma(\alpha)$ will have to be looked up in a special table. To give the reader some idea about the shape of the graphs of gamma densities, those for several special values of α and β are shown in Figure 6.1.

Special cases of the gamma distribution play important roles in statistics; for instance, for $\alpha = 1$ and $\beta = \theta$ we get

DEFINITION 6.3 A random variable **x** has an **exponential distribution**, and it is referred to as an exponential random variable, if and only if its probability density is given by

$$f(x) = \begin{cases} \dfrac{1}{\theta} e^{-x/\theta} & \text{for } x > 0 \\ 0 & \text{elsewhere} \end{cases}$$

where $\theta > 0$.

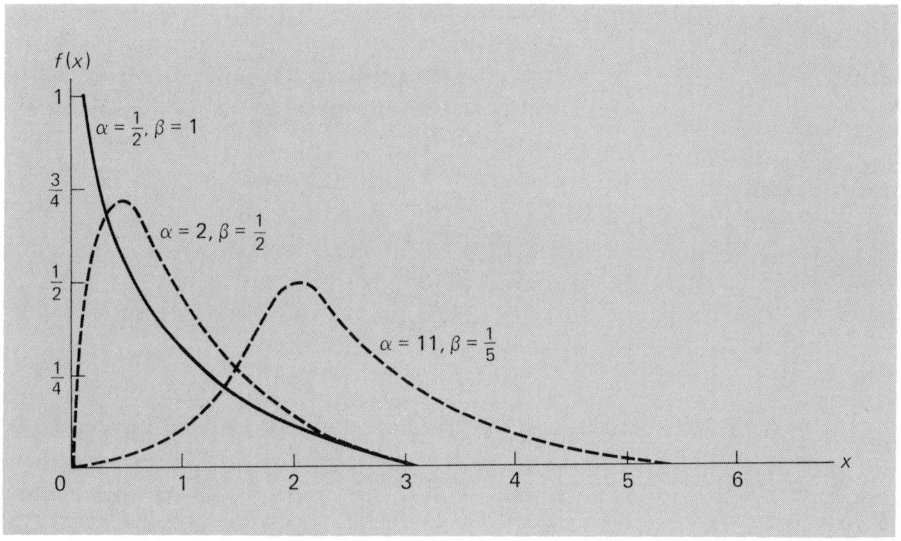

Figure 6.1 Graphs of gamma distributions.

To show how an exponential distribution might arise in practice, let us refer to the situation described in Exercise 14 on page 186, where we were interested in the probability of getting x successes during a time interval of length t when (i) the probability of a success during a very small time interval from t to $t + \Delta t$ is $\alpha \cdot \Delta t$, (ii) the probability of more than one success during such a time interval is negligible, and (iii) the probability of a success during such a time interval does not depend on what happened prior to time t. Now, let us determine the probability density of the random variable **y**, the **waiting time** until the first success. Making use of the result of that exercise and the formula for the Poisson distribution, we find that the probability of zero successes during a time interval of length y is $\frac{(\alpha y)^0 e^{-\alpha y}}{0!} = e^{-\alpha y}$, and, hence, that the probability of getting the first success prior to time y is

$$F(y) = 1 - e^{-\alpha y} \qquad \text{for } y > 0$$

and $F(y) = 0$ elsewhere. Having thus found the distribution function of **y**, differentiation with respect to y yields

$$f(y) = \alpha e^{-\alpha y} \qquad \text{for } y > 0$$

and, of course, $f(y) = 0$ elsewhere. Clearly, this is an exponential density with $\theta = \frac{1}{\alpha}$.

The exponential distribution applies not only to the occurrence of the first success in a **Poisson process**, which is what we call a situation like that described in Exercise 14 on page 186, but by virtue of condition (iii), see Exercise 12 on page 203, it applies also to the waiting times between successes.

EXAMPLE 6.1

If the number of speeders a radar unit spots per hour at a certain location on Route I-10 is a Poisson random variable with $\lambda = 8.4$, what is the probability of a waiting time less than 10 minutes between successive speeders?

Solution

If we treat the waiting time as a random variable having an exponential distribution with $\theta = \frac{1}{8.4}$, integration from 0 to $\frac{1}{6}$ (since the units are hours) yields

$$\int_0^{1/6} 8.4\, e^{-8.4x}\, dx = -e^{-8.4x}\Big|_0^{1/6} = -e^{-1.4} + 1$$

which is approximately 0.75.

Another special case of the gamma distribution arises when $\alpha = \frac{\nu}{2}$ and $\beta = 2$, where ν is the lowercase Greek letter *nu*.

DEFINITION 6.4 A random variable **x** has a **chi-square distribution**, and it is referred to as a chi-square random variable, if and only if its probability density is given by

$$f(x) = \begin{cases} \dfrac{1}{2^{\nu/2}\Gamma(\nu/2)} x^{\frac{\nu-2}{2}} e^{-\frac{x}{2}} & \text{for } x > 0 \\ 0 & \text{elsewhere} \end{cases}$$

The parameter ν is referred to as the **number of degrees of freedom**, or simply the **degrees of freedom**. The chi-square distribution plays a very important role in sampling theory, and it will be discussed in some detail in Chapter 8.

To derive formulas for the mean and the variance of the gamma distribution and, hence, the exponential and chi-square distributions, let us first prove the following theorem:

THEOREM 6.2 The rth moment about the origin of the gamma distribution is given by

$$\mu_r' = \frac{\beta^r \Gamma(\alpha + r)}{\Gamma(\alpha)}$$

Proof. By Definition 4.2,

$$\mu_r' = \int_0^\infty x^r \cdot \frac{1}{\beta^\alpha \Gamma(\alpha)} x^{\alpha-1} e^{-x/\beta}\, dx = \frac{\beta^r}{\Gamma(\alpha)} \cdot \int_0^\infty y^{\alpha+r-1} e^{-y}\, dy$$

where we let $y = \frac{x}{\beta}$. Since the integral on the right is $\Gamma(r + \alpha)$ according to the definition of the gamma function on page 195, this completes the proof.

Using Theorem 6.2, let us now prove the following results:

THEOREM 6.3 The mean and the variance of the gamma distribution are given by

$$\mu = \alpha\beta \quad \text{and} \quad \sigma^2 = \alpha\beta^2$$

Proof. From Theorem 6.2,

$$\mu_1' = \frac{\beta \Gamma(\alpha + 1)}{\Gamma(\alpha)} = \alpha\beta$$

and

$$\mu_2' = \frac{\beta^2 \Gamma(\alpha + 2)}{\Gamma(\alpha)} = \alpha(\alpha + 1)\beta^2,$$

so that $\mu = \alpha\beta$ and $\sigma^2 = \alpha(\alpha + 1)\beta^2 - (\alpha\beta)^2 = \alpha\beta^2$.

COROLLARY 1 The mean and the variance of the exponential distribution are given by

$$\mu = \theta \quad \text{and} \quad \sigma^2 = \theta^2$$

COROLLARY 2 The mean and the variance of the chi-square distribution are given by

$$\mu = \nu \quad \text{and} \quad \sigma^2 = 2\nu$$

To obtain these results from Theorem 6.3, we substitute $\alpha = 1$ and $\beta = \theta$ for the exponential distribution, and $\alpha = \frac{\nu}{2}$ and $\beta = 2$ for the chi-square distribution.

For future reference, let us give here also the moment-generating function of the gamma distribution.

THEOREM 6.4 The moment-generating function of the gamma distribution is given by

$$M_{\mathbf{x}}(t) = (1 - \beta t)^{-\alpha}$$

The reader will be asked to prove this result, and use it to find some of the lower moments of the gamma distribution, in Exercises 9 and 10 on page 203.

6.4 THE BETA DISTRIBUTION

The uniform density $f(x) = 1$ for $0 < x < 1$ and $f(x) = 0$ elsewhere is a special case of the **beta distribution**, which is defined in the following way:

DEFINITION 6.5 A random variable $\mathbf{x}$ has a **beta distribution**, and it is referred to as a beta random variable, if and only if its probability density is given by

$$f(x) = \begin{cases} \dfrac{\Gamma(\alpha + \beta)}{\Gamma(\alpha) \cdot \Gamma(\beta)} x^{\alpha - 1}(1 - x)^{\beta - 1} & \text{for } 0 < x < 1 \\ 0 & \text{elsewhere} \end{cases}$$

where $\alpha > 0$ and $\beta > 0$.

In recent years, the beta distribution has found important applications in **Bayesian inference**, where parameters are looked upon as random variables, and there is a need for a fairly "flexible" probability density for the parameter θ of the binomial distribution, which takes on non-zero values only on the interval from 0 to 1. By "flexible" we mean that the probability density can take on a great variety of different shapes, as the reader will be asked to verify for the beta distribution in Exercise 19 on page 204. This use of the beta distribution will be discussed in Chapter 10.

We shall not prove here that the total area under the curve of the beta distribution is equal to 1, but in the proof of the theorem which follows we shall make use of the fact that

$$\int_0^1 \frac{\Gamma(\alpha + \beta)}{\Gamma(\alpha) \cdot \Gamma(\beta)} x^{\alpha-1}(1 - x)^{\beta-1}\, dx = 1$$

and, hence, that

$$\int_0^1 x^{\alpha-1}(1 - x)^{\beta-1}\, dx = \frac{\Gamma(\alpha) \cdot \Gamma(\beta)}{\Gamma(\alpha + \beta)}$$

This integral defines the **beta function**, whose values are denoted $B(\alpha, \beta)$; in other words, $B(\alpha, \beta) = \frac{\Gamma(\alpha) \cdot \Gamma(\beta)}{\Gamma(\alpha + \beta)}$. Detailed discussion of the beta function may be found in any textbook on advanced calculus.

THEOREM 6.5 The mean and the variance of the beta distribution are given by

$$\mu = \frac{\alpha}{\alpha + \beta} \quad \text{and} \quad \sigma^2 = \frac{\alpha\beta}{(\alpha + \beta)^2(\alpha + \beta + 1)}$$

Proof. By definition,

$$\begin{aligned}
\mu &= \frac{\Gamma(\alpha + \beta)}{\Gamma(\alpha) \cdot \Gamma(\beta)} \cdot \int_0^1 x \cdot x^{\alpha-1}(1 - x)^{\beta-1}\, dx \\
&= \frac{\Gamma(\alpha + \beta)}{\Gamma(\alpha) \cdot \Gamma(\beta)} \cdot \frac{\Gamma(\alpha + 1) \cdot \Gamma(\beta)}{\Gamma(\alpha + \beta + 1)} \\
&= \frac{\alpha}{\alpha + \beta}
\end{aligned}$$

where we recognized the integral as $B(\alpha + 1, \beta)$ and made use of the fact that $\Gamma(\alpha + 1) = \alpha \cdot \Gamma(\alpha)$ and $\Gamma(\alpha + \beta + 1) = (\alpha + \beta) \cdot \Gamma(\alpha + \beta)$.

Similar steps, which will be left to the reader in Exercise 20 on page 204, yield

$$\mu_2' = \frac{(\alpha + 1)\alpha}{(\alpha + \beta + 1)(\alpha + \beta)}$$

and it follows that

$$\sigma^2 = \frac{(\alpha + 1)\alpha}{(\alpha + \beta + 1)(\alpha + \beta)} - \left(\frac{\alpha}{\alpha + \beta}\right)^2$$

$$= \frac{\alpha\beta}{(\alpha + \beta)^2(\alpha + \beta + 1)}$$

THEORETICAL EXERCISES

1. Show that if a random variable has a uniform density with the parameters α and β, the probability that it will take on a value less than $\alpha + p(\beta - \alpha)$ is equal to p.

2. Prove Theorem 6.1.

3. A random variable is said to have a **Cauchy distribution** if its density is given by

$$f(x) = \frac{\frac{\beta}{\pi}}{(x - \alpha)^2 + \beta^2} \qquad \text{for } -\infty < x < \infty$$

Show that for this distribution μ_1' and μ_2' do not exist.

4. Use integration by parts to show that $\Gamma(\alpha) = (\alpha - 1) \cdot \Gamma(\alpha - 1)$ for $\alpha > 1$.

5. Perform a suitable change of variable to show that the integral defining the gamma function can be written as

$$\Gamma(\alpha) = 2^{1-\alpha} \cdot \int_0^\infty z^{2\alpha-1} e^{-\frac{1}{2}z^2}\, dz \qquad \text{for } \alpha > 0$$

6. Using the form of the gamma function of Exercise 5, we can write

$$\Gamma(\tfrac{1}{2}) = \sqrt{2} \int_0^\infty e^{-\frac{1}{2}z^2}\, dz$$

and, hence,

$$[\Gamma(\tfrac{1}{2})]^2 = 2\left\{\int_0^\infty e^{-\frac{1}{2}x^2}\, dx\right\}\left\{\int_0^\infty e^{-\frac{1}{2}y^2}\, dy\right\} = 2\int_0^\infty \int_0^\infty e^{-\frac{1}{2}(x^2+y^2)}\, dx\, dy$$

Change to polar coordinates to evaluate this double integral, and thus show that $\Gamma(\frac{1}{2}) = \sqrt{\pi}$.

7. Find the probabilities that the value of a random variable will exceed 4, if it has a gamma distribution with

(a) $\alpha = 2$ and $\beta = 3$;
(b) $\alpha = 3$ and $\beta = 4$.

8. Show that a gamma distribution with $\alpha > 1$ has a relative maximum at $x = \beta(\alpha - 1)$. What happens when $0 < \alpha < 1$ and when $\alpha = 1$?

9. Prove Theorem 6.4, making the substitution $y = x\left(\frac{1}{\beta} - t\right)$ in the integral defining $M_{\mathbf{x}}(t)$.

10. Expand the moment-generating function of the gamma distribution as a binomial series, and read off the values of μ'_1, μ'_2, and μ'_3.

11. Show that if a random variable has an exponential density with the parameter θ, the probability that it will take on a value less than $-\theta \cdot \ln(1 - p)$ is equal to p for $0 \leqslant p < 1$.

12. If $\mathbf{x}$ has an exponential distribution, show that

$$P(\mathbf{x} \geqslant t + T | \mathbf{x} \geqslant T) = P(\mathbf{x} \geqslant t)$$

This property of an exponential random variable parallels that of a geometric random variable given in Exercise 6 on page 185.

13. A random variable $\mathbf{x}$ has a **Rayleigh distribution** if and only if its probability density is given by

$$f(x) = \begin{cases} 2\alpha x\, e^{-\alpha x^2} & \text{for } x > 0 \\ 0 & \text{elsewhere} \end{cases}$$

where $\alpha > 0$. Show that for this distribution

(a) $\mu = \frac{1}{2}\sqrt{\frac{\pi}{\alpha}}$;

(b) $\sigma^2 = \frac{1}{\alpha}\left(1 - \frac{\pi}{4}\right)$.

14. A random variable $\mathbf{x}$ has a **Pareto distribution** if and only if its probability density is given by

$$f(x) = \begin{cases} \dfrac{\alpha}{x^{\alpha+1}} & \text{for } x > 1 \\ 0 & \text{elsewhere} \end{cases}$$

Show that μ'_r exists only if $r < \alpha$.

15. A random variable $\mathbf{x}$ has a **Weibull distribution** if and only if its probability density is given by

$$f(x) = \begin{cases} kx^{\beta-1} e^{-\alpha x^{\beta}} & \text{for } x > 0 \\ 0 & \text{elsewhere} \end{cases}$$

where $\alpha > 0$ and $\beta > 0$.

(a) Express k in terms of α and β.

(b) Show that $\mu = \alpha^{-1/\beta}\Gamma\left(1 + \frac{1}{\beta}\right)$.

16. If the random variable $\mathbf{t}$ is the time to failure of a commercial product and the values of its probability density and distribution function at time t are, respectively, $f(t)$ and $F(t)$, then its **failure rate** at time t (see also Exercise 7 on page 185) is given by $\frac{f(t)}{1 - F(t)}$. Thus, the failure rate at time t is the probability density of failure at t given that failure does not occur prior to time t.

(a) Show that if $\mathbf{t}$ has an exponential distribution, the failure rate is constant.
(b) Show that if $\mathbf{t}$ has a Weibull distribution (see Exercise 15), the failure rate is given by $\alpha\beta t^{\beta-1}$.

17. Verify that the integral of the beta density, from 0 to 1, equals 1 for

(a) $\alpha = 2$ and $\beta = 4$;
(b) $\alpha = 3$ and $\beta = 3$.

18. Show that if $\alpha > 1$ and $\beta > 1$, the beta density has a relative maximum at

$$x = \frac{\alpha - 1}{\alpha + \beta - 2}.$$

19. Sketch the graphs of the beta densities having

(a) $\alpha = 2$ and $\beta = 2$;
(b) $\alpha = \frac{1}{2}$ and $\beta = 1$;
(c) $\alpha = 2$ and $\beta = \frac{1}{2}$;
(d) $\alpha = 2$ and $\beta = 5$.

[*Hint:* To evaluate $\Gamma(\frac{3}{2})$ and $\Gamma(\frac{5}{2})$ make use of the recursion formula $\Gamma(\alpha) = (\alpha - 1) \cdot \Gamma(\alpha - 1)$ and the result of Exercise 6.]

20. Verify the expression given for μ_2' in the proof of Theorem 6.5.

21. Show that the parameters of the beta distribution can be expressed as follows in terms of the mean and the variance of this distribution:

(a) $\alpha = \mu\left[\frac{\mu(1 - \mu)}{\sigma^2} - 1\right]$;

(b) $\beta = (1 - \mu)\left[\frac{\mu(1 - \mu)}{\sigma^2} - 1\right]$.

22. Karl Pearson, one of the founders of modern statistics, showed that the differential equation

$$\frac{1}{f(x)} \cdot \frac{d[f(x)]}{dx} = \frac{d - x}{a + bx + cx^2}$$

yields (for appropriate values of the constants a, b, c, and d) most of the important distributions of statistics. Verify that the differential equation gives

(a) the gamma distribution when $a = c = 0$, $b > 0$, and $d > -b$;
(b) the exponential distribution when $a = c = d = 0$ and $b > 0$;
(c) the beta distribution when $a = 0$, $b = -c$, $\frac{d-1}{b} < 1$, and $\frac{d}{b} > -1$.

APPLIED EXERCISES

23. A point X is chosen on the line AB, whose midpoint is C and whose length is a. If **x**, the distance from X to A, is a random variable having a uniform density with $\alpha = 0$ and $\beta = a$, what is the probability that AX, BX, and AC will form a triangle?

24. In certain experiments, the error made in determining the density of a substance is a random variable having a uniform density with $\alpha = -0.015$ and $\beta = 0.015$. Find the probabilities that

(a) such an error will be between 0.01 and 0.02;
(b) the size of such an error will exceed 0.005.

25. If a company employs n salespersons, its gross sales in thousands of dollars may be regarded as a random variable having a gamma distribution with $\alpha = 80\sqrt{n}$ and $\beta = 2$. If the sales cost is \$8,000 per salesperson, how many salespersons should the company employ to maximize the expected profit?

26. In a certain city, the daily consumption of electric power, in millions of kilowatt-hours, can be treated as a random variable having a gamma distribution with $\alpha = 3$ and $\beta = 2$. If the power plant of this city has a daily capacity of 12 million kilowatt-hours, what is the probability that this power supply will be inadequate on any given day?

27. The mileage (in thousands of miles) which car owners get with a certain kind of radial tire is a random variable having an exponential distribution with $\theta = 40$. Find the probabilities that one of these tires will last

(a) at least 20,000 miles;
(b) at most 30,000 miles.

28. A certain kind of appliance requires repairs on the average once every two years. Assuming that the times between repairs are exponentially distributed, what is the probability that such an appliance will work at least three years without requiring repairs?

29. If the annual proportion of erroneous income tax returns filed with the I.R.S. can be looked upon as a random variable having a beta distribution with $\alpha = 2$ and $\beta = 9$, what is the probability that in any given year there will be fewer than 10 percent erroneous returns?

30. If the annual proportion of new restaurants that fail in a given city may be looked upon as a random variable having a beta distribution with $\alpha = 1$ and $\beta = 4$, find

(a) the mean of this distribution, namely, the annual proportion of new restaurants that can be expected to fail in the given city;

(b) the probability that at least 25 percent of all new restaurants will fail in the given city in any one year.

31. Suppose that the service life, in hours, of a semiconductor is a random variable having a Weibull distribution (see Exercise 15) with $\alpha = 0.025$ and $\beta = 0.50$.

(a) How long can such a semiconductor be expected to last?

(b) What is the probability that such a semiconductor will still be in operating condition after 4,000 hours?

6.5 THE NORMAL DISTRIBUTION

The **normal distribution**, which we shall study in this section, is in many ways the cornerstone of modern statistical theory. It was investigated first in the eighteenth century when scientists observed an astonishing degree of regularity in errors of measurement. They found that the patterns (distributions) which they observed could be closely approximated by continuous curves which they referred to as "normal curves of errors" and attributed to the laws of chance. The mathematical properties of such normal curves were first studied by Abraham de Moivre (1667–1745), Pierre Laplace (1749–1827), and Karl Gauss (1777–1855).

DEFINITION 6.6 A random variable **x** has a **normal distribution**, and it is referred to as a normal random variable, if and only if its probability density is given by

$$n(x; \mu, \sigma) = \frac{1}{\sigma\sqrt{2\pi}} e^{-\frac{1}{2}\left(\frac{x-\mu}{\sigma}\right)^2} \qquad \text{for } -\infty < x < \infty$$

where $\sigma > 0$.

The graph of a normal distribution, shaped like the cross section of a bell, is shown in Figure 6.2.

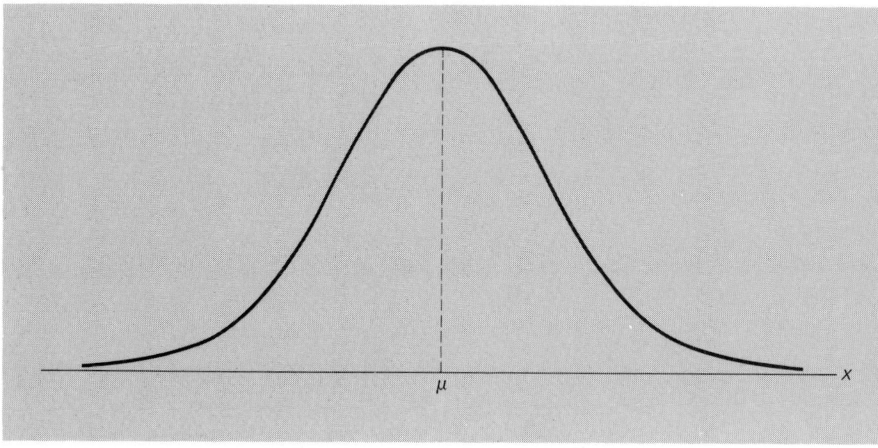

Figure 6.2 Graph of normal distribution.

The notation which we used here is similar to that used in connection with some of the distributions of Chapter 5; it shows explicitly that the two parameters of the normal distribution are μ and σ. It remains to be seen, however, whether the parameter μ is, in fact, $E(\mathbf{x})$ and the parameter σ is, in fact, $\sqrt{\text{var}(\mathbf{x})}$, where $\mathbf{x}$ is a random variable having a normal distribution with these two parameters.

First, though, let us show that the formula of Definition 6.6 can serve as a probability density. Since the values of $n(x; \mu, \sigma)$ are evidently positive so long as $\sigma > 0$, we must show that the total area under the curve is equal to 1. Integrating from $-\infty$ to ∞ and making the substitution $z = \dfrac{x - \mu}{\sigma}$, we get

$$\int_{-\infty}^{\infty} \frac{1}{\sigma\sqrt{2\pi}} e^{-\frac{1}{2}\left(\frac{x-\mu}{\sigma}\right)^2} dx = \frac{1}{\sqrt{2\pi}} \int_{-\infty}^{\infty} e^{-\frac{1}{2}z^2} dz = \frac{2}{\sqrt{2\pi}} \int_{0}^{\infty} e^{-\frac{1}{2}z^2} dz$$

Then, since the integral on the right equals $\dfrac{\Gamma(\frac{1}{2})}{\sqrt{2}} = \dfrac{\sqrt{\pi}}{\sqrt{2}}$ according to Exercise 6 on page 202, it follows that the total area under the curve is equal to $\dfrac{2}{\sqrt{2\pi}} \cdot \dfrac{\sqrt{\pi}}{\sqrt{2}} = 1$.

Next, let us show that

THEOREM 6.6 The moment-generating function of the normal distribution is given by

$$M_{\mathbf{x}}(t) = e^{\mu t + \frac{1}{2}\sigma^2 t^2}$$

Proof. By definition,

$$M_{\mathbf{x}}(t) = \int_{-\infty}^{\infty} e^{xt} \cdot \frac{1}{\sigma\sqrt{2\pi}} e^{-\frac{1}{2}\left(\frac{x-\mu}{\sigma}\right)^2} dx$$

$$= \frac{1}{\sigma\sqrt{2\pi}} \cdot \int_{-\infty}^{\infty} e^{-\frac{1}{2\sigma^2}[-2xt\sigma^2 + (x-\mu)^2]} dx$$

and if we complete the square, that is, use the identity

$$-2xt\sigma^2 + (x - \mu)^2 = [x - (\mu + t\sigma^2)]^2 - 2\mu t\sigma^2 - t^2\sigma^4$$

we get

$$M_{\mathbf{x}}(t) = e^{\mu t + \frac{1}{2}t^2\sigma^2}\left\{\frac{1}{\sigma\sqrt{2\pi}} \cdot \int_{-\infty}^{\infty} e^{-\frac{1}{2}\left[\frac{x-(\mu+t\sigma^2)}{\sigma}\right]^2} dx\right\}$$

Since the quantity inside the braces is the integral, from $-\infty$ to ∞, of a normal density with the parameters $\mu + t\sigma^2$ and σ, and hence equal to 1, it follows that

$$M_{\mathbf{x}}(t) = e^{\mu t + \frac{1}{2}\sigma^2 t^2}$$

Twice differentiating $M_{\mathbf{x}}(t)$ with respect to t, we get

$$M'_{\mathbf{x}}(t) = (\mu + \sigma^2 t) \cdot M_{\mathbf{x}}(t)$$

$$M''_{\mathbf{x}}(t) = [(\mu + \sigma^2 t)^2 + \sigma^2] \cdot M_{\mathbf{x}}(t)$$

so that $M'_{\mathbf{x}}(0) = \mu$ and $M''_{\mathbf{x}}(0) = \mu^2 + \sigma^2$. Thus, $E(\mathbf{x}) = \mu$ and $\text{var}(\mathbf{x}) = (\mu^2 + \sigma^2) - \mu^2 = \sigma^2$, which verifies that the parameters μ and σ are, indeed, the mean and the standard deviation of the normal distribution.

Since the normal distribution plays a basic role in statistics and its density cannot be integrated directly, its areas have been tabulated for the special case where $\mu = 0$ and $\sigma = 1$.

DEFINITION 6.7 The normal distribution with $\mu = 0$ and $\sigma = 1$ is referred to as the **standard normal distribution**.

The entries in Table III, represented by the white area of Figure 6.3, are the values of

$$\int_0^z \frac{1}{\sqrt{2\pi}} e^{-\frac{1}{2}x^2} dx$$

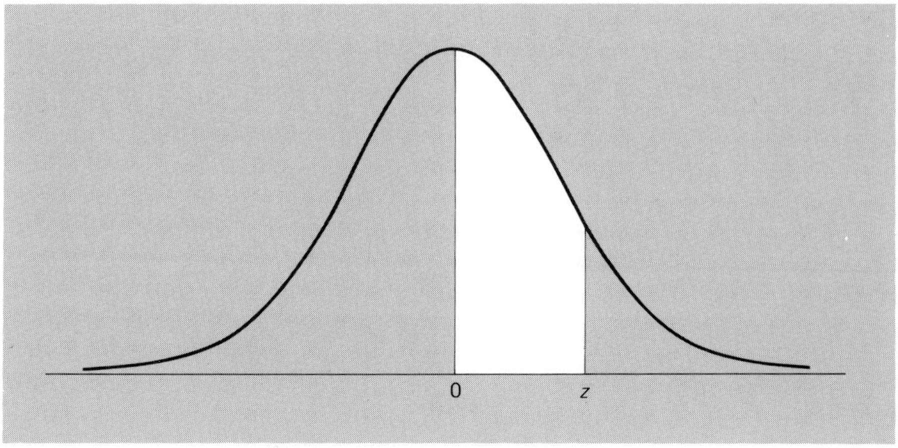

Figure 6.3 Tabulated areas under the standard normal distribution.

namely, the probabilities that a random variable having the standard normal distribution will take on a value on the interval from 0 to z for $z = 0.00, 0.01, 0.02, \ldots, 3.08$, and 3.09, and also $z = 4.0$, $z = 5.0$, and $z = 6.0$. By virtue of the symmetry of the normal distribution about its mean, it is unnecessary to extend Table III to negative values of z.

Occasionally, we are required to find a value of z corresponding to a specified probability that falls between values listed in Table III (see Exercises 17 and 19 on pages 222 and 223). For convenience, we shall always choose the z value corresponding to the tabular value that comes closest to the specified probability. However, if the given probability falls midway between tabular values, we shall choose for z the value falling midway between the corresponding values of z. For instance, to find the value of z which corresponds to a probability of 0.3512, which falls between 0.3508 and 0.3531 in Table III, we choose $z = 1.04$ since 0.3508 is closer to 0.3512. On the other hand, for a probability of 0.2533, which falls midway between 0.2517 and 0.2549, we take $z = 0.685$.

EXAMPLE 6.2

Find the probabilities that a random variable having the standard normal distribution will take on a value (a) less than 1.72, (b) less than -0.88, (c) between 1.30 and 1.75, and (d) between -0.25 and 0.45.

Solution

(a) We look up the entry corresponding to $z = 1.72$ in Table III, add 0.5000 (see Figure 6.4), and get $0.4573 + 0.5000 = 0.9573$; (b) we look up the entry

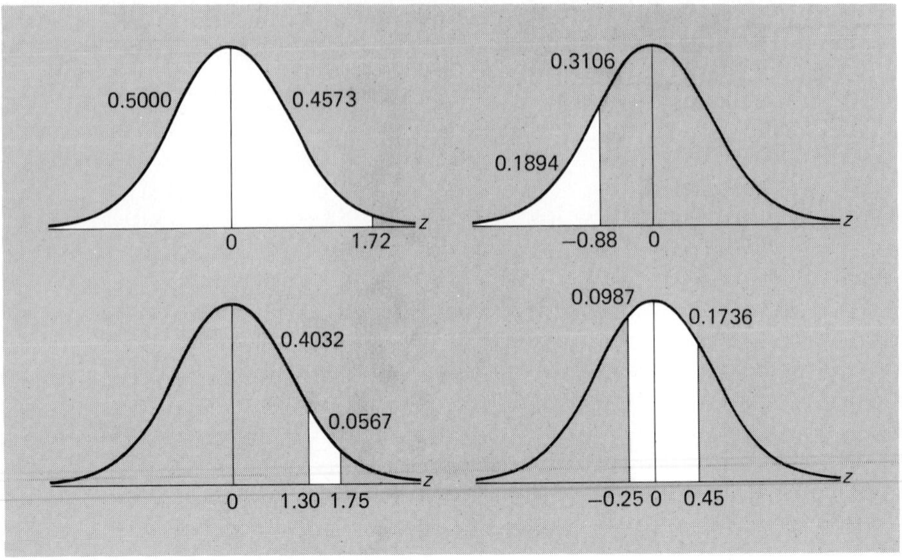

Figure 6.4 Diagrams for Example 6.2.

corresponding to $z = 0.88$ in Table III, subtract it from 0.5000 (see Figure 6.4), and get $0.5000 - 0.3106 = 0.1894$; (c) we look up the entries corresponding to $z = 1.75$ and $z = 1.30$ in Table III, subtract the second from the first (see Figure 6.4), and get $0.4599 - 0.4032 = 0.0567$; (d) we look up the entries corresponding to $z = 0.25$ and $z = 0.45$ in Table III, add them (see Figure 6.4), and get $0.0987 + 0.1736 = 0.2723$.

To determine probabilities relating to random variables having normal distributions other than the standard normal distribution, we make use of the following theorem:

THEOREM 6.7 If $\mathbf{x}$ has a normal distribution with the mean μ and the standard deviation σ, then

$$\mathbf{z} = \frac{\mathbf{x} - \mu}{\sigma}$$

has the standard normal distribution.

Proof. Since the relationship between the values of $\mathbf{x}$ and $\mathbf{z}$ is linear, $\mathbf{z}$ must take on a value between $z_1 = \dfrac{x_1 - \mu}{\sigma}$ and $z_2 = \dfrac{x_2 - \mu}{\sigma}$ when $\mathbf{x}$ takes on a value between x_1 and x_2. Hence, we can write

$$P(x_1 < \mathbf{x} < x_2) = \frac{1}{\sqrt{2\pi}\sigma} \int_{x_1}^{x_2} e^{-\frac{1}{2}\left(\frac{x-\mu}{\sigma}\right)^2} dx$$
$$= \frac{1}{\sqrt{2\pi}} \int_{z_1}^{z_2} e^{-\frac{1}{2}z^2} dz$$
$$= \int_{z_1}^{z_2} n(z; 0, 1)\, dz$$
$$= P(z_1 < \mathbf{z} < z_2)$$

where $\mathbf{z}$ is seen to be a random variable having the standard normal distribution.

Thus, to use Table III in connection with any random variable having a normal distribution, we perform the change of scale $z = \frac{x - \mu}{\sigma}$.

EXAMPLE 6.3

Suppose that the amount of cosmic radiation to which a person is exposed when flying by jet across the United States is a random variable having a normal distribution with a mean of 4.35 mrem and a standard deviation of 0.59 mrem. What is the probability that a person will be exposed to more than 5.20 mrem of cosmic radiation on such a flight?

Solution

We look up the entry corresponding to $z = \frac{5.20 - 4.35}{0.59} = 1.44$ in Table III, subtract from 0.5000 (see Figure 6.5), and get $0.5000 - 0.4251 = 0.0749$.

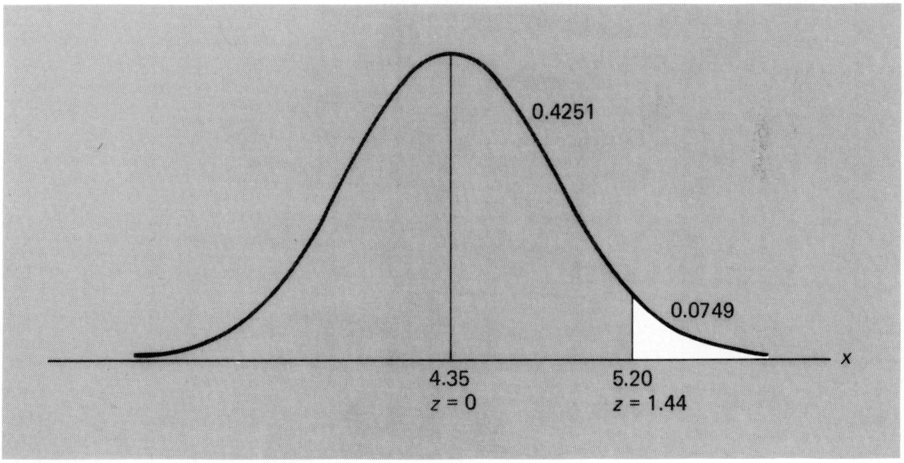

Figure 6.5 Diagram for Example 6.3.

6.6 THE NORMAL APPROXIMATION TO THE BINOMIAL DISTRIBUTION

The normal distribution is sometimes introduced as a continuous distribution which provides a close approximation to the binomial distribution when n, the number of trials, is very large and θ, the probability of a success on an individual trial, is close to $\frac{1}{2}$. Figure 6.6 shows the histograms of binomial distributions with $\theta = \frac{1}{2}$ and $n = 2, 5, 10$, and 25, and it can be seen that with increasing n these distributions approach the symmetrical bell-shaped pattern of the normal distribution.

To provide a theoretical foundation for this argument, let us first prove the following theorem:

THEOREM 6.8 If $\mathbf{x}$ is a random variable having a binomial distribution with the parameters n and θ, then the moment-generating function of

$$\mathbf{z} = \frac{\mathbf{x} - n\theta}{\sqrt{n\theta(1-\theta)}}$$

approaches that of the standard normal distribution when $n \to \infty$.

Proof. Making use of Theorems 4.10 and 5.4, we can write

$$M_{\mathbf{z}}(t) = M_{\frac{\mathbf{x}-\mu}{\sigma}}(t) = e^{-\mu t/\sigma} \cdot [1 + \theta(e^{t/\sigma} - 1)]^n$$

where $\mu = n\theta$ and $\sigma = \sqrt{n\theta(1-\theta)}$. Then, taking logarithms and substituting the Maclaurin's series of $e^{t/\sigma}$, we get

$$\begin{aligned}\ln M_{\frac{\mathbf{x}-\mu}{\sigma}}(t) &= -\frac{\mu t}{\sigma} + n \cdot \ln[1 + \theta(e^{t/\sigma} - 1)] \\ &= -\frac{\mu t}{\sigma} + n \cdot \ln\left[1 + \theta\left\{\frac{t}{\sigma} + \frac{1}{2}\left(\frac{t}{\sigma}\right)^2 + \frac{1}{6}\left(\frac{t}{\sigma}\right)^3 + \cdots\right\}\right]\end{aligned}$$

and, using the infinite series $\ln(1+x) = x - \frac{1}{2}x^2 + \frac{1}{3}x^3 - \cdots$, which converges for $|x| < 1$, to expand this logarithm, it follows that

$$\begin{aligned}\ln M_{\frac{\mathbf{x}-\mu}{\sigma}}(t) = -\frac{\mu t}{\sigma} &+ n\theta\left[\frac{t}{\sigma} + \frac{1}{2}\left(\frac{t}{\sigma}\right)^2 + \frac{1}{6}\left(\frac{t}{\sigma}\right)^3 + \cdots\right] \\ &- \frac{n\theta^2}{2}\left[\frac{t}{\sigma} + \frac{1}{2}\left(\frac{t}{\sigma}\right)^2 + \frac{1}{6}\left(\frac{t}{\sigma}\right)^3 + \cdots\right]^2 \\ &+ \frac{n\theta^3}{3}\left[\frac{t}{\sigma} + \frac{1}{2}\left(\frac{t}{\sigma}\right)^2 + \frac{1}{6}\left(\frac{t}{\sigma}\right)^3 + \cdots\right]^3 - \cdots\end{aligned}$$

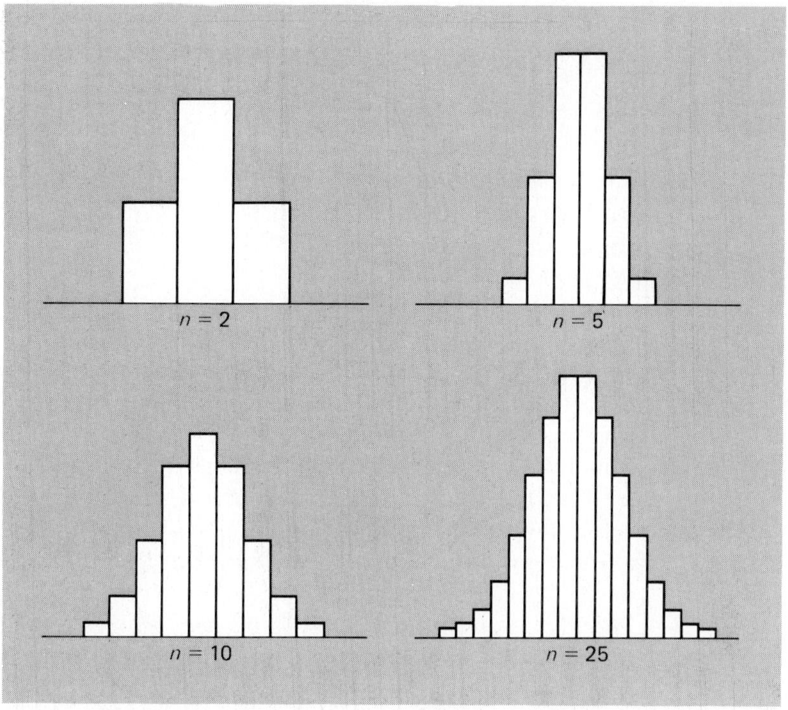

Figure 6.6 Binomial distributions with $\theta = \frac{1}{2}$.

Collecting powers of t, we obtain

$$\ln M_{\frac{\mathbf{x}-\mu}{\sigma}}(t) = \left(-\frac{\mu}{\sigma} + \frac{n\theta}{\sigma}\right)t + \left(\frac{n\theta}{2\sigma^2} - \frac{n\theta^2}{2\sigma^2}\right)t^2 + \left(\frac{n\theta}{6\sigma^3} - \frac{n\theta^2}{2\sigma^3} + \frac{n\theta^3}{3\sigma^3}\right)t^3 + \cdots$$

$$= \frac{1}{\sigma^2}\left(\frac{n\theta - n\theta^2}{2}\right)t^2 + \frac{n}{\sigma^3}\left(\frac{\theta - 3\theta^2 + 2\theta^3}{6}\right)t^3 + \cdots$$

since $\mu = n\theta$. Then, substituting $\sigma = \sqrt{n\theta(1-\theta)}$, we find that

$$\ln M_{\frac{\mathbf{x}-\mu}{\sigma}}(t) = \frac{1}{2}t^2 + \frac{n}{\sigma^3}\left(\frac{\theta - 3\theta^2 + 2\theta^3}{6}\right)t^3 + \cdots$$

where for $r > 2$ the coefficient of t^r is a constant times $\frac{n}{\sigma^r}$, which approaches 0 when $n \to \infty$. It follows that

$$\lim_{n \to \infty} \ln M_{\frac{x-\mu}{\sigma}}(t) = \tfrac{1}{2}t^2$$

and since the limit of a logarithm equals the logarithm of the limit (provided the two limits exist), we conclude that

$$\lim_{n \to \infty} M_{\frac{x-\mu}{\sigma}}(t) = e^{\frac{1}{2}t^2}$$

which is the moment-generating function of Theorem 6.6 with $\mu = 0$ and $\sigma = 1$.

This completes the proof of Theorem 6.8, but have we shown that when $n \to \infty$ the distribution of $\mathbf{z}$, the **standardized** binomial random variable, approaches the standard normal distribution? Not quite. To this end, we must refer to two theorems which we shall state here without proof:

1. *There is a one-to-one correspondence between moment-generating functions and probability distributions (densities) when the former exist.*
2. *If the moment-generating function of one random variable approaches that of another random variable, then the distribution (density) of the first random variable approaches that of the second random variable under the same limiting conditions.*

Strictly speaking, our results apply when $n \to \infty$, but the normal distribution is often used to approximate binomial probabilities even when n is fairly small. A good rule of thumb is to use this approximation only when $n\theta$ and $n(1 - \theta)$ are both greater than 5.

EXAMPLE 6.4

Find the probability of getting 6 heads and 10 tails in 16 tosses of a balanced coin, and also use the normal distribution to approximate this probability.

Solution

Substituting $x = 6$, $n = 16$, and $\theta = \frac{1}{2}$ into the formula for the binomial distribution, we get

$$b\left(6; 16, \frac{1}{2}\right) = \binom{16}{6}\left(\frac{1}{2}\right)^6\left(1 - \frac{1}{2}\right)^{16-6} = \frac{8{,}008}{65{,}536}$$

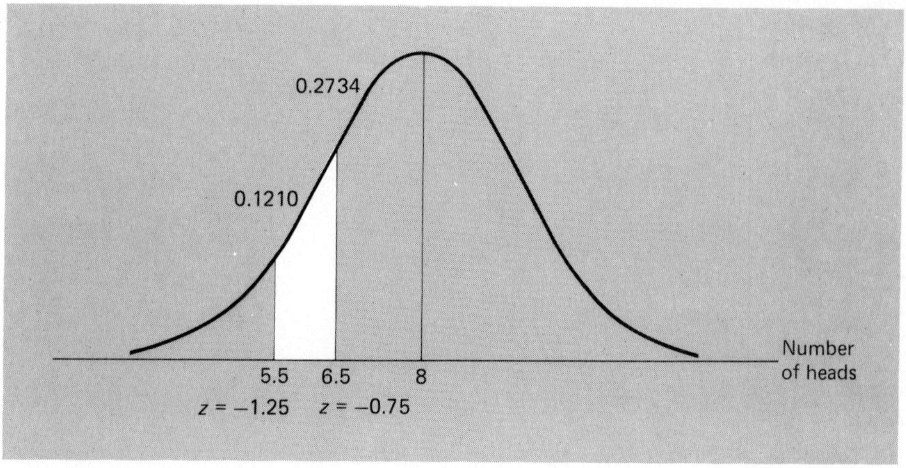

Figure 6.7 Diagram for Example 6.4.

or 0.1222 rounded to four decimals. To find the normal approximation to this probability, we must use the **continuity correction** according to which each non-negative integer k is represented by the interval from $k - \frac{1}{2}$ to $k + \frac{1}{2}$, and, hence, we must determine the area under the curve between 5.5 and 6.5 (see Figure 6.7). Since $\mu = 16 \cdot \frac{1}{2} = 8$ and $\sigma = \sqrt{16 \cdot \frac{1}{2} \cdot \frac{1}{2}} = 2$, we must determine the area under the curve between

$$z = \frac{5.5 - 8}{2} = -1.25 \quad \text{and} \quad z = \frac{6.5 - 8}{2} = -0.75$$

Since the entries in Table III corresponding to $z = 1.25$ and $z = 0.75$ are 0.3944 and 0.2734, we find that the normal approximation to the probability of "6 heads and 10 tails in 16 tosses of a balanced coin" is $0.3944 - 0.2734 = 0.1210$. This is very close, indeed, to the correct value of this probability rounded to four decimals.

EXAMPLE 6.5

Use the normal approximation to the binomial distribution to find the probability that at least 70 of 100 mosquitos will be killed by a new insect spray, when the probability is 0.75 that any one of them will be killed by the spray.

Solution

Using the same continuity correction as in Example 6.4, we must find the area under the curve to the right of 69.5 (see Figure 6.8), and since $\mu = 100(0.75) =$

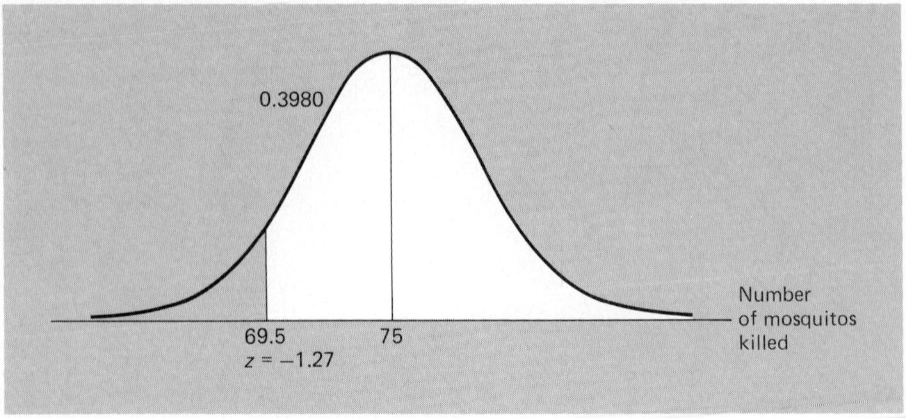

Figure 6.8 Diagram for Example 6.5.

75 and $\sigma = \sqrt{100(0.75)(0.25)} = 4.33$, we must find the area under the curve to the right of

$$z = \frac{69.5 - 75}{4.33} = -1.27$$

Since the entry in Table III corresponding to $z = 1.27$ is 0.3980, the answer is $0.3980 + 0.5000 = 0.8980$. Comparison with the correct value of this probability (looked up in the tables by Romig listed on page 193) shows that the error of the approximation is only 0.0018.

6.7 THE BIVARIATE NORMAL DISTRIBUTION

Among multivariate densities, of special importance is the **multivariate normal distribution**, which is a generalization of the normal distribution in one variable. As it is best (indeed, virtually necessary) to present this distribution in matrix notation, we shall give here only the **bivariate** case; discussions of the general case are referred to on page 224.

DEFINITION 6.8 A pair of random variables **x** and **y** have a **bivariate normal distribution**, and they are referred to as jointly normally distributed random variables, if and only if their joint probability density is given by

$$f(x, y) = \frac{e^{-\frac{1}{2(1-\rho^2)}\left[\left(\frac{x-\mu_1}{\sigma_1}\right)^2 - 2\rho\left(\frac{x-\mu_1}{\sigma_1}\right)\left(\frac{y-\mu_2}{\sigma_2}\right) + \left(\frac{y-\mu_2}{\sigma_2}\right)^2\right]}}{2\pi\sigma_1\sigma_2\sqrt{1-\rho^2}}$$

for $-\infty < x < \infty$ and $-\infty < y < \infty$, where $\sigma_1 > 0$, $\sigma_2 > 0$, and $-1 < \rho < 1$.

To study this distribution, let us first show that the parameters μ_1, μ_2, σ_1, and σ_2, are, respectively, the means and the standard deviations of the random variables **x** and **y**. Integrating on y, from $-\infty$ to ∞, to obtain the marginal density of **x**, we can write

$$g(x) = \frac{e^{-\frac{1}{2(1-\rho^2)}\left(\frac{x-\mu_1}{\sigma_1}\right)^2}}{2\pi\sigma_1\sigma_2\sqrt{1-\rho^2}} \int_{-\infty}^{\infty} e^{-\frac{1}{2(1-\rho^2)}\left[\left(\frac{y-\mu_2}{\sigma_2}\right)^2 - 2\rho\left(\frac{x-\mu_1}{\sigma_1}\right)\left(\frac{y-\mu_2}{\sigma_2}\right)\right]} dy$$

Temporarily making the substitution $u = \frac{x - \mu_1}{\sigma_1}$ to simplify the notation and changing the variable of integration by letting $v = \frac{y - \mu_2}{\sigma_2}$, we obtain

$$g(x) = \frac{e^{-\frac{1}{2(1-\rho^2)}u^2}}{2\pi\sigma_1\sqrt{1-\rho^2}} \int_{-\infty}^{\infty} e^{-\frac{1}{2(1-\rho^2)}(v^2 - 2\rho uv)} dv$$

and, after completing the square by letting $v^2 - 2\rho uv = (v - \rho u)^2 - \rho^2 u^2$ and collecting terms, this becomes

$$g(x) = \frac{e^{-\frac{1}{2}u^2}}{\sigma_1\sqrt{2\pi}} \left\{ \frac{1}{\sqrt{2\pi}\sqrt{1-\rho^2}} \int_{-\infty}^{\infty} e^{-\frac{1}{2}\left(\frac{v-\rho u}{\sqrt{1-\rho^2}}\right)^2} dv \right\}$$

Finally, identifying the quantity in parentheses as the integral of a normal density from $-\infty$ to ∞ and, hence, equalling 1, we get

$$g(x) = \frac{e^{-\frac{1}{2}u^2}}{\sigma_1\sqrt{2\pi}} = \frac{1}{\sigma_1\sqrt{2\pi}} e^{-\frac{1}{2}\left(\frac{x-\mu_1}{\sigma_1}\right)^2}$$

for $-\infty < x < \infty$. It follows by inspection that the marginal density of **x** is a normal distribution with the mean μ_1 and the standard deviation σ_1, and, by symmetry, that the marginal density of **y** is a normal distribution with the mean μ_2 and the standard deviation σ_2.

So far as the parameter ρ is concerned, where ρ is the lowercase Greek letter *rho*, it is called the **correlation coefficient**, and the necessary integration will show that $\text{cov}(\mathbf{x}, \mathbf{y}) = \rho\sigma_1\sigma_2$. Thus, the parameter ρ measures how the random variables **x** and **y** vary together, and its significance will be discussed further in Chapter 14.

When we deal with a pair of random variables having a bivariate normal distribution, their conditional densities are also of importance; so, let us prove the following theorem:

THEOREM 6.9 If **x** and **y** have a bivariate normal distribution, the conditional density of **y** given $\mathbf{x} = x$ is a normal distribution with the mean

$$\mu_{\mathbf{y}|x} = \mu_2 + \rho\frac{\sigma_2}{\sigma_1}(x - \mu_1)$$

and the variance

$$\sigma^2_{\mathbf{y}|x} = \sigma_2^2(1 - \rho^2)$$

and the conditional density of **x** given $\mathbf{y} = y$ is a normal distribution with the mean

$$\mu_{\mathbf{x}|y} = \mu_1 + \rho\frac{\sigma_1}{\sigma_2}(y - \mu_2)$$

and the variance

$$\sigma^2_{\mathbf{x}|y} = \sigma_1^2(1 - \rho^2)$$

Proof. Writing $w(y|x) = \dfrac{f(x, y)}{g(x)}$ in accordance with Definition 3.12, and letting $u = \dfrac{x - \mu_1}{\sigma_1}$ and $v = \dfrac{y - \mu_2}{\sigma_2}$ to simplify the notation, we get

$$w(y|x) = \frac{\dfrac{1}{2\pi\sigma_1\sigma_2\sqrt{1 - \rho^2}} e^{-\frac{1}{2(1-\rho^2)}[u^2 - 2\rho uv + v^2]}}{\dfrac{1}{\sqrt{2\pi}\sigma_1} e^{-\frac{1}{2}u^2}}$$

$$= \frac{1}{\sqrt{2\pi}\sigma_2\sqrt{1 - \rho^2}} e^{-\frac{1}{2(1-\rho^2)}[v^2 - 2\rho uv + \rho^2 u^2]}$$

$$= \frac{1}{\sqrt{2\pi}\sigma_2\sqrt{1 - \rho^2}} e^{-\frac{1}{2}\left[\frac{v - \rho u}{\sqrt{1 - \rho^2}}\right]^2}$$

Then, expressing this result in terms of the original variables, we obtain

$$w(y|x) = \frac{1}{\sigma_2\sqrt{2\pi}\sqrt{1-\rho^2}} e^{-\frac{1}{2}\left[\frac{y-\left\{\mu_2+\rho\frac{\sigma_2}{\sigma_1}(x-\mu_1)\right\}}{\sigma_2\sqrt{1-\rho^2}}\right]^2}$$

for $-\infty < y < \infty$, and it can be seen by inspection that this is a normal density with the mean $\mu_{\mathbf{y}|x} = \mu_2 + \rho\frac{\sigma_2}{\sigma_1}(x - \mu_1)$ and the variance $\sigma_{\mathbf{y}|x}^2 = \sigma_2^2(1 - \rho^2)$. The corresponding results for the conditional density of **x** given $\mathbf{y} = y$ follow by symmetry.

The bivariate normal distribution has many important properties, some statistical and some purely mathematical. Among the former, there is the following property, which the reader will be asked to prove in Exercise 9 on page 221:

THEOREM 6.10 If two random variables have a bivariate normal distribution, they are independent if and only if $\rho = 0$.

In connection with this, if $\rho = 0$ the random variables are said to be **uncorrelated**.

Also, we have shown that for two random variables having a bivariate normal distribution the two marginal densities are normal, but the converse is not necessarily true. In other words, the marginal distributions may both be normal without the joint distribution being a bivariate normal distribution. For instance, if the bivariate density of **x** and **y** is given by

$$f^*(x, y) = \begin{cases} 2f(x, y) & \text{inside squares 2 and 4 of Figure 6.9} \\ 0 & \text{inside squares 1 and 3 of Figure 6.9} \\ f(x, y) & \text{elsewhere} \end{cases}$$

where $f(x, y)$ is the value of the bivariate normal density with $\mu_1 = 0$, $\mu_2 = 0$, and $\rho = 0$ at (x, y), it is easy to see that the marginal densities of **x** and **y** are normal even though their joint density is not a bivariate normal distribution.

Many interesting properties of the bivariate normal density are obtained by studying the **bivariate normal surface**, pictured in Figure 6.10, whose equation is $z = f(x, y)$, where $f(x, y)$ is the value of the bivariate normal density at (x, y). As the reader will be asked to verify in the exercises that follow, the bivariate normal surface has a maximum at (μ_1, μ_2), any plane parallel to the z-axis intersects the surface in a curve having the shape of a normal distribution, and any plane parallel to the xy-plane which intersects the surface intersects it in an ellipse called a **contour of constant probability density**. When $\rho = 0$ and $\sigma_1 = \sigma_2$, the contours

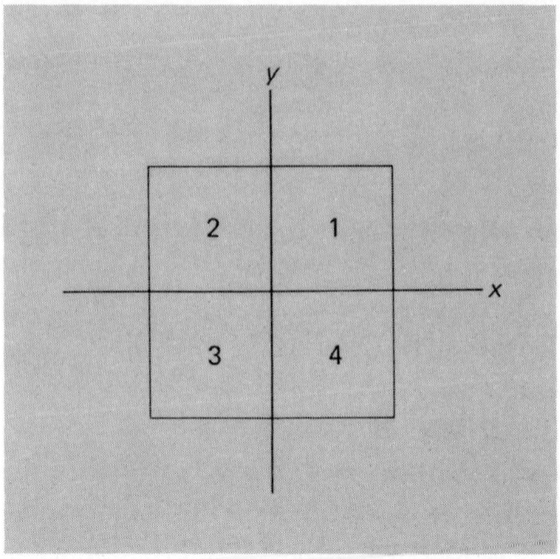

Figure 6.9 Sample space for the bivariate density given by $f^*(x, y)$.

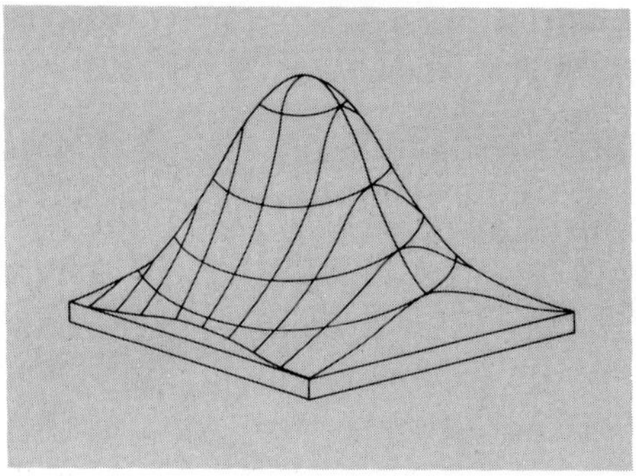

Figure 6.10 Bivariate normal surface.

of constant probability density are circles, and it is customary to refer to the corresponding joint density as a **circular normal distribution**.

THEORETICAL EXERCISES

1. Show that the normal distribution has a relative maximum at $x = \mu$ and inflection points at $x = \mu - \sigma$ and $x = \mu + \sigma$.
2. Show that the differential equation of Exercise 22 on page 205 yields a normal distribution when $b = c = 0$ and $a > 0$.
3. Twice more differentiating the moment-generating function of the normal distribution (see page 208), verify that $\mu_3 = 0$ and $\mu_4 = 3\sigma^4$.

4. Use the moment-generating function of the normal distribution given in Theorem 6.6 to show that for the normal distribution $\alpha_3 = 0$ and $\alpha_4 = 3$, where α_3 and α_4 are as defined in Exercises 8 and 9 on page 148.

5. Show that for the normal distribution

$$M_{\mathbf{x}-\mu}(t) = e^{\frac{1}{2}t^2\sigma^2}$$

6. If we let $K_{\mathbf{x}}(t) = \ln M_{\mathbf{x}-\mu}(t)$, the coefficient of $\frac{t^r}{r!}$ in the Maclaurin's series of $K_{\mathbf{x}}(t)$ is called the **rth cumulant** and it is denoted by κ_r. Equating coefficients of like powers, show that

(a) $\kappa_2 = \mu_2$; **(b)** $\kappa_3 = \mu_3$;
(c) $\kappa_4 = \mu_4 - 3\mu_2^2$; **(d)** $\kappa_5 = \mu_5 - 10\mu_3\mu_2$.

Also show that for the normal distribution $\kappa_2 = \sigma^2$ and all other cumulants are zero.

7. Show that when $\lambda \to \infty$, where λ is the parameter of the Poisson distribution, then the moment-generating function of

$$\mathbf{z} = \frac{\mathbf{x} - \lambda}{\sqrt{\lambda}}$$

namely, that of a standardized Poisson random variable, approaches the moment-generating function of the standard normal distribution.

8. Show that when $\alpha \to \infty$ and β remains constant, the moment-generating function of a standardized gamma random variable approaches the moment-generating function of the standard normal distribution.

9. Prove Theorem 6.10.

10. Show that any plane perpendicular to the xy-plane intersects the bivariate normal surface in a curve having the shape of a normal distribution.

11. If the exponent of e of a bivariate normal density is

$$\frac{-1}{102}[(x + 2)^2 - 2.8(x + 2)(y - 1) + 4(y - 1)^2]$$

find

(a) μ_1, μ_2, σ_1, σ_2, and ρ;
(b) $\mu_{\mathbf{y}|x}$ and $\sigma^2_{\mathbf{y}|x}$.

12. If the exponent of e of a bivariate normal density is

$$\frac{-1}{54}(x^2 + 4y^2 + 2xy + 2x + 8y + 4)$$

find σ_1, σ_2, and ρ, given that $\mu_1 = 0$ and $\mu_2 = -1$.

13. If **x** and **y** have a bivariate normal distribution, $\mathbf{u} = \mathbf{x} + \mathbf{y}$ and $\mathbf{v} = \mathbf{x} - \mathbf{y}$, find an expression for the correlation coefficient of **u** and **v**.

14. If **x** and **y** have a bivariate normal distribution, it can be shown that the joint moment-generating function (see Exercise 6 on page 161) of these random variables is given by

$$\begin{aligned} M_{\mathbf{x},\mathbf{y}}(t_1, t_2) &= E(e^{t_1\mathbf{x}+t_2\mathbf{y}}) \\ &= e^{t_1\mu_1+t_2\mu_2+\frac{1}{2}(\sigma_1^2t_1^2+2\rho\sigma_1\sigma_2t_1t_2+\sigma_2^2t_2^2)} \end{aligned}$$

Verify that

(a) the first partial derivative of this function with respect to t_1 at $t_1 = 0$ and $t_2 = 0$ is μ_1;

(b) the second partial derivative with respect to t_1 at $t_1 = 0$ and $t_2 = 0$ is $\sigma_1^2 + \mu_1^2$;

(c) the second partial derivative with respect to t_1 and t_2 at $t_1 = 0$ and $t_2 = 0$ is $\rho\sigma_1\sigma_2 + \mu_1\mu_2$.

APPLIED EXERCISES

15. If **z** is a random variable having the standard normal distribution, find the probabilities that this random variable will take on a value

(a) greater than 1.14; **(b)** less than −0.36;
(c) between −0.46 and −0.09; **(d)** between −0.58 and 1.12.

16. If **x** is a random variable having a normal distribution, what are the probabilities of getting a value

(a) within one standard deviation of the mean;
(b) within two standard deviations of the mean;
(c) within three standard deviations of the mean;
(d) within four standard deviations of the mean?

17. If z_α is defined by

$$\int_{z_\alpha}^{\infty} n(z; 0, 1)\, dz = \alpha$$

find its values for

(a) $\alpha = 0.05$; **(b)** $\alpha = 0.025$;
(c) $\alpha = 0.01$; **(d)** $\alpha = 0.005$.

18. Suppose that during periods of transcendental meditation the reduction of a person's oxygen consumption is a random variable having a normal distribution with $\mu = 37.6$ cc per minute and $\sigma = 4.6$ cc per minute. Find the probabilities that during a period of transcendental meditation a person's oxygen consumption will be reduced by

(a) at least 44.5 cc per minute;
(b) at most 35.0 cc per minute;
(c) anywhere from 30.0 to 40.0 cc per minute.

19. Suppose that the actual amount of instant coffee which a filling machine puts into "6-ounce" jars is a random variable having a normal distribution with $\sigma = 0.05$ ounce. If only 3 percent of the jars are to contain less than 6 ounces of coffee, what must be the mean fill of these jars?

20. Use the normal approximation to find the probability of getting 7 heads and 7 tails in 14 flips of a balanced coin, and compare the result with the exact value (rounded to four decimals) given in Table I.

21. If 23 percent of all patients with high blood pressure have bad side effects from a certain kind of medicine, use the normal approximation to find the probability that among 120 patients with high blood pressure treated with this medicine more than 32 will have bad side effects.

22. To illustrate the law of large numbers (see also Exercise 20 on page 175), use the normal approximation to the binomial distribution to find the probabilities that the proportion of heads will be anywhere from 0.49 to 0.51 when a balanced coin is flipped

(a) 100 times; **(b)** 1,000 times;
(c) 10,000 times.

23. The center of a target is taken as the origin of a rectangular system of coordinates, with reference to which the point of impact of a missile has the coordinates **x** and **y**. If **x** and **y** have a bivariate normal density with $\mu_1 = 0$, $\mu_2 = 0$, $\sigma_1 = 120$ feet, $\sigma_2 = 120$ feet, and $\rho = 0$, find the probabilities that the point of impact will be

(a) inside a square with sides of 180 feet, whose center is at the origin and whose sides are parallel to the coordinate axes;
(b) inside a circle with a radius of 75 feet with its center at the origin.

24. If **x** and **y** have the circular normal distribution with $\mu_1 = \mu_2 = 0$ and $\sigma_1 = \sigma_2 = 12$, find

(a) the probability of getting a point (x, y) inside the circle $x^2 + y^2 = 36$;
(b) the value of c for which the probability of getting a point (x, y) inside the circle $x^2 + y^2 = c^2$ is 0.80.

25. Suppose that **x** and **y**, the height and weight of certain animals, have a bivariate normal distribution with $\mu_1 = 18$ inches, $\mu_2 = 15$ pounds, $\sigma_1 = 3$ inches, $\sigma_2 = 2$ pounds, and $\rho = 0.75$. Find

(a) the expected weight of one of these animals that is 17 inches tall;
(b) the expected height of one of these animals that weighs 20 pounds.

References

Useful information about various special probability densities, in outline form, may be found in

HASTINGS, N. A. J., and PEACOCK, J. B., *Statistical Distributions.* London: Butterworth & Co. Ltd., 1975,

and

JOHNSON, N. L., and KOTZ, S., *Continuous Univariate Distributions*, Vols. 1 and 2. New York: John Wiley & Sons, Inc., 1970.

A direct proof that the standardized binomial distribution approaches the standard normal distribution when $n \to \infty$ is given in

KEEPING, E. S., *Introduction to Statistical Inference.* Princeton, N.J.: D. Van Nostrand Co., Inc., 1962.

A detailed treatment of the mathematical and statistical properties of the bivariate normal surface may be found in

YULE, G. U., and KENDALL, M. G., *An Introduction to the Theory of Statistics*, 14th ed. New York: Hafner Publishing Co., Inc., 1950.

The multivariate normal distribution is treated in matrix notation in

BICKEL, P. J., and DOKSUM, K. A., *Mathematical Statistics: Basic Ideas and Selected Topics.* San Francisco: Holden-Day, Inc., 1977,

LINDGREN, B. W., *Statistical Theory*, 2nd ed. New York: Macmillan Publishing Co., Inc., 1968.

functions of random variables

7

7.1 INTRODUCTION

In this chapter we shall concern ourselves with the problem of finding the probability distributions or densities of functions of one or more random variables. That is, given a set of random variables $\mathbf{x}_1, \mathbf{x}_2, \ldots, \mathbf{x}_n$, and their joint distribution or density, we shall be interested in finding the probability distribution or density of some random variable $\mathbf{y} = u(\mathbf{x}_1, \mathbf{x}_2, \ldots, \mathbf{x}_n)$. This means that the values of the random variable $\mathbf{y}$ are related to those of the $\mathbf{x}$'s by means of the equation $y = u(x_1, x_2, \ldots, x_n)$.

Several methods are available for solving this kind of problem. The ones we shall discuss in the next three sections are called the **distribution function technique**, the **transformation of variables technique**, and the **moment-generating function technique**. Although all three methods can be used in some situations, in most problems one technique will be preferable (easier to use than the others). This is true, for example, in some instances where the function in question is linear in the random variables $\mathbf{x}_1, \mathbf{x}_2, \ldots, \mathbf{x}_n$, and the moment-generating function technique yields the simplest derivations.

The various techniques we shall discuss in this chapter will be used again in Chapter 8 to derive several distributions that are of fundamental importance in statistical inference.

7.2 DISTRIBUTION FUNCTION TECHNIQUE

A straightforward method of obtaining the probability density of a function of continuous random variables consists of first finding its distribution function, and then its density by differentiation. Thus, if $\mathbf{x}_1, \mathbf{x}_2, \ldots, \mathbf{x}_n$ are continuous random variables with a given joint probability density, the probability density of $\mathbf{y} = u(\mathbf{x}_1, \mathbf{x}_2, \ldots, \mathbf{x}_n)$ is obtained by first determining an expression for the probabilities

$$F(y) = P(\mathbf{y} \leqslant y) = P[u(\mathbf{x}_1, \mathbf{x}_2, \ldots, \mathbf{x}_n) \leqslant y]$$

and then differentiating to get

$$f(y) = \frac{dF(y)}{dy}$$

according to Theorem 3.6.

EXAMPLE 7.1

If the probability density of $\mathbf{x}$ is given by

$$f(x) = \begin{cases} 6x(1-x) & \text{for } 0 < x < 1 \\ 0 & \text{elsewhere} \end{cases}$$

find the probability density of $\mathbf{y} = \mathbf{x}^3$.

Solution

Letting $G(y)$ denote the value of the distribution function of $\mathbf{y}$ at y, we can write

$$\begin{aligned} G(y) &= P(\mathbf{y} \leqslant y) \\ &= P(\mathbf{x}^3 \leqslant y) \\ &= P(\mathbf{x} \leqslant y^{1/3}) \\ &= \int_0^{y^{1/3}} 6x(1-x)\,dx \\ &= 3y^{2/3} - 2y \end{aligned}$$

and, hence,

$$g(y) = 2(y^{-1/3} - 1)$$

for $0 < y < 1$; elsewhere, $g(y) = 0$. In Exercise 5 on page 244 the reader will be asked to verify this result by a different technique.

EXAMPLE 7.2

If $\mathbf{y} = |\mathbf{x}|$, show that

$$g(y) = \begin{cases} f(y) + f(-y) & \text{for } y > 0 \\ 0 & \text{elsewhere} \end{cases}$$

where $f(x)$ is the value of the probability density of $\mathbf{x}$ at x and $g(y)$ is the value of the probability density of $\mathbf{y}$ at y. Also, use this result to find the probability density of $\mathbf{y} = |\mathbf{x}|$, where $\mathbf{x}$ has the standard normal distribution.

Solution

For $y > 0$ we have

$$\begin{aligned} G(y) &= P(\mathbf{y} \leqslant y) \\ &= P(|\mathbf{x}| \leqslant y) \\ &= P(-y \leqslant \mathbf{x} \leqslant y) \\ &= F(y) - F(-y) \end{aligned}$$

and, upon differentiation,

$$g(y) = f(y) + f(-y)$$

Since $|x|$ cannot be negative, $g(y) = 0$ for $y < 0$; the value of $g(0)$ is arbitrarily set equal to 0.

If $\mathbf{x}$ has the standard normal distribution and $\mathbf{y} = |\mathbf{x}|$, it follows that

$$\begin{aligned} g(y) &= n(y; 0, 1) + n(-y; 0, 1) \\ &= 2n(y; 0, 1) \end{aligned}$$

for $y > 0$; elsewhere, $g(y) = 0$. An important application of this result is given in Example 7.9.

EXAMPLE 7.3

If the joint density of $\mathbf{x}_1$ and $\mathbf{x}_2$ is given by

$$f(x_1, x_2) = \begin{cases} 6e^{-3x_1-2x_2} & \text{for } x_1 > 0, x_2 > 0 \\ 0 & \text{elsewhere} \end{cases}$$

find the density function of the random variable $\mathbf{y} = \mathbf{x}_1 + \mathbf{x}_2$.

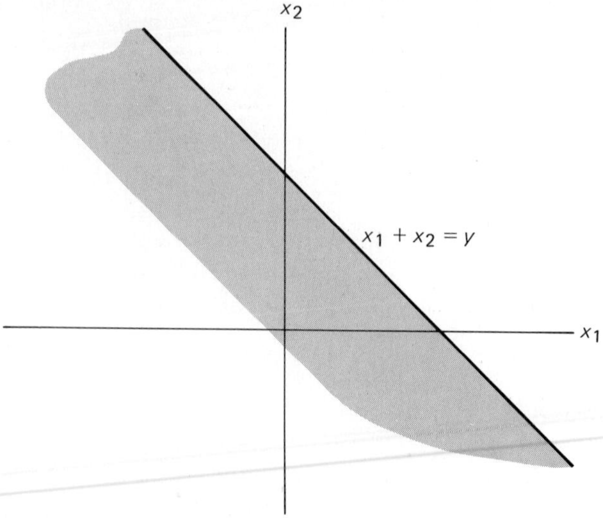

Figure 7.1 Diagram for Example 7.3.

Solution

Integrating the joint density over the shaded region of Figure 7.1, we get

$$F(y) = \int_0^y \int_0^{y-x_2} 6e^{-3x_1-2x_2}\, dx_1 dx_2$$
$$= 1 + 2e^{-3y} - 3e^{-2y}$$

and, differentiating with respect to y, we obtain

$$f(y) = 6(e^{-2y} - e^{-3y})$$

for $y > 0$; elsewhere, $f(y) = 0$.

THEORETICAL EXERCISES

1. If the probability density of $\mathbf{x}$ is given by

$$f(x) = \begin{cases} 2xe^{-x^2} & \text{for } x > 0 \\ 0 & \text{elsewhere} \end{cases}$$

and $\mathbf{y} = \mathbf{x}^2$, find

(a) the distribution function of $\mathbf{y}$;
(b) the probability density of $\mathbf{y}$.

2. If $\mathbf{x}$ has an exponential distribution with the parameter θ, use the distribution function technique to find the probability density of the random variable $\mathbf{y} = \ln \mathbf{x}$.

3. If $\mathbf{x}$ has the uniform density with the parameters $\alpha = 0$ and $\beta = 1$, use the method of Section 7.2 to find the probability density of the random variable $\mathbf{y} = \sqrt{\mathbf{x}}$.

4. If the joint density of $\mathbf{x}$ and $\mathbf{y}$ is given by

$$f(x, y) = \begin{cases} 4xye^{-(x^2+y^2)} & \text{for } x > 0, y > 0 \\ 0 & \text{elsewhere} \end{cases}$$

and $\mathbf{z} = \sqrt{\mathbf{x}^2 + \mathbf{y}^2}$, find

(a) the distribution function of $\mathbf{z}$;
(b) the probability density of $\mathbf{z}$.

5. If $\mathbf{x}_1$ and $\mathbf{x}_2$ are independent random variables having exponential densities with the parameters θ_1 and θ_2, use the method of Section 7.2 to find the probability density of $\mathbf{y} = \mathbf{x}_1 + \mathbf{x}_2$ when

(a) $\theta_1 \neq \theta_2$;
(b) $\theta_1 = \theta_2$.

Note that Example 7.3 is a special case of part (a) with $\theta_1 = \frac{1}{3}$ and $\theta_2 = \frac{1}{2}$.

6. If $\mathbf{x}_1$ and $\mathbf{x}_2$ are independent random variables having the uniform density with $\alpha = 0$ and $\beta = 1$, refer to Figure 7.2 to find expressions for the distribution function of $\mathbf{y} = \mathbf{x}_1 + \mathbf{x}_2$ which apply when

(a) $y \leqslant 0$; **(b)** $0 < y < 1$;
(c) $1 < y < 2$; **(d)** $y \geqslant 2$.

Also find the probability density of $\mathbf{y}$.

7. If the joint density of $\mathbf{x}$ and $\mathbf{y}$ is given by

$$f(x, y) = \begin{cases} e^{-(x+y)} & \text{for } x > 0, y > 0 \\ 0 & \text{elsewhere} \end{cases}$$

and $\mathbf{z} = \dfrac{\mathbf{x} + \mathbf{y}}{2}$, find the probability density of $\mathbf{z}$ by the method of Section 7.2.

APPLIED EXERCISES

8. In Exercise 23 on page 111, $\mathbf{p}$ is the price of a certain commodity (in dollars) and $\mathbf{s}$ its total sales (in 10,000 units). Use the joint density given in that exercise and the method of Section 7.2 to find the probability density of the random variable $\mathbf{v} = \mathbf{sp}$, the total amount of money spent on this commodity in units of \$10,000.

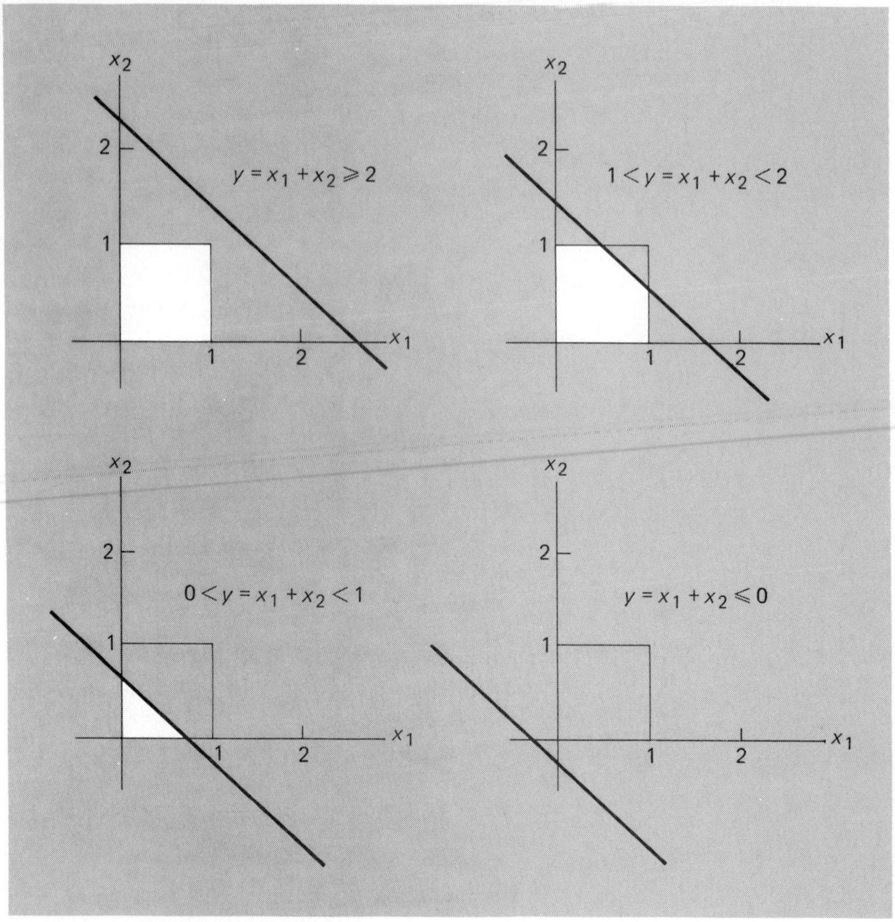

Figure 7.2 Diagram for Exercise 6.

9. In Exercise 19 on page 125, **x** is the amount (in dollars) a salesperson spends on gasoline and **y** is the amount (in dollars) for which he/she is reimbursed. Use the joint density given in that exercise and the method of Section 7.2 to find the probability density of the random variable $\mathbf{z} = \mathbf{x} - \mathbf{y}$, the amount for which he/she is not reimbursed.

7.3 TRANSFORMATION OF VARIABLE TECHNIQUE

Let us show how the probability distribution or density of a function of a random variable can be determined without first getting its distribution function. In the discrete case there is no real problem so long as the relationship between the values of **x** and $\mathbf{y} = u(\mathbf{x})$ is one-to-one; all we have to do is make the appropriate substitution.

EXAMPLE 7.4

If **x** is the number of heads obtained in four tosses of a balanced coin, find the probability distribution of $\mathbf{y} = \dfrac{1}{1 + \mathbf{x}}$.

Solution

Using the formula for the binomial distribution with $n = 4$ and $\theta = \frac{1}{2}$ (or referring to the result of Example 3.3 on page 75), we find that the probability distribution of **x** is given by

x	0	1	2	3	4
$f(x)$	$\frac{1}{16}$	$\frac{4}{16}$	$\frac{6}{16}$	$\frac{4}{16}$	$\frac{1}{16}$

Then, using the relationship $y = \dfrac{1}{1 + x}$ to substitute values of **y** for values of **x**, we find that the probability distribution of **y** is given by

y	1	$\frac{1}{2}$	$\frac{1}{3}$	$\frac{1}{4}$	$\frac{1}{5}$
$g(y)$	$\frac{1}{16}$	$\frac{4}{16}$	$\frac{6}{16}$	$\frac{4}{16}$	$\frac{1}{16}$

If we had wanted to make the substitution directly in the formula for the binomial distribution with $n = 4$ and $\theta = \frac{1}{2}$, we could have substituted $x = \dfrac{1}{y} - 1$ for x in

$$f(x) = \binom{4}{x}\left(\frac{1}{2}\right)^4 \quad \text{for } x = 0, 1, 2, 3, 4$$

getting

$$g(y) = f\left(\frac{1}{y} - 1\right) = \binom{4}{\frac{1}{y} - 1}\left(\frac{1}{2}\right)^4 \quad \text{for } y = 1, \tfrac{1}{2}, \tfrac{1}{3}, \tfrac{1}{4}, \tfrac{1}{5}$$

Note that in Example 7.4 the probabilities remained unchanged; the only difference is that in the result they are associated with the various values of **y** instead of the corresponding values of **x**. This is all there is to the **transformation of variable** (or **change of variable**) **technique** in the discrete case so long as the relationship is one-to-one. If it is not one-to-one, we may proceed as in the following example:

EXAMPLE 7.5

With reference to Example 7.4, find the probability distribution of the random variable $\mathbf{z} = (\mathbf{x} - 2)^2$.

Solution

Calculating the probabilities $h(z)$ associated with the various values of **z**, we get

$$h(0) = f(2) = \tfrac{6}{16}$$

$$h(1) = f(1) + f(3) = \tfrac{4}{16} + \tfrac{4}{16} = \tfrac{8}{16}$$

$$h(2) = f(0) + f(4) = \tfrac{1}{16} + \tfrac{1}{16} = \tfrac{2}{16}$$

and, hence,

z	0	1	2
$h(z)$	$\frac{3}{8}$	$\frac{4}{8}$	$\frac{1}{8}$

To perform a transformation of variable in the continuous case, we shall assume that the function given by $y = u(x)$ is differentiable and either increasing or decreasing for all values within the range of **x** for which $f(x) \neq 0$, so that the inverse function, given by $x = w(y)$, exists for all of the corresponding values of y, and is differentiable except where $u'(x) = 0$.† Under these conditions, we can prove the following theorem:

THEOREM 7.1 Let $f(x)$ be the value of the probability density of the continuous random variable **x** at x. If the function given by $y = u(x)$ is differentiable and either increasing or decreasing for all values within the range of **x** for which $f(x) \neq 0$, then, for these values of x, the equation

†Note that, to avoid points where $u'(x)$ might be 0, we generally did not include the endpoints of the intervals for which probability densities are non-zero. This is a practice we shall follow throughout this book.

$y = u(x)$ can be uniquely solved for x to give $x = w(y)$, and the probability density of $\mathbf{y}$ is given by

$$g(y) = f[w(y)] \cdot |w'(y)| \qquad \text{provided } u'(x) \neq 0$$

Elsewhere, $g(y) = 0$.

Proof. First let us prove the case where the function given by $y = u(x)$ is increasing. As can be seen from Figure 7.3, $\mathbf{x}$ must take on a value between $w(a)$ and $w(b)$ when $\mathbf{y}$ takes on a value between a and b. Hence,

$$\begin{aligned} P(a < \mathbf{y} < b) &= P[w(a) < \mathbf{x} < w(b)] \\ &= \int_{w(a)}^{w(b)} f(x)\, dx \\ &= \int_a^b f[w(y)]w'(y)\, dy \end{aligned}$$

where we performed the change of variable $y = u(x)$, or equivalently $x = w(y)$, in the integral. In accordance with Definition 3.4, the integrand gives the probability density of $\mathbf{y}$ so long as $w'(y)$ exists, and we can write

$$g(y) = f[w(y)]w'(y)$$

When the function given by $y = u(x)$ is decreasing, it can be seen from Figure 7.3 that $\mathbf{x}$ must take on a value between $w(b)$ and $w(a)$ when $\mathbf{y}$ takes on a value between a and b. Hence,

$$\begin{aligned} P(a < \mathbf{y} < b) &= P[w(b) < \mathbf{x} < w(a)] \\ &= \int_{w(b)}^{w(a)} f(x)\, dx \\ &= \int_b^a f[w(y)]w'(y)\, dy \\ &= -\int_a^b f[w(y)]w'(y)\, dy \end{aligned}$$

where we performed the same change of variable as before, and it follows that

$$g(y) = -f[w(y)]w'(y)$$

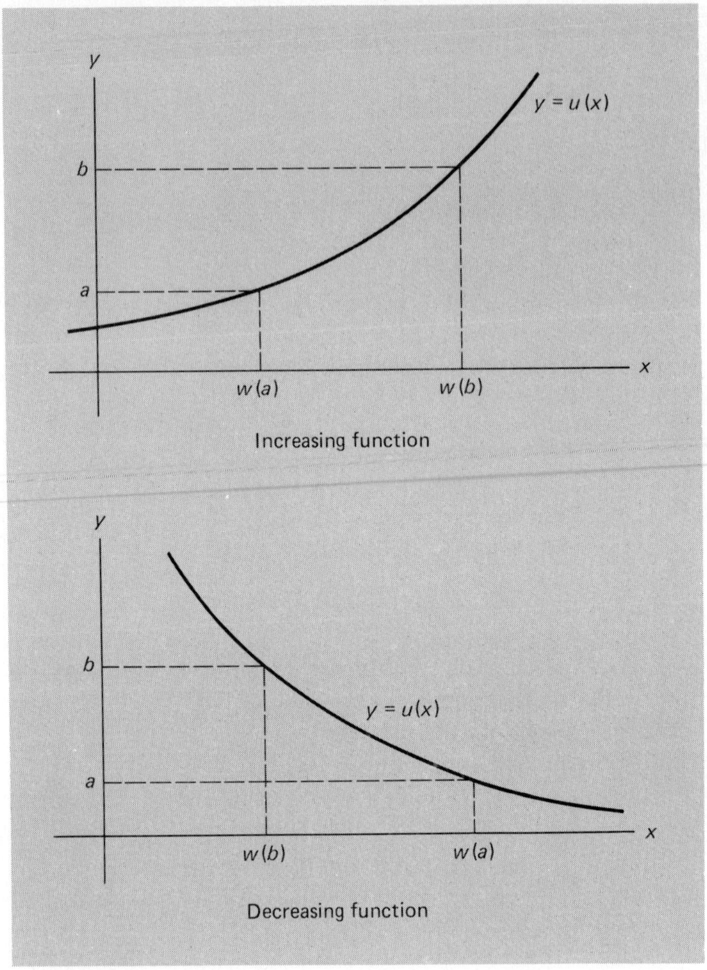

Figure 7.3 Diagrams for proof of Theorem 7.1.

Since $w'(y) = \dfrac{dx}{dy} = \dfrac{1}{\dfrac{dy}{dx}}$ is positive when the function given by $y = u(x)$ is increasing, and $-w'(y)$ is positive when the function given by $y = u(x)$ is decreasing, we can combine the two cases by writing

$$g(y) = f[w(y)] \cdot |w'(y)|$$

The result of Theorem 7.1 is easy to remember by writing the transformation of variable formula as

$$g(y) = f(x) \cdot \left|\frac{dx}{dy}\right|$$

where $f(x)$ and $\left|\frac{dx}{dy}\right|$ must both be expressed in terms of y.

EXAMPLE 7.6

If **x** has the exponential distribution given by

$$f(x) = \begin{cases} e^{-x} & \text{for } x > 0 \\ 0 & \text{elsewhere} \end{cases}$$

find the probability density of the random variable $\mathbf{y} = \sqrt{\mathbf{x}}$.

Solution

The equation $y = \sqrt{x}$, relating the values of **x** and **y**, has the unique inverse $x = y^2$, which yields $w'(y) = \dfrac{dx}{dy} = 2y$. Therefore,

$$g(y) = e^{-y^2}|2y| = 2ye^{-y^2}$$

for $y > 0$ in accordance with Theorem 7.1. Since the probability of getting a value of **y** less than or equal to 0, like the probability of getting a value of **x** less than or equal to 0, is 0, it follows that the probability density of **y** is given by

$$g(y) = \begin{cases} 2ye^{-y^2} & \text{for } y > 0 \\ 0 & \text{elsewhere} \end{cases}$$

Note that this is the Weibull distribution of Exercise 15 on page 204 with $\alpha = 1$ and $\beta = 2$.

The two diagrams of Figure 7.4 illustrate what happens in Example 7.6 when we transform from **x** to **y**. As in the discrete case (for instance, Example 7.4), probabilities remain the same, but they pertain to different values (intervals of values) of the respective random variables. In the diagram on the left, the 0.35

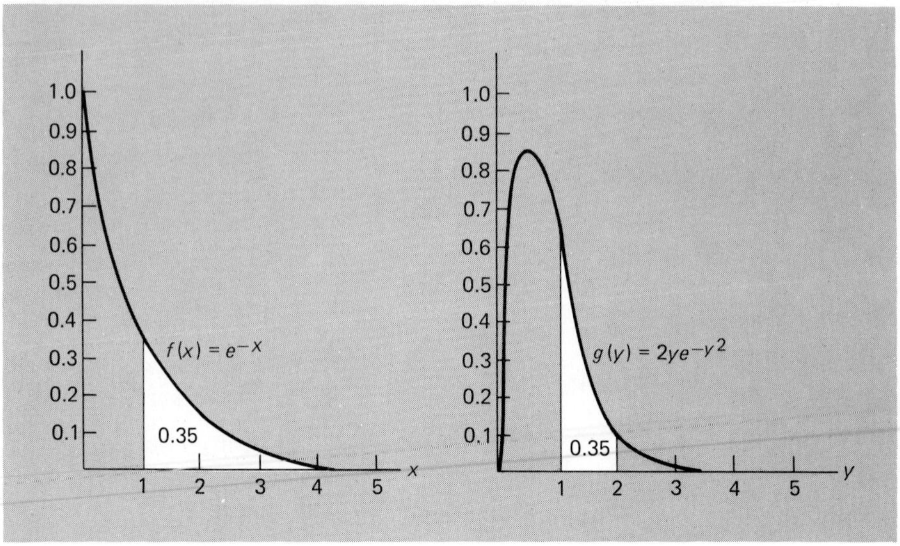

Figure 7.4 Diagrams for Example 7.6.

probability pertains to the event that **x** will take on a value on the interval from 1 to 4, and in the diagram on the right, the 0.35 probability pertains to the event that **y** will take on a value on the interval from 1 to 2.

EXAMPLE 7.7

If the double arrow of Figure 7.5 is spun so that the random variable θ has the uniform density

$$f(\theta) = \begin{cases} \dfrac{1}{\pi} & \text{for } -\dfrac{\pi}{2} < \theta < \dfrac{\pi}{2} \\ 0 & \text{elsewhere} \end{cases}$$

determine the probability density of **x**, the abscissa of the point on the x-axis to which the arrow will point.

Solution

As is apparent from the diagram, the relationship between x and θ is given by $x = a \cdot \tan \theta$, so that

$$\frac{d\theta}{dx} = \frac{a}{a^2 + x^2}$$

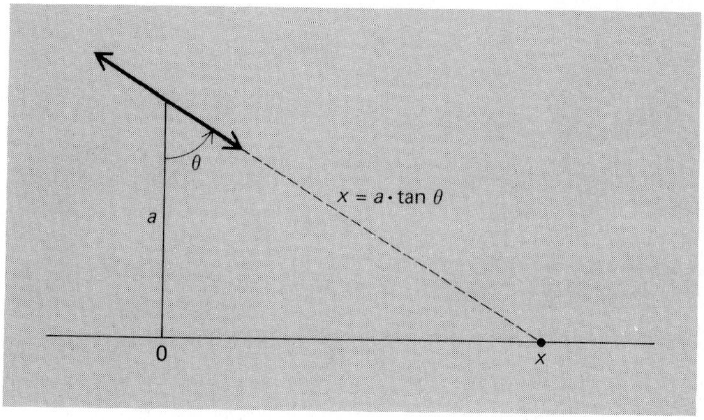

Figure 7.5 Diagram for Example 7.7.

and it follows that

$$g(x) = \frac{1}{\pi} \cdot \left| \frac{a}{a^2 + x^2} \right|$$

$$= \frac{1}{\pi} \cdot \frac{a}{a^2 + x^2} \qquad \text{for } -\infty < x < \infty$$

according to Theorem 7.1. Note that this is a special case of the Cauchy distribution of Exercise 3 on page 202.

EXAMPLE 7.8

If $F(x)$ is the value of the distribution function of the continuous random variable $\mathbf{x}$ at x, find the probability density of $\mathbf{y} = F(\mathbf{x})$.

Solution

As can be seen from Figure 7.6, the value of $\mathbf{y}$ corresponding to any particular value of $\mathbf{x}$ is given by the area under the graph of the density of $\mathbf{x}$ to the left of x. Differentiating $y = F(x)$ with respect to x, we get

$$\frac{dy}{dx} = F'(x) = f(x)$$

and, hence,

$$\frac{dx}{dy} = \frac{1}{\frac{dy}{dx}} = \frac{1}{f(x)}$$

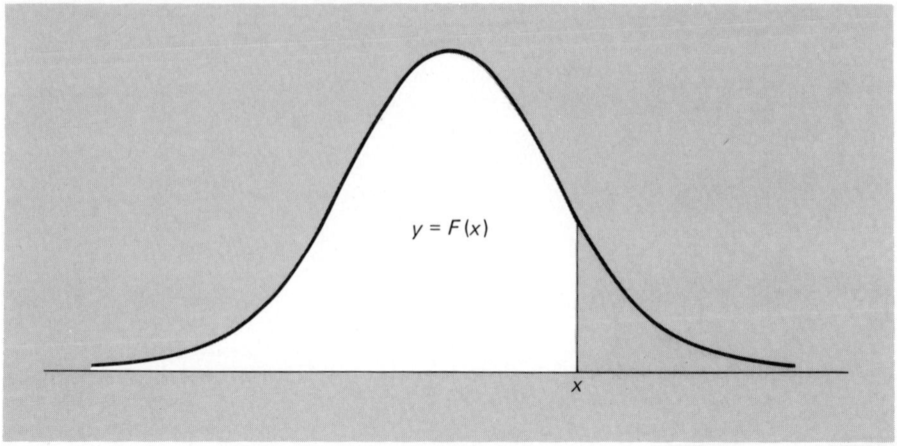

Figure 7.6 Diagram for Example 7.8 (Probability integral transformation).

provided $f(x) \neq 0$. It follows from Theorem 7.1 that

$$g(y) = f(x) \cdot \left| \frac{1}{f(x)} \right| = 1$$

for $0 < y < 1$, and we can say that y has the uniform density with $\alpha = 0$ and $\beta = 1$.

The transformation which we performed in Example 7.8 is called the **probability integral transformation**. The result is not only of theoretical importance, but it facilitates the **simulation** of observed values of continuous random variables. A reference to how this is done, especially in connection with the normal distribution, is given on page 252.

When the conditions underlying Theorem 7.1 are not met, we can be in serious difficulties, and we may have to use the method of Section 7.2 or a generalization of Theorem 7.1 referred to among the references on page 252; sometimes, there is an easy way out, as in the following example:

EXAMPLE 7.9

If $\mathbf{x}$ has the standard normal distribution, find the probability density of $\mathbf{z} = \mathbf{x}^2$.

Solution

Since the function given by $z = x^2$ is decreasing for negative values of x and increasing for positive values of x, the conditions of Theorem 7.1 are not met.

However, the transformation from $\mathbf{x}$ to $\mathbf{z}$ can be made in two steps: First we find the probability density of $\mathbf{y} = |\mathbf{x}|$, and then we find the probability density of $\mathbf{z} = \mathbf{y}^2 (= \mathbf{x}^2)$.

So far as the first step is concerned, we already studied the transformation $\mathbf{y} = |\mathbf{x}|$ in Example 7.2 on page 227; in fact, we showed there that if $\mathbf{x}$ has the standard normal distribution, then $\mathbf{y} = |\mathbf{x}|$ has the probability density

$$g(y) = 2n(y; 0, 1) = \frac{2}{\sqrt{2\pi}} e^{-\frac{1}{2}y^2}$$

for $y > 0$, and $g(y) = 0$ elsewhere. For the second step, the function given by $z = y^2$ is increasing for $y > 0$; that is, for all values of $\mathbf{y}$ for which $g(y) \neq 0$. Thus, we can use Theorem 7.1, and since

$$\frac{dy}{dz} = \tfrac{1}{2} z^{-\frac{1}{2}}$$

we get

$$h(z) = \frac{2}{\sqrt{2\pi}} e^{-\frac{1}{2}z} \left|\tfrac{1}{2} z^{-\frac{1}{2}}\right|$$

$$= \frac{1}{\sqrt{2\pi}} z^{-\frac{1}{2}} e^{-\frac{1}{2}z}$$

for $z > 0$, and $h(z) = 0$ elsewhere. Observe that since $\Gamma(\frac{1}{2}) = \sqrt{\pi}$, the distribution we have arrived at for $\mathbf{z}$ is a chi-square distribution (see Definition 6.4 on page 198) with $\nu = 1$.

The method of this section can also be used to find the distribution of a random variable which is a function of two or more random variables. Suppose, for instance, that we are given the joint distribution of two random variables $\mathbf{x}_1$ and $\mathbf{x}_2$, and that we want to determine the distribution of the random variable $\mathbf{y} = u(\mathbf{x}_1, \mathbf{x}_2)$. If the relationship between y and x_1 with x_2 held constant, or the relationship between y and x_2 with x_1 held constant, permits, we can proceed in the discrete case as in Example 7.4 to find the joint distribution of $\mathbf{y}$ and $\mathbf{x}_2$, or $\mathbf{x}_1$ and $\mathbf{y}$, and then sum on the values of the other random variable to get the marginal distribution of $\mathbf{y}$. In the continuous case, we first use Theorem 7.1 with the transformation formula written as

$$g(y, x_2) = f(x_1, x_2) \cdot \left|\frac{\partial x_1}{\partial y}\right|$$

or as

$$g(x_1, y) = f(x_1, x_2) \cdot \left|\frac{\partial x_2}{\partial y}\right|$$

where $f(x_1, x_2)$ and the partial derivative must be expressed in terms of y and x_2, or x_1 and y. Then we integrate out the other variable to get the marginal density of $\mathbf{y}$.

EXAMPLE 7.10

If $\mathbf{x}_1$ and $\mathbf{x}_2$ are independent random variables having Poisson distributions with the parameters λ_1 and λ_2, find the probability distribution of the random variable $\mathbf{y} = \mathbf{x}_1 + \mathbf{x}_2$.

Solution

Since $\mathbf{x}_1$ and $\mathbf{x}_2$ are independent, their joint distribution is given by

$$\begin{aligned} f(x_1, x_2) &= \frac{e^{-\lambda_1}(\lambda_1)^{x_1}}{x_1!} \cdot \frac{e^{-\lambda_2}(\lambda_2)^{x_2}}{x_2!} \\ &= \frac{e^{-(\lambda_1+\lambda_2)}(\lambda_1)^{x_1}(\lambda_2)^{x_2}}{x_1!x_2!} \end{aligned}$$

for $x_1 = 0, 1, 2, \ldots,$ and $x_2 = 0, 1, 2, \ldots$. Since $y = x_1 + x_2$ and, hence, $x_1 = y - x_2$, we can substitute $y - x_2$ for x_1, getting

$$g(y, x_2) = \frac{e^{-(\lambda_1+\lambda_2)}(\lambda_2)^{x_2}(\lambda_1)^{y-x_2}}{x_2!(y-x_2)!}$$

for $y = 0, 1, 2, \ldots,$ and $x_2 = 0, 1, \ldots, y$, for the joint distribution of $\mathbf{y}$ and $\mathbf{x}_2$. Then, summing on x_2, from 0 to y, we get

$$\begin{aligned} h(y) &= \sum_{x_2=0}^{y} \frac{e^{-(\lambda_1+\lambda_2)}(\lambda_2)^{x_2}(\lambda_1)^{y-x_2}}{x_2!(y-x_2)!} \\ &= \frac{e^{-(\lambda_1+\lambda_2)}}{y!} \cdot \sum_{x_2=0}^{y} \frac{y!}{x_2!(y-x_2)!}(\lambda_2)^{x_2}(\lambda_1)^{y-x_2} \end{aligned}$$

after factoring out $e^{-(\lambda_1+\lambda_2)}$ and multiplying and dividing by $y!$. Identifying the summation at which we arrived as the binomial expansion of $(\lambda_1 + \lambda_2)^y$, we finally get

$$h(y) = \frac{e^{-(\lambda_1+\lambda_2)}(\lambda_1 + \lambda_2)^y}{y!} \qquad \text{for } y = 0, 1, 2, \ldots$$

and we have, thus, shown that the sum of two independent random variables having Poisson distributions with the parameters λ_1 and λ_2 has a Poisson distribution with the parameter $\lambda = \lambda_1 + \lambda_2$.

EXAMPLE 7.11

If the joint density of $\mathbf{x}_1$ and $\mathbf{x}_2$ is given by

$$f(x_1, x_2) = \begin{cases} e^{-(x_1+x_2)} & \text{for } x_1 > 0, x_2 > 0 \\ 0 & \text{elsewhere} \end{cases}$$

find the density function of $\mathbf{y} = \dfrac{\mathbf{x}_1}{\mathbf{x}_1 + \mathbf{x}_2}$.

Solution

Since y decreases when x_2 increases and x_1 is held constant, we can use Theorem 7.1 (modified as indicated above) to find the joint density of $\mathbf{x}_1$ and $\mathbf{y}$. Since $y = \dfrac{x_1}{x_1 + x_2}$ yields $x_2 = x_1 \cdot \dfrac{1 - y}{y}$ and, hence,

$$\frac{\partial x_2}{\partial y} = -\frac{x_1}{y^2}$$

it follows that

$$g(x_1, y) = e^{-x_1/y}\left| -\frac{x_1}{y^2}\right| = \frac{x_1}{y^2} \cdot e^{-x_1/y}$$

for $x_1 > 0$ and $0 < y < 1$. Finally, integrating out x_1 and changing the variable of integration to $u = x_1/y$, we get

$$\begin{aligned} h(y) &= \int_0^\infty \frac{x_1}{y^2} \cdot e^{-x_1/y}\, dx_1 \\ &= \int_0^\infty u \cdot e^{-u}\, du \\ &= \Gamma(2) \\ &= 1 \end{aligned}$$

for $0 < y < 1$, and $h(y) = 0$ elsewhere. Thus, the random variable $\mathbf{y}$ has the uniform density with $\alpha = 0$ and $\beta = 1$.

EXAMPLE 7.12

If the joint density of $\mathbf{x}_1$ and $\mathbf{x}_2$ is given by

$$f(x_1, x_2) = \begin{cases} 1 & \text{for } 0 < x_1 < 1, 0 < x_2 < 1 \\ 0 & \text{elsewhere} \end{cases}$$

find the probability density of $\mathbf{y} = \mathbf{x}_1 + \mathbf{x}_2$.[†]

Solution

Since y increases when x_1 increases and x_2 is held constant, we can use Theorem 7.1 (modified as indicated on page 239) to find the joint density of $\mathbf{y}$ and $\mathbf{x}_2$. Since $x_1 = y - x_2$ and, hence, $\frac{\partial x_1}{\partial y} = 1$, it follows that

$$\begin{aligned} g(y, x_2) &= 1 \cdot |1| \\ &= 1 \end{aligned}$$

for all pairs of values of y and x_2 within the white region of Figure 7.7 (that is, for $0 < x_2 < 1$ and $x_2 < y < x_2 + 1$), and $g(y, x_2) = 0$ elsewhere.

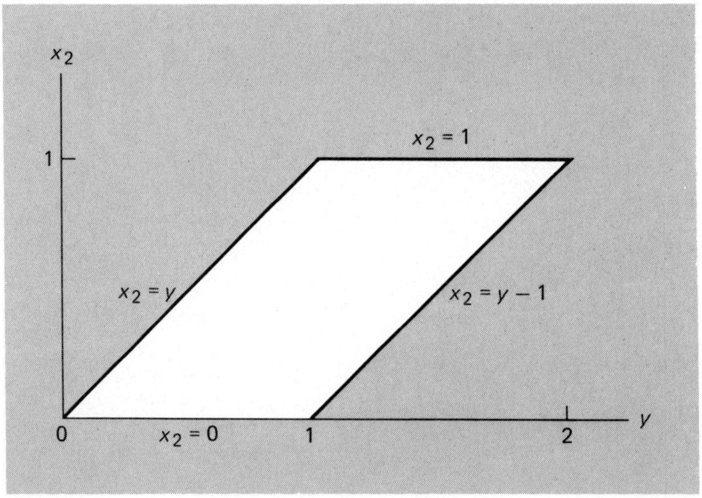

Figure 7.7 Sample space for Example 7.12.

[†]In Exercise 6 on page 229 the reader was asked to work the same problem by the method of Section 7.2.

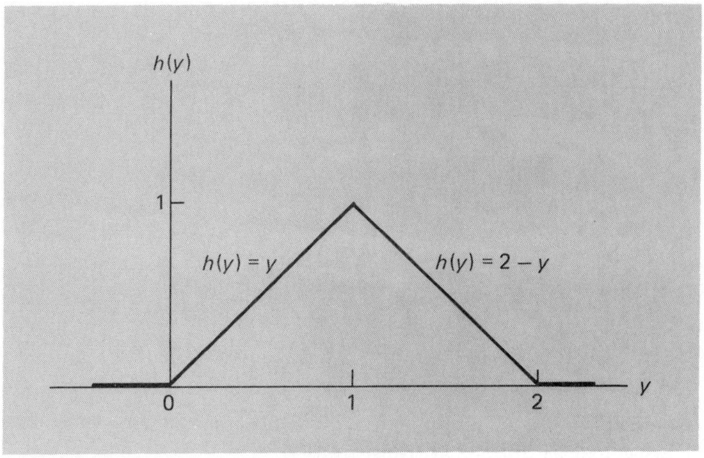

Figure 7.8 Triangular probability density.

Finally, integrating out x_2 separately for $y \leqslant 0$, $0 < y < 1$, $1 < y < 2$, and $y \geqslant 2$, we get

$$h(y) = \begin{cases} 0 & \text{for } y \leqslant 0 \\ \displaystyle\int_0^y 1\, dx_2 = y & \text{for } 0 < y < 1 \\ \displaystyle\int_{y-1}^1 1\, dx_2 = 2 - y & \text{for } 1 < y < 2 \\ 0 & \text{for } y \geqslant 2 \end{cases}$$

and to make the density function continuous, we let $h(1) = 1$. We have thus shown that the sum of the given random variables has the **triangular probability density**, whose graph is shown in Figure 7.8.

So far we have considered only functions of one or two random variables, but the method can easily be generalized. For instance, for a function of three random variables we perform a transformation to introduce the new random variable in place of one of the original variables, and then we eliminate (by summation or integration) the other two variables with which we began.

EXAMPLE 7.13

If the joint probability density of $\mathbf{x}_1$, $\mathbf{x}_2$, and $\mathbf{x}_3$ is given by

$$f(x_1, x_2, x_3) = \begin{cases} e^{-(x_1+x_2+x_3)} & \text{for } x_1 > 0, x_2 > 0, x_3 > 0 \\ 0 & \text{elsewhere} \end{cases}$$

find the probability density of the random variable $\mathbf{y} = \mathbf{x}_1 + \mathbf{x}_2 + \mathbf{x}_3$.

Solution

To find the joint distribution of **y**, $\mathbf{x}_2$, and $\mathbf{x}_3$, we make use of the formula

$$g(y, x_2, x_3) = f(x_1, x_2, x_3) \cdot \left|\frac{\partial x_1}{\partial y}\right|$$

and since $\dfrac{\partial x_1}{\partial y} = 1$, we get

$$\begin{aligned} g(y, x_2, x_3) &= e^{-y} \cdot |1| \\ &= e^{-y} \end{aligned}$$

for $x_2 > 0$, $x_3 > 0$, and $y > x_2 + x_3$, and $g(y, x_2, x_3) = 0$ elsewhere. Then, integrating out x_2 and x_3, we find that

$$h(y) = \int_0^y \int_0^{y-x_3} e^{-y}\, dx_2\, dx_3 = \tfrac{1}{2}y^2 e^{-y}$$

for $y > 0$, and $h(y) = 0$ elsewhere. Observe that we have shown that the sum of three independent random variables having the gamma distribution with $\alpha = 1$ and $\beta = 1$ is a random variable having the gamma distribution with $\alpha = 3$ and $\beta = 1$.

THEORETICAL EXERCISES

1. If **x** has a geometric distribution with $\theta = \frac{1}{3}$, find the probability distribution of $\mathbf{y} = 4 - 5\mathbf{x}$.

2. If **x** has a hypergeometric distribution with $k = 4$, $N = 15$, and $n = 3$, find the probability distribution of **y**, the number of successes minus the number of failures.

3. If **x** is the total we roll with a pair of dice, for which the probability distribution is given on page 74, find

(a) the probability distribution of **y** which takes on the values 0, 1, or 2, when the remainder is, respectively, 0, 1, or 2, when the value of **x** is divided by 3;

(b) the probability distribution of **z** which takes on the value -1 when the value of **x** is 2, 3, or 12, the value 1 when the value of **x** is 7 or 11, and the value 0 when the value of **x** is 4, 5, 6, 8, 9, or 10.

4. Use the transformation of variable technique to prove Theorem 6.7 on page 210.

5. Rework Example 7.1 on page 226 by the transformation of variable technique.

6. If $\mathbf{x} = \ln \mathbf{y}$ has a normal distribution with the mean μ and the variance σ^2, find the probability density of $\mathbf{y}$, which is said to have the **log-normal distribution**.

7. If the probability density of $\mathbf{x}$ is given by

$$f(x) = \begin{cases} \dfrac{x}{2} & \text{for } 0 < x < 2 \\ 0 & \text{elsewhere} \end{cases}$$

find the probability density of $\mathbf{y} = \mathbf{x}^3$. Also plot the graphs of these two probability densities and indicate the respective areas under the curves representing $P(\frac{1}{2} < \mathbf{x} < 1)$ and $P(\frac{1}{8} < \mathbf{y} < 1)$.

8. If the probability density of $\mathbf{x}$ is given by

$$f(x) = \begin{cases} \dfrac{kx^3}{(1 + 2x)^6} & \text{for } x > 0 \\ 0 & \text{elsewhere} \end{cases}$$

where k is an appropriate constant, find the probability density of the random variable $\mathbf{y} = \dfrac{2\mathbf{x}}{1 + 2\mathbf{x}}$. Also find the value of k by comparing the result with the probability density of Definition 6.5.

9. If $\mathbf{x}$ has a uniform density with $\alpha = 0$ and $\beta = 1$, show that the random variable $\mathbf{y} = -2 \cdot \ln \mathbf{x}$ has a gamma distribution. What are its parameters?

10. If the probability density of $\mathbf{x}$ is given by

$$f(x) = \begin{cases} \dfrac{3x^2}{2} & \text{for } -1 < x < 1 \\ 0 & \text{elsewhere} \end{cases}$$

find

(a) the probability density of $\mathbf{y} = |\mathbf{x}|$ using the result of Example 7.2 on page 227;

(b) the probability density of $\mathbf{z} = \mathbf{x}^2 (= \mathbf{y}^2)$.

11. If $\mathbf{x}$ has a uniform density with $\alpha = -1$ and $\beta = 3$, find

(a) the probability density of $\mathbf{y} = |\mathbf{x}|$ using the result of Example 7.2 on page 227;

(b) the probability density of $\mathbf{z} = \mathbf{x}^4 (= \mathbf{y}^4)$.

12. If the joint probability distribution of $\mathbf{x}_1$ and $\mathbf{x}_2$ is given by $f(x_1, x_2) = \dfrac{x_1 x_2}{36}$ for $x_1 = 1, 2, 3$, and $x_2 = 1, 2, 3$, find

(a) the probability distribution of $\mathbf{y} = \mathbf{x}_1 \mathbf{x}_2$;

(b) the probability distribution of $\mathbf{z} = \dfrac{\mathbf{x}_1}{\mathbf{x}_2}$.

13. With reference to Example 3.12 on page 98, find
 (a) the probability distribution of $\mathbf{u} = \mathbf{x} + \mathbf{y}$;
 (b) the probability distribution of $\mathbf{v} = \mathbf{xy}$;
 (c) the probability distribution of $\mathbf{w} = \mathbf{x} - \mathbf{y}$.

14. If $\mathbf{x}_1$ and $\mathbf{x}_2$ are independent random variables having binomial distributions with the respective parameters n_1 and θ and n_2 and θ, show that $\mathbf{y} = \mathbf{x}_1 + \mathbf{x}_2$ has the binomial distribution with the parameters $n_1 + n_2$ and θ. (*Hint*: Use Theorem 1.12.)

15. If $\mathbf{x}_1$ and $\mathbf{x}_2$ are independent random variables having the geometric distribution with the parameter θ, show that $\mathbf{y} = \mathbf{x}_1 + \mathbf{x}_2$ is a random variable having the negative binomial distribution with the parameters θ and $k = 2$.

16. If $\mathbf{x}$ and $\mathbf{y}$ are independent random variables having the standard normal distribution, show that $\mathbf{z} = \mathbf{x} + \mathbf{y}$ also has a normal distribution. (*Hint*: Complete the square in the exponent.) What are the mean and the variance of this normal distribution?

17. If the joint density of $\mathbf{x}$ and $\mathbf{y}$ is given by

$$f(x, y) = \begin{cases} 12xy(1 - y) & \text{for } 0 < x < 1, 0 < y < 1 \\ 0 & \text{elsewhere} \end{cases}$$

find the probability density of the random variable $\mathbf{z} = \mathbf{xy}^2$.

18. If $\mathbf{x}_1$ and $\mathbf{x}_2$ are independent random variables having the Cauchy distribution

$$f(x) = \frac{1}{\pi(1 + x^2)} \qquad \text{for } -\infty < x < \infty$$

find, and identify, the probability density of $\mathbf{y} = \mathbf{x}_1 + \mathbf{x}_2$. (*Hint*: Use partial fractions to perform the necessary integration.)

19. If $\mathbf{x}$ and $\mathbf{y}$ are two random variables whose joint density is given by

$$f(x, y) = \begin{cases} \frac{1}{2} & \text{for } x > 0, y > 0, x + y < 2 \\ 0 & \text{elsewhere} \end{cases}$$

find
 (a) the joint density of $\mathbf{u} = \mathbf{y} - \mathbf{x}$ and $\mathbf{y}$;
 (b) the probability density of $\mathbf{u}$.

20. In Example 7.12 we found the probability density of the sum of two independent random variables having the uniform density with $\alpha = 0$ and $\beta = 1$. Given a third random variable $\mathbf{x}_3$ which has the same uniform density and is independent of $\mathbf{x}_1$ and $\mathbf{x}_2$, show that if $\mathbf{u} = \mathbf{y} + \mathbf{x}_3 = \mathbf{x}_1 + \mathbf{x}_2 + \mathbf{x}_3$, then
 (a) the joint density of $\mathbf{u}$ and $\mathbf{y}$ is given by

$$g(u, y) = \begin{cases} y & \text{for Regions I and II of Figure 7.9} \\ 2 - y & \text{for Regions III and IV of Figure 7.9} \\ 0 & \text{elsewhere} \end{cases}$$

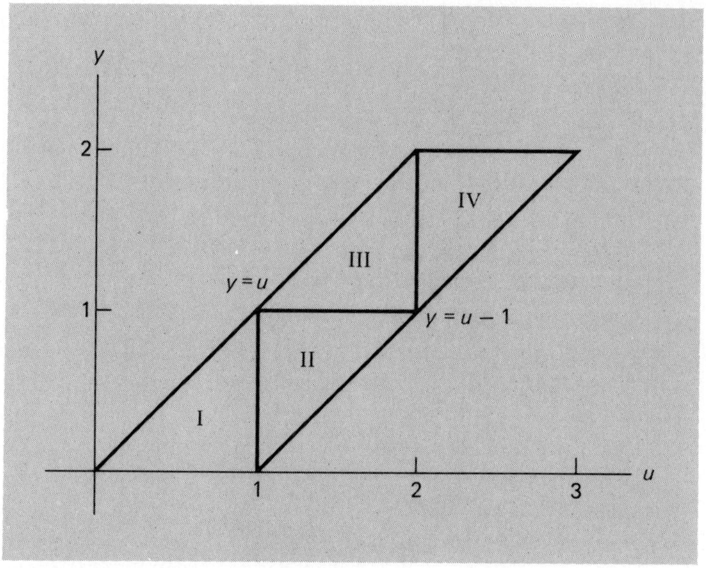

Figure 7.9 Diagram for Exercise 20.

(b) the probability density of **u** is given by

$$h(u) = \begin{cases} 0 & \text{for } u \leqslant 0 \\ \frac{1}{2}u^2 & \text{for } 0 < u < 1 \\ \frac{1}{2}u^2 - \frac{3}{2}(u-1)^2 & \text{for } 1 < u < 2 \\ \frac{1}{2}u^2 - \frac{3}{2}(u-1)^2 + \frac{3}{2}(u-2)^2 & \text{for } 2 < u < 3 \\ 0 & \text{for } u \geqslant 3 \end{cases}$$

Note that if we let $h(1) = h(2) = \frac{1}{2}$, this will make the probability density of u continuous.

APPLIED EXERCISES

21. According to the Maxwell–Boltzmann law of theoretical physics, the probability density of **v**, the velocity of a gas molecule, is

$$f(v) = \begin{cases} kv^2 e^{-\beta v^2} & \text{for } v > 0 \\ 0 & \text{elsewhere} \end{cases}$$

where β depends on its mass and the absolute temperature, and k is an appropriate constant. Show that the probability density of the kinetic energy **E**, whose values are related to those of v by means of the equation $E = \frac{1}{2}mv^2$, is a gamma distribution.

22. With reference to Exercise 22 on page 111, find the probability density of $\mathbf{z} = \frac{\mathbf{x} + \mathbf{y}}{2}$, the average proportion of correct answers a student will get on the two aptitude tests.

23. Use the method of this section to rework Exercise 8 on page 229, by determining first the joint density of $\mathbf{p}$ and $\mathbf{v} = \mathbf{sp}$, and then the probability density of $\mathbf{v}$.

7.4 MOMENT-GENERATING FUNCTION TECHNIQUE

Moment-generating functions can play an important role in determining the probability distribution or density of a function of random variables when the function is a linear combination of n *independent* random variables. We shall illustrate this technique here when such a linear combination is, in fact, the sum of n independent random variables, leaving it to the reader to generalize it in Exercises 5 and 6 on page 251.

The method is based on the theorem that the moment-generating function of the sum of n independent random variables equals the product of their moment-generating functions, namely,

THEOREM 7.2 If $\mathbf{x}_1, \mathbf{x}_2, \ldots,$ and $\mathbf{x}_n$ are independent random variables and $\mathbf{y} = \mathbf{x}_1 + \mathbf{x}_2 + \cdots + \mathbf{x}_n$, then

$$M_{\mathbf{y}}(t) = \prod_{i=1}^{n} M_{\mathbf{x}_i}(t)$$

where $M_{\mathbf{x}_i}(t)$ is the value of the moment-generating function of $\mathbf{x}_i$ at t.

Proof. Making use of the fact that the random variables are independent and, hence,

$$f(x_1, x_2, \ldots, x_n) = f_1(x_1) \cdot f_2(x_2) \cdot \ldots \cdot f_n(x_n)$$

according to Definition 3.14, we can write

$$\begin{aligned}
M_{\mathbf{y}}(t) &= E(e^{\mathbf{y}t}) = E[e^{(\mathbf{x}_1+\mathbf{x}_2+\cdots+\mathbf{x}_n)t}] \\
&= \int_{-\infty}^{\infty} \cdots \int_{-\infty}^{\infty} e^{(x_1+x_2+\cdots+x_n)t} f(x_1, x_2, \cdots, x_n)\, dx_1\, dx_2 \cdots dx_n \\
&= \int_{-\infty}^{\infty} e^{x_1 t} f_1(x_1)\, dx_1 \cdot \int_{-\infty}^{\infty} e^{x_2 t} f_2(x_2)\, dx_2 \cdots \int_{-\infty}^{\infty} e^{x_n t} f_n(x_n)\, dx_n \\
&= \prod_{i=1}^{n} M_{\mathbf{x}_i}(t)
\end{aligned}$$

which proves the theorem for the continuous case. To prove it for the discrete case, we have only to replace all of the integrals by sums.

Note that if we want to use Theorem 7.2 to find the probability distribution or density of the random variable $\mathbf{y} = \mathbf{x}_1 + \mathbf{x}_2 + \cdots + \mathbf{x}_n$, we must be able to identify whatever probability distribution or density corresponds to $M_{\mathbf{y}}(t)$ and rely on the first of the two theorems which we gave on page 214; namely, the uniqueness theorem about the correspondence between moment-generating functions and probability distributions or densities.

EXAMPLE 7.14

Find the probability distribution of the sum of n independent random variables $\mathbf{x}_1$, $\mathbf{x}_2, \ldots$, and $\mathbf{x}_n$, having Poisson distributions with the respective parameters $\lambda_1, \lambda_2, \ldots, \lambda_n$.

Solution

By Theorem 5.9, we have

$$M_{\mathbf{x}_i}(t) = e^{\lambda_i(e^t-1)}$$

and, hence,

$$M_{\mathbf{y}}(t) = \prod_{i=1}^{n} e^{\lambda_i(e^t-1)} = e^{(\lambda_1+\lambda_2+\cdots+\lambda_n)(e^t-1)}$$

which can readily be identified as the moment-generating function of the Poisson distribution with the parameter $\lambda = \lambda_1 + \lambda_2 + \cdots + \lambda_n$. Thus, the distribution of the sum of n independent random variables having Poisson distributions with the parameters λ_i is a Poisson distribution with the parameter $\lambda = \lambda_1 + \lambda_2 + \cdots + \lambda_n$. Note that in Example 7.10 we proved this for $n = 2$.

EXAMPLE 7.15

If $\mathbf{x}_1, \mathbf{x}_2, \ldots$, and $\mathbf{x}_n$ are independent random variables having exponential distributions with the same parameter θ, find the probability density of the random variable $\mathbf{y} = \mathbf{x}_1 + \mathbf{x}_2 + \cdots + \mathbf{x}_n$.

Solution

Since the exponential distribution is a gamma distribution with $\alpha = 1$ and $\beta = \theta$, we have

$$M_{\mathbf{x}_i}(t) = (1 - \theta t)^{-1}$$

by Theorem 6.4, and, hence,

$$M_{\mathbf{y}}(t) = \prod_{i=1}^{n} (1 - \theta t)^{-1} = (1 - \theta t)^{-n}$$

according to the second of the special rules for products in Appendix II. Identifying the moment-generating function of $\mathbf{y}$ as that of a gamma distribution with $\alpha = n$ and $\beta = \theta$, we conclude that the distribution of the sum of n independent random variables having exponential distributions with the same parameter θ is a gamma distribution with the parameters $\alpha = n$ and $\beta = \theta$. Note that this agrees with the result of Example 7.13, where we showed that the sum of three independent random variables having exponential distributions with the parameter $\theta = 1$ has a gamma distribution with $\alpha = 3$ and $\beta = 1$.

Theorem 7.2 also provides an easy and elegant way of deriving the moment-generating function of the binomial distribution. Suppose that $\mathbf{x}_1, \mathbf{x}_2, \ldots,$ and $\mathbf{x}_n$, are independent random variables having the same Bernoulli distribution $f(x; \theta) = \theta^x(1 - \theta)^{1-x}$ for $x = 0, 1$. By Definition 4.6,

$$M_{\mathbf{x}_i}(t) = e^{0 \cdot t}(1 - \theta) + e^{1 \cdot t}\theta = 1 + \theta(e^t - 1)$$

so that Theorem 7.2 yields

$$M_{\mathbf{y}}(t) = \prod_{i=1}^{n} [1 + \theta(e^t - 1)] = [1 + \theta(e^t - 1)]^n$$

which is readily identified as the moment-generating function of the binomial distribution with the parameters n and θ. Of course, $\mathbf{y} = \mathbf{x}_1 + \mathbf{x}_2 + \cdots + \mathbf{x}_n$ *is* the total number of successes in n trials, since $\mathbf{x}_1$ is the number of successes on the first trial, $\mathbf{x}_2$ is the number of successes on the second trial, . . . , and $\mathbf{x}_n$ is the number of successes on the nth trial. As we shall see later, this is a fruitful way of looking at the binomial distribution.

THEORETICAL EXERCISES

1. Use the moment-generating function technique to rework Exercise 14 on page 246.

2. Find the moment-generating function of the negative binomial distribution by making use of the fact that if k independent random variables have geometric distributions with the same parameter θ, their sum is a random variable having the negative binomial distribution with the parameters θ and k. (*Hint:* use the result of Exercise 4 on page 185.)

3. If n independent random variables have the same gamma distribution with the same parameters α and β, find the moment-generating function of their sum and, if possible, identify its distribution.

4. If n independent random variables $\mathbf{x}_i$ have normal distributions with the means μ_i and the standard deviations σ_i, find the moment-generating function of their sum and identify the corresponding distribution, its mean, and its variance.

5. Prove the following generalization of Theorem 7.2: If $\mathbf{x}_1, \mathbf{x}_2, \ldots,$ and $\mathbf{x}_n$ are independent random variables and $\mathbf{y} = a_1\mathbf{x}_1 + a_2\mathbf{x}_2 + \cdots + a_n\mathbf{x}_n$, then

$$M_{\mathbf{y}}(t) = \prod_{i=1}^{n} M_{\mathbf{x}_i}(a_i t)$$

where $M_{\mathbf{x}_i}(t)$ is the value of the moment-generating function of $\mathbf{x}_i$ at t.

6. Use the result of Exercise 5 to show that if n independent random variables $\mathbf{x}_i$ have normal distributions with the means μ_i and the standard deviations σ_i, then $\mathbf{y} = a_1\mathbf{x}_1 + a_2\mathbf{x}_2 + \cdots + a_n\mathbf{x}_n$ has a normal distribution. What are the mean and the standard deviation of this distribution?

APPLIED EXERCISES

7. If the number of fish a person catches per hour at Woods Canyon Lake is a random variable having a Poisson distribution with $\lambda = 1.6$, use Table II at the end of the book to find the probabilities that a person fishing there will catch

(a) 4 fish in 2 hours;
(b) at least 2 fish in 3 hours;
(c) at most 2 fish in 4 hours.

8. If the number of incoming airplanes per minute at a large metropolitan airport is a random variable having a Poisson distribution with $\lambda = 0.9$, use Table II at the end of the book to find the probabilities that there will be

(a) exactly 9 incoming planes during a period of 5 minutes;
(b) fewer than 10 incoming planes during a period of 8 minutes;
(c) at least 14 incoming planes during a period of 11 minutes.

9. If the number of minutes it takes a service station attendant to balance a tire is a random variable having an exponential distribution with the parameter $\theta = 5$, what are the probabilities that the attendant will take

(a) less than 8 minutes to balance 2 tires;
(b) at least 12 minutes to balance 3 tires?

10. If the number of minutes a doctor spends with a patient is a random variable having an exponential distribution with the parameter $\theta = 9$, what are the probabilities that it will take the doctor at least 20 minutes to treat

(a) two patients;
(b) three patients?

References

The use of the probability integral transformation in problems of simulation is discussed in

FREUND, J. E., and MILLER, I., *Probability and Statistics for Engineers*, 2nd ed. Englewood Cliffs, N.J.: Prentice-Hall, Inc., 1977.

A generalization of Theorem 7.1, which applies when the interval within the range of **x** for which $f(x) \neq 0$ can be partitioned into k sub-intervals so that the conditions of Theorem 7.1 apply separately for each of the sub-intervals, may be found in

WALPOLE, R. E., and MYERS, R. H., *Probability and Statistics for Engineers and Scientists*, 2nd ed. New York: Macmillan Publishing Co., Inc., 1978.

More detailed and more advanced treatments of the material in this chapter are given in many advanced texts on mathematical statistics; for instance, in

HOGG, R. V., and CRAIG, A. T., *Introduction to Mathematical Statistics*, 4th ed. New York: Macmillan Publishing Co., Inc., 1978,

ROUSSAS, G. G., *A First Course in Mathematical Statistics*. Reading, Mass.: Addison-Wesley Publishing Co., 1973,

WILKS S. S., *Mathematical Statistics*. New York: John Wiley & Sons, Inc., 1962.

sampling distributions

8

8.1 INTRODUCTION

Statistics concerns itself mainly with conclusions and predictions resulting from chance outcomes that occur in carefully planned experiments or investigations. In the finite case, these chance outcomes constitute a subset, or **sample**, of measurements or observations from a larger set of values called the **population**. In the continuous case they are usually values of identically distributed random variables, whose distribution we refer to as the **population distribution**, or the **infinite population** sampled. The word "infinite" implies that there is, logically speaking, no limit to the number of random variables whose values we could observe.

EXAMPLE 8.1

If a scientist must weigh five of 40 guinea pigs, we could say that the ones she selects constitute a sample from this set, or "population," of guinea pigs. It would be preferable, though, to look upon the weights of the five guinea pigs as a sample from the population which consists of the weights of all 40 guinea pigs. In this way, the population as well as the sample will be numbers, as prescribed by the definitions we shall give below.

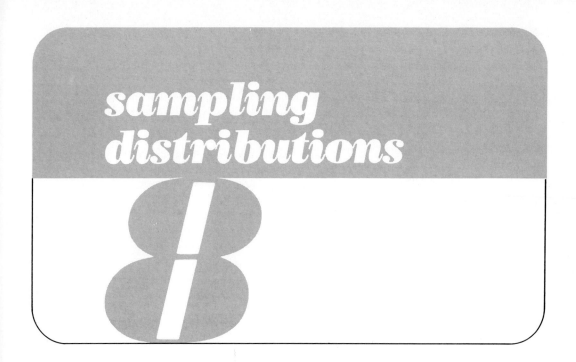

Also, to estimate the average useful life of a certain kind of transistor, an investigator selects a sample of ten of these transistors, tests them over a period of time, and records for each the time to failure. If these times to failure are values of random variables having the exponential distribution with the parameter θ, we say that the data constitute a sample from this exponential population.

As can well be imagined, not all samples lend themselves to valid generalizations about the populations from which they are obtained. In fact, most of the methods of inference discussed in this book are based on the assumption that we are dealing with **random samples**.

DEFINITION 8.1 If $\mathbf{x}_1, \mathbf{x}_2, \ldots,$ and $\mathbf{x}_n$ are independent and identically distributed random variables, we say that they constitute a **random sample** from the infinite population given by their common distribution.[†]

If $f(x_1, x_2, \ldots, x_n)$ is the value of the joint distribution of such a set of random variables at $(x_1, x_2, \ldots, x_n)$, we can write

$$f(x_1, x_2, \ldots, x_n) = \prod_{i=1}^{n} f(x_i)$$

where $f(x_i)$ is the value of the population distribution at x_i. Observe that Definition 8.1 applies also to sampling with replacement from finite populations; sampling without replacement from finite populations will be discussed in Section 8.3.

Statistical inferences are usually based on **statistics**, that is, on random variables which are functions of a set of random variables $\mathbf{x}_1, \mathbf{x}_2, \ldots,$ and $\mathbf{x}_n$, constituting a random sample. Typical of what we mean by "statistic" are the **sample mean** and the **sample variance**.

DEFINITION 8.2 If $\mathbf{x}_1, \mathbf{x}_2, \ldots,$ and $\mathbf{x}_n$ constitute a random sample, then

$$\bar{\mathbf{x}} = \frac{\sum_{i=1}^{n} \mathbf{x}_i}{n}$$

[†] In the past, it has been common practice to apply the term "random sample" to the values of the random variables instead of the random variables, themselves. Intuitively, this makes more sense, but it does not conform with current usage.

is called the **sample mean** and

$$\mathbf{s}^2 = \frac{\sum_{i=1}^{n} (\mathbf{x}_i - \bar{\mathbf{x}})^2}{n - 1}$$

is called the **sample variance.**†

As they are given here, these definitions apply only to random samples, but the sample mean and the sample variance can, similarly, be defined for any set of random variables $\mathbf{x}_1, \mathbf{x}_2, \ldots,$ and $\mathbf{x}_n$.

It is common practice to apply the terms "statistic," "sample mean," and "sample variance" also to values of the corresponding random variables. For instance, to compute the mean and the variance of a set of observed data, we substitute into the formulas

$$\bar{x} = \frac{\sum_{i=1}^{n} x_i}{n} \quad \text{and} \quad s^2 = \frac{\sum_{i=1}^{n} (x_i - \bar{x})^2}{n - 1}$$

where the x_i denote observed values of the corresponding random variables. Such values of $\bar{\mathbf{x}}$ and $\mathbf{s}^2$ are often used to estimate the mean μ and the variance σ^2 of the population from which the data were obtained. It must be understood, of course, that we have introduced $\bar{\mathbf{x}}$ and $\mathbf{s}^2$ as examples of statistics, and that there are many other statistics which can be used to estimate μ, σ^2, and other population parameters.

Since statistics are random variables, their values will vary from sample to sample, and it is customary to refer to their distributions as **sampling distributions**. Most of the remainder of this chapter will be devoted to the sampling distributions of statistics which play important roles in applications.

8.2 THE DISTRIBUTION OF THE MEAN

First let us study some theory about the sampling distribution of the mean, making only some very general assumptions about the nature of the population sampled.

† The reason for dividing by $n - 1$ rather than the seemingly more logical choice, n, will be explained in Section 10.3.

THEOREM 8.1 If $\mathbf{x}_1, \mathbf{x}_2, \ldots,$ and $\mathbf{x}_n$ constitute a random sample from an infinite population which has the mean μ and the variance σ^2, then

$$E(\bar{\mathbf{x}}) = \mu \quad \text{and} \quad \text{var}(\bar{\mathbf{x}}) = \frac{\sigma^2}{n}$$

Proof. Letting $\mathbf{y} = \bar{\mathbf{x}}$ in Theorem 4.14 and, hence, setting $a_i = \dfrac{1}{n}$, we get

$$E(\bar{\mathbf{x}}) = \sum_{i=1}^{n} \frac{1}{n} \cdot \mu = n\left(\frac{1}{n} \cdot \mu\right) = \mu$$

since $E(\mathbf{x}_i) = \mu$. Then, by the Corollary of Theorem 4.14 on page 157, we conclude that

$$\text{var}(\bar{\mathbf{x}}) = \sum_{i=1}^{n} \frac{1}{n^2} \cdot \sigma^2 = n\left(\frac{1}{n^2} \cdot \sigma^2\right) = \frac{\sigma^2}{n}$$

It is customary to write $E(\bar{\mathbf{x}})$ as $\mu_{\bar{\mathbf{x}}}$, $\text{var}(\bar{\mathbf{x}})$ as $\sigma^2_{\bar{\mathbf{x}}}$, and refer to $\sigma_{\bar{\mathbf{x}}}$ as the **standard error of the mean**. The formula for the standard error of the mean, $\sigma_{\bar{\mathbf{x}}} = \dfrac{\sigma}{\sqrt{n}}$, shows that the standard deviation of the distribution of $\bar{\mathbf{x}}$ decreases when n, the **sample size**, is increased. This means that when n becomes larger and we actually have more information (the values of more random variables), we can expect values of $\bar{\mathbf{x}}$ to be closer to μ, the quantity they are intended to estimate. If we refer to Chebyshev's theorem as it is formulated in Exercise 14 on page 149, we can express this more formally in the following way: *For any positive constant c, the probability that* $\bar{\mathbf{x}}$ *will take on a value between* $\mu - c$ *and* $\mu + c$ *is at least* $1 - \dfrac{\sigma^2}{nc^2}$; *when* $n \to \infty$, *this probability approaches 1.*

The result of the preceding paragraph, based on Chebyshev's theorem, is primarily of theoretical interest. Of much more practical value is the **central limit theorem**, one of the most important theorems of statistics, which concerns the limiting distribution of the **standardized mean** of n random variables when $n \to \infty$. We shall prove this theorem here only for the case where the n random variables are a random sample from a population whose moment-generating function exists. More general conditions under which the theorem holds are given in Exercises 8 and 9 on page 263, and the most general conditions under which it holds are referred to at the end of this chapter.

THEOREM 8.2 (Central limit theorem) If $\mathbf{x}_1, \mathbf{x}_2, \ldots,$ and $\mathbf{x}_n$ constitute a random sample from an infinite population having the mean μ, the variance σ^2, and the moment generating function $M_{\mathbf{x}}(t)$, then the limiting distribution of

$$\mathbf{z} = \frac{\bar{\mathbf{x}} - \mu}{\sigma/\sqrt{n}}$$

as $n \to \infty$, is the standard normal distribution.

Proof. First using part 3 of Theorem 4.10 and then part 2, we get

$$M_{\mathbf{z}}(t) = M_{\frac{\bar{\mathbf{x}}-\mu}{\sigma/\sqrt{n}}}(t) = e^{-\sqrt{n}\mu t/\sigma} \cdot M_{\bar{\mathbf{x}}}\left(\frac{\sqrt{n}t}{\sigma}\right)$$

$$= e^{-\sqrt{n}\mu t/\sigma} \cdot M_{n\bar{\mathbf{x}}}\left(\frac{t}{\sigma\sqrt{n}}\right)$$

Since $n\bar{\mathbf{x}} = \mathbf{x}_1 + \mathbf{x}_2 + \cdots + \mathbf{x}_n$, it follows from Theorem 7.2 that

$$M_{\mathbf{z}}(t) = e^{-\sqrt{n}\mu t/\sigma} \cdot \left[M_{\mathbf{x}}\left(\frac{t}{\sigma\sqrt{n}}\right)\right]^n$$

and, hence, that

$$\ln M_{\mathbf{z}}(t) = -\frac{\sqrt{n}\mu t}{\sigma} + n \cdot \ln M_{\mathbf{x}}\left(\frac{t}{\sigma\sqrt{n}}\right)$$

Expanding $M_{\mathbf{x}}\left(\frac{t}{\sigma\sqrt{n}}\right)$ as a power series in t, we obtain

$$\ln M_{\mathbf{z}}(t) = -\frac{\sqrt{n}\mu t}{\sigma} + n \cdot \ln\left[1 + \mu_1'\frac{t}{\sigma\sqrt{n}} + \mu_2'\frac{t^2}{2\sigma^2 n} + \mu_3'\frac{t^3}{6\sigma^3 n\sqrt{n}} + \cdots\right]$$

where $\mu_1', \mu_2', \mu_3', \ldots,$ are the moments of the population distribution, namely, those of the distribution of the original random variables $\mathbf{x}_i$.

If n is sufficiently large, we can use the expansion of $\ln(1 + x)$ as a power series in x (as on page 212), getting

$$\ln M_z(t) = -\frac{\sqrt{n}\mu t}{\sigma} + n\left\{\left[\mu_1'\frac{t}{\sigma\sqrt{n}} + \mu_2'\frac{t^2}{2\sigma^2 n} + \mu_3'\frac{t^3}{6\sigma^3 n\sqrt{n}} + \cdots\right] - \frac{1}{2}\left[\mu_1'\frac{t}{\sigma\sqrt{n}} + \mu_2'\frac{t^2}{2\sigma^2 n} + \mu_3'\frac{t^3}{6\sigma^3 n\sqrt{n}} + \cdots\right]^2 + \frac{1}{3}\left[\mu_1'\frac{t}{\sigma\sqrt{n}} + \mu_2'\frac{t^2}{2\sigma^2 n} + \mu_3'\frac{t^3}{6\sigma^3 n\sqrt{n}} + \cdots\right]^3 - \cdots\right\}$$

Then, collecting powers of t, we obtain

$$\ln M_z(t) = \left(-\frac{\sqrt{n}\mu}{\sigma} + \frac{\sqrt{n}\mu_1'}{\sigma}\right)t + \left(\frac{\mu_2'}{2\sigma^2} - \frac{\mu_1'^2}{2\sigma^2}\right)t^2 + \left(\frac{\mu_3'}{6\sigma^3\sqrt{n}} - \frac{\mu_1' \cdot \mu_2'}{2\sigma^3\sqrt{n}} + \frac{\mu_1'^3}{3\sigma^3\sqrt{n}}\right)t^3 + \cdots$$

and since $\mu_1' = \mu$ and $\mu_2' - \mu_1'^2 = \sigma^2$, this reduces to

$$\ln M_z(t) = \frac{1}{2}t^2 + \left(\frac{\mu_3'}{6} - \frac{\mu_1'\mu_2'}{2} + \frac{\mu_1'^3}{3}\right)\frac{t^3}{\sigma^3\sqrt{n}} + \cdots$$

Finally, observing that the coefficient of t^3 is a constant times $\frac{1}{\sqrt{n}}$ and in general the coefficient of t^r is a constant times $\frac{1}{\sqrt{n^{r-2}}}$, we get

$$\lim_{n\to\infty} \ln M_z(t) = \tfrac{1}{2}t^2$$

and, hence,

$$\lim_{n\to\infty} M_z(t) = e^{\frac{1}{2}t^2}$$

since the limit of a logarithm equals the logarithm of the limit (provided these limits exist). Identifying the limiting moment-generating function at which we have arrived as that of the standard normal distribution, we need only the two theorems stated on page 214 to complete the proof of Theorem 8.2.

Sometimes, the central limit theorem is interpreted incorrectly as implying that the distribution of $\bar{\mathbf{x}}$ approaches a normal distribution when $n \to \infty$. This is incorrect because $\text{var}(\bar{\mathbf{x}}) \to 0$ when $n \to \infty$; on the other hand, the central limit theorem justifies the approximation of the distribution of $\bar{\mathbf{x}}$ with a normal distribution having the mean μ and the variance $\frac{\sigma^2}{n}$ when n is large. In practice, this approximation is used when $n \geqslant 30$ regardless of the shape of the population sampled. For smaller values of n the approximation is questionable, but see Theorem 8.3 below.

EXAMPLE 8.2

A soft-drink vending machine is set so that the amount of drink dispensed is a random variable with a mean of 200 milliliters and a standard deviation of 15 milliliters. What is the probability that the average (mean) amount dispensed in a random sample of size 36 is at least 204 milliliters?

Solution

According to Theorem 8.1, the distribution of $\bar{\mathbf{x}}$ has the mean $\mu_{\bar{\mathbf{x}}} = 200$ and the standard deviation $\sigma_{\bar{\mathbf{x}}} = \frac{15}{\sqrt{36}} = 2.5$, and according to the central limit theorem, this distribution is approximately normal. Since $z = \frac{204 - 200}{2.5} = 1.6$, it follows from Table III that $P(\bar{\mathbf{x}} \geqslant 204) = P(\mathbf{z} \geqslant 1.6) = 0.5000 - 0.4452 = 0.0548$.

It is of interest to note that when the population we are sampling is normal, the distribution of $\bar{\mathbf{x}}$ *is* a normal distribution regardless of the size of n.

THEOREM 8.3 If $\bar{\mathbf{x}}$ is the mean of a random sample of size n from a normal population with the mean μ and the variance σ^2, its sampling distribution is a normal distribution with the mean μ and the variance σ^2/n.

Proof. According to Theorems 4.10 and 7.2 we can write

$$M_{\bar{\mathbf{x}}}(t) = \left[M_{\mathbf{x}}\left(\frac{t}{n}\right)\right]^n$$

and since the moment-generating function of a normal distribution with the mean μ and the variance σ^2 is given by

$$M_{\mathbf{x}}(t) = e^{\mu t + \frac{1}{2}\sigma^2 t^2}$$

according to Theorem 6.6, we get

$$M_{\bar{\mathbf{x}}}(t) = [e^{\mu \cdot \frac{t}{n} + \frac{1}{2}\left(\frac{t}{n}\right)^2 \sigma^2}]^n$$

$$= e^{\mu t + \frac{1}{2} t^2 \left(\frac{\sigma^2}{n}\right)}$$

This moment-generating function is readily seen to be that of a normal distribution with the mean μ and the variance σ^2/n, and to complete the proof of Theorem 8.3 we have only to refer to the two theorems on page 214.

8.3 THE DISTRIBUTION OF THE MEAN: FINITE POPULATIONS

If an experiment consists of selecting one or more values from a finite set of numbers $\{c_1, c_2, \ldots, c_N\}$, this set is referred to as a **finite population of size N**. If the selection is without replacement and $\mathbf{x}_1$ is the first number drawn, $\mathbf{x}_2$ is the second number drawn, . . . , and $\mathbf{x}_n$ is the nth number drawn, these random variables form a random sample of size n from this finite population provided their joint probability distribution is given by

$$f(x_1, x_2, \ldots, x_n) = \frac{1}{N(N-1)\cdots(N-n+1)}$$

for each ordered n-tuple of values selected from the set $\{c_1, c_2, \ldots, c_N\}$. It follows that the probability for each subset of n of the N elements of the finite population (regardless of the order in which the values are obtained) is

$$\frac{n!}{N(N-1)\cdots(N-n+1)} = \frac{1}{\binom{N}{n}}$$

and this is often given as the requirement for the selection of a random sample of size n from a finite population of size N.

It also follows from the above joint distribution that the marginal distribution of $\mathbf{x}_r$ is

$$f(x_r) = \frac{1}{N} \qquad \text{for } x_r = c_1, c_2, \ldots, c_N$$

for $r = 1, 2, \ldots, n$, and we refer to its mean and its variance as the mean and the variance of the finite population.

DEFINITION 8.3 The **mean** and the **variance** of the finite population $\{c_1, c_2, \ldots, c_N\}$ are

$$\mu = \sum_{i=1}^{N} c_i \cdot \frac{1}{N} \quad \text{and} \quad \sigma^2 = \sum_{i=1}^{N} (c_i - \mu)^2 \cdot \frac{1}{N}$$

The joint marginal distribution of any two of the random variables $\mathbf{x}_1, \mathbf{x}_2, \ldots,$ and $\mathbf{x}_n$ is given by

$$g(x_r, x_s) = \frac{1}{N(N-1)}$$

for each ordered pair of values of the finite population, and it follows that their covariance is

$$\begin{aligned}\operatorname{cov}(\mathbf{x}_r, \mathbf{x}_s) &= \sum_{i=1}^{N} \sum_{\substack{j=1 \\ i \neq j}}^{N} \frac{1}{N(N-1)}(c_i - \mu)(c_j - \mu) \\ &= \frac{\left[\sum_{i=1}^{N} c_i\right]^2 - \sum_{i=1}^{N} c_i^2}{N(N-1)} - \mu^2 \\ &= -\frac{\sigma^2}{N-1}\end{aligned}$$

Making use of all these results, let us now prove the following theorem:

THEOREM 8.4 If $\bar{\mathbf{x}}$ is the mean of a random sample of size n from a finite population of size N with the mean μ and the variance σ^2, then

$$E(\bar{\mathbf{x}}) = \mu \quad \text{and} \quad \operatorname{var}(\bar{\mathbf{x}}) = \frac{\sigma^2}{n} \cdot \frac{N-n}{N-1}$$

Proof. Substituting $a_i = \frac{1}{N}$, $\operatorname{var}(\mathbf{x}_i) = \sigma^2$, and $\operatorname{cov}(\mathbf{x}_i, \mathbf{x}_j) = -\frac{\sigma^2}{N-1}$ into Theorem 4.14, we get

$$E(\bar{\mathbf{x}}) = \sum_{i=1}^{n} \frac{1}{n} \cdot \mu = \mu$$

and

$$\begin{aligned}\text{var}(\bar{\mathbf{x}}) &= \sum_{i=1}^{n} \frac{1}{n^2}\cdot\sigma^2 + 2\cdot\sum_{i<j}\sum \frac{1}{n^2}\left(-\frac{\sigma^2}{N-1}\right)\\ &= \frac{\sigma^2}{n} + 2\cdot\frac{n(n-1)}{2}\cdot\frac{1}{n^2}\left(-\frac{\sigma^2}{N-1}\right)\\ &= \frac{\sigma^2}{n}\cdot\frac{N-n}{N-1}\end{aligned}$$

It is of interest to note that the formulas we obtained for $\text{var}(\bar{\mathbf{x}})$ in Theorems 8.1 and 8.4 differ only by the **finite population correction factor** $\frac{N-n}{N-1}$. Indeed, when N is large compared to n, the difference between the two formulas for $\text{var}(\bar{\mathbf{x}})$ is generally negligible, and the formula $\sigma_{\bar{\mathbf{x}}} = \frac{\sigma}{\sqrt{n}}$ is often used as an approximation when we are sampling from a large finite population. A general rule of thumb is to use this approximation so long as the sample does not constitute more than 5 percent of the population.

THEORETICAL EXERCISES

1. Verify the following computing formula for the value of a sample variance:

$$s^2 = \frac{n\cdot\sum_{i=1}^{n} x_i^2 - \left(\sum_{i=1}^{n} x_i\right)^2}{n(n-1)}$$

2. Use the Corollary of Theorem 4.15 to show that if $\mathbf{x}_1, \mathbf{x}_2, \ldots,$ and $\mathbf{x}_n$ constitute a random sample from an infinite population, then

$$\text{cov}(\mathbf{x}_r - \bar{\mathbf{x}}, \bar{\mathbf{x}}) = 0$$

for $r = 1, 2, \ldots, n$.

3. Use Theorem 4.14 and its Corollary to show that if $\mathbf{x}_{11}, \mathbf{x}_{12}, \ldots, \mathbf{x}_{1n_1}, \mathbf{x}_{21}, \mathbf{x}_{22}, \ldots, \mathbf{x}_{2n_2}$ are independent random variables, with the first n_1 constituting a random sample from an infinite population with the mean μ_1 and the variance σ_1^2, and the other n_2 constituting a random sample from an infinite population with the mean μ_2 and the variance σ_2^2, then

$$E(\bar{\mathbf{x}}_1 - \bar{\mathbf{x}}_2) = \mu_1 - \mu_2$$

and

$$\text{var}(\bar{\mathbf{x}}_1 - \bar{\mathbf{x}}_2) = \frac{\sigma_1^2}{n_1} + \frac{\sigma_2^2}{n_2}$$

4. Show that if the two samples of Exercise 3 come from normal populations, then $\bar{\mathbf{x}}_1 - \bar{\mathbf{x}}_2$ has a normal distribution with the mean $\mu_1 - \mu_2$ and the variance $\frac{\sigma_1^2}{n_1} + \frac{\sigma_2^2}{n_2}$. (*Hint:* Proceed as in the proof of Theorem 8.3.)

5. If $\mathbf{x}_1, \mathbf{x}_2, \ldots,$ and $\mathbf{x}_n$ are independent random variables having identical Bernoulli distributions with the parameter θ, then $\bar{\mathbf{x}}$ is the proportion of successes in n trials, which we denote by $\hat{\boldsymbol{\theta}}$. Verify that $E(\hat{\boldsymbol{\theta}}) = \theta$ and $\text{var}(\hat{\boldsymbol{\theta}}) = \frac{\theta(1-\theta)}{n}$.

6. If the first n_1 random variables of Exercise 3 have Bernoulli distributions with the parameter θ_1 and the other n_2 random variables have Bernoulli distributions with the parameter θ_2, show that

$$E(\hat{\boldsymbol{\theta}}_1 - \hat{\boldsymbol{\theta}}_2) = \theta_1 - \theta_2$$

and

$$\text{var}(\hat{\boldsymbol{\theta}}_1 - \hat{\boldsymbol{\theta}}_2) = \frac{\theta_1(1-\theta_1)}{n_1} + \frac{\theta_2(1-\theta_2)}{n_2}$$

7. Looking at the binomial distribution as on page 250, use the central limit theorem to prove Theorem 6.8.

8. The following is a sufficient condition for the central limit theorem: *If the random variables* $\mathbf{x}_1, \mathbf{x}_2, \ldots,$ *and* $\mathbf{x}_n$ *are independent and uniformly bounded (that is, there is a positive constant k such that the probability that any one of the* $\mathbf{x}_i$ *takes on a value greater than k or less than* $-k$ *is* 0), *then if the variance of* $\mathbf{y}_n = \mathbf{x}_1 + \mathbf{x}_2 + \cdots + \mathbf{x}_n$ *becomes infinite when* $n \to \infty$, *the distribution of the standardized mean of the* $\mathbf{x}_i$ *approaches the standard normal distribution.* Show that this sufficient condition holds for a sequence of independent random variables $\mathbf{x}_i$ having the probability distributions

$$f_i(x_i) = \begin{cases} \frac{1}{2} & \text{for } x_i = 1 - (\frac{1}{2})^i \\ \frac{1}{2} & \text{for } x_i = (\frac{1}{2})^i - 1 \end{cases}$$

9. The following is a sufficient condition, the Laplace–Liapounoff condition, for the central limit theorem: *If* $\mathbf{x}_1, \mathbf{x}_2, \mathbf{x}_3, \ldots,$ *is a sequence of independent random variables, each having an absolute third moment*

$$c_i = E(|\mathbf{x}_i - \mu_i|^3)$$

and if

$$\lim_{n\to\infty} [\text{var}(\mathbf{y}_n)]^{-\frac{3}{2}} \sum_{i=1}^{n} c_i = 0$$

where $\mathbf{y}_n = \mathbf{x}_1 + \mathbf{x}_2 + \cdots + \mathbf{x}_n$, *then the distribution of the standardized mean of the* $\mathbf{x}_i$ *approaches the standard normal distribution when* $n \to \infty$. Use this condition to show that the central limit theorem holds for the sequence of random variables of Exercise 8.

10. If $\mathbf{x}_1, \mathbf{x}_2, \mathbf{x}_3, \ldots$, is a sequence of independent random variables having the uniform densities

$$f_i(x_i) = \begin{cases} \dfrac{1}{2 - \dfrac{1}{i}} & \text{for } 0 < x_i < 2 - \dfrac{1}{i} \\ 0 & \text{elsewhere} \end{cases}$$

show that the central limit theorem holds using

(a) the conditions of Exercise 8;

(b) the conditions of Exercise 9.

11. Explain why the results of Theorem 8.1 apply instead of those of Theorem 8.4 when we sample with replacement from a finite population.

12. Explain the results of Exercise 11 on page 186 in the light of Theorem 8.4.

13. If a random sample of size n is selected from the finite population which consists of the first N positive integers, show that

(a) the mean of the distribution of $\bar{\mathbf{x}}$ is $\dfrac{N+1}{2}$;

(b) the variance of the distribution of $\bar{\mathbf{x}}$ is $\dfrac{(N+1)(N-n)}{12n}$;

(c) the mean and the variance of the distribution of $\mathbf{y} = n \cdot \bar{\mathbf{x}}$ are

$$E(\mathbf{y}) = \frac{n(N+1)}{2} \quad \text{and} \quad \text{var}(\mathbf{y}) = \frac{n(N+1)(N-n)}{12}$$

(*Hint:* Refer to Appendix II or the results of Exercise 1 on page 171.)

APPLIED EXERCISES

14. A random sample of size $n = 100$ is taken from an infinite population with the mean $\mu = 75$ and the variance $\sigma^2 = 256$. If we use Chebyshev's theorem, with what probability can we assert that the value we obtain for $\bar{\mathbf{x}}$ will fall between 67 and 83?

15. Rework Exercise 14, using the central limit theorem instead of Chebyshev's theorem.

16. A random sample of size $n = 81$ is taken from an infinite population with the mean $\mu = 128$ and the standard deviation $\sigma = 6.3$. With what probability can we assert that the value we obtain for $\bar{\mathbf{x}}$ will not fall between 126.6 and 129.4, if we use

(a) Chebyshev's theorem;
(b) the central limit theorem.

17. Rework part (b) of Exercise 16, assuming that the population is not infinite, but finite and of size $N = 400$.

18. A random sample of size 64 is taken from a normal population with $\mu = 51.4$ and $\sigma = 6.8$. What is the probability that the mean of the sample will

(a) exceed 52.9;
(b) fall between 50.5 and 52.3;
(c) be less than 50.6?

19. A random sample of size 100 is taken from a normal population with $\sigma = 25$. What is the probability that the mean of the sample will differ from the mean of the population by 3 or more either way?

20. Independent random samples of size 400 are taken from each of two populations having equal means and the standard deviations $\sigma_1 = 20$ and $\sigma_2 = 30$. Using Chebyshev's theorem and the results of Exercise 3, what can we assert with a probability of at least 0.99 about the value we will get for $\bar{\mathbf{x}}_1 - \bar{\mathbf{x}}_2$? (By "independent" we mean that the samples satisfy the conditions of Exercise 3.)

21. Assuming that the two populations of Exercise 20 are normal, use the result of Exercise 4 to find k such that $P(-k < \bar{\mathbf{x}}_1 - \bar{\mathbf{x}}_2 < k) = 0.99$.

22. Independent random samples of size $n_1 = 30$ and $n_2 = 50$ are taken from two normal populations having the means $\mu_1 = 78$ and $\mu_2 = 75$, and the variances $\sigma_1^2 = 150$ and $\sigma_2^2 = 200$. Use the results of Exercise 4 to find the probability that the mean of the first sample will exceed that of the second sample by at least 4.8.

23. The actual proportion of families in a certain city who own, rather than rent, their home is 0.70. If 84 families in this city are interviewed at random and their responses to the question whether they own or rent their home are looked upon as values of independent random variables having identical Bernoulli distributions with the parameter $\theta = 0.70$, with what probability can we assert that the value we obtain for the sample proportion, $\hat{\boldsymbol{\theta}}$, will fall between 0.64 and 0.76, using the result of Exercise 5 and

(a) Chebyshev's theorem;
(b) the central limit theorem?

24. The actual proportion of men who approve a certain tax proposal is 0.40 and the corresponding proportion for women is 0.25. If $n_1 = 500$ men and

$n_2 = 400$ women are interviewed at random and their individual responses are looked upon as the values of independent random variables having Bernoulli distributions with the respective parameters $\theta_1 = 0.40$ and $\theta_2 = 0.25$, what can we assert, according to Chebyshev's theorem, with a probability of at least 0.9975 about the value we will get for $\hat{\boldsymbol{\theta}}_1 - \hat{\boldsymbol{\theta}}_2$, the difference between the two sample proportions of favorable responses? Use the result of Exercise 6.

25. Find the values of the finite population correction factor for

(a) $n = 5$ and $N = 200$;
(b) $n = 50$ and $N = 300$;
(c) $n = 200$ and $N = 800$.

8.4 THE CHI-SQUARE DISTRIBUTION

In Example 7.9 we showed that if $\mathbf{x}$ has the standard normal distribution, then $\mathbf{x}^2$ has the special gamma distribution which we referred to as the chi-square distribution, and this accounts for the important role which the chi-square distribution plays in problems of sampling from normal populations. To review some of the results of Section 6.3, a random variable $\mathbf{x}$ has the chi-square distribution (also written χ^2 distribution) with ν degrees of freedom, if its density is given by

$$f(x) = \begin{cases} \dfrac{1}{2^{\nu/2}\Gamma(\nu/2)} x^{\frac{\nu-2}{2}} e^{-x/2} & \text{for } x > 0 \\ 0 & \text{elsewhere} \end{cases}$$

The mean and the variance of the chi-square distribution with ν degrees of freedom are ν and 2ν, and its moment-generating function is given by

$$M_{\mathbf{x}}(t) = (1 - 2t)^{-\nu/2}$$

The chi-square distribution has several important mathematical properties, which are given in Theorems 8.5 through 8.8. First, let us formally state the result of Example 7.9, which we referred to above.

THEOREM 8.5 If $\mathbf{x}$ has the standard normal distribution, then $\mathbf{x}^2$ has the chi-square distribution with $\nu = 1$ degree of freedom.

More generally, let us show that

THEOREM 8.6 If $\mathbf{x}_1, \mathbf{x}_2, \ldots,$ and $\mathbf{x}_n$ are independent random variables having standard normal distributions, then

$$\mathbf{y} = \sum_{i=1}^{n} \mathbf{x}_i^2$$

has the chi-square distribution with $\nu = n$ degrees of freedom.

Proof. By Theorem 8.5,

$$M_{\mathbf{x}_i^2}(t) = (1 - 2t)^{-\frac{1}{2}}$$

so that, by Theorem 7.2, it follows that

$$M_{\mathbf{y}}(t) = \prod_{i=1}^{n} (1 - 2t)^{-\frac{1}{2}} = (1 - 2t)^{-\frac{n}{2}}$$

This moment-generating function can be identified by inspection as that of the chi-square distribution with $\nu = n$ degrees of freedom.

Two further properties of the chi-square distribution are given in the following two theorems, which the reader will be asked to prove in Exercises 1 and 2 on page 276:

THEOREM 8.7 If $\mathbf{x}_1, \mathbf{x}_2, \ldots,$ and $\mathbf{x}_n$ are independent random variables having chi-square distributions with $\nu_1, \nu_2, \ldots,$ and ν_n degrees of freedom, then

$$\mathbf{y} = \sum_{i=1}^{n} \mathbf{x}_i$$

has the chi-square distribution with $\nu_1 + \nu_2 + \cdots + \nu_n$ degrees of freedom.

THEOREM 8.8 If $\mathbf{x}_1$ and $\mathbf{x}_2$ are independent random variables, $\mathbf{x}_1$ has a chi-square distribution with ν_1 degrees of freedom, and $\mathbf{x}_1 + \mathbf{x}_2$ has a chi-square distribution with $\nu > \nu_1$ degrees of freedom, then $\mathbf{x}_2$ has a chi-square distribution with $\nu - \nu_1$ degrees of freedom.

The chi-square distribution has many important applications, some of which will be discussed in Chapters 10 through 13. Foremost, there are those based, directly or indirectly, on the following theorem:

THEOREM 8.9 If $\bar{\mathbf{x}}$ and $\mathbf{s}^2$ are the mean and the variance of a random sample of size n from a normal population with the mean μ and the variance σ^2, then

1. $\bar{\mathbf{x}}$ and $\mathbf{s}^2$ are independent;
2. the random variable $\frac{(n-1)\mathbf{s}^2}{\sigma^2}$ has a chi-square distribution with $n-1$ degrees of freedom.

Proof. Since a detailed proof of part 1 would go beyond the scope of this text, we shall assume the independence of $\bar{\mathbf{x}}$ and $\mathbf{s}^2$ in our proof of part 2. (In addition to the references to proofs of part 1 on page 285, Exercise 8 on page 277outlines the major steps of a somewhat simpler proof based on the idea of a conditional moment-generating function, and in Exercise 7 on page 277 the reader will be asked to prove this independence for the special case where $n = 2$.)

To prove part 2, let us begin with the identity

$$\sum_{i=1}^{n} (\mathbf{x}_i - \mu)^2 = \sum_{i=1}^{n} (\mathbf{x}_i - \bar{\mathbf{x}})^2 + n(\bar{\mathbf{x}} - \mu)^2$$

which the reader will be asked to verify in Exercise 3 on page 276. Now, if we divide each term by σ^2 and substitute $(n-1)\mathbf{s}^2$ for $\sum_{i=1}^{n} (\mathbf{x}_i - \bar{\mathbf{x}})^2$, it follows that

$$\sum_{i=1}^{n} \left(\frac{\mathbf{x}_i - \mu}{\sigma}\right)^2 = \frac{(n-1)\mathbf{s}^2}{\sigma^2} + \left(\frac{\bar{\mathbf{x}} - \mu}{\sigma/\sqrt{n}}\right)^2$$

So far as the three terms of this identity are concerned, we know from Theorem 8.6 that $\sum_{i=1}^{n} \left(\frac{\mathbf{x}_i - \mu}{\sigma}\right)^2$ is a random variable having a chi-square distribution with n degrees of freedom. Also, according to Theorems 8.3 and 8.5, $\left(\frac{\bar{\mathbf{x}} - \mu}{\sigma/\sqrt{n}}\right)^2$ has a chi-square distribution with 1 degree of freedom. Now, since $\bar{\mathbf{x}}$ and $\mathbf{s}^2$ are independent, it follows that $\left(\frac{\bar{\mathbf{x}} - \mu}{\sigma/\sqrt{n}}\right)^2$ and $\frac{(n-1)\mathbf{s}^2}{\sigma^2}$ are

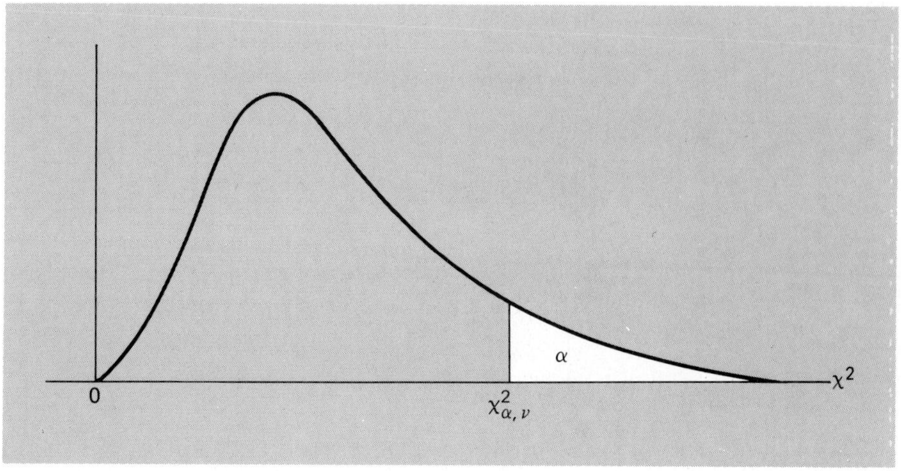

Figure 8.1 Chi-square distribution.

independent and, hence, we conclude by Theorem 8.8 that $\frac{(n-1)\mathbf{s}^2}{\sigma^2}$ has a chi-square distribution with $n - 1$ degrees of freedom.

Since the chi-square distribution arises in many important applications, integrals of its density have been extensively tabulated. Table V at the end of this book contains values of $\chi^2_{\alpha,\nu}$ for $\alpha = 0.995, 0.99, 0.975, 0.95, 0.05, 0.025, 0.01, 0.005$, and $\nu = 1, 2, \ldots, 30$, where $\chi^2_{\alpha,\nu}$ is such that the area to its right under the chi-square curve with ν degrees of freedom (see Figure 8.1) is equal to α. That is, $\chi^2_{\alpha,\nu}$ is such that

$$P(\boldsymbol{\chi}^2 \geqslant \chi^2_{\alpha,\nu}) = \alpha$$

When ν exceeds 30 so that Table V cannot be used, probabilities related to chi-square distributions are usually approximated with the use of normal distributions (see Exercises 5 and 6 on page 276).

EXAMPLE 8.3

Suppose that the thickness of a part used in a semiconductor is its critical dimension, and that the process of manufacturing these parts is considered to be under control if the true variation among the thicknesses of the parts is given by a standard deviation not greater than $\sigma = 0.60$ thousandth of an inch. To keep a check on the process, random samples of size $n = 20$ are taken periodically, and it is regarded to be "out of control" if the probability that $\mathbf{s}^2$ will take on a value greater than or equal to the observed sample value is 0.01 or less (even though $\sigma = 0.60$). What can one conclude about the process if the standard deviation of such a periodic random sample is $s = 0.84$ thousandth of an inch?

Solution

The process will be declared "out of control" if $\frac{(n-1)s^2}{\sigma^2}$ with $n = 20$ and $\sigma = 0.60$ exceeds $\chi^2_{.01,19} = 36.191$. Since

$$\frac{(n-1)s^2}{\sigma^2} = \frac{19(0.84)^2}{(0.60)^2} = 37.24$$

this is the case in our example. Of course, it is assumed in this analysis that the sample may be regarded as a random sample from a normal population.

8.5 THE *t* DISTRIBUTION

In Theorem 8.3 we showed that for random samples from a normal population with the mean μ and the variance σ^2, $\bar{\mathbf{x}}$ has a normal distribution with the mean μ and the variance $\frac{\sigma^2}{n}$; that is, $\frac{\bar{\mathbf{x}} - \mu}{\sigma/\sqrt{n}}$ has the standard normal distribution. The major difficulty in applying this result is that in actual practice σ is usually unknown, which makes it necessary to replace it with a value of the sample standard deviation $\mathbf{s}$ (or some other estimate). The theory which follows leads to the exact distribution of $\frac{\bar{\mathbf{x}} - \mu}{\mathbf{s}/\sqrt{n}}$ for random samples from normal populations.

To derive this sampling distribution, let us first study the more general problem stated in the following theorem:

THEOREM 8.10 If $\mathbf{y}$ and $\mathbf{z}$ are independent random variables, $\mathbf{y}$ has a chi-square distribution with ν degrees of freedom, and $\mathbf{z}$ has the standard normal distribution, then the distribution of

$$\mathbf{t} = \frac{\mathbf{z}}{\sqrt{\mathbf{y}/\nu}}$$

is given by

$$f(t) = \frac{\Gamma\left(\frac{\nu+1}{2}\right)}{\sqrt{\pi\nu}\,\Gamma\left(\frac{\nu}{2}\right)} \cdot \left(1 + \frac{t^2}{\nu}\right)^{-\frac{\nu+1}{2}} \qquad \text{for } -\infty < t < \infty$$

and it is called the ***t* distribution** with ν degrees of freedom.

Proof. Since **y** and **z** are independent, their joint density is given by

$$f(y, z) = \frac{1}{\sqrt{2\pi}} e^{-\frac{1}{2}z^2} \cdot \frac{1}{\Gamma\left(\frac{\nu}{2}\right)2^{\frac{\nu}{2}}} y^{\frac{\nu}{2}-1} e^{-\frac{y}{2}}$$

for $y > 0$ and $-\infty < z < \infty$, and $f(y, z) = 0$ elsewhere. Then, to use the change of variable technique of Section 7.3, we solve $t = \frac{z}{\sqrt{y/\nu}}$ for z getting $z = t\sqrt{y/\nu}$ and, hence, $\frac{\partial z}{\partial t} = \sqrt{y/\nu}$. Thus, by Theorem 7.1, the joint density of **y** and **t** is given by

$$g(y, t) = \begin{cases} \dfrac{1}{\sqrt{2\pi}\Gamma\left(\frac{\nu}{2}\right)2^{\frac{\nu}{2}}} y^{\frac{\nu-1}{2}} e^{-\frac{y}{2}\left(1+\frac{t^2}{\nu}\right)} & \text{for } y > 0 \text{ and } -\infty < t < \infty \\ 0 & \text{elsewhere} \end{cases}$$

and, integrating out y with the aid of the substitution $w = \frac{y}{2}\left(1 + \frac{t^2}{\nu}\right)$, we finally get

$$f(t) = \frac{\Gamma\left(\frac{\nu+1}{2}\right)}{\sqrt{\pi\nu}\,\Gamma\left(\frac{\nu}{2}\right)} \cdot \left(1 + \frac{t^2}{\nu}\right)^{-\frac{\nu+1}{2}} \qquad \text{for } -\infty < t < \infty$$

The t distribution was first obtained by W. S. Gosset, who published his research under the pen name "Student"; hence, the distribution is also known as the **Student-*t* distribution**, or **Student's-*t* distribution**.

In view of its importance, the t distribution has been tabulated extensively. Table IV, for example, contains values of $t_{\alpha,\nu}$ for $\alpha = 0.10, 0.05, 0.025, 0.01, 0.005$, and $\nu = 1, 2, \ldots, 29$, where $t_{\alpha,\nu}$ is such that the area to its right under the curve of the t distribution with ν degrees of freedom (see Figure 8.2) is equal to α. That is, $t_{\alpha,\nu}$ is such that

$$P(\mathbf{t} \geqslant t_{\alpha,\nu}) = \alpha$$

The table does not contain values of $t_{\alpha,\nu}$ for $\alpha > 0.50$, since the density is symmetrical about $t = 0$ and, hence, $t_{1-\alpha,\nu} = -t_{\alpha,\nu}$. When ν is 30 or more, probabilities related to the t distribution are usually approximated with the use of normal distributions (see Exercise 11 on page 278).

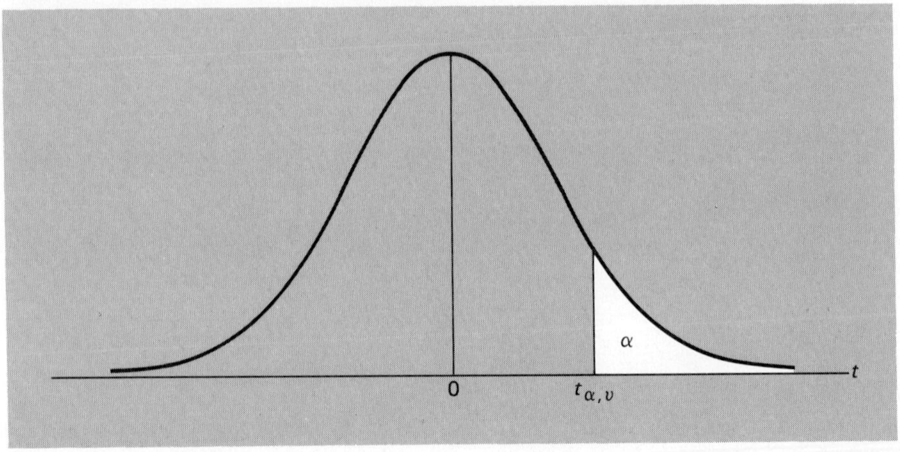

Figure 8.2 *t* distribution.

Among the many applications of the t distribution, some of which will be treated in Chapters 11 and 13, its major application (for which it was originally developed) is based on the following theorem:

THEOREM 8.11 If $\bar{\mathbf{x}}$ and $\mathbf{s}^2$ are the mean and the variance of a random sample of size n from a normal population with the mean μ and the variance σ^2, then

$$\mathbf{t} = \frac{\bar{\mathbf{x}} - \mu}{\mathbf{s}/\sqrt{n}}$$

has the t distribution with $n - 1$ degrees of freedom.

Proof. By Theorems 8.9 and 8.3, the random variables

$$\mathbf{y} = \frac{(n-1)\mathbf{s}^2}{\sigma^2} \quad \text{and} \quad \mathbf{z} = \frac{\bar{\mathbf{x}} - \mu}{\sigma/\sqrt{n}}$$

have, respectively, a chi-square distribution with $n - 1$ degrees of freedom and the standard normal distribution. Since they are, furthermore, independent by part 1 of Theorem 8.9, substitution into the formula for $\mathbf{t}$ of Theorem 8.10 yields

$$\mathbf{t} = \frac{\dfrac{\bar{\mathbf{x}} - \mu}{\sigma/\sqrt{n}}}{\sqrt{\mathbf{s}^2/\sigma^2}} = \frac{\bar{\mathbf{x}} - \mu}{\mathbf{s}/\sqrt{n}}$$

and this completes the proof.

EXAMPLE 8.4

In 16 one-hour test runs, the gasoline consumption of an engine averaged 16.4 gallons with a standard deviation of 2.1 gallons. Test the claim that the average gasoline consumption of this engine is 12.0 gallons per hour.

Solution

Substituting $n = 16$, $\mu = 12.0$, $\bar{x} = 16.4$, and $s = 2.1$ into the formula for t in Theorem 8.11, we get

$$t = \frac{\bar{x} - \mu}{s/\sqrt{n}} = \frac{16.4 - 12.0}{2.1/\sqrt{16}} = 8.38$$

Since Table IV shows that the probability of getting a value of t greater than 2.947 is 0.005 for 15 degrees of freedom, the probability of getting a value greater than 8 must be negligible. Thus, it would seem reasonable to conclude that the true average hourly gasoline consumption of the engine exceeds 12.0 gallons.

8.6 THE *F* DISTRIBUTION

Another distribution which plays an important role in connection with sampling from normal populations is the ***F* distribution**. We shall define this distribution as the sampling distribution of the ratio of two independent chi-square random variables, each divided by its respective degrees of freedom.

THEOREM 8.12 If $\mathbf{u}$ and $\mathbf{v}$ are independent random variables having chi-square distributions with ν_1 and ν_2 degrees of freedom, then the distribution of

$$\mathbf{F} = \frac{\mathbf{u}/\nu_1}{\mathbf{v}/\nu_2}$$

is given by

$$g(F) = \begin{cases} \dfrac{\Gamma\left(\dfrac{\nu_1 + \nu_2}{2}\right)}{\Gamma\left(\dfrac{\nu_1}{2}\right)\Gamma\left(\dfrac{\nu_2}{2}\right)} \left(\dfrac{\nu_1}{\nu_2}\right)^{\frac{\nu_1}{2}} \cdot F^{\frac{\nu_1}{2}-1} \left(1 + \dfrac{\nu_1}{\nu_2} \cdot F\right)^{-\frac{1}{2}(\nu_1+\nu_2)} & \text{for } F > 0 \\ 0 & \text{elsewhere} \end{cases}$$

and it is called the ***F* distribution** with ν_1 and ν_2 degrees of freedom.

Proof. The joint density of **u** and **v** is given by

$$f(u, v) = \frac{1}{2^{\nu_1/2}\Gamma\left(\frac{\nu_1}{2}\right)} \cdot u^{\frac{\nu_1}{2}-1} e^{-\frac{u}{2}} \cdot \frac{1}{2^{\nu_2/2}\Gamma\left(\frac{\nu_2}{2}\right)} \cdot v^{\frac{\nu_2}{2}-1} e^{-\frac{v}{2}}$$

$$= \frac{1}{2^{(\nu_1+\nu_2)/2}\Gamma\left(\frac{\nu_1}{2}\right)\Gamma\left(\frac{\nu_2}{2}\right)} \cdot u^{\frac{\nu_1}{2}-1} v^{\frac{\nu_2}{2}-1} e^{-\frac{u+v}{2}}$$

for $u > 0$ and $v > 0$, and $f(u, v) = 0$ elsewhere. Then, to use the change of variable technique of Section 7.3, we solve $F = \frac{u/\nu_1}{v/\nu_2}$ for u getting $u = \frac{\nu_1}{\nu_2} \cdot vF$ and, hence, $\frac{\partial u}{\partial F} = \frac{\nu_1}{\nu_2} \cdot v$. Thus, by Theorem 7.1, the joint density of **F** and **v** is given by

$$g(F, v) = \frac{\left(\frac{\nu_1}{\nu_2}\right)^{\nu_1/2}}{2^{(\nu_1+\nu_2)/2}\Gamma\left(\frac{\nu_1}{2}\right)\Gamma\left(\frac{\nu_2}{2}\right)} \cdot F^{\frac{\nu_1}{2}-1} v^{\frac{\nu_1+\nu_2}{2}-1} e^{-\frac{v}{2}\left(\frac{\nu_1 F}{\nu_2}+1\right)}$$

for $F > 0$ and $v > 0$, and $g(F, v) = 0$ elsewhere. Now, integrating out v by making the substitution $w = \frac{v}{2}\left(\frac{\nu_1 F}{\nu_2} + 1\right)$, we finally get

$$g(F) = \begin{cases} \dfrac{\Gamma\left(\frac{\nu_1+\nu_2}{2}\right)}{\Gamma\left(\frac{\nu_1}{2}\right)\Gamma\left(\frac{\nu_2}{2}\right)} \left(\frac{\nu_1}{\nu_2}\right)^{\frac{\nu_1}{2}} \cdot F^{\frac{\nu_1}{2}-1}\left(1 + \frac{\nu_1}{\nu_2}F\right)^{-\frac{1}{2}(\nu_1+\nu_2)} & \text{for } F > 0 \\ 0 & \text{elsewhere} \end{cases}$$

In view of its importance, the F distribution has been tabulated extensively. Table VI, for example, contains values of F_{α,ν_1,ν_2} for $\alpha = 0.05$ and 0.01, and for various values of ν_1 and ν_2, where F_{α,ν_1,ν_2} is such that the area to its right under the curve of the F distribution with ν_1 and ν_2 degrees of freedom (see Figure 8.3) is equal to α. That is, F_{α,ν_1,ν_2} is such that

$$P(\mathbf{F} \geqslant F_{\alpha,\nu_1,\nu_2}) = \alpha$$

Important applications of Theorem 8.12 arise in problems in which we are interested in comparing the variances σ_1^2 and σ_2^2 of two normal populations; for

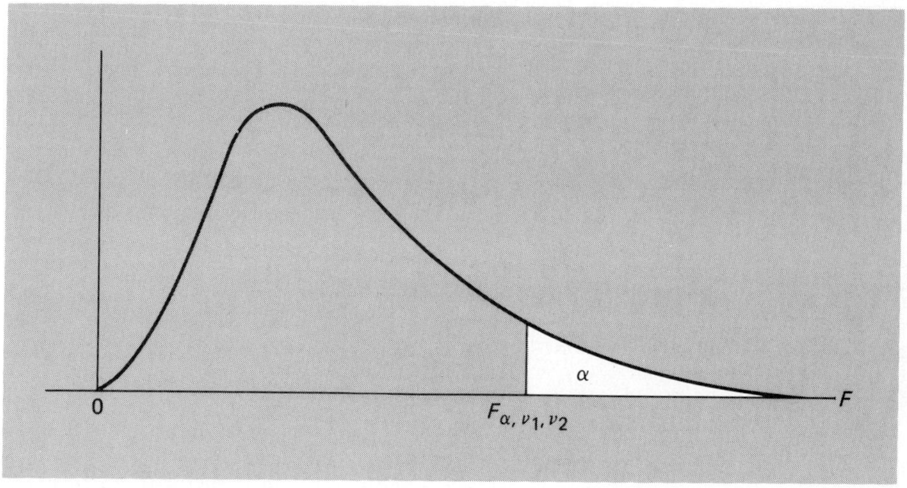

Figure 8.3 *F* distribution.

instance, in problems in which we want to estimate the ratio $\frac{\sigma_1^2}{\sigma_2^2}$, or perhaps test whether $\sigma_1^2 = \sigma_2^2$. We base such inferences on independent random samples of size n_1 and n_2 from the two populations and Theorem 8.9, according to which

$$\chi_1^2 = \frac{(n_1 - 1)\mathbf{s}_1^2}{\sigma_1^2} \quad \text{and} \quad \chi_2^2 = \frac{(n_2 - 1)\mathbf{s}_2^2}{\sigma_2^2}$$

are random variables having chi-square distributions with $n_1 - 1$ and $n_2 - 1$ degrees of freedom. By "independent random samples" we mean that the $n_1 + n_2$ random variables constituting the two random samples are all independent, so that χ_1^2 and χ_2^2 are also independent and their substitution into Theorem 8.12 gives the following result:

THEOREM 8.13 If $\mathbf{s}_1^2$ and $\mathbf{s}_2^2$ are the variances of independent random samples of size n_1 and n_2 from normal populations with the variances σ_1^2 and σ_2^2, then

$$\mathbf{F} = \frac{\mathbf{s}_1^2/\sigma_1^2}{\mathbf{s}_2^2/\sigma_2^2} = \frac{\sigma_2^2\mathbf{s}_1^2}{\sigma_1^2\mathbf{s}_2^2}$$

has an F distribution with $n_1 - 1$ and $n_2 - 1$ degrees of freedom.

In Chapter 11 we shall apply Theorem 8.13 to the problem of estimating the ratio $\frac{\sigma_1^2}{\sigma_2^2}$ when these two population variances are unknown; also, in Chapter 13 we shall demonstrate how to test whether $\sigma_1^2 = \sigma_2^2$. Still other tests based on the F distribution are presented in the analysis-of-variance procedures of Chapter 15. Since all these applications are based on the ratios of sample variances, the F distribution is also known as the **variance-ratio distribution**.

THEORETICAL EXERCISES

1. Prove Theorem 8.7.
2. Prove Theorem 8.8.
3. Verify the identity

$$\sum_{i=1}^{n} (\mathbf{x}_i - \mu)^2 = \sum_{i=1}^{n} (\mathbf{x}_i - \bar{\mathbf{x}})^2 + n(\bar{\mathbf{x}} - \mu)^2$$

which we used in the proof of Theorem 8.9.

4. Use Theorem 8.9 to show that for random samples of size n from a normal population with the variance σ^2, the sampling distribution of $\mathbf{s}^2$ has the mean σ^2 and the variance $\frac{2\sigma^4}{n-1}$. (A general formula for the variance of the sampling distribution of $\mathbf{s}^2$ for random samples from any population having finite second and fourth moments may be found in the book by H. Cramér listed on page 285.)
5. If the range of $\mathbf{x}$ is the set of all positive real numbers, show that
   (a) the probability that the random variable $\sqrt{2\mathbf{x}} - \sqrt{2n}$ takes on a value less than k equals the probability that the random variable $\frac{\mathbf{x} - n}{\sqrt{2n}}$ takes on a value less than $k + \frac{k^2}{2\sqrt{2n}}$;
   (b) if $\mathbf{x}$ has a chi-square distribution with n degrees of freedom, then for large n the distribution of $\sqrt{2\mathbf{x}} - \sqrt{2n}$ can be approximated with the standard normal distribution.

   Also use the result of part (b) and Theorem 8.9 to show that for large n the variance of the sampling distribution of $\mathbf{s}$ is approximately $\frac{\sigma^2}{2(n-1)}$.
6. Find approximate values for the probability that a random variable $\mathbf{x}$ having a chi-square distribution with 50 degrees of freedom will take on a value greater than 68.0
   (a) by treating $\frac{\mathbf{x} - \nu}{\sqrt{2\nu}}$ with $\nu = 50$ as a random variable having the standard normal distribution;

of strategies for which the corresponding entry in the payoff matrix is the smallest value of its row and the greatest value of its column. In Example 9.5 there is no saddle point since the smallest value of each row is also the smallest value of its column. On the other hand, in the game of Example 9.3 there is a saddle point corresponding to Strategies I and 2 since 8, the smallest value of the second row, is the greatest value of the first column. Also, the 3×2 game of Example 9.4 has a saddle point corresponding to Strategies II and 2 since 3, the smallest value of the second row, is the greatest value of the second column, and the 3×3 game on page 291 has a saddle point corresponding to Strategies I and 2 since 2, the smallest value of the second row, is the greatest value of the first column. In general, if a game has a saddle point it is said to be **strictly determined**, and the strategies corresponding to the saddle point are spyproof (and, hence, optimum) minimax strategies. The fact that there can be more than one saddle point in a game is illustrated in Exercise 1 on page 295; it also follows from that exercise that it does not matter in that case which of the saddle points is used to determine the optimum strategies of the two players.

If a game does not have a saddle point, minimax strategies are not spyproof, and each player can outsmart the other if he knows how his opponent will react in a given situation. To avoid this possibility, it suggests itself that each player should somehow mix up his behavior patterns intentionally, and the best way of doing this is by introducing an element of chance into the selection of his strategy.

EXAMPLE 9.6

With reference to the game of Example 9.5, suppose that Player A uses a gambling device (dice, cards, numbered slips of paper, a table of random numbers) which leads to the choice of Strategy I with probability x, and to the choice of Strategy II with probability $1 - x$. Find the value of x which will minimize Player A's maximum expected loss.

Solution

If Player B chooses Strategy 1, Player A can expect to lose

$$E = 8x - 5(1 - x)$$

dollars, and if Player B chooses Strategy 2, Player A can expect to lose

$$E = 2x + 6(1 - x)$$

dollars. Graphically, this situation is described in Figure 9.1, where we have plotted the lines whose equations are $E = 8x - 5(1 - x)$ and $E = 2x + 6(1 - x)$ for values of x from 0 to 1.

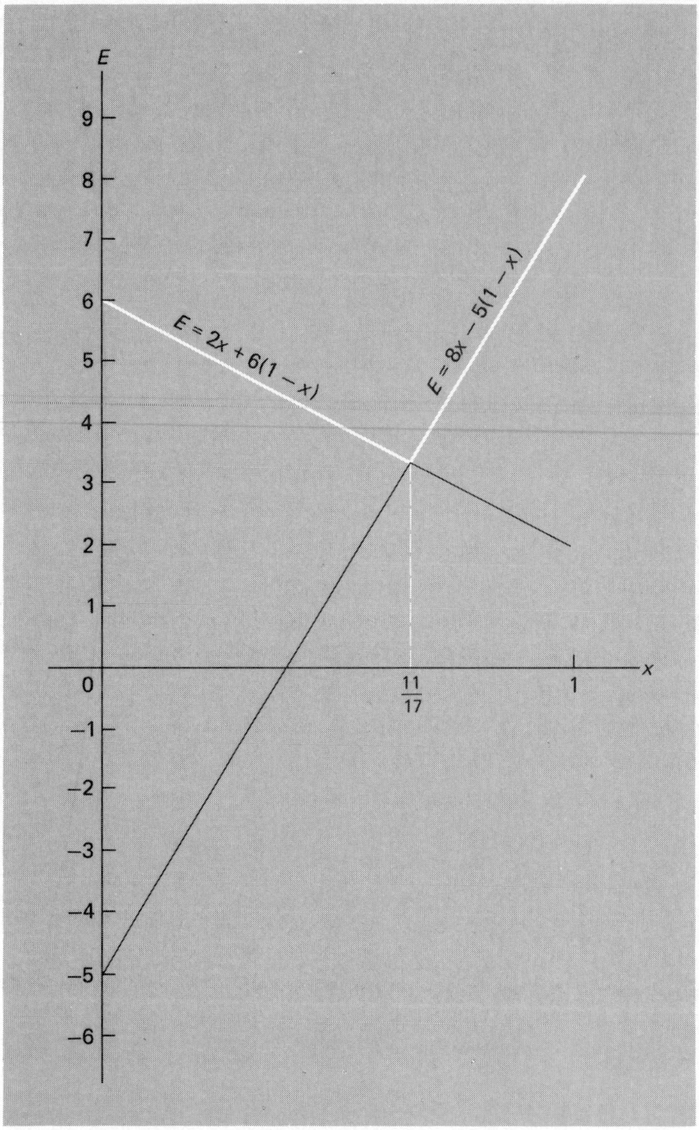

Figure 9.1 Diagram for Example 9.6.

Applying the minimax criterion to the expected losses of Player A, we find from Figure 9.1 that the greater of the two values of E for any given value of x is smallest where the two lines intersect, and to find the corresponding value of x we have only to solve the equation

$$8x - 5(1 - x) = 2x + 6(1 - x)$$

which yields $x = \frac{11}{17}$. This means that if Player A uses 11 slips of paper numbered I and 6 slips of paper numbered II, shuffles them thoroughly, and then acts according to which kind he randomly draws, he will be holding his expected loss down to $8 \cdot \frac{11}{17} - 5 \cdot \frac{6}{17} = 3\frac{7}{17}$ or \$3.41 to the nearest cent.

So far as Player B is concerned, in Exercise 14 on page 298 the reader will be asked to use a similar argument to show that Player B will maximize his minimum gain (which is the same as minimizing his maximum loss) by choosing between Strategies 1 and 2 with respective probabilities of $\frac{4}{17}$ and $\frac{13}{17}$, and that he will thus assure for himself an expected gain of $3\frac{7}{17}$ or \$3.41 to the nearest cent. Incidentally, the \$3.41 to which Player A can hold down his expected loss and Player B can raise his expected gain is called the value of this game. Also, if a player's ultimate choice is thus left to chance, his overall strategy is referred to as **randomized** or **mixed**, whereas the original Strategies I, II, 1, and 2 are referred to as **pure**.

The examples of this section were all given without any "physical" interpretation because we were interested only in introducing some of the basic concepts of the theory of games. If we apply these methods to Example 9.1, we find that the "game" has a saddle point and that the manufacturer's minimax strategy is to delay expanding the capacity of his plant. Of course, this assumes, questionably so, that Nature (which controls whether there is going to be a recession) is a malevolent opponent. Also, it would seem that in a situation like this the manager ought to have some idea about the chances for a recession, and, hence, that the problem should be solved by the first method of Section 9.1.

THEORETICAL EXERCISE

1. If a zero-sum two-person game has a saddle point corresponding to Strategies I and 4 and another corresponding to Strategies III and 2, show that

(a) there are also saddle points corresponding to Strategies I and 2, and Strategies III and 4;

(b) the payoff must be the same for all four of these saddle points.

APPLIED EXERCISES

2. With reference to Example 9.1 on page 286, what decision would minimize the manufacturer's expected loss if he felt that

(a) the odds for a recession are 3 to 2;

(b) the odds for a recession are 7 to 4?

3. With reference to Example 9.1 on page 286, would the manufacturer's decision remain the same if

(a) the \$164,000 profit is replaced by a \$200,000 profit and the odds are 2 to 1 that there will be a recession;

(b) the \$40,000 loss is replaced by a \$60,000 loss and the odds are 3 to 2 that there will be a recession?

4. Ms. Cooper is planning to attend a convention in Honolulu, and she must send in her room reservation immediately. The convention is so large that the activities are held partly in Hotel X and partly in Hotel Y, and Ms. Cooper does not know whether the particular session she wants to attend will be held at Hotel X or Hotel Y. She is planning to stay only one night, which would cost her \$36.00 at Hotel X and \$32.40 at Hotel Y, and it will cost her an extra \$6.00 for cab fare if she stays at the wrong hotel.
   (a) If Ms. Cooper feels that the odds are 3 to 1 that the session she wants to attend will be held at Hotel X, where should she make her reservation so as to minimize her expected cost?
   (b) If Ms. Cooper feels that the odds are 5 to 1 that the session she wants to attend will be held at Hotel X, where should she make her reservation so as to minimize her expected cost?

5. A truck driver has to deliver a load of lumber to one of two construction sites, which are, respectively, 27 and 33 miles from the lumberyard, but he has misplaced the order telling him where the load of lumber should go. The two construction sites are 12 miles apart, and, to complicate matters, the telephone at the lumberyard is out of order. Where should he go first if he wants to minimize the distance he can expect to drive and he feels that
   (a) the odds are 5 to 1 that the lumber should go to the construction site which is 33 miles from the lumberyard;
   (b) the odds are 2 to 1 that the lumber should go to the construction site which is 33 miles from the lumberyard;
   (c) the odds are 3 to 1 that the lumber should go to the construction site which is 33 miles from the lumberyard?

6. Basing their decisions on pessimism as in Example 9.2, where should
   (a) Ms. Cooper of Exercise 4 make her reservation;
   (b) the truck driver of Exercise 5 go first?

7. Basing their decisions on optimism (that is, maximizing maximum gains or minimizing minimum losses), what decisions should be reached by
   (a) the manufacturer of Example 9.1 on page 286;
   (b) Ms. Cooper of Exercise 4;
   (c) the truck driver of Exercise 5?

8. Suppose that the manufacturer of Example 9.1 on page 286 is the kind of person who always worries about losing out on a good deal. For instance, he finds that if he delays expansion and economic conditions remain good, he will lose out by \$84,000 (the difference between the \$164,000 profit he would have made if he had decided to expand right away, and the \$80,000 profit he will actually make). Referring to this quantity as an **opportunity loss**, or **regret**, find
   (a) the opportunity losses corresponding to the other three possibilities;

(b) which decision would minimize the manufacturer's maximum loss of opportunity.

9. With reference to the definition of Exercise 8, find which decisions will minimize the maximum opportunity loss of

(a) Ms. Cooper of Exercise 4;
(b) the truck driver of Exercise 5.

10. With reference to Example 9.1 on page 286, suppose that the manufacturer has the option of hiring an infallible forecaster for \$15,000 to find out for certain whether there will be a recession. Based on the original 2 to 1 odds that there will be a recession, would it be worthwhile for the manufacturer to spend this \$15,000?

11. Each of the following is the payoff matrix (the payments Player A makes to Player B) for a zero-sum two-person game. Eliminate all dominated strategies and determine the optimum strategy for each player as well as the value of the game:

(a)

3	−2
5	7

(b)

14	11
16	−2

(c)

−5	0	3
−6	−3	−3
−12	−1	1

(d)

7	10	8
8	8	11
7	5	9

12. Each of the following is the payoff matrix of a zero-sum two-person game. Find the saddle point (or saddle points) and the value of each game:

(a)

−1	5	−2
0	3	1
−2	−4	5

(b)

3	2	4	9
4	4	4	3
5	6	5	6
5	7	5	9

13. A small town has two service stations, which share the town's market for gasoline. The owner of Station A is debating whether or not to give away free glasses to her customers as part of a promotional scheme, and the owner of

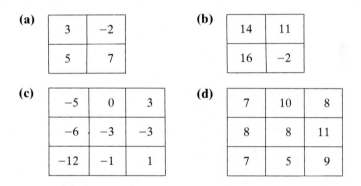

Station B is debating whether or not to give away free steak knives. They know (from similar situations elsewhere) that if Station A gives away free glasses and Station B does not give away free steak knives, Station A's share of the market will increase by 6 percent; if Station B gives away free steak knives and Station A does not give away free glasses, Station B's share of the market will increase by 8 percent; and if both stations give away the respective items, Station B's share of the market will increase by 3 percent.

(a) Present this information in the form of a payoff table, in which the entries are Station A's losses in its share of the market.

(b) Find optimum strategies for the owners of the two stations.

14. Verify the probabilities of $\frac{4}{17}$ and $\frac{13}{17}$ given on page 295 for the randomized strategy of Player B.

15. The following is the payoff matrix of a 2×2 zero-sum two-person game:

3	−4
−3	1

(a) What randomized strategy should Player A use so as to minimize his maximum expected loss?

(b) What randomized strategy should Player B use so as to maximize his minimum expected gain?

(c) What is the value of the game?

16. With reference to Exercise 4, which randomized strategy will minimize Ms. Cooper's maximum expected cost?

17. A country has two airfields with installations worth \$2,000,000 and \$10,000,000, respectively, of which it can defend only one against an attack by its enemy. The enemy, on the other hand, can attack only one of these airfields and take it successfully only if it is left undefended. Considering the "payoff" to the country to be the total value of the installations it holds after the attack, find the optimum strategy of the country as well as that of its enemy, and the value of the "game."

18. Two persons agree to play the following game: The first writes either 1 or 4 on a slip of paper and at the same time the second writes either 0 or 3 on another slip of paper. If the sum of the two numbers is odd, the first wins this amount in dollars; otherwise, the second wins \$2.

(a) Construct the payoff matrix in which the payoffs are the first person's losses.

(b) What randomized decision procedure should the first person use so as to minimize her maximum expected loss?

(c) What randomized decision procedure should the second person use so as to maximize his minimum expected gain?

19. There are two gas stations in a certain block, and the owner of the first station knows that if neither station lowers its prices, he can expect a net profit of \$100 on any given day. If he lowers his prices while the other station does not, he can expect a net profit of \$140; if he does not lower his prices but the other station does, he can expect a net profit of \$70; and if both stations participate in this "price war," he can expect a net profit of \$80. The owners of the two gas stations decide independently what prices to charge on any given day, and it is assumed that they cannot change their prices after they discover those charged by the other.

(a) Should the owner of the first gas station charge his regular prices or should he lower them, if he wants to maximize his minimum net profit?

(b) Assuming that the above profit figures apply also to the second gas station, how might the owners of the gas stations collude so that each could expect a net profit of \$105?

Note that this "game" is not zero-sum, so that the possibility of collusion opens entirely new possibilities.

9.3 STATISTICAL GAMES

In statistical inference we base decisions about populations on sample data, and it is by no means far-fetched to look upon such an inference as a game between Nature, which controls the relevant feature (or features) of the population, and the person (scientist, or statistician) who must arrive at some decision about Nature's choice. For instance, if we want to estimate the mean μ of a normal population on the basis of a random sample of size n, we could say that Nature has control over the "true" value of μ. On the other hand, we might estimate μ in terms of the value of the sample mean or that of the sample median, and presumably there is some penalty or reward which depends on the size of our error.

In spite of the obvious similarity between this problem and the ones of the preceding section, there are essentially two features in which **statistical games** are different. First, there is the question which we already met when we tried to apply the theory of games to the decision problem of Example 9.1, namely, the question of whether it is reasonable to treat Nature as a malevolent opponent. Obviously not, but this does not simplify matters; if we could treat Nature as a rational opponent, we would know, at least, what to expect.

The other distinction is that in the games of Section 9.2 each player had to choose his strategy without any knowledge of what his opponent had done or was planning to do, whereas in a statistical game the statistician is supplied with sample data which provide him with some information about Nature's choice. This also complicates matters, but it merely amounts to the fact that we are dealing with more complicated kinds of games. To illustrate, let us consider the following decision problem: *We are told that a coin is either balanced with heads on one side and tails on the other or two-headed. We cannot inspect the coin, but we can flip it once and observe whether it comes up heads or tails. Then we must decide whether or not it is two-headed, keeping in mind that there is a penalty of \$1 if our decision is wrong, and no penalty (or reward) if our decision is right.* If we ignored the fact that we can observe one flip of the coin, we could treat the problem as the following game:

Player A (The Statistician)

		a_1	a_2
Player B (Nature)	θ_1	$L(a_1, \theta_1) = 0$	$L(a_2, \theta_1) = 1$
	θ_2	$L(a_1, \theta_2) = 1$	$L(a_2, \theta_2) = 0$

which should remind the reader of the scheme on page 288. Now, θ_1 is the "state of Nature" that the coin is two-headed, θ_2 is the "state of Nature" that the coin is balanced with heads on one side and tails on the other, a_1 is the statistician's decision that the coin is two-headed, and a_2 is the statistician's decision that the coin is balanced with heads on one side and tails on the other. The entries in the table are the corresponding values of the given loss function.

Now let us consider also the fact that we (Player A, or the statistician) know what happened in the flip of the coin; that is, we know whether a random variable $\mathbf{x}$ has taken on the value $x = 0$ (heads) or $x = 1$ (tails). Since we shall want to make use of this information in choosing between a_1 and a_2, we need a function, a **decision function**, which tells us what action to take when $x = 0$ and what action to take when $x = 1$. One possibility is to choose a_1 when $x = 0$ and a_2 when $x = 1$, and we can express this symbolically by writing

$$d_1(x) = \begin{cases} a_1 & \text{when } x = 0 \\ a_2 & \text{when } x = 1 \end{cases}$$

or more simply $d_1(0) = a_1$ and $d_1(1) = a_2$. The purpose of the subscript is to distinguish this decision function from others, for instance, from

$$d_2(0) = a_1 \quad \text{and} \quad d_2(1) = a_1$$

which tells us to choose a_1 regardless of the outcome of the experiment, and

$$d_3(0) = a_2 \quad \text{and} \quad d_3(1) = a_2$$

which tells us to choose a_2 regardless of the outcome of the experiment, and

$$d_4(0) = a_2 \quad \text{and} \quad d_4(1) = a_1$$

which tells us to choose a_2 when $x = 0$ and a_1 when $x = 1$.

To compare the merits of all these decision functions, let us first determine the expected losses to which they lead for the various strategies of Nature, namely, the values of the **risk function**

$$R(d_i, \theta_j) = E\{L[d_i(\mathbf{x}), \theta_j]\}$$

where the expectation is taken with respect to the random variable $\mathbf{x}$. Since the probabilities for $x = 0$ and $x = 1$ are, respectively, 1 and 0 for θ_1, and $\frac{1}{2}$ and $\frac{1}{2}$ for θ_2, we get

$$\begin{aligned}
R(d_1, \theta_1) &= 1 \cdot L(a_1, \theta_1) + 0 \cdot L(a_2, \theta_1) = 1 \cdot 0 + 0 \cdot 1 = 0 \\
R(d_1, \theta_2) &= \tfrac{1}{2} \cdot L(a_1, \theta_2) + \tfrac{1}{2} \cdot L(a_2, \theta_2) = \tfrac{1}{2} \cdot 1 + \tfrac{1}{2} \cdot 0 = \tfrac{1}{2} \\
R(d_2, \theta_1) &= 1 \cdot L(a_1, \theta_1) + 0 \cdot L(a_1, \theta_1) = 1 \cdot 0 + 0 \cdot 0 = 0 \\
R(d_2, \theta_2) &= \tfrac{1}{2} \cdot L(a_1, \theta_2) + \tfrac{1}{2} \cdot L(a_1, \theta_2) = \tfrac{1}{2} \cdot 1 + \tfrac{1}{2} \cdot 1 = 1 \\
R(d_3, \theta_1) &= 1 \cdot L(a_2, \theta_1) + 0 \cdot L(a_2, \theta_1) = 1 \cdot 1 + 0 \cdot 1 = 1 \\
R(d_3, \theta_2) &= \tfrac{1}{2} \cdot L(a_2, \theta_2) + \tfrac{1}{2} \cdot L(a_2, \theta_2) = \tfrac{1}{2} \cdot 0 + \tfrac{1}{2} \cdot 0 = 0 \\
R(d_4, \theta_1) &= 1 \cdot L(a_2, \theta_1) + 0 \cdot L(a_1, \theta_1) = 1 \cdot 1 + 0 \cdot 0 = 1 \\
R(d_4, \theta_2) &= \tfrac{1}{2} \cdot L(a_2, \theta_2) + \tfrac{1}{2} \cdot L(a_1, \theta_2) = \tfrac{1}{2} \cdot 0 + \tfrac{1}{2} \cdot 1 = \tfrac{1}{2}
\end{aligned}$$

where the values of the loss function were obtained from the table on page 300.

We have thus arrived at the following 4×2 zero-sum two-person game, in which the payoffs are the corresponding values of the risk function:

Player A (The Statistician)

		d_1	d_2	d_3	d_4
Player B (Nature)	θ_1	0	0	1	1
	θ_2	$\frac{1}{2}$	1	0	$\frac{1}{2}$

As can be seen by inspection, d_2 is dominated by d_1 and d_4 is dominated by d_3, so that d_2 and d_4 can be discarded—in decision theory we say that they are **inadmissible**. Actually, this should not come as a surprise, since in d_2 as well as d_4 we accept alternative a_1 (that the coin is two-headed) even though it came up tails.

This leaves us with the 2×2 zero-sum two-person game in which Player A has to choose between d_1 and d_3. It can easily be verified that if Nature is looked upon as a malevolent opponent, the optimum strategy is to randomize between d_1 and d_3 with respective probabilities of $\frac{2}{3}$ and $\frac{1}{3}$, and the value of the game (the expected risk) is $\frac{1}{3}$ of a dollar. If Nature is not looked upon as a malevolent opponent, some other criterion will have to be used for choosing between d_1 and d_3, and this will be discussed in the sections which follow. Incidentally, we formulated this problem with reference to a two-headed coin and an ordinary coin, but we could just as well have formulated it more abstractly as a decision problem in which we must decide on the basis of a single observation whether a random variable has the Bernoulli distribution (see Section 5.3) with the parameter $\theta = 0$ or the parameter $\theta = \frac{1}{2}$.

To illustrate further the concepts of a loss function and a risk function, let us consider the following example, in which Nature as well as the statistician has a continuum of strategies:

EXAMPLE 9.7

A random variable has the uniform density

$$f(x) = \begin{cases} \dfrac{1}{\theta} & \text{for } 0 < x < \theta \\ 0 & \text{elsewhere} \end{cases}$$

and we want to estimate the parameter θ (the "move" of Nature) on the basis of a single observation. If the decision function is to be of the form $d(x) = kx$, where $k \geqslant 1$, and the lossses are proportional to the absolute value of the errors, that is,

$$L(kx, \theta) = c|kx - \theta|$$

where c is a positive constant, find the value of k which will minimize the risk.

Solution

For the risk function we get

$$\begin{aligned} R(d, \theta) &= \int_0^{\theta/k} c(\theta - kx) \cdot \frac{1}{\theta}\, dx + \int_{\theta/k}^{\theta} c(kx - \theta) \cdot \frac{1}{\theta}\, dx \\ &= c\theta\left(\frac{k}{2} - 1 + \frac{1}{k}\right) \end{aligned}$$

and there is nothing we can do about the factor θ, but it can easily be verified that $k = \sqrt{2}$ will minimize $\frac{k}{2} - 1 + \frac{1}{k}$. Thus, if we actually took the observation and got $x = 5$, our estimate of θ would be $5\sqrt{2}$ or approximately 7.07.

9.4 DECISION CRITERIA

In Example 9.7 we were able to find a decision function which minimized the risk regardless of the true state of Nature (that is, regardless of the true value of the parameter θ), but this is the exception rather than the rule. Had we not limited ourselves to decision functions of the form $d(x) = kx$, then the decision function given by $d(x) = \theta_1$ would be best when θ happens to equal θ_1, the one given by $d(x) = \theta_2$ would be best when θ happens to equal $\theta_2, \ldots$, and it is obvious that there can be no decision function which is best for all values of θ.

In general, we thus have to be satisfied with decision functions that are best only with respect to some criterion, and the two criteria which we shall study in this chapter are: (1) the **minimax criterion**, according to which we choose the decision function d for which $R(d, \theta)$, maximized with respect to θ, is a minimum; and (2) the **Bayes criterion**, according to which we choose the decision function d for which the **Bayes risk** $E[R(d, \boldsymbol{\theta})]$ is a minimum, where the expectation is taken with respect to $\boldsymbol{\theta}$. This requires that we look upon $\boldsymbol{\theta}$ as a random variable having a given distribution.

It is of interest to note that in the example of Section 9.1 we used both of these criteria. When we quoted odds for a recession, we assigned probabilities to the two states of Nature, θ_1 and θ_2, and when we suggested that the manufacturer minimize his expected loss, we suggested, in fact, that he use the Bayes criterion. Also, when we asked on page 288 what the manufacturer might do if he were a confirmed pessimist, we suggested that he would protect himself against the worst that can happen by using the minimax criterion.

9.5 THE MINIMAX CRITERION

If we apply the minimax criterion to the illustration of Section 9.3, dealing with the coin which is either two-headed or balanced with heads on one side and tails on the other, we find from the table on page 301 with d_2 and d_4 deleted that for d_1 the maximum risk is $\frac{1}{2}$, for d_3 the maximum risk is 1, and, hence, the one that minimizes the maximum risk is d_1.

EXAMPLE 9.8

Use the minimax criterion to estimate the parameter θ of a binomial distribution on the basis of a value of the random variable $\mathbf{x}$, the observed number of successes in n trials, when the decision function is of the form

$$d(x) = \frac{x + a}{n + b}$$

where a and b are constants, and the loss function is given by

$$L\left(\frac{x+a}{n+b}, \theta\right) = c\left(\frac{x+a}{n+b} - \theta\right)^2$$

where c is a positive constant.

Solution

The problem is to find the values of a and b which will minimize the corresponding risk function after it has been maximized with respect to θ. After all, we have control over the choice of a and b, while Nature (our presumed opponent) has control over the choice of θ.

Since $E(\mathbf{x}) = n\theta$ and $E(\mathbf{x}^2) = n\theta(1 - \theta + n\theta)$, as we saw on page 169, it follows that

$$R(d, \theta) = E\left[c\left(\frac{\mathbf{x}+a}{n+b} - \theta\right)^2\right]$$

$$= \frac{c}{(n+b)^2}[\theta^2(b^2 - n) + \theta(n - 2ab) + a^2]$$

and, using calculus, we could find the value of θ which maximizes this expression, and then minimize $R(d, \theta)$ for this value of θ with respect to a and b. This is not particularly difficult, but it will be left to the reader in Exercise 4 on page 307 as it involves some tedious algebraic detail.

Instead, we shall use the **equalizer principle**, according to which (under fairly general conditions) the risk function of a minimax decision rule is a constant; in this case it would not depend on the value of θ.[†] To justify this principle, at least intuitively, observe that in Example 9.6 the minimax strategy of Player A leads to an expected loss of \$3.41 regardless of whether Player B chooses Strategy 1 or Strategy 2.

To make the risk function independent of θ in our example, the coefficients of θ and θ^2 will both have to equal 0 in the expression obtained above. This yields $b^2 - n = 0$ and $n - 2ab = 0$, and, hence, $a = \frac{1}{2}\sqrt{n}$ and $b = \sqrt{n}$. Thus, the minimax decision function is given by

$$d(x) = \frac{x + \frac{1}{2}\sqrt{n}}{n + \sqrt{n}}$$

and if we actually obtained 39 successes in 100 trials, we would estimate the parameter of this binomial distribution as $\frac{39 + 5}{100 + 10} = 0.40$.

[†] The exact conditions under which the equalizer principle holds are given in the book by T. S. Ferguson listed among the references at the end of this chapter.

9.6 THE BAYES CRITERION

To apply the Bayes criterion in the illustration of Section 9.3, the one dealing with the coin which is either two-headed or balanced with heads on one side and tails on the other, we will have to assign probabilities to the two strategies of Nature, θ_1 and θ_2. If we assign θ_1 and θ_2, respectively, the probabilities p and $1 - p$, it can be seen from the table on page 301 that for d_1 the Bayes risk is

$$0 \cdot p + \tfrac{1}{2} \cdot (1 - p) = \tfrac{1}{2} \cdot (1 - p)$$

and that for d_3 the Bayes risk is

$$1 \cdot p + 0 \cdot (1 - p) = p$$

It follows that the Bayes risk of d_1 is less than that of d_3 (and d_1 is to be preferred to d_3) when $p > \frac{1}{3}$, and that the Bayes risk of d_3 is less than that of d_1 (and d_3 is to be preferred to d_1) when $p < \frac{1}{3}$. When $p = \frac{1}{3}$, the two Bayes risks are equal, and we can use either d_1 or d_3.

EXAMPLE 9.9

With reference to Example 9.7, suppose that the parameter of the uniform density is looked upon as a random variable with the probability density

$$h(\theta) = \begin{cases} \theta \cdot e^{-\theta} & \text{for } \theta > 0 \\ 0 & \text{elsewhere} \end{cases}$$

If there is no restriction on the form of the decision function and the loss function is quadratic, that is, its values are given by

$$L[d(x), \theta] = c\{d(x) - \theta\}^2$$

find the decision function which minimizes the Bayes risk.

Solution

Since $\boldsymbol{\theta}$ is now a random variable, we look upon the original density function as the conditional density

$$f(x|\theta) = \begin{cases} \dfrac{1}{\theta} & \text{for } 0 < x < \theta \\ 0 & \text{elsewhere} \end{cases}$$

and, letting $f(x, \theta) = f(x|\theta) \cdot h(\theta)$ in accordance with Definition 3.13 on page 117, we get

$$f(x, \theta) = \begin{cases} e^{-\theta} & \text{for } 0 < x < \theta \\ 0 & \text{elsewhere} \end{cases}$$

As the reader will be asked to verify in Exercise 6 on page 308, this yields

$$g(x) = \begin{cases} e^{-x} & \text{for } x > 0 \\ 0 & \text{elsewhere} \end{cases}$$

for the marginal density of **x** and

$$\varphi(\theta|x) = \begin{cases} e^{x-\theta} & \theta > x \\ 0 & \text{elsewhere} \end{cases}$$

for the conditional density of $\boldsymbol{\theta}$ given $\mathbf{x} = x$.

Now, the Bayes risk $E[R(d, \boldsymbol{\theta})]$ which we shall want to minimize is given by the double integral

$$\int_0^\infty \left\{ \int_0^\theta c[d(x) - \theta]^2 f(x|\theta)\, dx \right\} h(\theta)\, d\theta$$

which can also be written as

$$\int_0^\infty \left\{ \int_x^\infty c[d(x) - \theta]^2 \varphi(\theta|x)\, d\theta \right\} g(x)\, dx$$

making use of the fact that $f(x|\theta) \cdot h(\theta) = \varphi(\theta|x) \cdot g(x)$ and changing the order of integration. To minimize this double integral, we must choose $d(x)$ for each x so that the integral

$$\int_x^\infty c[d(x) - \theta]^2 \varphi(\theta|x)\, d\theta = \int_x^\infty c[d(x) - \theta]^2\, e^{x-\theta}\, d\theta$$

is as small as possible. Differentiating with respect to $d(x)$ and putting the derivative equal to 0, we get

$$2c\, e^x \cdot \int_x^\infty [d(x) - \theta]\, e^{-\theta}\, d\theta = 0$$

This yields

$$d(x) \cdot \int_x^\infty e^{-\theta}\, d\theta - \int_x^\infty \theta\, e^{-\theta}\, d\theta = 0$$

and, finally,

$$d(x) = \frac{\int_x^\infty \theta\, e^{-\theta}\, d\theta}{\int_x^\infty e^{-\theta}\, d\theta} = \frac{(x+1)\, e^{-x}}{e^{-x}} = x + 1$$

Thus, if the observation we get is $x = 5$ (as on page 302), this decision function gives the Bayes estimate $5 + 1 = 6$ for the parameter of the original uniform density.

THEORETICAL EXERCISES

1. With reference to the illustration on page 300, show that even if the coin is flipped n times, there are only two admissible decision functions. Also construct a table showing the values of the risk function corresponding to these two decision functions and the two states of Nature.

2. With reference to Example 9.7, show that if the losses are proportional to the squared errors instead of their absolute values, the risk function becomes

$$R(d, \theta) = \frac{c\theta^2}{3}(k^2 - 3k + 3)$$

and its minimum is at $k = \frac{3}{2}$.

3. A statistician has to decide on the basis of a single observation whether the parameter θ of the density

$$f(x) = \begin{cases} \dfrac{2x}{\theta^2} & \text{for } 0 < x < \theta \\ 0 & \text{elsewhere} \end{cases}$$

equals θ_1 or θ_2, where $\theta_1 < \theta_2$. If he decides on θ_1 when the observed value is less than the constant k, on θ_2 when the observed value is greater than or equal to the constant k, and he is fined C dollars for making the wrong decision, which value of k will minimize the maximum risk?

4. Find the value of θ which maximizes the risk function of Example 9.8, and then find the values of a and b which minimize the risk function for that value of θ. Compare the results with those given on page 304.

5. If we assume in Example 9.8 that $\boldsymbol{\theta}$ is a random variable having the uniform density of Section 6.2 with $\alpha = 0$ and $\beta = 1$, show that the Bayes risk is given by

$$\frac{c}{(n+b)^2}\left[\tfrac{1}{3}(b^2 - n) + \tfrac{1}{2}(n - 2ab) + a^2\right]$$

Also show that this Bayes risk is a minimum when $a = 1$ and $b = 2$, so that the optimum Bayes decision rule is given by $d(x) = \dfrac{x+1}{n+2}$.

6. Verify the results given on page 306 for the marginal density of $\mathbf{x}$ and the conditional density of $\boldsymbol{\theta}$ given $\mathbf{x} = x$.

7. Suppose that we want to estimate the parameter θ of the geometric distribution of Section 5.5 on the basis of a single observation. If the loss function is given by

$$L[d(x), \theta] = c\{d(x) - \theta\}^2$$

and $\boldsymbol{\theta}$ is looked upon as a random variable having the uniform density $h(\theta) = 1$ for $0 < \theta < 1$ and $h(\theta) = 0$ elsewhere, duplicate the steps in Example 9.9 to show that

(a) the conditional density of $\boldsymbol{\theta}$ given x is

$$\varphi(\theta|x) = \begin{cases} x(x+1)\theta(1-\theta)^{x-1} & \text{for } 0 < \theta < 1 \\ 0 & \text{elsewhere} \end{cases}$$

(b) the Bayes risk is minimized by the decision function

$$d(x) = \frac{2}{x+2}$$

(*Hint:* Make use of the fact that the integral of any beta density is equal to 1.)

APPLIED EXERCISES

8. A statistician has to decide on the basis of one observation whether the parameter θ of a Bernoulli distribution is 0, $\frac{1}{2}$, or 1; his loss in dollars (a penalty which is deducted from his fee) is 100 times the absolute value of his error.
   (a) Construct a table showing the nine possible values of the loss function.
   (b) List the nine possible decision functions and construct a table showing all the values of the corresponding risk function.
   (c) Show that five of the decision functions are not admissible, and that according to the minimax criterion the remaining decision functions are all equally good.
   (d) Which decision function is best according to the Bayes criterion, if the three possible values of the parameter θ are regarded as equally likely?

9. A statistician has to decide on the basis of two observations whether the parameter θ of a binomial distribution is $\frac{1}{4}$ or $\frac{1}{2}$; his loss (a penalty which is deducted from his fee) is \$160 if he is wrong.

(a) Construct a table showing the four possible values of the loss function.

(b) List the eight possible decision functions and construct a table showing all the values of the corresponding risk function.

(c) Show that three of the decision functions are not admissible.

(d) Find the decision function which is best according to the minimax criterion.

(e) Find the decision function which is best according to the Bayes criterion, if the probabilities assigned to $\theta = \frac{1}{4}$ and $\theta = \frac{1}{2}$ are, respectively, $\frac{2}{3}$ and $\frac{1}{3}$.

10. A manufacturer produces an item consisting of two components, which must both work for the item to function properly. The cost of returning one of the items to the manufacturer for repairs is α dollars, the cost of inspecting one of the components is β dollars, and the cost of repairing a faulty component is φ dollars. He can ship each item without inspection with the guarantee that it will be put into perfect working condition at his factory in case it does not work; he can inspect both components and repair them if necessary; or he can randomly select one of the components and ship the item with the original guarantee if it works, or repair it and also check the other component.

(a) Construct a table showing the manufacturer's expected losses corresponding to his three "strategies" and the three "states" of Nature that 0, 1, or 2 of the components do not work.

(b) What should the manufacturer do if $\alpha = \$25.00$, $\varphi = \$10.00$, and he wants to minimize his maximum expected losses?

(c) What should the manufacturer do to minimize his Bayes risk if $\alpha = \$10.00$, $\beta = \$12.00$, $\varphi = \$30.00$, and he feels that the probabilities for 0, 1, and 2 defective components are, respectively, 0.70, 0.20, and 0.10.

References

Some fairly elementary material on the theory of games and decision theory can be found in

CHERNOFF, H., and MOSES, L. E., *Elementary Decision Theory.* New York: John Wiley & Sons, Inc., 1959,

DRESHER, M., *Games of Strategy: Theory and Applications.* Englewood Cliffs, N.J.: Prentice-Hall, Inc., 1961,

MCKINSEY, J. C. C., *Introduction to the Theory of Games.* New York: McGraw-Hill Book Company, 1952,

OWEN, G., *Game Theory.* Philadelphia: W. B. Saunders Company, 1968,

WILLIAMS, J. D., *The Compleat Strategyst.* New York: McGraw-Hill Book Company, 1954,

and more advanced treatments in

BICKEL, P. J., and DOKSUM, K. A., *Mathematical Statistics: Basic Ideas and Selected Topics.* San Francisco: Holden-Day, Inc., 1977,

FERGUSON, T. S., *Mathematical Statistics: A Decision Theoretic Approach.* New York: Academic Press, Inc., 1967,

WALD, A., *Statistical Decision Functions.* New York: John Wiley & Sons, Inc., 1950.

point estimation

10

10.1 INTRODUCTION

Traditionally, statistical inference has been divided into **problems of estimation** and the **testing of hypotheses**, and we shall continue to make this distinction mainly to facilitate the organization of the material which we want to present in this book. Let us point out, though, that all problems of statistical inference, namely, all problems in which we make generalizations on the basis of sample data, are essentially decision problems, and, hence, can be handled by a unified approach like that presented in Chapter 9. The main distinction is that in problems of estimation we must choose a value of a parameter (that is, we must choose one particular strategy of Nature) from a possible continuum of alternatives, while in the testing of hypotheses we must decide whether to accept or reject one specified value or a set of values of a parameter.

As we already pointed out on page 303, perfect decision functions do not exist, and this is another reason for distinguishing between problems of estimation and the testing of hypotheses—the methods which we are willing to accept as "second best" or the restrictions which we must impose to obtain optimum decision functions differ somewhat for the two kinds of problems.

10.2 POINT ESTIMATION

If we use the value of a statistic to estimate a population parameter, this value is a **point estimate** of the parameter. For example, if we use a sample mean to estimate the mean of a population, a sample proportion to estimate the parameter θ of a binomial population, or a sample variance to estimate the variance of a population, we are in each case using a point estimate of the parameter in question. These estimates are called point estimates because they are single numbers, or points on the real axis, used, respectively, to estimate μ, θ, and σ^2.

The statistic, whose value is used as the point estimate of a parameter, is called an **estimator**. Therefore, the statistic $\bar{\mathbf{x}}$ is an estimator of μ, and its value $\bar{x}$ is the point estimate. Similarly, the statistic $\mathbf{s}^2$ is an estimator of σ^2, and its value s^2 is the point estimate. Clearly then, an estimator is a decision function and the point estimate is the decision or action resulting from the values of a particular sample.

Since estimators are random variables, one of the key problems of point estimation is to study their sampling distributions. For instance, when we estimate the variance of a population on the basis of a random sample, we can hardly expect that the value of $\mathbf{s}^2$ we get will actually equal σ^2, but it would be reassuring, at least, to know whether we can expect it to be close. Also, if we must decide whether to use a sample mean or a sample median to estimate the parameter of a population, it would be important to know, among other things, whether $\bar{\mathbf{x}}$ or $\tilde{\mathbf{x}}$ is more likely to yield a value which is actually close.

Various statistical properties of estimators can, thus, be used to decide which estimator is most appropriate in a given situation, which will expose us to the smallest risk, which will give us the most information at the lowest cost, and so forth. The particular properties of estimators which we shall discuss in Sections 10.3 through 10.5 are **unbiasedness**, **minimum variance**, **consistency**, **relative efficiency**, and **sufficiency**.

10.3 UNBIASED ESTIMATORS

Since there can be no perfect estimator which always gives the right answer, it would seem reasonable that an estimator should do so at least on the average. In other words, it would seem desirable that the expected value of an estimator equal the parameter which it is supposed to estimate. If this is the case, the estimator is said to be **unbiased**; otherwise, it is said to be **biased**. Formally,

DEFINITION 10.1 A statistic $\hat{\boldsymbol{\theta}}$ is an **unbiased estimator** of the parameter θ if and only if $E(\hat{\boldsymbol{\theta}}) = \theta$.

According to this definition, it follows from Theorems 8.1 and 8.4 that $\bar{\mathbf{x}}$ is an unbiased estimator of μ for any population whose mean exists. The following are further examples of unbiased as well as biased estimators:

EXAMPLE 10.1

If $\mathbf{x}$ has a binomial distribution, show that $\mathbf{x}/n$, the observed proportion of successes, is an unbiased estimator of the parameter θ.

Solution

Since $E(\mathbf{x}) = n\theta$, it follows that

$$E\left(\frac{\mathbf{x}}{n}\right) = \frac{1}{n} \cdot E(\mathbf{x}) = \frac{1}{n} \cdot n\theta = \theta$$

EXAMPLE 10.2

Show that the minimax estimator of the binomial parameter θ given on page 304 is biased.

Solution

Since $E(\mathbf{x}) = n\theta$, it follows that in general

$$E\left(\frac{\mathbf{x} + \frac{1}{2}\sqrt{n}}{n + \sqrt{n}}\right) = \frac{E(\mathbf{x} + \frac{1}{2}\sqrt{n})}{n + \sqrt{n}} = \frac{n\theta + \frac{1}{2}\sqrt{n}}{n + \sqrt{n}} \neq \theta$$

We can now explain why we divided by $n - 1$ and not by n when we defined the sample variance—it makes $\mathbf{s}^2$ an unbiased estimator of σ^2 for random samples from infinite populations.

THEOREM 10.1 If $\mathbf{s}^2$ is the variance of a random sample from an infinite population, then $E(\mathbf{s}^2) = \sigma^2$.

Proof. By Definition 8.2,

$$E(\mathbf{s}^2) = E\left[\frac{1}{n-1} \cdot \sum_{i=1}^{n} (\mathbf{x}_i - \bar{\mathbf{x}})^2\right]$$

$$= \frac{1}{n-1} \cdot E\left[\sum_{i=1}^{n} \{(\mathbf{x}_i - \mu) - (\bar{\mathbf{x}} - \mu)\}^2\right]$$

$$= \frac{1}{n-1}\left[\sum_{i=1}^{n} E\{(\mathbf{x}_i - \mu)^2\} - n \cdot E\{(\bar{\mathbf{x}} - \mu)^2\}\right]$$

Then, since $E[(\mathbf{x}_i - \mu)^2] = \sigma^2$ and $E[(\bar{\mathbf{x}} - \mu)^2] = \dfrac{\sigma^2}{n}$, it follows that

$$E(\mathbf{s}^2) = \frac{1}{n-1}\left[\sum_{i=1}^{n} \sigma^2 - n \cdot \frac{\sigma^2}{n}\right] = \sigma^2$$

Although $\mathbf{s}^2$ is an unbiased estimator of the variance of an infinite population, it is not an unbiased estimator of the variance of a finite population, and in neither case is **s** an unbiased estimator of σ. The bias of **s** is discussed in the book by E. S. Keeping referred to on page 340.

If we have to choose one of several unbiased estimators, we usually take the one whose sampling distribution has the smallest variance. We already indicated this on page 282, although we did not mention at the time that the sample median is also an unbiased estimator of the mean μ of a normal population. To check whether a given unbiased estimator has the smallest possible variance, namely, whether it is a **minimum variance unbiased estimator**, we make use of the fact that if $\hat{\boldsymbol{\theta}}$ is an unbiased estimator of θ, it can be shown under very general conditions (referred to on page 340) that the variance of $\hat{\boldsymbol{\theta}}$ must satisfy the inequality

$$\text{var}(\hat{\boldsymbol{\theta}}) \geq \frac{1}{n \cdot E\left[\left(\dfrac{\partial \ln f(\mathbf{x})}{\partial \theta}\right)^2\right]}$$

where $f(x)$ is the value of the population density at x, and n is the size of the random sample. This inequality, the **Cramér–Rao inequality**, leads to the following result:

THEOREM 10.2 If $\hat{\boldsymbol{\theta}}$ is an unbiased estimator of θ and

$$\text{var}(\hat{\boldsymbol{\theta}}) = \frac{1}{n \cdot E\left[\left(\dfrac{\partial \ln f(\mathbf{x})}{\partial \theta}\right)^2\right]}$$

then $\hat{\boldsymbol{\theta}}$ is a minimum variance unbiased estimator of θ.

EXAMPLE 10.3

Show that $\bar{\mathbf{x}}$ is a minimum variance unbiased estimator of the mean μ of a normal population.

Solution

Since

$$f(x) = \frac{1}{\sigma\sqrt{2\pi}} \cdot e^{-\frac{1}{2}\left(\frac{x-\mu}{\sigma}\right)^2} \quad \text{for } -\infty < x < \infty$$

it follows that

$$\ln f(x) = -\ln \sigma\sqrt{2\pi} - \frac{1}{2}\left(\frac{x-\mu}{\sigma}\right)^2$$

so that

$$\frac{\partial \ln f(x)}{\partial \mu} = \frac{1}{\sigma}\left(\frac{x-\mu}{\sigma}\right)$$

and, hence,

$$E\left[\left(\frac{\partial \ln f(\mathbf{x})}{\partial \mu}\right)^2\right] = \frac{1}{\sigma^2} \cdot E\left[\left(\frac{\mathbf{x}-\mu}{\sigma}\right)^2\right] = \frac{1}{\sigma^2} \cdot 1 = \frac{1}{\sigma^2}$$

Thus,

$$\frac{1}{n \cdot E\left[\left(\frac{\partial \ln f(\mathbf{x})}{\partial \mu}\right)^2\right]} = \frac{1}{n \cdot \frac{1}{\sigma^2}} = \frac{\sigma^2}{n}$$

and since $\bar{\mathbf{x}}$ is unbiased and $\text{var}(\bar{\mathbf{x}}) = \frac{\sigma^2}{n}$ according to Theorem 8.1, it follows that $\bar{\mathbf{x}}$ *is* a minimum variance unbiased estimator of μ.

It would be erroneous to conclude from this example that $\bar{\mathbf{x}}$ is a minimum variance unbiased estimator of the mean μ of every population. To show that this is not so, the reader will be asked to verify in Exercise 15 on page 324 that for random samples of size $n = 3$ from the continuous uniform population

$$f(x) = \begin{cases} 1 & \text{for } \theta - \frac{1}{2} < x < \theta + \frac{1}{2} \\ 0 & \text{elsewhere} \end{cases}$$

the variance of the **mid-range**, which is the mean of the largest value and the smallest, is less than the variance of $\bar{\mathbf{x}}$.

As we have indicated, unbiased estimators are usually compared in terms of their variances. If $\hat{\boldsymbol{\theta}}_1$ and $\hat{\boldsymbol{\theta}}_2$ are two unbiased estimators of a parameter θ and the variance of $\hat{\theta}_1$ is less than the variance of $\hat{\theta}_2$, we say that $\hat{\theta}_1$ is **relatively more efficient**. Also, we use the ratio

$$\frac{\text{var}(\hat{\theta}_1)}{\text{var}(\hat{\theta}_2)}$$

to measure the **efficiency** of $\hat{\theta}_2$ relative to $\hat{\theta}_1$.

EXAMPLE 10.4

In estimating the mean μ of a normal population on the basis of a random sample of size $2n + 1$, what is the efficiency of the median relative to the mean?

Solution

From Theorem 8.1 we know that $\bar{\mathbf{x}}$ is unbiased and that

$$\text{var}(\bar{\mathbf{x}}) = \frac{\sigma^2}{2n + 1}$$

So far as $\tilde{\mathbf{x}}$ is concerned, it is unbiased by virtue of the symmetry of the normal distribution about its mean, and we know from the discussion following Theorem 8.15 that for large samples

$$\text{var}(\tilde{\mathbf{x}}) = \frac{\pi\sigma^2}{4n}$$

Thus, for large samples, the efficiency of the median relative to the mean is approximately

$$\frac{\text{var}(\bar{\mathbf{x}})}{\text{var}(\tilde{\mathbf{x}})} = \frac{\dfrac{\sigma^2}{2n + 1}}{\dfrac{\pi\sigma^2}{4n}} = \frac{4n}{\pi(2n + 1)}$$

and the **asymptotic efficiency** of the median with respect to the mean is

$$\lim_{n\to\infty} \frac{4n}{\pi(2n + 1)} = \frac{2}{\pi}$$

or about 64 percent. This result may be interpreted as follows: For large samples, the mean requires only 64 percent as many observations as the median to estimate μ with the same reliability.

It is important to note that we have limited our discussion of relative efficiency to unbiased estimators. If we included biased estimators, we could always assure ourselves of an estimator with zero variance by letting its values equal the same constant regardless of the data which we may obtain. Thus, if $\hat{\boldsymbol{\theta}}$ is not an unbiased estimator of the parameter θ, it is preferable to judge its merits and make efficiency comparisons on the basis of the **mean square error**

$$E[(\boldsymbol{\theta} - \theta)^2]$$

instead of the variance of $\hat{\boldsymbol{\theta}}$.

10.4 CONSISTENT ESTIMATORS

The idea of choosing a minimum variance unbiased estimator is closely related to that of minimizing the risk function when the loss function is quadratic (as in Example 9.8). Of course, there are other kinds of loss functions and other ways of measuring the chance fluctuations of statistics. The fact that the variance may not even provide a good criterion for this purpose is illustrated by the following example: Suppose that we want to estimate on the basis of one observation the parameter θ of the population

$$f(x) = w \cdot \frac{1}{\sigma\sqrt{2\pi}} \cdot e^{-\frac{1}{2}\left(\frac{x-\theta}{\sigma}\right)^2} + (1 - w) \cdot \frac{1}{\pi} \cdot \frac{1}{1 + (x - \theta)^2}$$

for $-\infty < x < \infty$ and $0 < w < 1$. Evidently, this density is a **weighted mean** of a normal density with the mean θ and the variance σ^2 and a Cauchy density (see Exercise 3 on page 202) with $\beta = 1$ and $\alpha = \theta$. Now, if w is very close to 1, say, $w = 1 - 10^{-100}$, and σ is very small, say, $\sigma = 10^{-100}$, the probability that a random variable having this distribution will take on a value which is very close to θ, and hence is a very good estimate of θ, is practically 1. Yet, the variance of this estimator exceeds any bound, which follows from the fact that the variance of the Cauchy distribution does not exist.

The preceding example is somewhat out of the ordinary, but it suggests that we pay more attention to the probabilities of estimators taking on values close to the parameters they are supposed to estimate. The reader may recall that we already touched upon questions concerning the "closeness" of estimates in Sections 5.4 and 8.2. Basing our argument on Chebyshev's theorem, we showed in Section 5.4 that when $n \to \infty$ the probability approaches 0 that the sample proportion $\frac{\mathbf{x}}{n}$ will take on a value which differs from the binomial parameter θ by more than any arbitrary constant $c > 0$. Also using Chebyshev's theorem, we showed in Section 8.2 that when $n \to \infty$ the probability approaches 0 that $\bar{\mathbf{x}}$ will take on a value which differs from μ, the mean of the population sampled, by more than any arbitrary constant $c > 0$.

In both of the examples of the preceding paragraph we were practically assured that, at least for large n, the estimators will take on values which are very close to the respective parameters. This concept of closeness is generalized in the following definition of **consistency**:

DEFINITION 10.2 The statistic $\hat{\boldsymbol{\theta}}$ is a **consistent estimator** of the parameter θ if and only if for each positive constant c,

$$\lim_{n\to\infty} P(|\hat{\boldsymbol{\theta}} - \theta| \geqslant c) = 0$$

or, equivalently, if and only if

$$\lim_{n\to\infty} P(|\hat{\boldsymbol{\theta}} - \theta| < c) = 1$$

In accordance with this definition, we showed in Section 5.4 that $\frac{\mathbf{x}}{n}$ is a consistent estimator of the binomial parameter θ, and in Section 8.2 that $\bar{\mathbf{x}}$ is a consistent estimator of the mean μ of a population which has a finite variance.

Note that consistency is an **asymptotic property**, namely, a limiting property of an estimator; informally, Definition 10.2 says that when n is sufficiently large, we can be practically certain that the error made with a consistent estimator will be less than any small preassigned positive constant.

In actual practice, we can often judge whether an estimator is consistent by using the following **sufficient conditions** (though not necessary conditions), which are an immediate consequence of Chebyshev's theorem:

THEOREM 10.3 The statistic $\hat{\boldsymbol{\theta}}$ is a consistent estimator of the parameter θ if

1. $\hat{\boldsymbol{\theta}}$ is unbiased;
2. $\text{var}(\hat{\boldsymbol{\theta}}) \to 0$ as $n \to \infty$.

To demonstrate that these conditions are not necessary conditions, we have only to show that an estimator can be consistent without being unbiased. An example of such an estimator is the one of Exercise 5 on page 307, and in Exercise 16 on page 324 the reader will be asked to verify that this estimator, $\frac{\mathbf{x} + 1}{n + 2}$, is, indeed, a consistent estimator of the binomial parameter θ. We should add, though, that a biased estimator can be consistent only if it is asymptotically unbiased, namely, that it becomes unbiased when $n \to \infty$. For instance, the minimax estimator on page 304 is asymptotically unbiased since

$$\lim_{n\to\infty} E\left(\frac{\mathbf{x} + \frac{1}{2}\sqrt{n}}{n + \sqrt{n}}\right) = \lim_{n\to\infty} \frac{n\theta + \frac{1}{2}\sqrt{n}}{n + \sqrt{n}} = \theta$$

EXAMPLE 10.5

Show that the sample variance $\mathbf{s}^2$ is a consistent estimator of σ^2 for random samples from normal populations.

Solution

Since $\mathbf{s}^2$ is unbiased according to Theorem 10.1, it remains to be shown that the variance of $\mathbf{s}^2$ approaches 0 as $n \to \infty$. Referring to the result of Exercise 4 on page 276 (or Theorem 8.9 on which this exercise is based), we find that

$$\text{var}(\mathbf{s}^2) = \frac{2\sigma^4}{n - 1}$$

and, hence, that $\text{var}(\mathbf{s}^2) \to 0$ as $n \to \infty$, so long as σ is finite.

10.5 SUFFICIENT ESTIMATORS

An estimator $\hat{\boldsymbol{\theta}}$ is said to be **sufficient** if it utilizes all the information in a sample relevant to the estimation of the population parameter θ; that is, if all the knowledge we can gain about θ by actually specifying the individual sample values and their order can just as well be obtained by observing only the value of the statistic $\hat{\boldsymbol{\theta}}$.

Formally, this can be expressed in terms of the conditional distribution (probability distribution or density) of the sample values given $\hat{\boldsymbol{\theta}} = \hat{\theta}$. This quantity is given by

$$f(x_1, x_2, \ldots, x_n | \hat{\theta}) = \frac{f(x_1, x_2, \ldots, x_n, \hat{\theta})}{g(\hat{\theta})}$$

If it depends on θ, particular values of $\mathbf{x}_1, \mathbf{x}_2, \ldots,$ and $\mathbf{x}_n$ yielding $\hat{\boldsymbol{\theta}} = \hat{\theta}$ are more probable for some values of θ than for others, and the knowledge of these sample values will help in the estimation of θ. If it does not depend on θ, particular values of $\mathbf{x}_1, \mathbf{x}_2, \ldots,$ and $\mathbf{x}_n$ yielding $\hat{\boldsymbol{\theta}} = \hat{\theta}$ are just as likely to occur for any value of θ, and the knowledge of these sample values will not help in the estimation of θ.

DEFINITION 10.3 The statistic $\hat{\boldsymbol{\theta}}$ is a **sufficient estimator** of the parameter θ if and only if for each value of $\hat{\boldsymbol{\theta}}$ the conditional distribution of the random sample $\mathbf{x}_1, \mathbf{x}_2, \ldots,$ and $\mathbf{x}_n$ given $\hat{\boldsymbol{\theta}} = \hat{\theta}$ is independent of θ.

EXAMPLE 10.6

If $\mathbf{x}_1, \mathbf{x}_2, \ldots,$ and $\mathbf{x}_n$ are independent Bernoulli random variables with the same parameter θ, show that the statistic $\hat{\boldsymbol{\theta}} = \mathbf{x}/n$, where $\mathbf{x} = \mathbf{x}_1 + \mathbf{x}_2 + \cdots + \mathbf{x}_n$, is a sufficient estimator of θ.

Solution

By Definition 5.2,

$$f(x_i; \theta) = \theta^{x_i}(1-\theta)^{1-x_i} \qquad \text{for } x_i = 0, 1$$

so that

$$\begin{aligned} f(x_1, x_2, \ldots, x_n) &= \prod_{i=1}^{n} \theta^{x_i}(1-\theta)^{1-x_i} \\ &= \theta^{\sum_{i=1}^{n} x_i}(1-\theta)^{n-\sum_{i=1}^{n} x_i} \\ &= \theta^{x}(1-\theta)^{n-x} \\ &= \theta^{n\hat{\theta}}(1-\theta)^{n-n\hat{\theta}} \end{aligned}$$

for $x_i = 0, 1$, and $i = 1, 2, \ldots, n$. Also, since $\mathbf{x}$ is a binomial random variable with the parameters n and θ, its distribution is given by

$$b(x; n, \theta) = \binom{n}{x} \theta^{x}(1-\theta)^{n-x}$$

and the transformation of variable technique of Section 7.3 yields

$$g(\hat{\theta}) = \binom{n}{n\hat{\theta}}\theta^{n\hat{\theta}}(1-\theta)^{n-n\hat{\theta}} \qquad \text{for } \hat{\theta} = 0, \frac{1}{n}, \ldots, 1$$

Now, substituting into the formula for $f(x_1, x_2, \ldots, x_n \mid \hat{\theta})$ on page 318 we get

$$\frac{f(x_1, x_2, \ldots, x_n, \hat{\theta})}{g(\hat{\theta})} = \frac{f(x_1, x_2, \ldots, x_n)}{g(\hat{\theta})}$$

$$= \frac{\theta^{n\hat{\theta}}(1-\theta)^{n-n\hat{\theta}}}{\binom{n}{n\hat{\theta}}\theta^{n\hat{\theta}}(1-\theta)^{n-n\hat{\theta}}}$$

$$= \frac{1}{\binom{n}{n\hat{\theta}}}$$

$$= \frac{1}{\binom{n}{x}}$$

for $x_i = 0, 1$, and $i = 1, 2, \ldots, n$, which evidently does not depend on θ. We conclude, therefore, that $\hat{\boldsymbol{\theta}} = \dfrac{\mathbf{x}}{n}$ is a sufficient estimator of θ.

EXAMPLE 10.7

Show that the statistic $\mathbf{y} = \frac{1}{6}(\mathbf{x}_1 + 2\mathbf{x}_2 + 3\mathbf{x}_3)$ is not sufficient for estimating the parameter θ of a Bernoulli population.

Solution

We must show that

$$f(x_1, x_2, x_3 | y) = \frac{f(x_1, x_2, x_3, y)}{g(y)}$$

is not independent of θ for some values of $\mathbf{x}_1$, $\mathbf{x}_2$, and $\mathbf{x}_3$. Thus, let us consider the particular case where $x_1 = 1$, $x_2 = 1$, and $x_3 = 0$, so that

$$f(1, 1, 0 | \mathbf{y} = \tfrac{1}{2}) = \frac{P(\mathbf{x}_1 = 1, \mathbf{x}_2 = 1, \mathbf{x}_3 = 0, \ \mathbf{y} = \frac{1}{2})}{P(\mathbf{y} = \frac{1}{2})}$$

$$= \frac{f(1, 1, 0)}{f(1, 1, 0) + f(0, 0, 1)}$$

where

$$f(x_1, x_2, x_3) = \theta^{x_1+x_2+x_3}(1-\theta)^{3-(x_1+x_2+x_3)}$$

for $x_i = 0, 1$, and $i = 1, 2, 3$. Since $f(1, 1, 0) = \theta^2(1 - \theta)$ and $f(0, 0, 1) = \theta(1 - \theta)^2$, it follows that

$$f(1, 1, 0|\mathbf{y} = \tfrac{1}{2}) = \frac{\theta^2(1 - \theta)}{\theta^2(1 - \theta) + \theta(1 - \theta)^2} = \theta$$

which depends on θ. Therefore, the statistic $\mathbf{y} = \frac{1}{6}(\mathbf{x}_1 + 2\mathbf{x}_2 + 3\mathbf{x}_3)$ is not a sufficient estimator of θ.

As it can be quite tedious to check with Definition 10.3 whether a statistic is a sufficient estimator of a given parameter, it is usually easier to use the following factorization theorem:

THEOREM 10.4 The statistic $\hat{\boldsymbol{\theta}}$ is a sufficient estimator of the parameter θ if and only if the joint density or probability distribution of the random sample can be factored so that

$$f(x_1, x_2, \ldots, x_n; \theta) = g(\hat{\theta}, \theta) \cdot h(x_1, x_2, \ldots, x_n)$$

where $g(\hat{\theta}, \theta)$ depends only on $\hat{\theta}$ and θ, and $h(x_1, x_2, \ldots, x_n)$ does not depend on θ.

To illustrate the use of this theorem, a proof of which is given in the book by R. V. Hogg and A. T. Craig referred to on page 340, let us consider the following example:

EXAMPLE 10.8

Show that the statistic $\bar{\mathbf{x}}$ is a sufficient estimator of the mean μ of a normal population with the known variance σ^2.

Solution

Making use of the fact that

$$f(x_1, x_2, \ldots, x_n; \mu) = \left(\frac{1}{\sigma\sqrt{2\pi}}\right)^n \cdot e^{-\frac{1}{2}\cdot\sum_{i=1}^{n}\left(\frac{x_i - \mu}{\sigma}\right)^2}$$

and that

$$\begin{aligned}\sum_{i=1}^{n}(x_i-\mu)^2 &= \sum_{i=1}^{n}[(x_i-\bar{x})-(\mu-\bar{x})]^2\\ &= \sum_{i=1}^{n}(x_i-\bar{x})^2+\sum_{i=1}^{n}(\bar{x}-\mu)^2\\ &= \sum_{i=1}^{n}(x_i-\bar{x})^2+n(\bar{x}-\mu)^2\end{aligned}$$

we get

$$f(x_1, x_2, \ldots, x_n; \mu) = \left\{\frac{\sqrt{n}}{\sigma\sqrt{2\pi}}\cdot e^{-\frac{1}{2}\left(\frac{\bar{x}-\mu}{\sigma/\sqrt{n}}\right)^2}\right\}\times\left\{\frac{1}{\sqrt{n}}\left(\frac{1}{\sigma\sqrt{2\pi}}\right)^{n-1}\cdot e^{-\frac{1}{2}\cdot\sum_{i=1}^{n}\left(\frac{x_i-\bar{x}}{\sigma}\right)^2}\right\}$$

where the first factor on the right-hand side depends only on the estimate $\bar{x}$ and the population mean μ, and the second factor does not involve μ. Therefore, according to Theorem 10.4, $\bar{\mathbf{x}}$ is a sufficient estimator of the mean μ of a normal population with the known variance σ^2.

With Definition 10.3 and Theorem 10.4 we have presented two ways of checking whether a statistic $\hat{\boldsymbol{\theta}}$ is a sufficient estimator of a given parameter θ. Usually, the factorization criterion of Theorem 10.4 leads to the easier solution, but to show that $\hat{\boldsymbol{\theta}}$ is not sufficient, it is almost always simpler to proceed by Definition 10.3, as illustrated in Example 10.7.

Let us conclude this section with a very important property of sufficient estimators. If $\hat{\boldsymbol{\theta}}$ is a sufficient estimator of θ, then every single-valued function $\mathbf{y} = u(\hat{\boldsymbol{\theta}})$, not involving θ, is also a sufficient estimator of θ, and therefore of $u(\theta)$, provided $\mathbf{y} = u(\hat{\boldsymbol{\theta}})$ can be solved to give the single-valued inverse $\hat{\boldsymbol{\theta}} = w(\mathbf{y})$. This result follows from Theorem 10.4, since we can write

$$f(x_1, x_2, \ldots, x_n; \theta) = g[w(y), \theta]\cdot h(x_1, x_2, \ldots, x_n)$$

where $g[w(y), \theta]$ depends only on y and θ. If we apply this result to Example 10.6, where we showed that $\hat{\boldsymbol{\theta}} = \dfrac{\mathbf{x}}{n}$ is a sufficient estimator of the Bernoulli parameter θ, it follows that the estimator $\mathbf{x} = \mathbf{x}_1 + \mathbf{x}_2 + \cdots + \mathbf{x}_n$ is also a sufficient estimator of the binomial mean $\mu = n\theta$.

THEORETICAL EXERCISES

1. Use the formula for the sampling distribution of $\tilde{\mathbf{x}}$ on page 281 to show that for random samples of size 3 the median is an unbiased estimator of the parameter θ of the uniform density with $\alpha = \theta - \frac{1}{2}$ and $\theta + \frac{1}{2}$.

2. Refer to Example 8.5 to show that for random samples of size $n = 3$ the median is a biased estimator of the parameter θ of an exponential population.

3. Given a random sample of size n from a population which has the known mean μ and the finite variance σ^2, show that $\frac{1}{n} \cdot \sum_{i=1}^{n} (\mathbf{x}_i - \mu)^2$ is an unbiased estimator of σ^2.

4. Show that if $\hat{\boldsymbol{\theta}}$ is an unbiased estimator of θ and $\text{var}(\hat{\boldsymbol{\theta}})$ does not equal 0, then $\hat{\boldsymbol{\theta}}^2$ is not an unbiased estimator of θ^2.

5. Use the results of Theorem 8.1 to show that $\bar{\mathbf{x}}^2$ is an asymptotically unbiased estimator of μ^2.

6. Show that the Bayes estimator of Exercise 5 on page 307 is biased.

7. If $\hat{\boldsymbol{\theta}}$ is an estimator of a parameter θ, its bias is given by $b = E(\hat{\boldsymbol{\theta}}) - \theta$. Show that $E[(\hat{\boldsymbol{\theta}} - \theta)^2] = \text{var}(\hat{\boldsymbol{\theta}}) + b^2$.

8. For what value of k is $\hat{\boldsymbol{\theta}} = k\mathbf{x}$ an unbiased estimator of the parameter θ of the population given by

$$f(x) = \begin{cases} \frac{1}{\theta} & \text{for } 0 < x < \theta \\ 0 & \text{elsewhere?} \end{cases}$$

9. Show that the sample proportion $\frac{\mathbf{x}}{n}$ is a minimum variance unbiased estimator of the binomial parameter θ.

10. Show that the mean of a random sample of size n is a minimum variance unbiased estimator of the parameter λ of a Poisson population.

11. If $\bar{\mathbf{x}}_1$ is the mean of a random sample of size n from a normal population with the mean μ and the variance σ_1^2, and $\bar{\mathbf{x}}_2$ is the mean of a random sample of size n from a normal population with the mean μ and the variance σ_2^2, show that

(a) $w \cdot \bar{\mathbf{x}}_1 + (1 - w) \cdot \bar{\mathbf{x}}_2$, where $0 \leq w \leq 1$, is an unbiased estimator of μ;

(b) the variance of this estimator is a minimum when

$$w = \frac{\sigma_2^2}{\sigma_1^2 + \sigma_2^2}$$

12. If $\bar{\mathbf{x}}_1$ and $\bar{\mathbf{x}}_2$ are the means of independent random samples of size n_1 and n_2 from a normal population with the mean μ and the variance σ^2, show that the variance of the unbiased estimator $w \cdot \bar{\mathbf{x}}_1 + (1 - w) \cdot \bar{\mathbf{x}}_2$ is a minimum when

$$w = \frac{n_1}{n_1 + n_2}.$$

13. If $\mathbf{x}_1$, $\mathbf{x}_2$, and $\mathbf{x}_3$ are a random sample from a normal population with the mean μ and the variance σ^2, what is the relative efficiency of the estimator $\hat{\boldsymbol{\mu}} = \frac{\mathbf{x}_1 + 2\mathbf{x}_2 + \mathbf{x}_3}{4}$ with respect to $\bar{\mathbf{x}}$?

14. If $\mathbf{x}$ is a binomial random variable and $\hat{\boldsymbol{\theta}}_1 = \dfrac{\mathbf{x}}{n}$ and $\hat{\boldsymbol{\theta}}_2 = \dfrac{\mathbf{x}+1}{n+2}$ are estimators of the parameter θ, for what values of θ is $E[(\hat{\boldsymbol{\theta}}_2 - \theta)^2]$ less than $E[(\hat{\boldsymbol{\theta}}_1 - \theta)^2]$?

15. Since the variances of the mean and the mid-range are not affected if the same constant is added to each observation, we can determine these variances for random samples of size 3 from the uniform population on page 314 by referring instead to the uniform population

$$f(x) = \begin{cases} 1 & \text{for } 0 < x < 1 \\ 0 & \text{elsewhere} \end{cases}$$

(a) Show that $E(\mathbf{x}) = \frac{1}{2}$, $E(\mathbf{x}^2) = \frac{1}{3}$, and $\text{var}(\mathbf{x}) = \frac{1}{12}$ for this distribution, so that for a random sample of size 3, $\text{var}(\bar{\mathbf{x}}) = \frac{1}{36}$.

(b) Use the results of Exercise 2 on page 282 and part (b) of Exercise 6 on page 283 (or derive the necessary densities and joint density) to show that for a random sample of size three from this distribution the order statistics $\mathbf{y}_1$ and $\mathbf{y}_3$ have $E(\mathbf{y}_1) = \frac{1}{4}$, $E(\mathbf{y}_1^2) = \frac{1}{10}$, $E(\mathbf{y}_3) = \frac{3}{4}$, $E(\mathbf{y}_3^2) = \frac{3}{5}$, and $E(\mathbf{y}_1\mathbf{y}_3) = \frac{1}{5}$, so that $\text{var}(\mathbf{y}_1) = \frac{3}{80}$, $\text{var}(\mathbf{y}_3) = \frac{3}{80}$, and $\text{cov}(\mathbf{y}_1, \mathbf{y}_3) = \frac{1}{80}$.

(c) Use the results of part (b) and Theorem 4.14 to show that

$$E\left(\frac{\mathbf{y}_1 + \mathbf{y}_3}{2}\right) = \frac{1}{2} \quad \text{and} \quad \text{var}\left(\frac{\mathbf{y}_1 + \mathbf{y}_3}{2}\right) = \frac{1}{40}$$

16. Show that the estimator of Exercise 5 on page 307 is a consistent estimator of the parameter θ of a binomial population.

17. Show that the estimator of Exercise 11 is consistent.

18. Show that the mean of a random sample of size n from an exponential population is a consistent estimator of its parameter θ.

19. Suppose we use the largest value of a random sample of size n (namely, the order statistic $\mathbf{y}_n$) to estimate the parameter θ of the population

$$f(x) = \begin{cases} \dfrac{1}{\theta} & \text{for } 0 < x < \theta \\ 0 & \text{elsewhere} \end{cases}$$

Check whether this estimator is (a) unbiased, and (b) consistent.

20. If $\mathbf{x}_1$, $\mathbf{x}_2$, and $\mathbf{x}_3$ are independent random variables having Bernoulli distributions with the same parameter θ, show that $\mathbf{y} = \mathbf{x}_1 + 2\mathbf{x}_2 + \mathbf{x}_3$ is not a sufficient estimator of θ. (*Hint*: Consider special values of $\mathbf{x}_1$, $\mathbf{x}_2$, and $\mathbf{x}_3$.)

21. If $\mathbf{x}_1$ and $\mathbf{x}_2$ are independent random variables having, respectively, binomial distributions with the parameters θ and n, and θ and m, show that $\dfrac{\mathbf{x}_1 + \mathbf{x}_2}{n+m}$ is a sufficient estimator of θ.

22. If $\mathbf{x}_1$ and $\mathbf{x}_2$ are independent random variables having Poisson distributions with the same parameter λ, show that their mean is a sufficient estimator of λ.
23. If $\mathbf{x}_1, \mathbf{x}_2, \ldots,$ and $\mathbf{x}_n$ is a random sample of size n from a geometric population, show that $\mathbf{y} = \sum_{i=1}^{n} \mathbf{x}_i$ is a sufficient estimator of its parameter θ.
24. Show that the estimator of Exercise 3 is a sufficient estimator of the variance of a normal population with the known mean μ.
25. With reference to Exercise 18 show that the sample mean is a sufficient estimator of the exponential parameter θ.

10.6 THE METHOD OF MOMENTS

As we have seen in this chapter and in Chapter 9, there can be many different ways of estimating a parameter of a population. In decision theory, we will generally get different estimators depending on the form of the loss function and the criterion which we apply to the risk function in order to choose a decision function (that is, an estimator) which is thus considered "best." In actual practice, where we seldom have enough information to calculate a Bayes risk (and where it may be difficult even to decide on the functional form of the loss function), it is desirable to have some general methods which yield estimators with as many as possible of the properties discussed in the preceding sections. In this section and the next we shall treat two such methods, the **method of moments**, historically one of the oldest methods of estimation, and the **method of maximum likelihood**. Some further discussion of Bayesian estimation will be given in Section 10.8, and another method, the **method of least squares**, will be taken up in Chapter 14.

The method of moments consists of equating the first few moments of a population to the corresponding moments of a sample, thus getting as many equations as are needed to solve for the unknown parameters of the population.

DEFINITION 10.4 The kth **sample moment** of a set of observations $x_1, x_2, \ldots,$ and x_n is the mean of their kth powers and it is denoted by m'_k; symbolically,

$$m'_k = \frac{\sum_{i=1}^{n} x_i^k}{n}$$

Thus, the method of moments consists of solving the system of equations

$$m'_k = \mu'_k \qquad k = 1, 2, \ldots, p$$

for the p parameters of the population.

EXAMPLE 10.9

Given a random sample of size n from a gamma population, use the method of moments to estimate its parameters α and β.

Solution

The system of equations we shall have to solve is

$$m'_1 = \mu'_1 \quad \text{and} \quad m'_2 = \mu'_2$$

Since $\mu'_1 = \alpha\beta$ and $\mu'_2 = \alpha(\alpha + 1)\beta^2$ according to Theorem 6.2, we get

$$m'_1 = \alpha\beta \quad \text{and} \quad m'_2 = \alpha(\alpha + 1)\beta^2$$

and, solving these two equations for α and β, we find that the estimates of the two parameters of the gamma distribution are

$$\hat{\alpha} = \frac{(m'_1)^2}{m'_2 - (m'_1)^2} \quad \text{and} \quad \hat{\beta} = \frac{m'_2 - (m'_1)^2}{m'_1}$$

where $m'_1 = \bar{x}$ and $m'_2 = \frac{\sum_{i=1}^{n} x_i^2}{n}$. Thus, in terms of the original observations

$$\hat{\alpha} = \frac{n\bar{x}^2}{\sum_{i=1}^{n} (x_i - \bar{x})^2} \quad \text{and} \quad \hat{\beta} = \frac{\sum_{i=1}^{n} (x_i - \bar{x})^2}{n\bar{x}}$$

and we get the corresponding estimators by substituting $\mathbf{x}_i$ for x_i and $\bar{\mathbf{x}}$ for $\bar{x}$.

In the above example we estimated the parameters of a specific population. It is important to note, however, that when the parameters to be estimated are the moments of the population, the method of moments can be used without knowledge of the exact functional form of the population.

10.7 THE METHOD OF MAXIMUM LIKELIHOOD

In two papers published in the early 1920's, R. A. Fisher, one of the foremost statisticians of our time, proposed a general method of estimation called the **method of maximum likelihood**. He also demonstrated the advantages of this method by showing that it yields sufficient estimators whenever they exist, and that the estimators are asymptotically minimum variance unbiased estimators.

To help understand the principle on which the method of maximum likelihood is based, suppose we are told that four letters arrived in the morning mail, but one of them has been misplaced. If, among the other three, two contain credit card billings and the other an invitation to a church social, what might be a good estimate of the parameter k, the total number of credit card billings in the four letters received? Clearly, it must be two or three, and assuming the missing letter to have been a random selection, we obtain probabilities of

$$\frac{\binom{2}{2}\binom{2}{1}}{\binom{4}{3}} = \frac{1}{2} \quad \text{and} \quad \frac{\binom{3}{2}\binom{1}{1}}{\binom{4}{3}} = \frac{3}{4}$$

respectively, for getting the observed data. Therefore, if we choose as our estimate of the total number of credit card billings the value which maximizes the probability of the observed data provided by the three letters, we get $k = 3$. We call this estimate a **maximum likelihood estimate**, and the method by which it was obtained the method of maximum likelihood.

Thus, the essential feature of the method of maximum likelihood is that we look at the values of a random sample and then choose as our estimate of the unknown population parameter the value for which the probability of obtaining the observed data is a maximum. If the observed sample values are $x_1, x_2, \ldots,$ and x_n, we can write in the discrete case

$$P(\mathbf{x}_1 = x_1, \mathbf{x}_2 = x_2, \ldots, \mathbf{x}_n = x_n) = f(x_1, x_2, \ldots, x_n; \theta)$$

which is just the value of the joint probability distribution of the random variables $\mathbf{x}_1, \mathbf{x}_2, \ldots,$ and $\mathbf{x}_n$ at the sample point $(x_1, x_2, \ldots, x_n)$. Since the sample values have been observed and are therefore fixed numbers, we regard $f(x_1, x_2, \ldots, x_n; \theta)$ as the value of a function of the parameter θ, referred to as the **likelihood function**. A similar definition applies when the random sample comes from a continuous population, but in that case $f(x_1, x_2, \ldots, x_n; \theta)$ is the value of the joint probability density at the sample point $(x_1, x_2, \ldots, x_n)$.

DEFINITION 10.5 If $x_1, x_2, \ldots,$ and x_n are the values of a random sample from a population with the parameter θ, the **likelihood function** of the sample is given by

$$L(\theta) = f(x_1, x_2, \ldots, x_n;\ \theta)$$

for values of θ within a given domain. Here $f(x_1, x_2, \ldots, x_n; \theta)$ is the value of the joint probability distribution or joint density function of the random variables $\mathbf{x}_1, \mathbf{x}_2, \ldots,$ and $\mathbf{x}_n$ at the observed sample point.

Thus, the method of maximum likelihood consists of maximizing the likelihood function with respect to θ, and we refer to the value of θ which maximizes the likelihood function as the maximum likelihood estimate of θ.

EXAMPLE 10.10

Given x "successes" in n trials, find the maximum likelihood estimator of the parameter θ of the binomial distribution.

Solution

To find the value of θ which maximizes

$$L(\theta) = b(x; n, \theta) = \binom{n}{x}\theta^x(1 - \theta)^{n-x}$$

it will be convenient to make use of the fact that the value of θ which maximizes $L(\theta)$ will also maximize

$$\ln L(\theta) = \ln\binom{n}{x} + x \cdot \ln \theta + (n - x) \cdot \ln(1 - \theta)$$

Thus, we get

$$\frac{d[\ln L(\theta)]}{d\theta} = \frac{x}{\theta} - \frac{n - x}{1 - \theta}$$

and, equating this derivative to 0 and solving for θ, we find that the likelihood function has a maximum at $\theta = \frac{x}{n}$. Hence, the maximum likelihood estimator of the parameter θ of the binomial distribution is $\hat{\boldsymbol{\theta}} = \frac{\mathbf{x}}{n}$.

EXAMPLE 10.11

If $x_1, x_2, \ldots,$ and x_n are the values of a random sample from an exponential population, find the maximum likelihood estimate of its parameter θ.

Solution

Since the likelihood function is given by

$$L(\theta) = f(x_1, x_2, \ldots, x_n; \theta)$$
$$= \prod_{i=1}^{n} f(x_i; \theta)$$
$$= \left(\frac{1}{\theta}\right)^n \cdot e^{-\frac{1}{\theta}\left(\sum_{i=1}^{n} x_i\right)}$$

differentiation of $\ln L(\theta)$ with respect to θ yields

$$\frac{d[\ln L(\theta)]}{d\theta} = -\frac{n}{\theta} + \frac{1}{\theta^2} \cdot \sum_{i=1}^{n} x_i$$

Equating this derivative to zero and solving for θ, we get the maximum likelihood estimate

$$\hat{\theta} = \frac{1}{n} \cdot \sum_{i=1}^{n} x_i = \bar{x}$$

Let us also consider an example in which the methods of elementary calculus cannot be used to find the maximum value of the likelihood function.

EXAMPLE 10.12

If $x_1, x_2, \ldots,$ and x_n are the values of a random sample from a continuous uniform population with $\alpha = 0$ and $\beta = \theta$, find the maximum likelihood estimator of θ.

Solution

Since the likelihood function is given by

$$L(\theta) = \prod_{i=1}^{n} f(x_i; \theta) = \left(\frac{1}{\theta}\right)^n$$

it can be seen that the values of this function increase when θ decreases, so that we must make θ as small as possible. Evidently, θ cannot be less than any values of the sample, so that the likelihood function attains its maximum when we set θ equal to the largest value of the sample. Thus, the maximum likelihood estimator of θ is $\hat{\boldsymbol{\theta}} = \mathbf{y}_n$, the nth order statistic.

The method of maximum likelihood can also be used for the simultaneous estimation of several parameters of a given population; in that case we must find the values of the parameters which together maximize the likelihood function.

EXAMPLE 10.13

Given a random sample of size n from a normal population with the mean μ and the variance σ^2, find joint maximum likelihood estimates of these two parameters.

Solution

Since the likelihood function is given by

$$L(\mu, \sigma^2) = \prod_{i=1}^{n} n(x_i; \mu, \sigma)$$

$$= \left(\frac{1}{\sigma\sqrt{2\pi}}\right)^n \cdot e^{-\frac{1}{2\sigma^2}\cdot \sum_{i=1}^{n}(x_i-\mu)^2}$$

partial differentiation of $\ln L(\mu, \sigma^2)$ with respect to μ and σ^2 yields

$$\frac{\partial[\ln L(\mu, \sigma^2)]}{\partial\mu} = \frac{1}{\sigma^2}\cdot\sum_{i=1}^{n}(x_i - \mu)$$

and

$$\frac{\partial[\ln L(\mu, \sigma^2)]}{\partial\sigma^2} = -\frac{n}{2\sigma^2} + \frac{1}{2\sigma^4}\cdot\sum_{i=1}^{n}(x_i - \mu)^2$$

Equating the first of these two partial derivatives to zero and solving for μ, we get

$$\hat{\mu} = \frac{1}{n}\cdot\sum_{i=1}^{n} x_i = \bar{x}$$

and equating the second of these partial derivatives to zero and solving for σ^2 after substituting $\mu = \bar{x}$, we get

$$\hat{\sigma}^2 = \frac{1}{n}\cdot\sum_{i=1}^{n}(x_i - \bar{x})^2$$

Since $\hat{\sigma}^2$ does not equal s^2 in this example, we find that maximum likelihood estimators need not be unbiased. It should also be observed that we did not prove that $\hat{\sigma}$ is a maximum likelihood estimate of σ, but only that $\hat{\sigma}^2$ is a maximum likelihood estimate of σ^2. However, it can be shown (see reference on page 340) that maximum likelihood estimators have the invariance property that if $\hat{\boldsymbol{\theta}}$ is a maximum likelihood estimator of θ and the function given by $g(\theta)$ is continuous, then $g(\hat{\boldsymbol{\theta}})$ is also a maximum likelihood estimator of $g(\theta)$. From this we can conclude that

$$\hat{\sigma} = \sqrt{\frac{1}{n} \cdot \sum_{i=1}^{n} (x_i - \bar{x})^2}$$

is also a maximum likelihood estimate of σ.

Although we maximized the logarithm of the likelihood function instead of the likelihood function, itself, in all of the examples in which we were able to use calculus, this is by no means necessary; it so happened that it was convenient in each case.

THEORETICAL EXERCISES

1. Use the method of moments to find an estimator for the parameter θ of the uniform density of Example 10.12.

2. If $x_1, x_2, \ldots,$ and x_n are the values of a random sample of size n from a population having the density

$$f(x; \theta) = \begin{cases} \dfrac{2(\theta - x)}{\theta^2} & \text{for } 0 < x < \theta \\ 0 & \text{elsewhere} \end{cases}$$

find an estimator for θ by the method of moments.

3. If $x_1, x_2, \ldots,$ and x_n are the values of a random sample of size n from a Poisson population with the parameter λ, find an estimate of λ using

(a) the method of moments;

(b) the method of maximum likelihood.

4. If $x_1, x_2, \ldots,$ and x_n are the values of a random sample of size n from a beta population with $\beta = 1$, find an estimate of the parameter α using

(a) the method of moments;

(b) the method of maximum likelihood.

5. Given a random sample of size n from a normal population with the known mean μ, find the maximum likelihood estimator of σ.

6. If $x_1, x_2, \ldots,$ and x_n are the values of a random sample of size n from a geometric population, find an estimate of its parameter θ using

(a) the method of moments;

(b) the method of maximum likelihood.

7. Given a random sample of size n from a population having the density

$$f(x;\theta) = \begin{cases} e^{-(x-\theta)} & \text{for } x > \theta \\ 0 & \text{elsewhere} \end{cases}$$

find an estimator of the parameter θ by the method of maximum likelihood.

8. Given a random sample of size n from a population having the density

$$f(x;\theta) = \begin{cases} (\theta+1)x^{\theta} & \text{for } 0 < x < 1 \\ 0 & \text{elsewhere} \end{cases}$$

find an estimator of θ using

(a) the method of moments;

(b) the method of maximum likelihood.

9. Among N independent random variables having identical binomial distributions with the parameters θ and $n = 2$, n_0 take on the value 0, n_1 take on the value 1, and n_2 take on the value 2. Find an estimate of θ using

(a) the method of moments;

(b) the method of maximum likelihood.

10. Given a random sample of size n from a gamma population with the known parameter α, find

(a) the maximum likelihood estimator of β;

(b) the maximum likelihood estimator of $\tau = (2\beta - 1)^2$.

11. Given a random sample of size n from a population having a uniform density, find simultaneous maximum likelihood estimators of the parameters α and β.

12. Given a random sample of size n from a population having the density

$$f(x;\delta,\varphi) = \begin{cases} \dfrac{1}{\varphi}\cdot e^{-\frac{x-\delta}{\varphi}} & \text{for } x > \delta \\ 0 & \text{elsewhere} \end{cases}$$

where $-\infty < \delta < \infty$ and $0 < \varphi < \infty$, find simultaneous maximum likelihood estimates of δ and φ.

13. Given independent random samples $\mathbf{x}_1, \mathbf{x}_2, \ldots, \mathbf{x}_n$, and $\mathbf{y}_1, \mathbf{y}_2, \ldots, \mathbf{y}_n$ from two normal populations having the means $\mu_1 = \alpha + \beta$ and $\mu_2 = \alpha - \beta$ and the common variance $\sigma^2 = 1$, find simultaneous maximum likelihood estimators for α and β.

10.8 BAYESIAN ESTIMATORS

So far we have assumed in this chapter that the parameters which we want to estimate are unknown constants; in Bayesian estimation the parameters are looked upon as random variables having **prior distributions**, usually reflecting the strength of one's belief about the possible values they can assume. In Section 9.6, we already met a problem of Bayesian estimation—the parameter was that of the uniform density whose values are $\frac{1}{\theta}$ for the interval from 0 to θ, and 0 elsewhere, and its prior distribution was a gamma distribution with $\alpha = 2$ and $\beta = 1$.

The main problem of Bayesian estimation is that of combining prior feelings about a parameter with direct sample evidence, and in Example 9.9 we accomplished this by determining $\varphi(\theta|x)$, the conditional density of $\boldsymbol{\theta}$ given $\mathbf{x} = x$. In contrast to the prior distribution of $\boldsymbol{\theta}$, this conditional distribution which also reflects the direct sample evidence is called the **posterior distribution** of $\boldsymbol{\theta}$. In general, if $h(\theta)$ is the value of the prior distribution of $\boldsymbol{\theta}$ at θ, and we want to combine the information which it conveys with direct sample evidence about $\boldsymbol{\theta}$, say, the value of a statistic $\mathbf{w} = u(\mathbf{x}_1, \mathbf{x}_2, \ldots, \mathbf{x}_n)$, we determine the posterior distribution of $\boldsymbol{\theta}$ by means of the formula

$$\varphi(\theta|w) = \frac{f(\theta, w)}{g(w)} = \frac{h(\theta) \cdot f(w|\theta)}{g(w)}$$

Here $f(w|\theta)$ is a value of the sampling distribution of $\mathbf{w}$ given $\boldsymbol{\theta} = \theta$, $f(\theta, w)$ is a value of the joint distribution of $\boldsymbol{\theta}$ and $\mathbf{w}$, and $g(w)$ is a value of the marginal distribution of $\mathbf{w}$. Note that the above formula for $\varphi(\theta|w)$ is, in fact, an extension of Bayes' theorem, Theorem 2.13, to the continuous case; hence, the term "Bayesian estimation."

Once the posterior distribution of a parameter has been obtained, it can be used to make estimates, as in Example 9.9, or it can be used to make probability statements about the parameter, as will be illustrated in Example 10.15. Although the method we have described has extensive applications, we shall limit our discussion here to inferences about the parameter θ of a binomial population and the parameter μ of a normal population; inferences about the parameter λ of a Poisson population are treated in Exercise 4 on page 338.

THEOREM 10.5 If $\mathbf{x}$ is a binomial random variable and the prior distribution of $\boldsymbol{\theta}$ is a beta distribution with given values of α and β, then the posterior distribution of $\boldsymbol{\theta}$ given $\mathbf{x} = x$ is a beta distribution with the parameters $x + \alpha$ and $n - x + \beta$.

Proof. For $\boldsymbol{\theta} = \theta$ we have

$$f(x|\theta) = \binom{n}{x}\theta^x(1-\theta)^{n-x} \qquad \text{for } x = 0, 1, 2, \ldots, n$$

$$h(\theta) = \begin{cases} \dfrac{\Gamma(\alpha+\beta)}{\Gamma(\alpha)\cdot\Gamma(\beta)} \cdot \theta^{\alpha-1}(1-\theta)^{\beta-1} & \text{for } 0 < \theta < 1 \\ 0 & \text{elsewhere} \end{cases}$$

and, hence,

$$f(\theta, x) = \frac{\Gamma(\alpha+\beta)}{\Gamma(\alpha)\cdot\Gamma(\beta)} \cdot \theta^{\alpha-1}(1-\theta)^{\beta-1} \times \binom{n}{x}\theta^x(1-\theta)^{n-x}$$

$$= \binom{n}{x} \cdot \frac{\Gamma(\alpha+\beta)}{\Gamma(\alpha)\cdot\Gamma(\beta)} \cdot \theta^{x+\alpha-1}(1-\theta)^{n-x+\beta-1}$$

for $0 < \theta < 1$ and $x = 0, 1, 2, \ldots, n$, and $f(\theta, x) = 0$ elsewhere. To obtain the marginal density of **x**, let us make use of the fact that the integral of the beta density from 0 to 1 equals 1, namely, that

$$\int_0^1 x^{\alpha-1}(1-x)^{\beta-1}\, dx = \frac{\Gamma(\alpha)\cdot\Gamma(\beta)}{\Gamma(\alpha+\beta)}$$

Thus, we get

$$g(x) = \binom{n}{x} \cdot \frac{\Gamma(\alpha+\beta)}{\Gamma(\alpha)\cdot\Gamma(\beta)} \cdot \frac{\Gamma(\alpha+x)\cdot\Gamma(n-x+\beta)}{\Gamma(n+\alpha+\beta)}$$

for $x = 0, 1, \ldots, n$ and, hence,

$$\varphi(\theta|x) = \frac{\Gamma(n+\alpha+\beta)}{\Gamma(\alpha+x)\cdot\Gamma(n-x+\beta)} \cdot \theta^{x+\alpha-1}(1-\theta)^{n-x+\beta-1}$$

for $0 < \theta < 1$, and $\varphi(\theta|x) = 0$ elsewhere. As can be seen by inspection, this is a beta density with the parameters $x + \alpha$ and $n - x + \beta$.

To make use of this theorem, let us refer to the result that (under very general conditions) the mean of the posterior distribution minimizes the Bayes risk when the loss function is quadratic, namely, when the loss function is given by

$$L[d(x), \theta] = c[d(x) - \theta]^2$$

where c is a positive constant. Note that this is the kind of loss function which we used in Example 9.9. Since the posterior distribution of $\boldsymbol{\theta}$ of Theorem 10.5 is a beta distribution with the parameters $x + \alpha$ and $n - x + \beta$, it follows from Theorem 6.5 that

$$E(\boldsymbol{\theta}|x) = \frac{x + \alpha}{\alpha + \beta + n}$$

is a value of an estimator of $\boldsymbol{\theta}$ which minimizes the Bayes risk when the loss function is quadratic and the prior distribution of $\boldsymbol{\theta}$ is of the given form.

EXAMPLE 10.14

Find a point estimate of the "true" probability of a success, if 42 successes are obtained in 120 binomial trials and the prior distribution of $\boldsymbol{\theta}$ is a beta distribution with $\alpha = \beta = 40$.

Solution

Substituting $x = 42$, $n = 120$, $\alpha = 40$, and $\beta = 40$ into the above formula for $E(\boldsymbol{\theta}|x)$, we get

$$E(\boldsymbol{\theta}|42) = \frac{42 + 40}{40 + 40 + 120} = 0.41$$

Note that without knowledge of the prior distribution of $\boldsymbol{\theta}$, the minimum variance unbiased estimate of θ (see Exercise 9 on page 323) would be the sample proportion

$$\hat{\theta} = \frac{x}{n} = \frac{42}{120} = 0.35$$

THEOREM 10.6 If $\bar{\mathbf{x}}$ is the mean of a random sample of size n from a normal population with the known variance σ^2, and the prior distribution of $\boldsymbol{\mu}$ is a normal distribution with the mean μ_0 and the variance σ_0^2, then the posterior distribution of $\boldsymbol{\mu}$ given $\bar{\mathbf{x}} = \bar{x}$ is a normal distribution with the mean μ_1 and the variance σ_1^2, where

$$\mu_1 = \frac{n\bar{x}\sigma_0^2 + \mu_0\sigma^2}{n\sigma_0^2 + \sigma^2} \quad \text{and} \quad \frac{1}{\sigma_1^2} = \frac{n}{\sigma^2} + \frac{1}{\sigma_0^2}$$

Proof. For $\boldsymbol{\mu} = \mu$ we have

$$f(\bar{x}|\mu) = \frac{\sqrt{n}}{\sigma\sqrt{2\pi}} \cdot e^{-\frac{1}{2}\left(\frac{\bar{x}-\mu}{\sigma/\sqrt{n}}\right)^2} \quad \text{for } -\infty < \bar{x} < \infty$$

according to Theorem 8.3, and

$$h(\mu) = \frac{1}{\sigma_0\sqrt{2\pi}} \cdot e^{-\frac{1}{2}\left(\frac{\mu-\mu_0}{\sigma_0}\right)^2} \quad \text{for } -\infty < \mu < \infty$$

so that

$$\varphi(\mu|\bar{x}) = \frac{h(\mu)\cdot f(\bar{x}|\mu)}{g(\bar{x})} = \frac{\sqrt{n}}{2\pi\sigma\sigma_0 g(\bar{x})} \cdot e^{-\frac{1}{2}\left(\frac{\bar{x}-\mu}{\sigma/\sqrt{n}}\right)^2 - \frac{1}{2}\left(\frac{\mu-\mu_0}{\sigma_0}\right)^2}$$

$$\text{for } -\infty < \mu < \infty$$

Now, if we collect powers of μ in the exponent of e, we get

$$-\frac{1}{2}\left(\frac{n}{\sigma^2} + \frac{1}{\sigma_0^2}\right)\mu^2 + \left(\frac{n\bar{x}}{\sigma^2} + \frac{\mu_0}{\sigma_0^2}\right)\mu - \frac{1}{2}\left(\frac{n\bar{x}^2}{\sigma^2} + \frac{\mu_0^2}{\sigma_0^2}\right)$$

and if we let

$$\frac{1}{\sigma_1^2} = \frac{n}{\sigma^2} + \frac{1}{\sigma_0^2} \quad \text{and} \quad \mu_1 = \frac{n\bar{x}\sigma_0^2 + \mu_0\sigma^2}{n\sigma_0^2 + \sigma^2}$$

factor out $-\dfrac{1}{2\sigma_1^2}$, and complete the square, the exponent of e in the expression for $\varphi(\mu|\bar{x})$ becomes

$$-\frac{1}{2\sigma_1^2}(\mu - \mu_1)^2 + R$$

where R involves n, $\bar{x}$, μ_0, σ, and σ_0, but not μ. Thus, the posterior distribution of $\boldsymbol{\mu}$ becomes

$$\varphi(\mu|\bar{x}) = \frac{\sqrt{n}\cdot e^R}{2\pi\sigma\sigma_0 g(\bar{x})} \cdot e^{-\frac{1}{2\sigma_1^2}(\mu-\mu_1)^2} \quad \text{for } -\infty < \mu < \infty$$

which is easily identified as a normal distribution with the mean μ_1 and the variance σ_1^2. Hence, it can be written as

$$\varphi(\mu|\bar{x}) = \frac{1}{\sigma_1\sqrt{2\pi}} \cdot e^{-\frac{1}{2}\left(\frac{\mu-\mu_1}{\sigma_1}\right)^2} \quad \text{for } -\infty < \mu < \infty$$

where μ_1 and σ_1 are defined above. Note that we did not have to determine $g(\bar{x})$ as it was absorbed in the constant in the final result.

EXAMPLE 10.15

A distributor of soft drink vending machines feels that in a supermarket one of his machines will sell on the average $\mu_0 = 738$ drinks per week. Of course the mean will vary somewhat from market to market, and the distributor feels that this variation is measured by the standard deviation $\sigma_0 = 13.4$. So far as a machine placed in a particular market is concerned, the number of drinks sold will vary from week to week, and this variation is measured by the standard deviation $\sigma = 42.5$. If one of the distributor's machines put into a new supermarket averaged $\bar{x} = 692$ during the first ten weeks, what is the probability (the distributor's personal probability) that for this market the value of μ is actually between 700 and 720?

Solution

Assuming that the population sampled is approximately normal and that it is reasonable to treat the prior distribution of μ as a normal distribution with the mean $\mu_0 = 738$ and the standard deviation $\sigma_0 = 13.4$, substitution into the two formulas of Theorem 10.6 yields

$$\mu_1 = \frac{10 \cdot 692(13.4)^2 + 738(42.5)^2}{10(13.4)^2 + (42.5)^2} = 715$$

and

$$\frac{1}{\sigma_1^2} = \frac{10}{(42.5)^2} + \frac{1}{(13.4)^2} = 0.0111$$

so that $\sigma_1^2 = 90.0$ and $\sigma_1 = 9.5$. Now, the answer to our question is given by the area of the white region of Figure 10.1, namely, the area under the standard

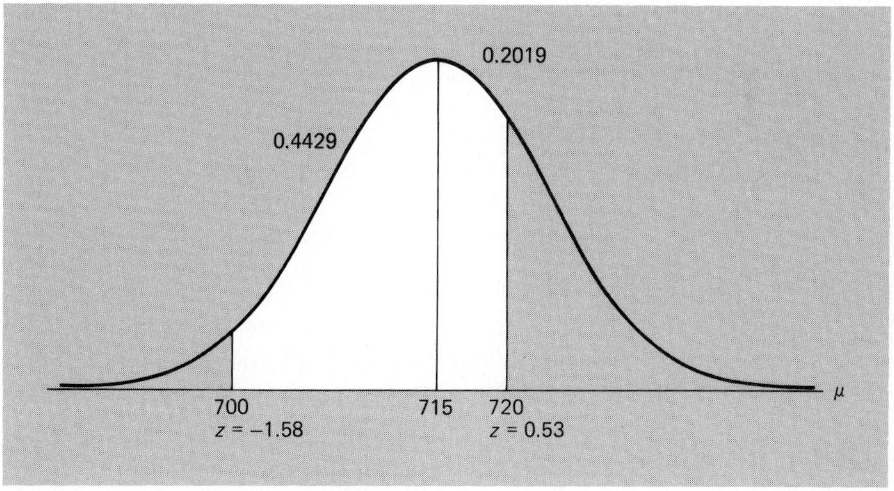

Figure 10.1 Posterior distribution of μ.

normal curve between

$$z = \frac{700 - 715}{9.5} = -1.58 \quad \text{and} \quad z = \frac{720 - 715}{9.5} = 0.53$$

Thus, the probability that $\boldsymbol{\mu}$ is between 700 and 720 is $0.4429 + 0.2019 = 0.6448$, or approximately 0.64.

THEORETICAL EXERCISES

1. Making use of the results of Exercise 21 on page 204, show that the mean of the posterior distribution of $\boldsymbol{\theta}$ given on page 334 can be written as

$$E(\boldsymbol{\theta}|x) = w \cdot \frac{x}{n} + (1 - w) \cdot \theta_0$$

namely, as a weighted mean of $\frac{x}{n}$ and θ_0, where θ_0 and σ_0^2 are the mean and the variance of the prior beta distribution of $\boldsymbol{\theta}$, and

$$w = \frac{n}{n + \frac{\theta_0(1 - \theta_0)}{\sigma_0^2} - 1}$$

2. In Example 10.14, the prior distribution of the parameter $\boldsymbol{\theta}$ of the binomial distribution was a beta distribution with the parameters $\alpha = \beta = 40$. Use Theorem 6.5 to find the mean and the variance of this prior distribution and describe its shape.

3. Show that the mean of the posterior distribution of $\boldsymbol{\mu}$ given in Theorem 10.6 can be written as

$$\mu_1 = w \cdot \bar{x} + (1 - w) \cdot \mu_0$$

namely, as a weighted mean of $\bar{x}$ and μ_0, where

$$w = \frac{n}{n + \frac{\sigma^2}{\sigma_0^2}}$$

4. If $\mathbf{x}$ has a Poisson distribution and the prior distribution of its parameter $\boldsymbol{\lambda}$ is a gamma distribution with the parameters α and β, show that

(a) the posterior distribution of $\boldsymbol{\lambda}$ given $\mathbf{x} = x$ is a gamma distribution with the parameters $\alpha + x$ and $\frac{\beta}{\beta + 1}$;

(b) the mean of this posterior distribution of $\boldsymbol{\lambda}$ is

$$\mu_1 = \frac{\beta(\alpha + x)}{\beta + 1}$$

APPLIED EXERCISES

5. The output of a certain transistor production line is checked daily by inspecting a sample of 100 units. Over a long period of time, the process has maintained a yield of 80 percent, namely, a proportion defective of 20 percent, and the variation of the proportion defective from day to day is measured by a standard deviation of 0.04. If on a certain day the sample contains 38 defectives, find the mean of the posterior distribution of $\boldsymbol{\theta}$ as an estimate of that day's proportion defective. Assume that the prior distribution of $\boldsymbol{\theta}$ is a beta distribution.

6. Records of a university (collected over many years) show that on the average 74 percent of all incoming freshmen have I.Q.'s of at least 115. Of course, the percentage varies somewhat from year to year and this variation is measured by a standard deviation of 3 percent. If a sample check of 30 freshmen entering the university in 1979 showed that only 18 of them have I.Q.'s of at least 115, estimate the true proportion of students with I.Q.'s of at least 115 in that freshman class using

(a) only the prior information;
(b) only the direct information;
(c) the results of Exercise 1 to combine the prior information with the direct information.

7. With reference to Example 10.15, find $P(712 < \boldsymbol{\mu} < 725 | \bar{x} = 692)$.

8. A history professor is making up a final examination which is to be given to a very large group of students. His feelings about the average grade they should get is expressed subjectively by a normal distribution with the mean $\mu_0 = 65.2$ and the standard deviation $\sigma_0 = 1.5$.

(a) What prior probability does the professor assign to the actual average grade being somewhere on the interval from 63.0 to 68.0?
(b) What posterior probability would he assign to this event if the examination is tried on a random sample of 40 students whose grades have a mean of 72.9 and a standard deviation of 7.4? Use $s = 7.4$ as an estimate of σ.

9. An office manager knows that for a certain kind of business the daily number of incoming telephone calls is a random variable having a Poisson distribution, whose parameter has a prior gamma distribution with $\alpha = 50$ and $\beta = 2$. Being told that one such business had 112 incoming calls on a given day, what would be his estimate of that particular business' average daily number of incoming calls if he considers

(a) only the prior information;
(b) only the direct information;
(c) both kinds of information and the theory of Exercise 4?

References

Various properties of sufficient estimators are discussed in

WILKS, S. S., *Mathematical Statistics.* New York: John Wiley & Sons, Inc., 1962,

and a proof of Theorem 10.4 may be found in

HOGG, R. V., and CRAIG, A. T., *Introduction to Mathematical Statistics*, 4th ed. New York: Macmillan Publishing Co., Inc., 1978.

Important properties of maximum likelihood estimators are discussed in

KEEPING, E. S., *Introduction to Statistical Inference.* Princeton, N.J.: D. Van Nostrand Co., Inc., 1962,

and a derivation of the Cramér–Rao inequality, as well as the most general conditions under which it applies, may be found in

RAO, C. R., *Advanced Statistical Methods in Biometric Research.* New York: John Wiley & Sons, Inc., 1952.

11 interval estimation

11.1 INTRODUCTION

Since we can rarely expect point estimates to equal the parameters they are supposed to estimate, it is generally desirable to give ourselves some leeway by using **interval estimates**. An interval estimate of a parameter θ is an interval of the form $\hat{\theta}_1 < \theta < \hat{\theta}_2$, where $\hat{\theta}_1$ and $\hat{\theta}_2$ depend on the value taken on by the estimator $\hat{\boldsymbol{\theta}}$ in a given sample and also on the sampling distribution of $\hat{\boldsymbol{\theta}}$. For instance, if we are asked to estimate the average I.Q. of a very large group of students on the basis of a random sample, we might arrive at the interval estimate $109 < \mu < 117$ on the basis of the sample mean $\bar{x} = 113$ as well as knowledge about the sampling distribution of $\bar{\mathbf{x}}$.

Since different samples will generally yield different values of $\hat{\boldsymbol{\theta}}$ and, hence, different values of $\hat{\theta}_1$ and $\hat{\theta}_2$, these endpoints of the interval are values of corresponding random variables $\hat{\boldsymbol{\theta}}_1$ and $\hat{\boldsymbol{\theta}}_2$. Based on the sampling distribution of $\hat{\boldsymbol{\theta}}$ we can thus assert with a given probability whether such an interval will actually contain the parameter it is supposed to estimate. In other words, we can use the sampling distribution of $\hat{\boldsymbol{\theta}}$ to choose $\hat{\theta}_1$ and $\hat{\theta}_2$ such that for any specified probability $1 - \alpha$, where $0 < \alpha < 1$,

$$P(\hat{\boldsymbol{\theta}}_1 < \theta < \hat{\boldsymbol{\theta}}_2) = 1 - \alpha$$

Such an interval $\hat{\theta}_1 < \theta < \hat{\theta}_2$, computed for a particular sample, is called a

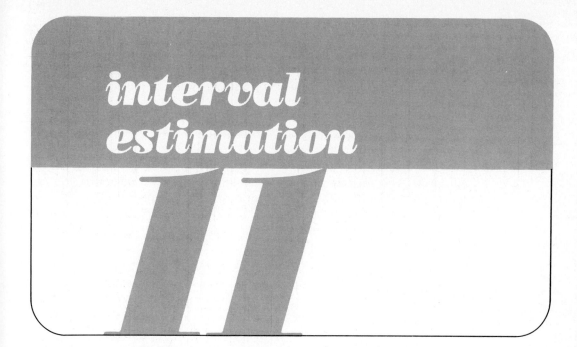

$(1 - \alpha)100\%$ **confidence interval**, the fraction $1 - \alpha$ is called the **confidence coefficient** or the **degree of confidence**, and the endpoints $\hat{\theta}_1$ and $\hat{\theta}_2$ are called the lower and upper **confidence limits**. For instance, when $\alpha = 0.05$ the degree of confidence is 0.95 and we get a 95% confidence interval.

Let us make it clear at this point that confidence intervals for given parameters are not unique. This is illustrated by Exercises 2 and 3 on page 350, and also in Section 11.2, where we show that, based on a single random sample, there exist numerous confidence intervals for μ all having the same degree of confidence. As in the case of point estimation, methods of obtaining confidence intervals must, thus, be judged by their various statistical properties. For instance, one desirable property is to have the length of a $(1 - \alpha)100\%$ confidence interval as short as possible. Another desirable property is to have the expected length, $E(\hat{\boldsymbol{\theta}}_2 - \hat{\boldsymbol{\theta}}_1)$, as small as possible.

11.2 CONFIDENCE INTERVALS FOR MEANS

Since $\bar{\mathbf{x}}$ is a sufficient estimator of the mean of a normal population with the known variance σ^2 (see Example 10.8 on page 321), let us use it to derive a confidence interval for μ from such a population. By Theorem 8.3, the sampling distribution of $\bar{\mathbf{x}}$ for a random sample of size n from a normal population with the mean μ and the variance σ^2 is a normal distribution with the mean $\mu_{\bar{\mathbf{x}}} = \mu$ and the variance $\sigma_{\bar{\mathbf{x}}}^2 = \dfrac{\sigma^2}{n}$. Thus, we can write

$$P(-z_{\alpha/2} < \mathbf{z} < z_{\alpha/2}) = 1 - \alpha$$

where

$$\mathbf{z} = \frac{\bar{\mathbf{x}} - \mu}{\sigma/\sqrt{n}}$$

and $z_{\alpha/2}$ is such that the integral of the standard normal density from $z_{\alpha/2}$ to ∞ equals $\alpha/2$. It follows that

$$P\left(-z_{\alpha/2} < \frac{\bar{\mathbf{x}} - \mu}{\sigma/\sqrt{n}} < z_{\alpha/2}\right) = 1 - \alpha$$

or, equivalently,

$$P\left(\bar{\mathbf{x}} - z_{\alpha/2} \cdot \frac{\sigma}{\sqrt{n}} < \mu < \bar{\mathbf{x}} + z_{\alpha/2} \cdot \frac{\sigma}{\sqrt{n}}\right) = 1 - \alpha$$

and it can be seen that the random variables $\hat{\boldsymbol{\theta}}_1$ and $\hat{\boldsymbol{\theta}}_2$, defined in Section 11.1, are

$$\hat{\boldsymbol{\theta}}_1 = \bar{\mathbf{x}} - z_{\alpha/2} \cdot \frac{\sigma}{\sqrt{n}} \quad \text{and} \quad \hat{\boldsymbol{\theta}}_2 = \bar{\mathbf{x}} + z_{\alpha/2} \cdot \frac{\sigma}{\sqrt{n}}$$

For a given value of α, they depend only on known constants and the observed sample, and the lower and upper confidence limits can now be obtained by letting $\bar{\mathbf{x}}$ take on its sample value $\bar{x}$. Thus, we have shown that

THEOREM 11.1 *(Confidence interval for μ, σ known)* If $\bar{x}$ is the value of the mean of a random sample of size n from a normal population with the known variance σ^2, a $(1 - \alpha)100\%$ confidence interval for μ is given by

$$\bar{x} - z_{\alpha/2} \cdot \frac{\sigma}{\sqrt{n}} < \mu < \bar{x} + z_{\alpha/2} \cdot \frac{\sigma}{\sqrt{n}}$$

By virtue of the central limit theorem, this result can also be used for random samples from non-normal populations with the known variance σ^2, provided n is sufficiently large; that is, when $n \geqslant 30$.

EXAMPLE 11.1

If a random sample of size $n = 20$ from a normal population with the variance $\sigma^2 = 225$ has the mean $\bar{x} = 64.3$, construct a 95% confidence interval for the population mean μ.

Solution

For $\alpha = 0.05$, we find from Table III that $z_{.025} = 1.96$. Therefore, the 95% confidence interval for μ is

$$64.3 - 1.96 \cdot \frac{15}{\sqrt{20}} < \mu < 64.3 + 1.96 \cdot \frac{15}{\sqrt{20}}$$

which reduces to

$$57.7 < \mu < 70.9$$

The interval which we obtained in this example is a 95% confidence interval for the mean of the population. Of course, the double inequality $57.7 < \mu <$

70.9 must be either true or false, but if we had to bet, 95 to 5 (or 19 to 1) would be **fair odds** that it is true. These odds are fair because the method by which it was obtained works, so to speak, 95% of the time.

The fact that confidence intervals for given parameters are not unique is readily seen by writing the $(1 - \alpha)100\%$ confidence interval of Theorem 11.1 as

$$\bar{x} - z_{2\alpha/3} \cdot \frac{\sigma}{\sqrt{n}} < \mu < \bar{x} + z_{\alpha/3} \cdot \frac{\sigma}{\sqrt{n}}$$

or as the **one-sided** $(1 - \alpha)100\%$ confidence interval

$$\mu < \bar{x} + z_{\alpha} \cdot \frac{\sigma}{\sqrt{n}}$$

We could also have based the confidence interval, say, on the sample median instead of the sample mean.

In order to construct an approximate $(1 - \alpha)100\%$ confidence interval for μ when σ is unknown but $n \geqslant 30$, we replace σ by the value of the sample standard deviation s and proceed as above. However, when we are dealing with a small sample from a normal population and $n < 30$, a $(1 - \alpha)100\%$ confidence interval for μ can be constructed by making use of the fact that the random variable

$$\mathbf{t} = \frac{\bar{\mathbf{x}} - \mu}{\mathbf{s}/\sqrt{n}}$$

has a t distribution with $n - 1$ degrees of freedom (see Theorem 8.11). Hence,

$$P(-t_{\alpha/2,n-1} < \mathbf{t} < t_{\alpha/2,n-1}) = 1 - \alpha$$

where $t_{\alpha/2,n-1}$ is as defined on page 271, and substituting for $\mathbf{t}$ we get

$$P\left(-t_{\alpha/2,n-1} < \frac{\bar{\mathbf{x}} - \mu}{\mathbf{s}/\sqrt{n}} < t_{\alpha/2,n-1}\right) = 1 - \alpha$$

or, equivalently,

$$P\left(\bar{\mathbf{x}} - t_{\alpha/2,n-1} \cdot \frac{\mathbf{s}}{\sqrt{n}} < \mu < \bar{\mathbf{x}} + t_{\alpha/2,n-1} \cdot \frac{\mathbf{s}}{\sqrt{n}}\right) = 1 - \alpha$$

This leads to the following small-sample confidence interval for μ:

THEOREM 11.2 *(Confidence interval for μ, σ unknown)* If $\bar{x}$ and s are the values of the mean and the standard deviation of a random sample of size n from a normal population with the unknown variance σ^2, a $(1 - \alpha)100\%$ confidence interval for μ is given by

$$\bar{x} - t_{\alpha/2,n-1} \cdot \frac{s}{\sqrt{n}} < \mu < \bar{x} + t_{\alpha/2,n-1} \cdot \frac{s}{\sqrt{n}}$$

For $n \geq 30$, this confidence interval formula and the one of Theorem 11.1 with σ replaced by s will generally yield nearly the same results.

EXAMPLE 11.2

A paint manufacturer wants to determine the average drying time of a new interior wall paint. If for 12 test areas of equal size he obtained a mean drying time of 66.3 minutes and a standard deviation of 8.4 minutes, construct a 95% confidence interval for the true mean μ.

Solution

Substituting $\bar{x} = 66.3$, $s = 8.4$, and $t_{.025,11} = 2.201$ (from Table IV), the 95% confidence interval for μ becomes

$$66.3 - 2.201 \cdot \frac{8.4}{\sqrt{12}} < \mu < 66.3 + 2.201 \cdot \frac{8.4}{\sqrt{12}}$$

or simply

$$61.0 < \mu < 71.6$$

This means that we can assert with a 95% degree of confidence that the interval from 61.0 minutes to 71.6 minutes contains the true average drying time of the paint.

The method by which we constructed confidence intervals in this section consisted essentially of finding a suitable random variable whose values are determined by the sample data as well as the population parameters, yet whose distribution does not involve the parameter we are trying to estimate. This was the case, for example, when we used the random variable $\mathbf{z} = \dfrac{\bar{\mathbf{x}} - \mu}{\sigma/\sqrt{n}}$, whose values cannot be calculated without knowledge of μ, but whose distribution, the standard normal distribution, does not involve μ. Although this method of obtaining confidence intervals, sometimes called the **pivotal method**, is very widely used, and we shall use it again in the next few sections, there exist more general methods, for instance, the one discussed in the book by Mood, Graybill, and Boes referred to on page 360.

11.3 CONFIDENCE INTERVALS FOR DIFFERENCES BETWEEN MEANS

In Exercise 4 on page 263 the reader was asked to show that if $\bar{\mathbf{x}}_1$ and $\bar{\mathbf{x}}_2$ are the means of independent random samples of size n_1 and n_2 from normal populations having the means μ_1 and μ_2 and the variances σ_1^2 and σ_2^2, then $\bar{\mathbf{x}}_1 - \bar{\mathbf{x}}_2$ is a random variable having a normal distribution with the mean

$$\mu_{\bar{\mathbf{x}}_1 - \bar{\mathbf{x}}_2} = \mu_1 - \mu_2$$

and the variance

$$\sigma^2_{\bar{\mathbf{x}}_1 - \bar{\mathbf{x}}_2} = \frac{\sigma_1^2}{n_1} + \frac{\sigma_2^2}{n_2}$$

It follows that

$$\mathbf{z} = \frac{(\bar{\mathbf{x}}_1 - \bar{\mathbf{x}}_2) - (\mu_1 - \mu_2)}{\sqrt{\dfrac{\sigma_1^2}{n_1} + \dfrac{\sigma_2^2}{n_2}}}$$

has a standard normal distribution. Substituting this expression for $\mathbf{z}$ into

$$P(-z_{\alpha/2} < \mathbf{z} < z_{\alpha/2}) = 1 - \alpha$$

the pivotal method leads to

$$P\left[(\bar{\mathbf{x}}_1 - \bar{\mathbf{x}}_2) - z_{\alpha/2} \cdot \sqrt{\frac{\sigma_1^2}{n_1} + \frac{\sigma_2^2}{n_2}} < \mu_1 - \mu_2 < (\bar{\mathbf{x}}_1 - \bar{\mathbf{x}}_2) + z_{\alpha/2} \cdot \sqrt{\frac{\sigma_1^2}{n_1} + \frac{\sigma_2^2}{n_2}}\right] = 1 - \alpha$$

and, hence, to the following confidence interval for $\mu_1 - \mu_2$:

THEOREM 11.3 *(Confidence interval for $\mu_1 - \mu_2$, σ_1 and σ_2 known)* If $\bar{x}_1$ and $\bar{x}_2$ are the values of the means of independent random samples of size n_1 and n_2 from normal populations with the known variances σ_1^2 and σ_2^2, a $(1 - \alpha)100\%$ confidence interval for $\mu_1 - \mu_2$ is given by

$$(\bar{x}_1 - \bar{x}_2) - z_{\alpha/2} \cdot \sqrt{\frac{\sigma_1^2}{n_1} + \frac{\sigma_2^2}{n_2}} < \mu_1 - \mu_2 < (\bar{x}_1 - \bar{x}_2) + z_{\alpha/2} \cdot \sqrt{\frac{\sigma_1^2}{n_1} + \frac{\sigma_2^2}{n_2}}$$

By virtue of the central limit theorem, this result can also be used for independent random samples from non-normal populations with the known variances σ_1^2 and σ_2^2, provided n_1 and n_2 are sufficiently large, that is, when n_1 and $n_2 \geqslant 30$.

EXAMPLE 11.3

Construct a 94% confidence interval for the actual difference between the average lifetimes of two kinds of light bulbs, given that a random sample of 40 light bulbs of one kind lasted on the average 418 hours of continuous use and 50 light bulbs of another kind lasted on the average 402 hours. The population standard deviations are known to be $\sigma_1 = 26$ and $\sigma_2 = 22$.

Solution

For $\alpha = 0.06$, we find from Table III that $z_{.03} = 1.88$. Therefore, the 94% confidence interval for $\mu_1 - \mu_2$ is

$$(418 - 402) - 1.88 \cdot \sqrt{\frac{26^2}{40} + \frac{22^2}{50}} < \mu_1 - \mu_2 <$$

$$(418 - 402) + 1.88 \cdot \sqrt{\frac{26^2}{40} + \frac{22^2}{50}}$$

which reduces to

$$6.3 < \mu_1 - \mu_2 < 25.7$$

Hence, we are 94% confident that the interval from 6.3 to 25.7 contains the true difference between the average lifetimes of the two kinds of light bulbs. The fact that both confidence limits are positive suggests that on the average the first kind of bulb is superior to the second kind.

In order to construct a $(1-\alpha)100\%$ confidence interval for $\mu_1-\mu_2$ when σ_1 and σ_2 are unknown but n_1 and $n_2 \geqslant 30$, we replace σ_1 and σ_2 by the values of the sample standard deviations s_1 and s_2 and proceed as above. The procedure for estimating the difference between two means when σ_1 and σ_2 are unknown and the sample sizes are small is not straightforward unless the unknown standard deviations of the two normal populations are equal. If $\sigma_1 = \sigma_2 = \sigma$, then

$$\mathbf{z} = \frac{(\bar{\mathbf{x}}_1 - \bar{\mathbf{x}}_2) - (\mu_1 - \mu_2)}{\sigma\sqrt{\dfrac{1}{n_1} + \dfrac{1}{n_2}}}$$

is a random variable having a standard normal distribution and σ^2 can be estimated by **pooling** the squared deviations from the means of the two samples. In Exercise 6 on page 351 the reader will be asked to show that the resulting **pooled estimator**

$$\mathbf{s}_p^2 = \frac{(n_1-1)\mathbf{s}_1^2 + (n_2-1)\mathbf{s}_2^2}{n_1+n_2-2}$$

is, indeed, an unbiased estimator of σ^2. Now, by Theorems 8.9 and 8.7, the independent random variables $\dfrac{(n_1-1)\mathbf{s}_1^2}{\sigma^2}$ and $\dfrac{(n_2-1)\mathbf{s}_2^2}{\sigma^2}$ have chi-square distributions with n_1-1 and n_2-1 degrees of freedom, and their sum

$$\mathbf{y} = \frac{(n_1-1)\mathbf{s}_1^2}{\sigma^2} + \frac{(n_2-1)\mathbf{s}_2^2}{\sigma^2} = \frac{(n_1+n_2-2)\mathbf{s}_p^2}{\sigma^2}$$

has a chi-square distribution with n_1+n_2-2 degrees of freedom. As it can be shown that the above random variables $\mathbf{z}$ and $\mathbf{y}$ are independent (see references on page 360), it follows from Theorem 8.10 that

$$\mathbf{t} = \frac{\mathbf{z}}{\sqrt{\dfrac{\mathbf{y}}{n_1+n_2-2}}}$$

$$= \frac{(\bar{\mathbf{x}}_1 - \bar{\mathbf{x}}_2) - (\mu_1 - \mu_2)}{\mathbf{s}_p\sqrt{\dfrac{1}{n_1} + \dfrac{1}{n_2}}}$$

has a t distribution with n_1+n_2-2 degrees of freedom. Substituting this expression for t into

$$P(-t_{\alpha/2,n_1+n_2-2} < \mathbf{t} < t_{\alpha/2,n_1+n_2-2}) = 1-\alpha$$

and algebraically simplifying the result, we arrive at the following $(1-\alpha)100\%$ confidence interval for $\mu_1 - \mu_2$:

THEOREM 11.4 *(Confidence interval for $\mu_1 - \mu_2, \sigma_1 = \sigma_2$ unknown)* If $\bar{x}_1$ and $\bar{x}_2$ are the values of the means of independent random samples of size n_1 and n_2 from normal populations with unknown but equal variances, a $(1-\alpha)100\%$ confidence interval for $\mu_1 - \mu_2$ is given by

$$(\bar{x}_1 - \bar{x}_2) - t_{\alpha/2, n_1+n_2-2} \cdot s_p\sqrt{\frac{1}{n_1} + \frac{1}{n_2}} < \mu_1 - \mu_2 < (\bar{x}_1 - \bar{x}_2) + t_{\alpha/2, n_1+n_2-2} \cdot s_p\sqrt{\frac{1}{n_1} + \frac{1}{n_2}}$$

where s_p is the square root of the value of the pooled estimator of the population variance given on page 348.

EXAMPLE 11.4

A study has been made to compare the nicotine contents of two brands of cigarettes. Ten cigarettes of Brand A had an average nicotine content of 3.1 milligrams with a standard deviation of 0.5 milligram, while eight cigarettes of Brand B had an average nicotine content of 2.7 milligrams with a standard deviation of 0.7 milligram. Assuming that the two sets of data are random samples from normal populations with equal variances, construct a 95% confidence interval for the true difference in the average nicotine content of the two brands of cigarettes.

Solution

Let us summarize the data as follows:

Brand A	*Brand B*
$n_1 = 10$	$n_2 = 8$
$\bar{x}_1 = 3.1$	$\bar{x}_2 = 2.7$
$s_1 = 0.5$	$s_2 = 0.7$

For $\alpha = 0.05$ and $n_1 + n_2 - 2 = 16$ degrees of freedom, we find from Table IV that $t_{.025,16} = 2.120$. The value of s_p is

$$s_p = \sqrt{\frac{9(0.25) + 7(0.49)}{16}} = 0.596$$

and, therefore, the 95% confidence interval for $\mu_1 - \mu_2$ is

$$(3.1 - 2.7) - 2.120(0.596)\sqrt{\tfrac{1}{10} + \tfrac{1}{8}} < \mu_1 - \mu_2 < (3.1 - 2.7) + 2.120(0.596)\sqrt{\tfrac{1}{10} + \tfrac{1}{8}}$$

which reduces to

$$-0.20 < \mu_1 - \mu_2 < 1.00$$

This is the required result, but observe that since the actual difference could thus be zero, we cannot conclude that there is a real difference in the nicotine contents of the two brands.

THEORETICAL EXERCISES

1. If x is a value of a random variable having an exponential distribution, find k so that the interval from 0 to kx is a $(1 - \alpha)100\%$ confidence interval for the parameter θ.

2. If x_1 and x_2 are the values of a random sample of size 2 from a population having a uniform density with $\alpha = 0$ and $\beta = \theta$, find k so that

$$0 < \theta < k(x_1 + x_2)$$

is a $(1 - \alpha)100\%$ confidence interval for θ. (*Hint:* Make use of the fact that $\mathbf{x}_1 + \mathbf{x}_2$ has a triangular density similar to that pictured in Figure 7.8.)

3. By using the methods of Section 8.7, it can be shown that for a random sample of size 2 from the population of Exercise 2 the distribution of the sample range is given by

$$f(R) = \begin{cases} \dfrac{2}{\theta^2}(\theta - R) & \text{for } 0 < R < \theta \\ 0 & \text{elsewhere} \end{cases}$$

Use this result to find c so that

$$R < \theta < cR$$

is a $(1 - \alpha)100\%$ confidence interval for θ.

4. Show that the $(1 - \alpha)100\%$ confidence interval

$$\bar{x} - z_{\alpha/2} \cdot \frac{\sigma}{\sqrt{n}} < \mu < \bar{x} + z_{\alpha/2} \cdot \frac{\sigma}{\sqrt{n}}$$

is shorter than the corresponding interval given by

$$\bar{x} - z_{2\alpha/3} \cdot \frac{\sigma}{\sqrt{n}} < \mu < \bar{x} + z_{\alpha/3} \cdot \frac{\sigma}{\sqrt{n}}$$

5. If $\bar{x}$ is used as an estimate of μ, show that we can be $(1 - \alpha)100\%$ confident that $|\bar{x} - \mu|$, the absolute value of our error, will be less than a specified amount e when the sample size is $n = \left(z_{\alpha/2} \cdot \frac{\sigma}{e}\right)^2$.

6. Show that $\mathbf{s}_p^2$ is an unbiased estimator of σ^2 and find its variance.

7. Verify the result on page 348 which expresses $\mathbf{t}$ in terms of $\bar{\mathbf{x}}_1$, $\bar{\mathbf{x}}_2$, and $\mathbf{s}_p$.

APPLIED EXERCISES

8. Measurements of the blood pressure of 25 elderly women have a mean of $\bar{x} = 140$ mm of mercury. If these data can be looked upon as a random sample from a normal population with $\sigma = 10$ mm of mercury, construct a 95% confidence interval for the population mean μ.

9. For several years, a mathematics placement test had been administered to all incoming freshmen at a certain college. If 64 students, randomly selected over this period of time, took on the average 28.5 minutes to complete the test with a variance of 9.3 minutes, construct a 99% confidence interval for the true average time it takes a freshman to complete the test.

10. An efficiency expert wants to determine the average time it takes a pit crew to change a set of four tires on a race car. Use the result of Exercise 5 to determine the sample size required to be able to assert with 95% confidence that the sample mean will differ from the true mean by less than 2 seconds. It is known from previous studies that the population standard deviation is 12 seconds.

11. The length of the skulls of 10 fossil skeletons of an extinct species of birds has a mean of 5.68 cm and a standard deviation of 0.29 cm. Assuming that such measurements are normally distributed, find a 95% confidence interval for the mean length of the skulls of this species of birds.

12. A food inspector, examining 12 jars of a certain brand of peanut butter, obtained the following percentages of impurities: 2.3, 1.9, 2.1, 2.8, 2.3, 3.6, 1.4, 1.8, 2.1, 3.2, 2.0, and 1.9. Assuming that such determinations are normally distributed, construct a 99% confidence interval for the average percentage of impurities in this brand of peanut butter.

13. A random sample of size $n_1 = 16$ from a normal population with $\sigma_1 = 4.8$ has the mean $\bar{x}_1 = 18$, and a random sample of size $n_2 = 25$ from a different normal population with $\sigma_2 = 3.5$ has the mean $\bar{x}_2 = 23$. Find a 90% confidence interval for $\mu_1 - \mu_2$.

14. A study of two kinds of photocopying equipment shows that 60 failures of the first kind of equipment took on the average 80.7 minutes to repair with a standard deviation of 19.4 minutes, while 60 failures of the second kind of equipment took on the average 88.1 minutes to repair with a standard deviation of 18.8 minutes. Find a 99% confidence interval for the difference between the true average times it takes to repair failures of the two kinds of photocopying equipment.

15. Twelve randomly selected mature citrus trees of one variety have a mean height of 13.8 feet with a standard deviation of 1.2 feet, and fifteen randomly selected mature citrus trees of another variety have a mean height of 12.9 feet with a standard deviation of 1.5 feet. Assuming that the random samples were selected from normal populations with equal variances, construct a 95% confidence interval for the difference between the true average heights of the two kinds of citrus trees.

16. The following are the heat-producing capacities of coal from two mines (in millions of calories per ton):

Mine A: 8,500, 8,330, 8,480, 7,960, 8,030
Mine B: 7,710, 7,890, 7,920, 8,270, 7,860

Assuming that the data constitute independent random samples from normal populations with equal variances, construct a 99% confidence interval for the difference between the true average heat-producing capacities of coal from the two mines.

11.4 CONFIDENCE INTERVALS FOR PROPORTIONS

There are many problems in which we must estimate proportions, probabilities, percentages, or rates, such as the proportion of defectives in a large shipment of transistors, the probability that a car stopped at a road block will have faulty lights, the percentage of school children with I.Q.'s over 115, or the mortality rate of a disease. In many of these it is reasonable to assume that we are sampling a binomial population, and, hence, that our problem is to estimate the binomial parameter θ. Making use of the fact that for large n the binomial distribution can be approximated with a normal distribution, namely, that the random variable

$$\mathbf{z} = \frac{\mathbf{x} - n\theta}{\sqrt{n\theta(1 - \theta)}}$$

can be treated as if it had the standard normal distribution, we can write

$$P\left(-z_{\alpha/2} < \frac{\mathbf{x} - n\theta}{\sqrt{n\theta(1-\theta)}} < z_{\alpha/2}\right) = 1 - \alpha$$

and obtain a $(1 - \alpha)100\%$ confidence interval for θ by solving the two inequalities

$$-z_{\alpha/2} < \frac{x - n\theta}{\sqrt{n\theta(1-\theta)}} \quad \text{and} \quad \frac{x - n\theta}{\sqrt{n\theta(1-\theta)}} < z_{\alpha/2}$$

for the corresponding confidence limits. Leaving the details of this to the reader in Exercise 1 on page 358, let us instead give here a large-sample approximation by first writing the above probability as

$$P\left(\hat{\boldsymbol{\theta}} - z_{\alpha/2} \cdot \sqrt{\frac{\theta(1-\theta)}{n}} < \theta < \hat{\boldsymbol{\theta}} + z_{\alpha/2} \cdot \sqrt{\frac{\theta(1-\theta)}{n}}\right) = 1 - \alpha$$

where $\hat{\boldsymbol{\theta}} = \frac{\mathbf{x}}{n}$. Then, substituting $\hat{\theta}$ for θ inside the radicals, which is a further approximation, we get[†]

THEOREM 11.5 *(Large-sample confidence interval for θ)* An approximate $(1 - \alpha)100\%$ confidence interval for the binomial parameter θ is given by

$$\hat{\theta} - z_{\alpha/2} \cdot \sqrt{\frac{\hat{\theta}(1-\hat{\theta})}{n}} < \theta < \hat{\theta} + z_{\alpha/2} \cdot \sqrt{\frac{\hat{\theta}(1-\hat{\theta})}{n}}$$

where $\hat{\theta} = \frac{x}{n}$.

EXAMPLE 11.5

A study is being made to estimate the proportion of voters in a sizeable community who favor the construction of a nuclear power plant. If it is found that only 140 of 400 voters selected at random favor the project, find a 95%

[†] In Exercise 3 on page 359 the reader will be asked to show how this result will have to be modified if we use the continuity correction when approximating the binomial distribution with a normal distribution.

confidence interval for the proportion of all voters in this community who favor the project.

Solution

Substituting $\hat{\theta} = \frac{140}{400} = 0.35$ and $z_{.025} = 1.96$ into the above large-sample confidence interval for θ, we obtain

$$0.35 - 1.96 \cdot \sqrt{\frac{(0.35)(0.65)}{400}} < \theta < 0.35 + 1.96 \cdot \sqrt{\frac{(0.35)(0.65)}{400}}$$

which reduces to

$$0.303 < \theta < 0.397$$

In actual practice, confidence limits for the binomial parameter θ are often obtained by means of the specially constructed tables referred to on page 360. This not only saves a good deal of arithmetic, but it makes it possible to obtain confidence intervals for this parameter when n is small.

11.5 CONFIDENCE INTERVALS FOR DIFFERENCES BETWEEN PROPORTIONS

Problems frequently arise where it is desirable to estimate the difference between the binomial parameters θ_1 and θ_2 on the basis of independent random samples from two binomial populations. Such is the case, for example, if we wish to estimate the difference between the proportions of voters in two different districts that favor Candidate X for election to the United States Senate.

If the respective numbers of successes are $\mathbf{x}_1$ and $\mathbf{x}_2$ and the corresponding sample proportions are $\hat{\boldsymbol{\theta}}_1 = \frac{\mathbf{x}_1}{n_1}$ and $\hat{\boldsymbol{\theta}}_2 = \frac{\mathbf{x}_2}{n_2}$, let us investigate the sampling distribution of $\hat{\boldsymbol{\theta}}_1 - \hat{\boldsymbol{\theta}}_2$, as a potential estimator of $\theta_1 - \theta_2$. For large values of n_1 and n_2, the distributions of $\mathbf{x}_1$ and $\mathbf{x}_2$ can again be approximated with normal distributions having the means $n_1\theta_1$ and $n_2\theta_2$ and the variances $n_1\theta_1(1 - \theta_1)$ and $n_2\theta_2(1 - \theta_2)$. Therefore, according to Exercise 6 on page 263, it follows that $\hat{\boldsymbol{\theta}}_1 - \hat{\boldsymbol{\theta}}_2$ has approximately a normal distribution with the mean

$$\mu_{\hat{\boldsymbol{\theta}}_1 - \hat{\boldsymbol{\theta}}_2} = \theta_1 - \theta_2$$

and the variance

$$\sigma^2_{\hat{\boldsymbol{\theta}}_1 - \hat{\boldsymbol{\theta}}_2} = \frac{\theta_1(1 - \theta_1)}{n_1} + \frac{\theta_2(1 - \theta_2)}{n_2}$$

and the random variable

$$\mathbf{z} = \frac{(\hat{\boldsymbol{\theta}}_1 - \hat{\boldsymbol{\theta}}_2) - (\theta_1 - \theta_2)}{\sqrt{\dfrac{\theta_1(1-\theta_1)}{n_1} + \dfrac{\theta_2(1-\theta_2)}{n_2}}}$$

has approximately the standard normal distribution. Substituting this expression for $\mathbf{z}$ into

$$P(-z_{\alpha/2} < \mathbf{z} < z_{\alpha/2}) = 1 - \alpha$$

and proceeding as in Section 11.4, we arrive at the following result:[†]

THEOREM 11.6 *(Large-sample confidence interval for $\theta_1 - \theta_2$)* An approximate $(1-\alpha)100\%$ confidence interval for $\theta_1 - \theta_2$, the difference between two binomial parameters, is given by

$$(\hat{\theta}_1 - \hat{\theta}_2) - z_{\alpha/2} \cdot \sqrt{\frac{\hat{\theta}_1(1-\hat{\theta}_1)}{n_1} + \frac{\hat{\theta}_2(1-\hat{\theta}_2)}{n_2}} < \theta_1 - \theta_2 <$$
$$(\hat{\theta}_1 - \hat{\theta}_2) + z_{\alpha/2} \cdot \sqrt{\frac{\hat{\theta}_1(1-\hat{\theta}_1)}{n_1} + \frac{\hat{\theta}_2(1-\hat{\theta}_2)}{n_2}}$$

where $\hat{\theta}_1 = x_1/n_1$ and $\hat{\theta}_2 = x_2/n_2$.

EXAMPLE 11.6

If it is found that 132 of 200 voters in District A favor a given candidate for election to the United States Senate and 90 of 150 voters in District B favor this same candidate, find a 99% confidence interval for $\theta_1 - \theta_2$, the difference between the actual proportions of voters from the two districts who favor the candidate.

Solution

Substituting $\hat{\theta}_1 = \frac{132}{200} = 0.66$, $\hat{\theta}_2 = \frac{90}{150} = 0.60$, and $z_{.005} = 2.575$ into the large-sample confidence interval of Theorem 11.6, we obtain

$$(0.66 - 0.60) - 2.575\sqrt{\frac{(0.66)(0.34)}{200} + \frac{(0.60)(0.40)}{150}} < \theta_1 - \theta_2 <$$
$$(0.66 - 0.60) + 2.575\sqrt{\frac{(0.66)(0.34)}{200} + \frac{(0.60)(0.40)}{150}}$$

[†] In Exercise 4 on page 359 the reader will be asked to fill the details.

which reduces to

$$-0.074 < \theta_1 - \theta_2 < 0.194$$

Thus, we are 99% confident that the interval from −0.074 to 0.194 contains the difference between the actual proportions of voters from the two districts who favor the candidate. Observe that this includes the possibility of a zero difference between the two proportions.

11.6 CONFIDENCE INTERVALS FOR VARIANCES

Given a random sample of size n from a normal population, we can obtain a $(1 - \alpha)100\%$ confidence interval for σ^2 by making use of Theorem 8.9, according to which

$$\frac{(n-1)\mathbf{s}^2}{\sigma^2}$$

is a random variable having a chi-square distribution with $n - 1$ degrees of freedom. Thus, we have

$$P\left[\chi^2_{1-\alpha/2,n-1} < \frac{(n-1)\mathbf{s}^2}{\sigma^2} < \chi^2_{\alpha/2,n-1}\right] = 1 - \alpha$$

or

$$P\left[\frac{(n-1)\mathbf{s}^2}{\chi^2_{\alpha/2,n-1}} < \sigma^2 < \frac{(n-1)\mathbf{s}^2}{\chi^2_{1-\alpha/2,n-1}}\right] = 1 - \alpha$$

where $\chi^2_{\alpha/2,n-1}$ and $\chi^2_{1-\alpha/2,n-1}$ are as defined on page 269, and we get

THEOREM 11.7 *(Confidence interval for σ^2)* If s^2 is the value of the variance of a random sample of size n from a normal population, a $(1 - \alpha)100\%$ confidence interval for σ^2 is given by

$$\frac{(n-1)s^2}{\chi^2_{\alpha/2,n-1}} < \sigma^2 < \frac{(n-1)s^2}{\chi^2_{1-\alpha/2,n-1}}$$

Corresponding $(1 - \alpha)100\%$ confidence limits for σ can be obtained by taking the square roots of the confidence limits for σ^2.

EXAMPLE 11.7

In 16 test runs the gasoline consumption of an experimental engine had a standard deviation of 2.2 gallons. Construct a 99% confidence interval for σ^2, measuring the true variability of the gasoline consumption of this engine.

Solution

Assuming that the observed data can be looked upon as a random sample from a normal population, we substitute $n = 16$ and $s = 2.2$, along with $\chi^2_{.005,15} = 32.801$ and $\chi^2_{.995,15} = 4.601$, obtained from Table V, into the confidence interval of Theorem 11.7, and we get

$$\frac{15(2.2)^2}{32.801} < \sigma^2 < \frac{15(2.2)^2}{4.601}$$

or

$$2.21 < \sigma^2 < 15.78$$

Taking square roots, we find that the corresponding 99% confidence interval for σ is given by

$$1.49 < \sigma < 3.97$$

11.7 CONFIDENCE INTERVALS FOR RATIOS OF TWO VARIANCES

If $\mathbf{s}_1^2$ and $\mathbf{s}_2^2$ are the sample variances of independent random samples of size n_1 and n_2 from normal populations, then according to Theorem 8.13

$$\mathbf{F} = \frac{\sigma_2^2\mathbf{s}_1^2}{\sigma_1^2\mathbf{s}_2^2}$$

is a random variable having an F distribution with $n_1 - 1$ and $n_2 - 1$ degrees of freedom. Substituting this expression for $\mathbf{F}$ into

$$P(F_{1-\alpha/2,n_1-1,n_2-1} < \mathbf{F} < F_{\alpha/2,n_1-1,n_2-1}) = 1 - \alpha$$

where $F_{1-\alpha/2,n_1-1,n_2-1}$ and $F_{\alpha/2,n_1-1,n_2-1}$ are as defined on page 274, making use of the fact that $F_{1-\alpha/2,n_1-1,n_2-1} = 1/F_{\alpha/2,n_2-1,n_1-1}$ (see Exercise 15 on page 278), and proceeding as in Section 11.4, we arrive at the following result:†

† In Exercise 6 on page 359 the reader will be asked to fill in the details.

THEOREM 11.8 $\left(\textit{Confidence interval for } \frac{\sigma_1^2}{\sigma_2^2}\right)$ If s_1^2 and s_2^2 are the values of the variances of independent random samples of size n_1 and n_2 from two normal populations, a $(1-\alpha)100\%$ confidence interval for $\frac{\sigma_1^2}{\sigma_2^2}$ is given by

$$\frac{s_1^2}{s_2^2} \cdot \frac{1}{F_{\alpha/2,n_1-1,n_2-1}} < \frac{\sigma_1^2}{\sigma_2^2} < \frac{s_1^2}{s_2^2} \cdot F_{\alpha/2,n_2-1,n_1-1}$$

Corresponding $(1-\alpha)100\%$ confidence limits for $\frac{\sigma_1}{\sigma_2}$ can be obtained by taking the square roots of the confidence limits for $\frac{\sigma_1^2}{\sigma_2^2}$.

EXAMPLE 11.8

With reference to Example 11.4, find a 98% confidence interval for $\frac{\sigma_1^2}{\sigma_2^2}$.

Solution

From Example 11.4 we have $n_1 = 10$, $n_2 = 8$, $s_1 = 0.5$, and $s_2 = 0.7$, and from Table VIb we find that $F_{.01,9,7} = 6.72$ and $F_{.01,7,9} = 5.61$. Thus, substitution into the confidence interval of Theorem 11.8 yields

$$\frac{0.25}{0.49} \cdot \frac{1}{6.72} < \frac{\sigma_1^2}{\sigma_2^2} < \frac{0.25}{0.49} \cdot 5.61$$

or

$$0.076 < \frac{\sigma_1^2}{\sigma_2^2} < 2.862$$

Since this includes the possibility that the ratio is 1, we cannot conclude that the assumption of equal population variances in Example 11.4 was unjustified.

THEORETICAL EXERCISES

1. By solving the inequalities on page 353, namely,

$$-z_{\alpha/2} < \frac{x - n\theta}{\sqrt{n\theta(1-\theta)}} \quad \text{and} \quad \frac{x - n\theta}{\sqrt{n\theta(1-\theta)}} < z_{\alpha/2}$$

show that the $(1 - \alpha)100\%$ confidence limits for θ are

$$\frac{x + \frac{1}{2} \cdot z_{\alpha/2}^2 \pm z_{\alpha/2}\sqrt{\frac{x(n-x)}{n} + \frac{1}{4} \cdot z_{\alpha/2}^2}}{n + z_{\alpha/2}^2}$$

2. Use the large-sample confidence interval for θ to show that we can be at least $(1 - \alpha)100\%$ confident that $|\hat{\theta} - \theta|$, the absolute value of the error which we make when we use the sample proportion $\hat{\theta} = \frac{x}{n}$ as an estimate of θ, will be less than a specified amount e when the sample size is $n = \frac{z_{\alpha/2}^2}{4e^2}$.
3. Modify the large-sample confidence interval of Theorem 11.5 to account for the continuity correction which we use when we approximate a binomial distribution with a normal distribution.
4. Fill in the details which lead from the $\mathbf{z}$ statistic on page 355 and the probability $P(-z_{\alpha/2} < \mathbf{z} < z_{\alpha/2}) = 1 - \alpha$ to the confidence interval of Theorem 11.6.
5. For large n, the sampling distribution of $\mathbf{s}$ is sometimes approximated with a normal distribution having the mean σ and the variance $\frac{\sigma^2}{2n}$ (see Exercise 5 on page 276). Show that this approximation leads to the following large-sample $(1 - \alpha)100\%$ confidence interval for σ:

$$\frac{s}{1 + \frac{z_{\alpha/2}}{\sqrt{2n}}} < \sigma < \frac{s}{1 - \frac{z_{\alpha/2}}{\sqrt{2n}}}$$

6. Fill in the details which lead from the $\mathbf{F}$ statistic on page 357 and the probability $P(F_{1-\alpha/2,n_1-1,n_2-1} < \mathbf{F} < F_{\alpha/2,n_1-1,n_2-1}) = 1 - \alpha$ to the confidence interval of Theorem 11.8.

APPLIED EXERCISES

7. A sample survey at a supermarket showed that 204 of 300 shoppers regularly use cents-off coupons. Use the large-sample confidence interval of Theorem 11.5 to find a 99% confidence interval for the corresponding true proportion.
8. In a random sample of 250 television viewers in a certain area, 190 had seen a certain controversial program. Construct a 95% confidence interval for the corresponding true proportion, using
   - **(a)** the large-sample confidence interval of Theorem 11.5;
   - **(b)** the confidence limits of Exercise 1.

9. Use the theory of Exercise 2 to find the minimum sample size which will enable us to assert with a degree of confidence of at least 95% that a sample proportion (which is used to estimate the parameter θ of a binomial population) is "off" by less than 0.03.

10. Use the theory of Exercise 2 to find the minimum sample size which will enable us to assert with a degree of confidence of at least 99% that a sample proportion (which is used to estimate the parameter θ of a binomial population) is "off" by less than 0.02.

11. In a random sample of visitors to a famous tourist attraction, 84 of 250 men and 156 of 250 women bought souvenirs. Construct a 95% confidence interval for the difference between the true proportions of men and women who buy souvenirs at this tourist attraction.

12. Among 500 marriage license applications, chosen at random in 1971, there were 48 in which the women were at least one year older than the men, and among 400 marriage license applications, chosen at random in 1977, there were 68 in which the women were at least one year older than the men. Construct a 99% confidence interval for the difference between the corresponding true proportions of marriage license applications in which the women were at least one year older than the men.

13. With reference to Exercise 11 on page 351, construct a 95% confidence interval for the true variance of the length of the skulls of the given species of birds.

14. Use the data of Exercise 9 on page 351 and the large-sample confidence interval of Exercise 5 to construct a 99% confidence interval for the true standard deviation of the time it takes students to complete the test.

15. With reference to Exercise 15 on page 352, construct a 98% confidence interval for the ratio of the two population variances.

16. With reference to Exercise 16 on page 352, construct a 90% confidence interval for the ratio of the two population standard deviations.

References

A general method for obtaining confidence intervals is given in

MOOD, A. M., GRAYBILL, F. A., and BOES, D. C., *Introduction to the Theory of Statistics.* New York: McGraw-Hill Book Company, 1974,

and further criteria for judging the relative merits of confidence intervals may be found in

LEHMANN, E. L., *Testing Statistical Hypotheses.* New York: John Wiley & Sons, Inc., 1959,

and in other advanced texts on mathematical statistics. Special tables for constructing 95% and 99% confidence intervals for proportions are given in the *Biometrika Tables* referred to on page 285. For a proof of the independence of the random variables **z** and **y** on page 348, see

BRUNK, H. D., *An Introduction to Mathematical Statistics*, 3rd ed. Lexington, Mass.: Xerox College Publishing, 1975.

hypothesis testing: theory

12

12.1 STATISTICAL HYPOTHESES

If an engineer has to decide on the basis of sample data whether the true average lifetime of a certain kind of tire is at least 22,000 miles, if an agronomist has to decide on the basis of experiments whether one kind of fertilizer produces a higher yield of soybeans than another, and if a manufacturer of pharmaceutical products has to decide on the basis of samples whether 90 percent of all patients given a new medication will recover from a certain disease, these problems can all be translated into the language of **statistical tests of hypotheses**. In the first case we might say that the engineer has to test the hypothesis that θ, the parameter of an exponential population, is at least 22,000; in the second case we might say that the agronomist has to decide whether $\mu_1 > \mu_2$, where μ_1 and μ_2 are the means of two normal populations; and in the third case we might say that the manufacturer has to decide whether θ, the parameter of a binomial population, equals 0.90. In each case it must be assumed, of course, that the chosen distribution correctly describes the experimental conditions, namely, that the distribution provides the correct **statistical model**.

As in the above examples, most tests of statistical hypotheses concern the parameters of distributions, but sometimes they also concern the type, or nature, of the distributions, themselves. For instance, in the first of our three examples the engineer may also have to decide whether he is actually dealing with a sample from an exponential population, or whether his data are values of random variables having, say, the Weibull distribution of Exercise 15 on page 204.

DEFINITION 12.1 A **statistical hypothesis** is an assertion or conjecture about the distribution of one or more random variables. If a statistical hypothesis completely specifies the distribution, it is referred to as a **simple hypothesis**; if not, it is referred to as a **composite hypothesis**.

A simple hypothesis must therefore not only specify the functional form of the underlying distribution, but also the values of all parameters. Thus, in the third of the above examples, the one dealing with the effectiveness of the new medication, the hypothesis $\theta = 0.90$ is simple, assuming, of course, that we specify the sample size and that the population is binomial. However, in the first of the above examples the hypothesis is composite since $\theta \geqslant 22{,}000$ does not assign a specific value to the parameter θ.

To be able to construct suitable criteria for testing statistical hypotheses, it is necessary that we also formulate **alternative hypotheses**. For instance, in the example dealing with the lifetimes of the tires we might formulate the alternative hypothesis that the parameter θ of the exponential population is less than 22,000; in the example dealing with the two kinds of fertilizer we might formulate the alternative hypothesis $\mu_1 = \mu_2$; and in the example dealing with the new medication we might formulate the alternative hypothesis that the parameter θ of the given binomial population is only 0.60, which is the disease's recovery rate without the new medication.

The concept of simple and composite hypotheses applies also to alternative hypotheses, and in the first example we can now say that we are testing the composite hypothesis $\theta \geqslant 22{,}000$ against the **composite alternative** $\theta < 22{,}000$, where θ is the parameter of an exponential population. Similarly, in the second example we are testing the composite hypothesis $\mu_1 > \mu_2$ against the composite alternative $\mu_1 = \mu_2$, where μ_1 and μ_2 are the means of two normal populations, and in the third example we are testing the simple hypothesis $\theta = 0.90$ against the **simple alternative** $\theta = 0.60$, where θ is the parameter of a binomial population for which n is given.

Frequently, statisticians state as their hypotheses the opposite of what they believe to be true, with the hope that the test procedures will lead to their rejection. For instance, if we want to show that the students in one school have a higher average I.Q. then those of another, we might formulate the hypothesis that there is no difference, namely, that $\mu_1 = \mu_2$. Similarly, if we want to show that one kind of ore has a higher percentage content of uranium than another kind of ore, we might formulate the hypothesis that the two percentages are the same; and if we want to show that there is a greater variability in the quality of one product than there is in the quality of another, we might formulate the hypothesis that there is no difference, namely, that $\sigma_1 = \sigma_2$. In view of the assumptions of no difference, hypotheses like these led to the term **null hypothesis**, although nowadays this term is applied to any hypothesis we may wish to test.

Symbolically, we shall use the symbol H_0 for the null hypothesis we want to test and H_1 for the alternative hypothesis. Problems involving more than two hypotheses, that is, problems involving several alternative hypotheses, tend to be quite complicated, and will not be studied in this book.

12.2 TESTING A STATISTICAL HYPOTHESIS

The testing of a statistical hypothesis is the application of an explicit set of rules for deciding whether to accept the null hypothesis or to reject it in favor of the alternative hypothesis. Suppose, for example, that a statistician wants to test the null hypothesis $\theta = \theta_0$ against the alternative hypothesis $\theta = \theta_1$. In order to make a choice, he will generate sample data by conducting an experiment and then compute the value of a **test statistic** $\hat{\boldsymbol{\theta}}$, which will tell him for each possible outcome of the sample space what action to take. The test procedure, therefore, partitions the sample space, and hence also the possible values of the test statistic, into two subsets: an **acceptance region** for H_0 and a **rejection region** for H_0.

The procedure just described can lead to two kinds of errors. For instance, if the true value of the parameter θ is θ_0 and the statistician incorrectly concludes that $\theta = \theta_1$, he is committing an error referred to as a **type I error**. On the other hand, if the true value of the parameter θ is θ_1 and the statistician incorrectly concludes that $\theta = \theta_0$, he is committing a second kind of error referred to as a **type II error**.

DEFINITION 12.2

1. Rejection of the null hypothesis when it is true is called a **type I error**; the probability of committing a type I error is denoted by α.
2. Acceptance of the null hypothesis when it is false is called a **type II error**; the probability of committing a type II error is denoted by β.

It is customary to refer to the rejection region for H_0 as the **critical region** of the test, and to the probability of obtaining a value of the test statistic inside the critical region when H_0 is true as the **size** of the critical region. Thus, the size of a critical region is just the probability α of committing a type I error. This probability is also called the **level of significance** of the test (see discussion on page 377).

EXAMPLE 12.1

With reference to the third illustration on page 361, suppose the manufacturer of the new medication wants to test the null hypothesis $\theta = 0.90$ against the alternative hypothesis $\theta = 0.60$. His test statistic is $\mathbf{x}$, the observed number of successes in $n = 20$ trials, and he will accept the null hypothesis if $x \geqslant 15$; otherwise, he will conclude that $\theta = 0.60$. Evaluate the probabilities α and β.

Solution

The acceptance region for H_0 is given by $x = 15, 16, 17, 18, 19$, and 20, and, correspondingly, the rejection or critical region is given by $x = 0, 1, 2, \ldots, 14$. Therefore, from Table I,

$$\begin{aligned}\alpha &= P(\text{type I error})\\ &= P(\mathbf{x} < 15;\ \theta = 0.90)\\ &= 0.0114\end{aligned}$$

and

$$\begin{aligned}\beta &= P(\text{type II error})\\ &= P(\mathbf{x} \geqslant 15;\ \theta = 0.60)\\ &= 0.1255\end{aligned}$$

A good test procedure is one in which both α and β are small, thereby giving us a good chance of making the correct decision. The probability of a type II error in Example 12.1 is rather high, but this can be reduced by appropriately changing the critical region. For instance, if we use the acceptance region $x \geqslant 16$ in Example 12.1, so that the critical region is $x < 16$, it can easily be checked that this would make $\alpha = 0.0433$ and $\beta = 0.0509$. Thus, while the probability of a type II error has become smaller, the probability of a type I error has become larger. The only way in which we can reduce the probabilities of both types of errors is to increase the size of the sample, but so long as n is held fixed, this inverse relationship between the probabilities of type I and type II errors is typical of statistical decision procedures. In other words, if the probability of one type of error is reduced, that of the other type of error is increased.

EXAMPLE 12.2

Suppose we want to test the null hypothesis that the mean μ of a normal population with $\sigma^2 = 1$ is μ_0 against the alternative hypothesis that it is μ_1, where $\mu_1 > \mu_0$. Find the value of k such that $\bar{x} > k$ provides a critical region of size

$\alpha = 0.05$ for a random sample of size n. Also determine the minimum sample size needed for testing the null hypothesis $\mu = 10$ against the alternative hypothesis $\mu = 11$ by this procedure so that $\beta \leqslant 0.05$.

Solution

Referring to Figure 12.1 and Table III, we find that $z = 1.645$ corresponds to an entry of 0.4500. Thus,

$$1.645 = \frac{k - \mu_0}{1/\sqrt{n}}$$

and

$$k = \mu_0 + \frac{1.645}{\sqrt{n}}$$

so that the desired critical region of size $\alpha = 0.05$ is $\bar{x} > \mu_0 + \dfrac{1.645}{\sqrt{n}}$.

When testing the null hypothesis $\mu = \mu_0$ against the alternative hypothesis $\mu = \mu_1$, where $\mu_1 > \mu_0$, and $\alpha = 0.05$, the probability β of a type II error is given by the area of the ruled region of Figure 12.1. For the null hypothesis $\mu = 10$, the alternative hypothesis $\mu = 11$, and $\alpha = 0.05$, we thus get

$$\beta = P\left(\bar{\mathbf{x}} < 10 + \frac{1.645}{\sqrt{n}};\ \mu = 11\right)$$

$$= P\left[\mathbf{z} < \frac{\left(10 + \frac{1.645}{\sqrt{n}}\right) - 11}{1/\sqrt{n}}\right]$$

$$= P(\mathbf{z} < -\sqrt{n} + 1.645)$$

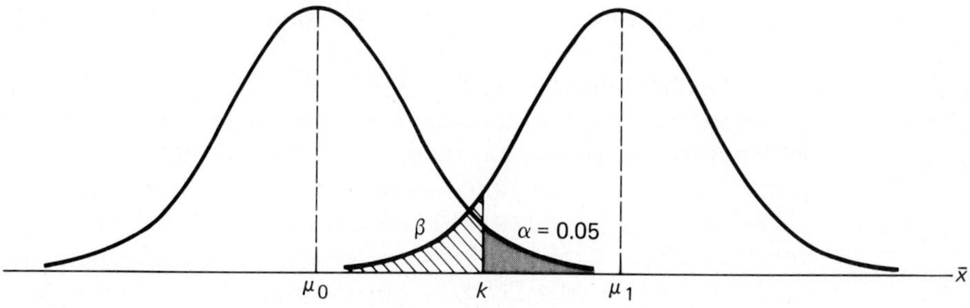

Figure 12.1 Critical region for testing $\mu = \mu_0$ against $\mu = \mu_1$.

Now, since $P(\mathbf{z} < -z_{0.05}) = 0.05$, we set

$$-\sqrt{n} + 1.645 = -1.645$$

from which we obtain $n = 10.8$. Therefore, the minimum sample size needed to keep $\beta \leqslant 0.05$ in this example is $n = 11$.

12.3 LOSSES AND RISKS

The concepts of loss functions and risk functions that were introduced in Chapter 9 also play an important part in the theory of hypothesis testing. In the decision theory approach to testing the null hypothesis that a population parameter θ equals θ_0 against the alternative that it equals θ_1, the statistician either takes the action a_0 and accepts the null hypothesis, or he takes the action a_1 and accepts the alternative hypothesis. Depending on the true "state of Nature" and the action which he takes, his losses are shown in the following table:

		Statistician a_0	*Statistician* a_1
Nature	θ_0	$L(a_0, \theta_0)$	$L(a_1, \theta_0)$
Nature	θ_1	$L(a_0, \theta_1)$	$L(a_1, \theta_1)$

These losses can be positive or negative (reflecting penalties or rewards), and the only condition which we shall impose is that

$$L(a_0, \theta_0) < L(a_1, \theta_0) \quad \text{and} \quad L(a_1, \theta_1) < L(a_0, \theta_1)$$

namely, that in either case the right decision is more profitable than the wrong one.

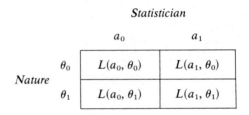

As in the statistical games of Section 9.3, the statistician's choice will depend on the outcome of an experiment and the decision function d, which tells him for each possible outcome what action to take. If the null hypothesis is true and the statistician accepts the alternative hypothesis, namely, if the value of the parameter is θ_0 and the statistician takes action a_1, he commits a type I error; correspondingly, if the value of the parameter is θ_1 and the statistician takes action a_0, he commits a type II error. For the decision function d, we shall let $\alpha(d)$ denote the probability of committing a type I error and $\beta(d)$ the probability of committing a type II error. The values of the risk function (defined on page 301) are thus

$$\begin{aligned}R(d, \theta_0) &= [1 - \alpha(d)]L(a_0, \theta_0) + \alpha(d)L(a_1, \theta_0)\\ &= L(a_0, \theta_0) + \alpha(d)[L(a_1,\theta_0) - L(a_0, \theta_0)]\end{aligned}$$

and

$$\begin{aligned}R(d, \theta_1) &= \beta(d)L(a_0, \theta_1) + [1 - \beta(d)]L(a_1, \theta_1)\\ &= L(a_1, \theta_1) + \beta(d)[L(a_0, \theta_1) - L(a_1, \theta_1)]\end{aligned}$$

where, by assumption, the quantities in brackets are both positive. It is apparent from this (and should, perhaps, have been obvious from the beginning) that to minimize the risks the statistician must choose a decision function which, in some way, keeps the probabilities of both types of errors as small as possible.

If we could assign prior probabilities to θ_0 and θ_1 and if we knew the exact values of all the losses $L(a_j, \theta_i)$ in the table on page 366, we could calculate the Bayes risk (defined on page 303) and look for the decision function which minimizes this risk. Alternatively, if we looked upon Nature as a malevolent opponent we could use the minimax criterion and choose the decision function which minimizes the maximum risk, but as must have been apparent from the applied exercises on page 296, this is not a very realistic approach in most practical situations.

12.4 THE NEYMAN–PEARSON LEMMA

In the theory of hypothesis testing which is nowadays referred to as "classical" or "traditional," namely, the **Neyman–Pearson theory**, we circumvent the dependence between probabilities of type I and type II errors by limiting ourselves to test statistics for which the probability of a type I error is less than or equal to some constant α. In other words, we restrict ourselves to critical regions of size less than or equal to α. We must allow for the critical region to be of size less than α to take care of discrete random variables, where it may be impossible to find a test statistic for which the size of the critical region is exactly equal to α. For all practical purposes, then, we hold the probability of a type I error fixed, and look for the test statistic which minimizes the probability of a type II error, or equivalently, which maximizes the quantity $1 - \beta$. When testing the null hypothesis $\theta = \theta_0$ against the alternative hypothesis $\theta = \theta_1$, the quantity $1 - \beta$ is referred to as the **power** of the test at $\theta = \theta_1$.

A critical region for testing a simple null hypothesis $\theta = \theta_0$ against a simple alternative hypothesis $\theta = \theta_1$ is said to be **best** or **most powerful**, if the power of the test at $\theta = \theta_1$ is a maximum. To construct a most powerful critical region in this kind of situation, we refer to the likelihoods (see page 327) of a random sample of size n from the population under consideration when $\theta = \theta_0$ and $\theta = \theta_1$. Denoting these likelihoods by L_0 and L_1, we thus have

$$L_0 = \prod_{i=1}^{n} f(x_i; \theta_0) \quad \text{and} \quad L_1 = \prod_{i=1}^{n} f(x_i; \theta_1)$$

Intuitively speaking, it stands to reason that $\frac{L_0}{L_1}$ should be small for sample points inside the critical region, which lead to type I errors when $\theta = \theta_0$ and to correct decisions when $\theta = \theta_1$; similarly, it stands to reason that $\frac{L_0}{L_1}$ should be large for sample points outside the critical region, which lead to correct decisions when $\theta = \theta_0$ and type II errors when $\theta = \theta_1$. The fact that this argument does, indeed, guarantee a most powerful critical region is proved by the following theorem:

THEOREM 12.1 *(Neyman–Pearson Lemma)* If C is a critical region of size α and k is a constant such that

$$\frac{L_0}{L_1} \leqslant k \qquad \text{inside } C$$

and

$$\frac{L_0}{L_1} \geqslant k \qquad \text{outside } C$$

then C is a most powerful critical region of size α for testing $\theta = \theta_0$ against $\theta = \theta_1$.

Proof. Suppose that C is a critical region satisfying the conditions of the theorem and that D is some other critical region of size α. Thus,

$$\int_C \cdots \int L_0 \, dx = \int_D \cdots \int L_0 \, dx = \alpha$$

where dx stands for $dx_1 \, dx_2 \, \ldots \, dx_n$, and the two multiple integrals are taken over the respective n-dimensional regions C and D. Now, making use of the fact that C is the union of the disjoint sets $C \cap D$ and $C \cap D'$ while D is the union of the disjoint sets $C \cap D$ and $C' \cap D$, we can write

$$\int_{C\cap D} \cdots \int L_0\, dx + \int_{C\cap D'} \cdots \int L_0\, dx = \int_{C\cap D} \cdots \int L_0\, dx + \int_{C'\cap D} \cdots \int L_0\, dx = \alpha$$

and, hence,

$$\int_{C\cap D'} \cdots \int L_0\, dx = \int_{C'\cap D} \cdots \int L_0\, dx$$

Then, since $L_1 \geqslant L_0/k$ inside C and $L_1 \leqslant L_0/k$ outside C, it follows that

$$\int_{C\cap D'} \cdots \int L_1\, dx \geqslant \int_{C\cap D'} \cdots \int \frac{L_0}{k}\, dx = \int_{C'\cap D} \cdots \int \frac{L_0}{k}\, dx \geqslant \int_{C'\cap D} \cdots \int L_1\, dx$$

and, hence, that

$$\int_{C\cap D'} \cdots \int L_1\, dx \geqslant \int_{C'\cap D} \cdots \int L_1\, dx$$

Finally,

$$\int_{C} \cdots \int L_1\, dx = \int_{C\cap D} \cdots \int L_1\, dx + \int_{C\cap D'} \cdots \int L_1\, dx \geqslant$$

$$\int_{C\cap D} \cdots \int L_1\, dx + \int_{C'\cap D} \cdots \int L_1\, dx = \int_{D} \cdots \int L_1\, dx$$

so that

$$\int_{C} \cdots \int L_1\, dx \geqslant \int_{D} \cdots \int L_1\, dx$$

and this completes the proof of Theorem 12.1. The final inequality states that for the critical region C the probability of *not* committing a type II error is greater than or equal to the corresponding probability for any other critical region of size α. (For the discrete case the proof is the same, with summations taking the place of integrals.)

EXAMPLE 12.3

A random sample of size n from a normal population with $\sigma^2 = 1$ is to be used to test the null hypothesis $\mu = \mu_0$ against the alternative hypothesis $\mu = \mu_1$, where $\mu_1 > \mu_0$. Use the Neyman–Pearson lemma to find the most powerful critical region of size α.

Solution

The two likelihoods are

$$L_0 = \left(\frac{1}{\sqrt{2\pi}}\right)^n \cdot e^{-\frac{1}{2}\Sigma(x_i-\mu_0)^2} \quad \text{and} \quad L_1 = \left(\frac{1}{\sqrt{2\pi}}\right)^n \cdot e^{-\frac{1}{2}\cdot\Sigma(x_i-\mu_1)^2}$$

where the summations extend from $i = 1$ to $i = n$, and after some simplifications their ratio becomes

$$\frac{L_0}{L_1} = e^{\frac{n}{2}(\mu_1^2-\mu_0^2)+(\mu_0-\mu_1)\cdot\Sigma x_i}$$

Thus, we must find a constant k and a region C of the sample space such that

$$e^{\frac{n}{2}(\mu_1^2-\mu_0^2)+(\mu_0-\mu_1)\cdot\Sigma x_i} \leq k \qquad \text{inside } C$$

$$e^{\frac{n}{2}(\mu_1^2-\mu_0^2)+(\mu_0-\mu_1)\cdot\Sigma x_i} \geq k \qquad \text{outside } C$$

and after taking logarithms, subtracting $\frac{n}{2}(\mu_1^2 - \mu_0^2)$, and dividing by the *negative* quantity $n(\mu_0 - \mu_1)$, these two inequalities become

$$\bar{x} \geq K \qquad \text{inside } C$$

$$\bar{x} \leq K \qquad \text{outside } C$$

where K is an expression in k, n, μ_0, and μ_1. Thus, in this example $\bar{x} \geq K$ is a most powerful critical region.

In actual practice, constants like K are determined by making use of the size of the critical region and appropriate statistical theory. In our case (see Example 12.2) we obtain $K = \mu_0 + z_\alpha \cdot \frac{1}{\sqrt{n}}$, where z_α is as defined on page 342. Thus, the most powerful critical region of size α for testing the null hypothesis $\mu = \mu_0$ against the alternative $\mu = \mu_1$ (with $\mu_1 > \mu_0$) for the given normal population is

$$\bar{x} \geq \mu_0 + z_\alpha \cdot \frac{1}{\sqrt{n}}$$

and it should be noted that it does not depend on μ_1. This is an important property, to which we shall refer again in Section 12.5.

Note that in Example 12.3 we derived the critical region without first mentioning that the test statistic is to be $\bar{\mathbf{x}}$. Since the specification of a critical region thus defines a corresponding test statistic and vice versa, the two terms are used interchangeably in the language of statistics.

THEORETICAL EXERCISES

1. Decide in each case whether the given hypothesis is simple or composite:
- **(a)** the hypothesis that a random variable has a gamma distribution with $\alpha = 3$ and $\beta = 2$;
- **(b)** the hypothesis that a random variable has a gamma distribution with $\alpha = 3$ and $\beta \neq 2$;
- **(c)** the hypothesis that a random variable has an exponential density;
- **(d)** the hypothesis that a random variable has a beta distribution with the mean $\mu = 0.50$;
- **(e)** the hypothesis that a random variable has a Poisson distribution with $\lambda = 1.25$;
- **(f)** the hypothesis that a random variable has a Poisson distribution with $\lambda > 1.25$;
- **(g)** the hypothesis that a random variable has a normal distribution with the mean $\mu = 100$;
- **(h)** the hypothesis that a random variable has a negative binomial distribution with $k = 3$ and $\theta < 0.60$.

2. A bowl contains seven marbles of which θ are red while the others are blue. In order to test the null hypothesis $\theta = 2$ against the alternative $\theta = 4$, two of the marbles are randomly drawn without replacement and the null hypothesis is rejected if and only if both are red. Find the probabilities of committing type I and type II errors with this criterion.

3. With reference to Example 12.1 on page 364, what would have been the probabilities of type I and type II errors if the acceptance region had been $x \geqslant 17$ and the corresponding rejection region had been $x < 17$?

4. Let $\mathbf{x}_1$ and $\mathbf{x}_2$ constitute a random sample of size 2 from a normal population with $\sigma^2 = 1$. If the null hypothesis $\mu = \mu_0$ is to be rejected in favor of the alternative hypothesis $\mu = \mu_1$, where $\mu_1 > \mu_0$, when $\bar{x} > \mu_0 + 1$, what is the size of this critical region?

5. A single observation of a random variable having an exponential distribution is used to test the null hypothesis that the mean of the distribution is $\theta = 2$ against the alternative that it is $\theta = 5$. If the null hypothesis is accepted if and only if the observed value of the random variable is less than 3, find the probabilities of type I and type II errors.

6. Show that if $\mu_1 < \mu_0$ in Example 12.3, the Neyman–Pearson lemma yields the critical region $\bar{x} \leqslant \mu_0 - z_\alpha \cdot \frac{1}{\sqrt{n}}$.
7. Let $\mathbf{x}_1$ and $\mathbf{x}_2$ constitute a random sample of size 2 from the population given by

$$f(x; \theta) = \begin{cases} \theta x^{\theta-1} & \text{for } 0 < x < 1 \\ 0 & \text{elsewhere} \end{cases}$$

If the critical region $x_1 x_2 \geqslant \frac{3}{4}$ is used to test the null hypothesis $\theta = 1$ against the alternative hypothesis $\theta = 2$, what is the power of this test at $\theta = 2$?
8. A random sample of size n from an exponential population is used to test the null hypothesis that its parameter is θ_0 against the alternative that its parameter is θ_1, where $\theta_1 > \theta_0$. Use the Neyman–Pearson lemma to find the most powerful critical region of size α, and use the result of Example 7.15 on page 249 to indicate how to evaluate the constant.
9. Use the Neyman–Pearson lemma to indicate how to construct the most powerful critical region of size α to test the null hypothesis that θ, the parameter of a binomial distribution with a given value of n, equals θ_0 against the alternative that it equals $\theta_1 < \theta_0$.
10. If $n = 100$, $\theta_0 = 0.40$, $\theta_1 = 0.30$, and α is as large as possible without exceeding 0.05, use the normal approximation to the binomial distribution to find the probability of committing a type II error with the criterion constructed in Exercise 9.
11. A single observation of a random variable having a geometric distribution is to be used to test the null hypothesis that its parameter equals θ_0 against the alternative that it equals $\theta_1 > \theta_0$. Use the Neyman–Pearson lemma to find the best critical region of size α.
12. Given a random sample of size n from a normal population with $\mu = 0$, use the Neyman–Pearson lemma to construct the most powerful critical region of size α to test the null hypothesis $\sigma = \sigma_0$ against the alternative $\sigma = \sigma_1 > \sigma_0$.

APPLIED EXERCISES

13. An airline wants to test the null hypothesis that 60 percent of its passengers object to smoking inside the plane. Explain under what conditions they would be committing a type I error and under what conditions they would be committing a type II error.
14. A doctor is asked to give an executive a thorough physical checkup to test the null hypothesis that he will be able to take on additional responsibilities. Explain under what conditions the doctor would be committing a type I error and under what conditions he would be committing a type II error.

15. Suppose that in Example 12.1 on page 364 the manufacturer of the new medication feels that the odds are 4 to 1 that with this medication the recovery rate from the disease is 0.90 rather than 0.60. With these odds, what are the probabilities that he will make a wrong decision if he uses the decision function

(a) $d_1(x) = \begin{cases} a_0 & \text{for } x \geqslant 15 \\ a_1 & \text{for } x < 15; \end{cases}$

(b) $d_2(x) = \begin{cases} a_0 & \text{for } x \geqslant 16 \\ a_1 & \text{for } x < 16; \end{cases}$

(c) $d_3(x) = \begin{cases} a_0 & \text{for } x \geqslant 14 \\ a_1 & \text{for } x < 14. \end{cases}$

12.5 THE POWER FUNCTION OF A TEST

In Example 12.1 we were able to give unique values for the probabilities of committing type I and type II errors because we were testing a simple hypothesis against a simple alternative. In actual practice, it is relatively rare, however, that simple hypotheses are tested against simple alternatives; usually one or the other, or both, are composite. For instance, in Example 12.1 it might well have been more realistic to test the null hypothesis that the recovery rate from the disease is $\theta \geqslant 0.90$ against the alternative that $\theta < 0.90$, namely, the alternative that the new medication is not as effective as claimed.

When we deal with composite hypotheses, the problem of evaluating the merits of a test criterion, or critical region, becomes much more difficult. In that case we have to consider the probabilities $\alpha(\theta)$ of committing a type I error for all values of θ within the domain specified under the null hypothesis H_0, and the probabilities $\beta(\theta)$ of committing a type II error for all values of θ within the domain specified under the alternative hypothesis H_1. It is customary to combine the two sets of probabilities in the following way:

DEFINITION 12.3 The **power function** of a test of a statistical hypothesis H_0 against an alternative hypothesis H_1 is given by

$$\pi(\theta) = \begin{cases} \alpha(\theta) & \text{for values of } \theta \text{ assumed under } H_0 \\ 1 - \beta(\theta) & \text{for values of } \theta \text{ assumed under } H_1 \end{cases}$$

Thus, the values of the power function are the probabilities of rejecting the null hypothesis H_0 for various values of the parameter θ. Observe also that for values of θ assumed under H_0, the power function gives the probability of committing a type I error, and for values of θ assumed under H_1, it gives the probability of *not* committing a type II error.

EXAMPLE 12.4

With reference to Example 12.1, suppose that we had wanted to test the null hypothesis

$$H_0: \quad \theta \geqslant 0.90$$

against the alternative hypothesis

$$H_1: \quad \theta < 0.90$$

Investigate the power function for the test criterion according to which we reject H_0 when $x < 15$, and otherwise we accept it.

Solution

Choosing select values of θ, we find from Table I the probabilities $\alpha(\theta)$ of getting fewer than 15 successes for $\theta = 0.90$ and 0.95, and the probabilities $\beta(\theta)$ of getting 15 or more successes for $\theta = 0.85, 0.80, \ldots,$ and 0.50. These probabilities and the corresponding values of the power function are shown in the following table:

θ	*Probability of type I error* $\alpha(\theta)$	*Probability of type II error* $\beta(\theta)$	*Probability of rejecting* H_0 $\pi(\theta)$
0.95	0.0003		0.0003
0.90	0.0114		0.0114
0.85		0.9326	0.0674
0.80		0.8042	0.1958
0.75		0.6171	0.3829
0.70		0.4163	0.5837
0.65		0.2455	0.7545
0.60		0.1255	0.8745
0.55		0.0553	0.9447
0.50		0.0207	0.9793

The graph of this power function is shown in Figure 12.2. Of course, it applies only to the critical region $x < 15$ of Example 12.1, but it is of interest to note that the power function of an ideal test criterion for this problem would be given by the dashed lines of Figure 12.2.

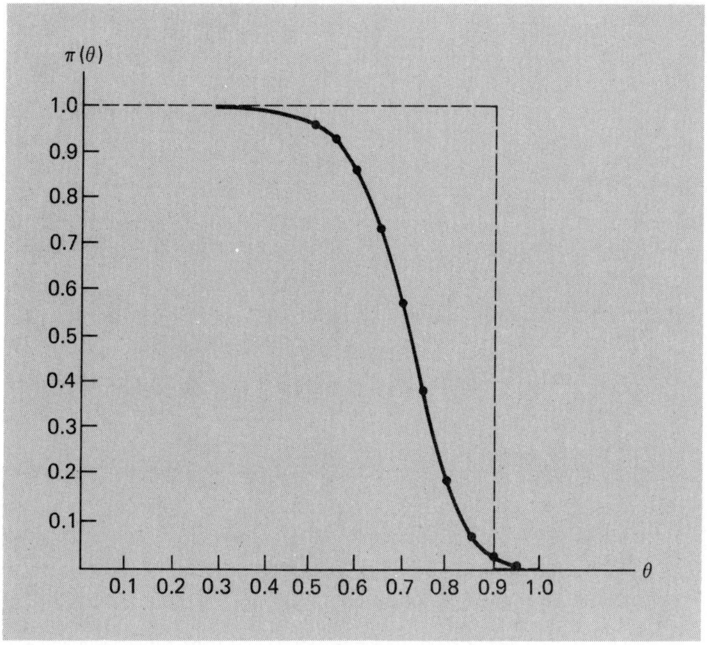

Figure 12.2 Power function of Example 12.4.

Power functions play a very important role in the evaluation of statistical tests, particularly in the comparison of several critical regions which might all be used to test a given null hypothesis against a given alternative. Incidentally, if we had plotted in Figure 12.2 the probabilities of accepting H_0 (instead of those of rejecting H_0), we would have obtained the **operating characteristic curve**, or simply the **OC-curve**, of the given critical region. In other words, the values of the operating characteristic function, used mainly in industrial applications, are given by $1 - \pi(\theta)$.

On page 367 we indicated that in the Neyman–Pearson theory of testing hypotheses we hold α, the probability of a type I error, fixed, and this requires that the null hypothesis H_0 be a simple hypothesis, say, $\theta = \theta_0$. As a result, the power function of any test of this null hypothesis will pass through the point (θ_0, α), the only point at which the value of a power function is the probability of making an error. This facilitates the comparison of the power functions of several critical regions, which are all designed to test the simple null hypothesis $\theta = \theta_0$ against a composite alternative, say, the alternative hypothesis $\theta \neq \theta_0$. To illustrate, consider Figure 12.3, giving the power functions of three different critical regions, or test criteria, designed for this purpose. Since for each value of θ except θ_0 the values of power functions are probabilities of making correct decisions, it is

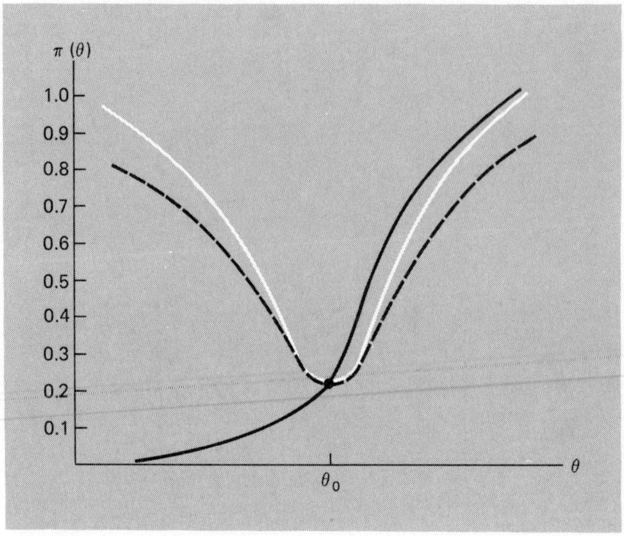

Figure 12.3 Power functions.

desirable to have them as close to 1 as possible. Thus, it can be seen by inspection that the critical region whose power function is given by the white curve of Figure 12.3 is preferable to the critical region whose power function is given by the curve which is dashed. The probability of not committing a type II error with the first of these critical regions always exceeds that of the second, and we say that the first critical region is **uniformly more powerful** than the second; also, the other critical region is said to be **inadmissible**.

The same clear-cut distinction is not possible if we attempt to compare the critical regions whose power functions are given by the white and solid curves of Figure 12.3—in this case the first one is preferable for $\theta < \theta_0$ while the other is preferable for $\theta > \theta_0$. In situations like this we need further criteria for comparing power functions, for instance that of Exercise 16 on page 386. Note that if the alternative hypothesis had been $\theta > \theta_0$, the critical region whose power function is given by the solid curve would have been uniformly more powerful than the critical region whose power function is given by the white curve.

In general, when testing a simple hypothesis against a composite alternative, we specify α, the probability of committing a type I error, and refer to one critical region of size α as uniformly more powerful than another if the values of its power function are always greater than or equal to those of the other, with the strict inequality holding for at least one value of the parameter under consideration. If, for a given problem, a critical region of size α is uniformly more powerful than any other critical region of size α, it is said to be **uniformly most powerful**;

unfortunately, uniformly most powerful critical regions rarely exist when we test a simple hypothesis against a composite alternative. Of course, when we test a simple hypothesis against a simple alternative, a most powerful critical region of size α, as defined on page 367, is, in fact, uniformly most powerful.

Until now we have always assumed that the acceptance of H_0 is equivalent to the rejection of H_1, and vice versa, but this is not the case, for example, in **multi-stage** or **sequential tests**, where the alternatives are to accept H_0, to accept H_1, or to defer the decision until more data have been obtained. It is also not the case in so-called **tests of significance**, where the alternative to rejecting H_0 is reserving judgment instead of accepting H_0. For instance, if we want to test the null hypothesis that a coin is perfectly balanced against the alternative that this is not the case, and 100 tosses yield 57 heads and 43 tails, this will not enable us to reject the null hypothesis when $\alpha = 0.05$ (see Exercise 6 on page 384). However, since we obtained quite a few more heads than the 50 which we can expect for a balanced coin, we may well be reluctant to accept the null hypothesis as true. To avoid this, we can say that the difference between 50 and 57, the number of heads which we expected and the number of heads which we obtained, may reasonably be attributed to chance—or we can say that this difference is not large enough to reject the null hypothesis. In either case, we do not really commit ourselves one way or the other, and so long as we do not actually accept the null hypothesis, we cannot commit a type II error. It is mainly in connection with tests of this kind that we refer to the probability of a type I error as the **level of significance**.

12.6 LIKELIHOOD RATIO TESTS

The Neyman–Pearson lemma provides a means of constructing most powerful critical regions for testing a simple null hypothesis against a simple alternative hypothesis, but it does not apply to composite hypotheses. We shall now present a general method for constructing critical regions for tests of composite hypotheses which in most cases have very satisfactory properties. The resulting tests, called **likelihood ratio tests**, are based on a generalization of the method of Section 12.4, but they are not necessarily uniformly most powerful. We shall discuss this method here with reference to tests concerning one parameter θ and continuous populations, but all of our arguments can easily be extended to the multi-parameter case and to discrete populations.

To illustrate the likelihood ratio technique, let us suppose that $\mathbf{x}_1, \mathbf{x}_2, \ldots,$ and $\mathbf{x}_n$ constitute a random sample of size n from a population whose density at x is $f(x; \theta)$, and that Ω is the set of values which can be taken on by the parameter θ. We often refer to Ω as the **parameter space** for θ. The null hypothesis we shall want to test is

$$H_0\colon \quad \theta \in \omega$$

and the alternative hypothesis is

$$H_1: \quad \theta \in \omega'$$

where ω is a subset of Ω and ω' is the complement of ω with respect to Ω. Thus, the parameter space for θ is partitioned into the disjoint sets ω and ω'; according to the null hypothesis θ is an element of the first set, and according to the alternative hypothesis it is an element of the second set. In most problems Ω is either the set of all real numbers, the set of all positive real numbers, some interval of real numbers, or a discrete set of real numbers.

When H_0 and H_1 are both simple hypotheses, ω and ω' each have only one element, and in Section 12.4 we constructed tests by comparing the likelihoods L_0 and L_1. In the general case, where at least one of the two hypotheses is composite, we compare instead the two quantities max L_0 and max L, where max L_0 is the maximum value of the likelihood function (see page 327) for all values of θ in ω, and max L is the maximum value of the likelihood function for all values of θ in Ω. In other words, if we have a random sample of size n from a population whose density at x is $f(x; \theta)$, $\hat{\theta}$ is the maximum likelihood estimate of θ subject to the restriction that θ must be an element of ω, and $\hat{\hat{\theta}}$ is the maximum likelihood estimate of θ for all values of θ in Ω, then

$$\max L_0 = \prod_{i=1}^{n} f(x_i; \hat{\theta})$$

and

$$\max L = \prod_{i=1}^{n} f(x_i; \hat{\hat{\theta}})$$

These quantities are both values of random variables since they depend on the observed values $x_1, x_2, \ldots, x_n$, and their ratio

$$\lambda = \frac{\max L_0}{\max L}$$

is referred to as a value of the **likelihood ratio statistic λ**.

Since max L_0 and max L are both values of a likelihood function, and therefore never negative, it follows that $\lambda \geqslant 0$; also, since ω is a subset of the parameter space Ω, it follows that $\lambda \leqslant 1$. When the null hypothesis is false, we would expect max L_0 to be small compared to max L, in which case λ would be close to zero. On the other hand, when the null hypothesis is true and $\theta \in \omega$, we would expect max L_0 to be close to max L, in which case λ would be close to 1. A likelihood ratio test states, therefore, that the null hypothesis H_0 is rejected if and only if λ falls in a critical region of the form $\lambda \leq k$, where $0 < k < 1$. To summarize,

DEFINITION 12.4 If ω and ω' are complementary subsets of the parameter space Ω, and if

$$\lambda = \frac{\max L_0}{\max L}$$

where $\max L_0$ and $\max L$ are the maximum values of the likelihood function for all values of θ in ω and Ω, respectively, then the critical region

$$\lambda \leq k$$

where $0 < k < 1$, defines a **likelihood ratio test** of the null hypothesis $\theta \in \omega$ against the alternative hypothesis $\theta \in \omega'$.

If H_0 is a simple hypothesis, k is chosen so that the size of the critical region equals α; if H_0 is composite, k is chosen so that the probability of a type I error is less than or equal to α for all θ in ω, and equal to α, if possible, for at least one value of θ in ω. Thus, if H_0 is a simple hypothesis and $g(\lambda)$ is the density of $\boldsymbol{\lambda}$ at λ when H_0 is true, then k must be such that

$$P(\boldsymbol{\lambda} \leq k) = \int_0^k g(\lambda)d\lambda = \alpha$$

In the discrete case, the integral is replaced by a sum and k is taken to be the largest value for which the sum is less than or equal to α.

EXAMPLE 12.5

Find the critical region of the likelihood ratio test for testing the null hypothesis

$$H_0\colon \quad \mu = \mu_0$$

against the composite alternative

$$H_1\colon \quad \mu \neq \mu_0$$

on the basis of a random sample of size n from a normal population with the known variance σ^2.

Solution

Since ω contains only μ_0, it follows that $\hat{\mu} = \mu_0$, and since Ω is the set of all real numbers, it follows by the method of Section 10.7 that $\hat{\hat{\mu}} = \bar{x}$. Thus,

$$\max L_0 = \left(\frac{1}{\sigma\sqrt{2\pi}}\right)^n \cdot e^{-\frac{1}{2\sigma^2}\cdot\Sigma(x_i-\mu_0)^2}$$

and

$$\max L = \left(\frac{1}{\sigma\sqrt{2\pi}}\right)^n \cdot e^{-\frac{1}{2\sigma^2}\cdot\Sigma(x_i-\bar{x})^2}$$

where the summations extend from $i = 1$ to $i = n$, and the value of the likelihood ratio statistic becomes

$$\lambda = \frac{e^{-\frac{1}{2\sigma^2}\cdot\Sigma(x_i-\mu_0)^2}}{e^{-\frac{1}{2\sigma^2}\cdot\Sigma(x_i-\bar{x})^2}} = e^{-\frac{n}{2\sigma^2}(\bar{x}-\mu_0)^2}$$

after suitable simplifications which the reader will be asked to verify in Exercise 7 on page 384. Hence, the critical region of the likelihood ratio test is

$$e^{-\frac{n}{2\sigma^2}(\bar{x}-\mu_0)^2} \leqslant k$$

and, after taking logarithms and dividing by $-\frac{n}{2\sigma_2}$, it becomes

$$(\bar{x}-\mu_0)^2 \geqslant -\frac{2\sigma^2}{n}\cdot\ln k$$

or

$$|\bar{x}-\mu_0| \geqslant K$$

where K will have to be determined so that the size of the critical region is α. Note that $\ln k$ is negative in view of the fact that $0 < k < 1$.

Since $\bar{\mathbf{x}}$ has a normal distribution with the mean μ_0 and the variance $\frac{\sigma^2}{n}$ (see Theorem 8.3 on page 259), we find that the critical region of this likelihood ratio test is

$$|\bar{x}-\mu_0| \geqslant z_{\alpha/2}\cdot\frac{\sigma}{\sqrt{n}}$$

or, equivalently,

$$|z| \geqslant z_{\alpha/2}$$

where

$$z = \frac{\bar{x} - \mu_0}{\sigma/\sqrt{n}}$$

In other words, the null hypothesis must be rejected when $\mathbf{z}$ takes on a value greater than or equal to $z_{\alpha/2}$, or a value less than or equal to $-z_{\alpha/2}$.

In the preceding example it was easy to find the constant that made the size of the critical region equal to α, because we were able to refer to the known distribution of $\bar{\mathbf{x}}$, and did not have to derive the distribution of the likelihood ratio statistic $\boldsymbol{\lambda}$, itself. Since the distribution of $\boldsymbol{\lambda}$ is generally quite complicated, which makes it difficult to evaluate k, it is often preferable to use the following approximation, whose proof is referred to on page 386:

THEOREM 12.2 For large n, the distribution of $-2 \cdot \ln \boldsymbol{\lambda}$ approaches, under very general conditions, the chi-square distribution with 1 degree of freedom.

We should add that this theorem applies only to the one-parameter case; if the population involves more than one unknown parameter, upon which the null hypothesis imposes r restrictions, the number of degrees of freedom in the chi-square approximation of the distribution of $-2 \cdot \ln \boldsymbol{\lambda}$ is equal to r. Thus, if we want to test the null hypothesis that the unknown mean and variance of a normal population are, respectively, $\mu = \mu_0$ and $\sigma^2 = \sigma_0^2$ against the alternative hypothesis that $\mu \neq \mu_0$ and $\sigma^2 \neq \sigma_0^2$, the number of degrees of freedom in the chi-square approximation of the distribution of $-2 \cdot \ln \boldsymbol{\lambda}$ would be 2; the two restrictions are $\mu = \mu_0$ and $\sigma^2 = \sigma_0^2$.

Since small values of λ correspond to large values of $-2 \cdot \ln \lambda$, we can use Theorem 12.2 to write the critical region of this approximate likelihood ratio test as

$$-2 \cdot \ln \lambda \geqslant \chi^2_{\alpha,1}$$

where $\chi^2_{\alpha,1}$ is as defined on page 269. In connection with Example 12.5 we find that

$$-2 \cdot \ln \lambda = \frac{n}{\sigma^2}(\bar{x} - \mu_0)^2 = \left(\frac{\bar{x} - \mu_0}{\sigma/\sqrt{n}}\right)^2$$

which actually *is* a value of a random variable having the chi-square distribution with 1 degree of freedom.

As we indicated on page 377, the likelihood ratio technique will generally produce satisfactory results. That this is not always the case is illustrated by the following example, which is somewhat out of the ordinary:

EXAMPLE 12.6

On the basis of a single observation, we want to test the simple null hypothesis that the probability distribution of **x** is

x	1	2	3	4	5	6	7
$f(x)$	$\frac{1}{12}$	$\frac{1}{12}$	$\frac{1}{12}$	$\frac{1}{4}$	$\frac{1}{6}$	$\frac{1}{6}$	$\frac{1}{6}$

against the composite alternative that the probability distribution is

x	1	2	3	4	5	6	7
$g(x)$	$\frac{a}{3}$	$\frac{b}{3}$	$\frac{c}{3}$	$\frac{2}{3}$	0	0	0

where $a + b + c = 1$. Show that the critical region obtained by means of the likelihood ratio technique is inadmissible.

Solution

The composite alternative hypothesis includes all the probability distributions which we get by assigning different values from 0 to 1 to a, b, and c, subject only to the restriction that $a + b + c = 1$. To determine λ for each value of x, we first let $x = 1$. For this value we get max $L_0 = \frac{1}{12}$, max $L = \frac{1}{3}$ (corresponding to $a = 1$), and, hence, $\lambda = \frac{1}{4}$. Determining λ for the other values of x in the same way, we get the results shown in the following table:

x	1	2	3	4	5	6	7
λ	$\frac{1}{4}$	$\frac{1}{4}$	$\frac{1}{4}$	$\frac{3}{8}$	1	1	1

If the size of the critical region is to be $\alpha = 0.25$, we find that the likelihood ratio technique yields the critical region for which the null hypothesis is rejected when

$\lambda = \frac{1}{4}$, namely, when $x = 1$, $x = 2$, or $x = 3$; clearly, $f(1) + f(2) + f(3) = \frac{1}{12} + \frac{1}{12} + \frac{1}{12} = 0.25$. The corresponding probability of a type II error is given by $g(4) + g(5) + g(6) + g(7)$, and hence, it equals $\frac{2}{3}$.

Now let us consider the critical region for which the null hypothesis is rejected only when $x = 4$. Its size is also $\alpha = 0.25$ since $f(4) = \frac{1}{4}$, but the corresponding probability of a type II error is

$$g(1) + g(2) + g(3) + g(5) + g(6) + g(7) = \frac{a}{3} + \frac{b}{3} + \frac{c}{3} + 0 + 0 + 0$$
$$= \frac{1}{3}$$

and since this is less than $\frac{2}{3}$, the critical region obtained by means of the likelihood ratio technique is inadmissible. Of course, as we pointed out at the beginning, this example is somewhat out of the ordinary.

THEORETICAL EXERCISES

1. A bowl contains 7 marbles of which θ are red while the others are blue. In order to test the null hypothesis $\theta \leqslant 2$ against the alternative $\theta > 2$, two of the marbles are randomly drawn without replacement and the null hypothesis is rejected if and only if both are red.

(a) Find the probabilities of committing type I errors when $\theta = 0, 1$, and 2.
(b) Find the probabilities of committing type II errors when $\theta = 3, 4, 5, 6$, and 7.

Also plot the graph of the power function.

2. Suppose that in Example 12.4 on page 374 we accept the null hypothesis $\theta \geqslant 0.90$ if $x \geqslant 16$ and reject it in favor of the alternative hypothesis $\theta < 0.90$ if $x < 16$. Construct the power function of this test criterion by calculating $\pi(\theta)$ for the same values of θ as in the table on page 374.

3. A single observation is to be used to test the null hypothesis that the parameter of an exponential distribution equals 10 against the alternative hypothesis that it does not equal 10. If the null hypothesis is to be rejected if and only if the observed value is less than 8 or greater than 12, find

(a) the probability of a type I error;
(b) the probabilities of type II errors when $\theta = 2, 4, 8, 16$, and 20.

Also plot the graph of the power function.

4. A random sample of size 64 is to be used to test the null hypothesis that the mean of a normal population with the variance $\sigma^2 = 256$ is less than or equal to 40 against the alternative hypothesis that it is greater than 40. If the null hypothesis is to be rejected if and only if the mean of the random sample exceeds 43, find

(a) the probabilities of type I errors when $\mu = 37, 38, 39$, and 40;
(b) the probabilities of type II errors when $\mu = 41, 42, 43, 44, 45, 46, 47$, and 48.

Also plot the graph of the power function.

5. The sum of the values obtained in a random sample of size 5 from a Poisson population is to be used to test the null hypothesis that the mean of the population is greater than 2 against the alternative hypothesis that it is less than or equal to 2. If the null hypothesis is to be rejected if and only if the sum of the observations is 5 or less, find

(a) the probabilities of type I errors when the mean of the population is 2.2, 2.4, 2.6, 2.8, and 3.0;
(b) the probabilities of type II errors when the mean of the population is 2.0, 1.5, 1.0, and 0.5.

Also plot the graph of the power function. (*Hint:* Use the result obtained in Example 7.14 on page 249.)

6. Verify the statement on page 377 that 57 heads and 43 tails in 100 flips of a coin does not enable us to reject the null hypothesis that the coin is perfectly balanced (against the alternative that it is not perfectly balanced) at the level of significance $\alpha = 0.05$. (*Hint:* Use the normal approximation to the binomial distribution.)

7. Verify the final step on page 380 which led to

$$\lambda = e^{-\frac{n}{2\sigma^2}(\bar{x}-\mu_0)^2}$$

8. The number of successes in n trials is to be used to test the null hypothesis that the parameter θ of a binomial population equals $\frac{1}{2}$ against the alternative that it does not equal $\frac{1}{2}$.

(a) Find an expression for the likelihood ratio statistic.
(b) Use the result of part (a) to show that the critical region of the likelihood ratio test can be written as

$$x \cdot \ln x + (n - x) \cdot \ln (n - x) \geq k$$

where x is the observed number of successes.

(c) Studying the graph of $f(x) = x \cdot \ln x + (n - x) \cdot \ln (n - x)$, its minimum, and its symmetry, show that the critical region of this likelihood ratio test can also be written as $\left| x - \frac{n}{2} \right| \geq c$, where c is a constant which depends on the size of the critical region.

9. A random sample of size n is to be used to test the null hypothesis that the parameter θ of an exponential population equals θ_0 against the alternative that it does not equal θ_0.

(a) Find an expression for the likelihood ratio statistic.

(b) Use the result of part (a) to show that the critical region of the likelihood ratio test can be written as

$$\bar{x} \cdot e^{-\bar{x}/\theta_0} \leqslant K$$

10. A random sample of size n from a normal population with unknown mean and variance is to be used to test the null hypothesis $\mu = \mu_0$ against the alternative $\mu \neq \mu_0$. Using the simultaneous maximum likelihood estimates of μ and σ^2 obtained in Example 10.13, show that the values of the likelihood ratio statistic can be written in the form

$$\lambda = \left(1 + \frac{t^2}{n-1}\right)^{-n/2}$$

where $t = \dfrac{\bar{x} - \mu_0}{s/\sqrt{n}}$. Note that the likelihood ratio test can, thus, be based on the t distribution of Section 8.5.

11. For the likelihood ratio statistic of Exercise 10, show that $-2 \cdot \ln \lambda$ approaches t^2 as $n \to \infty$. [*Hint:* Use the infinite series expansion of $\ln(1 + x)$ given on page 212.]

12. Given a random sample of size n from a normal population with unknown mean and variance, find an expression for the likelihood ratio statistic for testing the null hypothesis $\sigma = \sigma_0$ against the alternative hypothesis $\sigma \neq \sigma_0$. (*Hint:* See Example 10.13 on page 330.)

13. Independent random samples of size $n_1, n_2, \ldots,$ and n_k from k normal populations with unknown means and variances are to be used to test the null hypothesis $\sigma_1^2 = \sigma_2^2 = \cdots = \sigma_k^2$ against the alternative that these variances are not all equal.

(a) Show that under the null hypothesis the maximum likelihood estimates of the means μ_i and the variances σ_i^2 are

$$\hat{\mu}_i = \bar{x}_i \quad \text{and} \quad \hat{\sigma}_i^2 = \sum_{i=1}^{k} \frac{(n_i - 1)s_i^2}{n}$$

where $n = \sum_{i=1}^{k} n_i$, while without restrictions the maximum likelihood estimates of the means μ_i and the variances σ_i^2 are

$$\hat{\hat{\mu}}_i = \bar{x}_i \quad \text{and} \quad \hat{\hat{\sigma}}_i^2 = \frac{(n_i - 1)s_i^2}{n_i}$$

This follows directly from the results obtained in Section 10.7.

(b) Using the results of part (a), show that the likelihood ratio statistic can be written as

$$\lambda = \frac{\prod_{i=1}^{k} \left[\frac{(n_i - 1)s_i^2}{n_i}\right]^{n_i/2}}{\left[\sum_{i=1}^{k} \frac{(n_i - 1)s_i^2}{n}\right]^{n/2}}$$

(c) If $n_1 = 8$, $s_1^2 = 16$, $n_2 = 10$, $s_2^2 = 25$, $n_3 = 6$, $s_3^2 = 12$, $n_4 = 8$, and $s_4^2 = 24$ are the sample sizes and the variances of four independent random samples from four normal populations, use the result of part (b) to calculate $-2 \cdot \ln \lambda$ and then test the null hypothesis stated at the beginning of this exercise. (Note that the number of degrees of freedom for this approximate chi-square test is 3, since $\sigma_1^2 = \sigma_2^2 = \sigma_3^2 = \sigma_4^2$ imposes 3 restrictions on the parameters.)

14. Show that for $k = 2$ the likelihood ratio statistic of Exercise 13 can be expressed in terms of the ratio of the two sample variances and that the likelihood ratio test can, therefore, be based on the F distribution.

15. If 15, 28, 3, 12, 42, 19, 20, 2, 25, 30, 62, 12, 18, 16, 44, 65, 33, 51, 4, and 28 are the values of a random sample from an exponential population, use part (a) of Exercise 9 and Theorem 12.2 to test the null hypothesis that the mean of the population is 15 against the composite alternative that it is not equal to 15. Let α, the size of the critical region, be 0.05.

16. When we test a simple null hypothesis against a composite alternative, a critical region is said to be **unbiased** if the corresponding power function takes on its minimum value at the value of the parameter assumed under the null hypothesis. In other words, a critical region is unbiased if the probability of rejecting the null hypothesis is least when the null hypothesis is true. Given a single observation of the random variable **x** having the density

$$f(x) = \begin{cases} 1 + \theta^2(\frac{1}{2} - x) & \text{for } 0 < x < 1 \\ 0 & \text{elsewhere} \end{cases}$$

where $-1 \leqslant \theta \leqslant 1$, show that the critical region $x \leqslant \alpha$ provides an unbiased and uniformly most powerful critical region of size α for testing the null hypothesis $\theta = 0$ against the alternative hypothesis $\theta \neq 0$.

References

Discussions of various properties of likelihood ratio tests, particularly their large-sample properties, and a proof of Theorem 12.2 may be found in most advanced textbooks on the theory of statistics, for example, in the book by S. S. Wilks referred to on page 252. Much of the original research done in this area is reproduced in

Selected Papers in Statistics and Probability by Abraham Wald. Stanford, Calif.: Stanford University Press, 1957.

hypothesis testing: applications

13

13.1 INTRODUCTION

In Chapter 12 we discussed some of the theory which underlies statistical tests, and in this chapter we shall present some of the standard tests that are most widely used in applications. Most of these tests, at least those based on known population distributions, can be obtained by the likelihood ratio technique.

To explain the terminology which we shall use, let us consider a situation in which we want to test the null hypothesis H_0: $\theta = \theta_0$ against the **two-sided alternative** hypothesis H_1: $\theta \neq \theta_0$. Since it appears reasonable to accept the null hypothesis when our point estimate $\hat{\theta}$ of θ is close to θ_0 and to reject it when $\hat{\theta}$ is much larger or much smaller than θ_0, it would be logical to let the critical region consist of both tails of the sampling distribution of our test statistic $\hat{\boldsymbol{\theta}}$. Such a test is referred to as a **two-tailed test**.

On the other hand, if we are testing the null hypothesis H_0: $\theta = \theta_0$ against the **one-sided alternative** H_1: $\theta < \theta_0$, it would seem reasonable to reject H_0 only when $\hat{\theta}$ is much smaller than θ_0. Therefore, in this case it would be logical to let the critical region consist only of the left tail of the sampling distribution of $\hat{\boldsymbol{\theta}}$. Likewise, in testing H_0: $\theta = \theta_0$ against the one-sided alternative H_1: $\theta > \theta_0$, we reject H_0 only for large values of $\hat{\theta}$ and the critical region consists only of the right tail of the sampling distribution of $\hat{\boldsymbol{\theta}}$. Any test where the critical region consists only of one tail of the sampling distribution of the test statistic is called a **one-tailed test**.

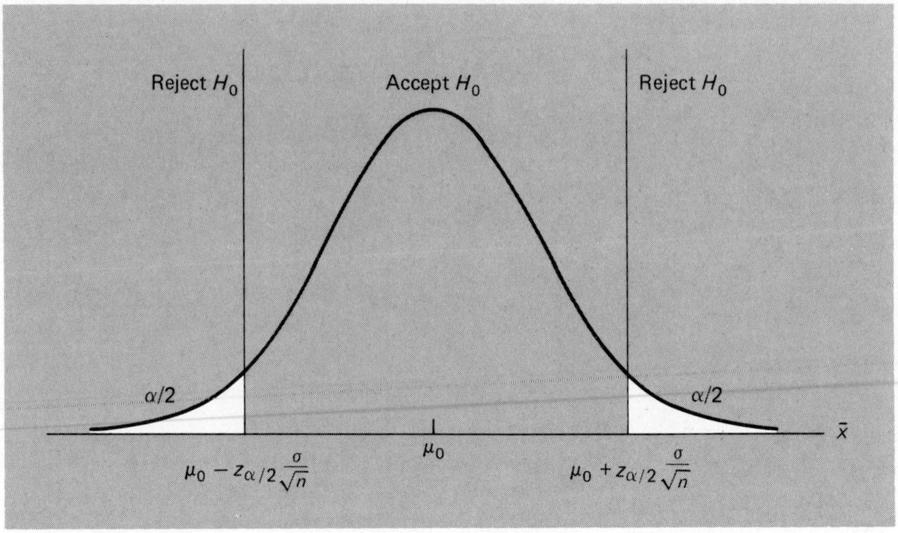

Figure 13.1 Critical region for the alternative hypothesis $\mu \neq \mu_0$.

For instance, for the two-sided alternative $\mu \neq \mu_0$ in Example 12.5 on page 379, the likelihood ratio technique led to a two-tailed test with the critical region

$$|\bar{x} - \mu_0| \geqslant z_{\alpha/2} \cdot \frac{\sigma}{\sqrt{n}}$$

or

$$\bar{x} \leqslant \mu_0 - z_{\alpha/2} \cdot \frac{\sigma}{\sqrt{n}} \quad \text{and} \quad \bar{x} \geqslant \mu_0 + z_{\alpha/2} \cdot \frac{\sigma}{\sqrt{n}}$$

As is pictured in Figure 13.1, the null hypothesis $\mu = \mu_0$ is rejected if $\bar{\mathbf{x}}$ takes on a value falling in either tail of its sampling distribution. In terms of the statistic $\mathbf{z}$, the critical region can be stated in the form $z \leqslant -z_{\alpha/2}$ and $z \geqslant z_{\alpha/2}$, where

$$z = \frac{\bar{x} - \mu_0}{\sigma/\sqrt{n}}$$

Had we used the one-sided alternative $\mu > \mu_0$, the likelihood ratio technique would have led to the one-tailed test pictured in Figure 13.2, and if we had used the one-sided alternative $\mu < \mu_0$, the likelihood ratio technique would have led to the one-tailed test pictured in Figure 13.3. It certainly stands to reason that

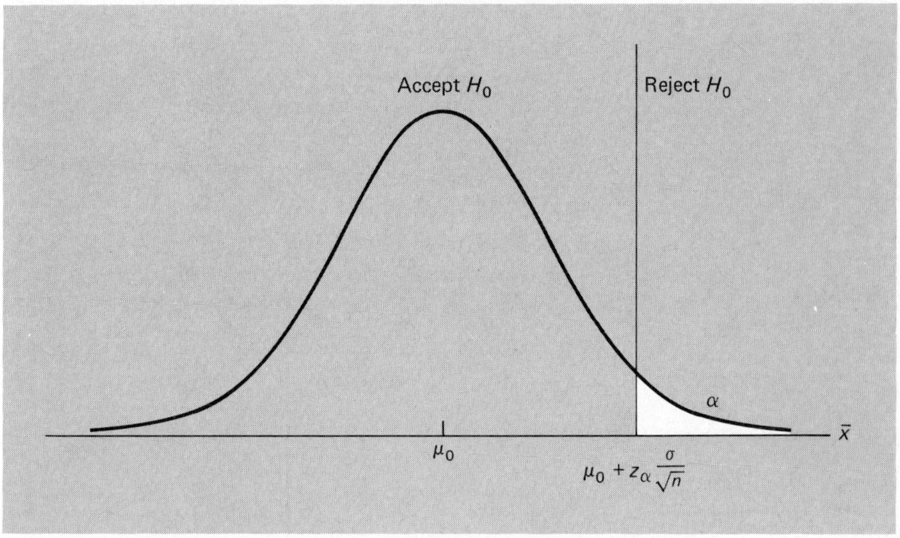

Figure 13.2 Critical region for the alternative hypothesis $\mu > \mu_0$.

in the first case we would reject the null hypothesis and accept the alternative only when $\bar{x}$ is large, namely, when it falls into the right tail of the sampling distribution, and that the opposite is true when the alternative hypothesis is $\mu < \mu_0$. Although there are exceptions to this rule (see Exercise 1 on page 396), two-sided

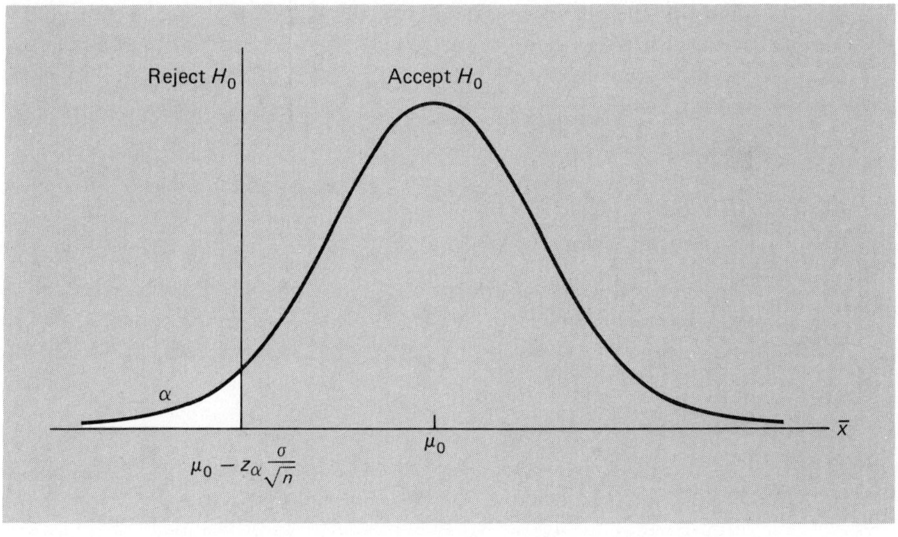

Figure 13.3 Critical region for the alternative hypothesis $\mu < \mu_0$.

alternatives usually lead to two-tailed tests and one-sided alternatives usually lead to one-tailed tests.

In the remainder of this chapter, the outline of each test procedure will consists of the following four steps:

1. *State the null hypothesis H_0 and an appropriate alternative hypothesis H_1.*
2. *Using the sampling distribution of an appropriate test statistic, determine a critical region of size α, where α is specified.*
3. *Compute the value of the test statistic from sample data.*
4. *Decide whether to reject the null hypothesis, whether to accept it, or whether to reserve judgment.*

13.2 TESTS CONCERNING MEANS

In this section we shall discuss the most commonly used tests concerning the mean of a population, and in Section 13.3 we shall discuss the corresponding tests concerning the means of two populations. Tests concerning the means of more than two populations will be taken up later in Chapter 15. All the tests in this section are based on normal distribution theory, assuming either that the samples come from normal populations or that they are large enough to justify normal approximations; some nonparametric alternatives to these tests, which do not require knowledge about the population or populations from which the samples are obtained, will be taken up in Chapter 16.

Suppose that we want to test the null hypothesis $\mu = \mu_0$ against one of the alternatives $\mu \neq \mu_0$, $\mu > \mu_0$, or $\mu < \mu_0$ on the basis of a random sample of size n from a normal population with the known variance σ^2. This, of course, is the test that was considered in Example 12.5 to illustrate the likelihood ratio technique and the critical regions for the respective alternatives are $|z| \geqslant z_{\alpha/2}$, $z \geqslant z_\alpha$, and $z \leqslant -z_\alpha$, where

$$z = \frac{\bar{x} - \mu_0}{\sigma/\sqrt{n}}$$

The most frequently used values of α, the probability of a type I error, are 0.05 and 0.01, and as the reader was asked to show in Exercise 17 on page 222, the corresponding values of z_α and $z_{\alpha/2}$ are $z_{.05} = 1.645$, $z_{.01} = 2.33$, $z_{.025} = 1.96$, and $z_{.005} = 2.575$.

EXAMPLE 13.1

Suppose that it is known from experience that the standard deviation of the weight of 8-ounce packages of cookies made by a certain bakery is 0.16 ounce. To check whether its production is under control on a given day, namely, to check whether the true average weight of the packages is 8 ounces, they select a random sample of 25 packages and find that their mean weight is $\bar{x} = 8.112$ ounces. Since the bakery stands to lose money when $\mu > 8$ and the customer loses out when $\mu < 8$, test the null hypothesis $\mu = 8$ against the alternative $\mu \neq 8$ using $\alpha = 0.01$.

Solution

1. H_0: $\mu = 8$
H_1: $\mu \neq 8$

2. Critical region: $|z| \geqslant z_{.005} = 2.575$, where $z = \dfrac{\bar{x} - \mu_0}{\sigma/\sqrt{n}}$

3. Computations: $\bar{x} = 8.112$, $n = 25$, and, hence,

$$z = \frac{8.112 - 8}{0.16/\sqrt{25}} = 3.5$$

4. Decision: Reject the null hypothesis and make suitable adjustments in the production process.

It should be noted that the critical region $z \geqslant z_\alpha$ can also be used to test the null hypothesis $\mu = \mu_0$ against the simple alternative $\mu = \mu_1 > \mu_0$, or to test the composite hypothesis $\mu \leqslant \mu_0$ against the composite alternative $\mu > \mu_0$. In the first case we would be testing a simple hypothesis against a simple alternative as in Section 12.4 (see Example 12.3 on page 370, where we studied this test for $\sigma = 1$), and in the second case α would be the maximum probability of committing a type I error for any value of μ assumed under the null hypothesis. Of course, similar arguments apply to the critical region $z \leqslant -z_\alpha$.

When we are dealing with a large sample of size $n \geqslant 30$ from a population which need not be normal but has a finite variance, we can use the central limit theorem to justify using the test for normal populations, and even when σ^2 is unknown we can approximate its value with s^2 in the computation of the test statistic. To illustrate the use of such an approximate **large-sample test**, consider the following example:

EXAMPLE 13.2

Suppose that 100 tires of a certain brand lasted on the average 21,431 miles with a standard deviation of 1,295 miles. Using $\alpha = 0.05$, test the null hypothesis $\mu = 22{,}000$ miles against the alternative hypothesis $\mu < 22{,}000$.

Solution

1. H_0: $\mu = 22{,}000$
   H_1: $\mu < 22{,}000$
2. Critical region: $z \leqslant -z_{.05} = -1.645$, where $z = \dfrac{\bar{x} - \mu_0}{\sigma/\sqrt{n}}$
3. Computations: $\bar{x} = 21{,}431$, $n = 100$, $\sigma \approx s = 1{,}295$, and, hence,

$$z = \frac{21{,}431 - 22{,}000}{1{,}295/\sqrt{100}} = -4.39$$

4. Decision: Reject H_0 and conclude that the tires are not as good as claimed.

When $n < 30$ and σ^2 is unknown, the test which we have been discussing in this section cannot be used. However, in Exercise 10 on page 385 we saw that for random samples from normal populations the likelihood ratio technique yields a corresponding test based on

$$t = \frac{\bar{x} - \mu_0}{s/\sqrt{n}}$$

which, according to Theorem 8.11, is a value of a random variable having the t distribution with $n - 1$ degrees of freedom. Thus, critical regions of size α for testing the null hypothesis $\mu = \mu_0$ against the alternatives $\mu \neq \mu_0$, $\mu > \mu_0$, or $\mu < \mu_0$, are, respectively, $|t| \geqslant t_{\alpha/2,n-1}$, $t \geqslant t_{\alpha,n-1}$, and $t \leqslant -t_{\alpha,n-1}$. Note that the comments made on page 391 in connection with the alternative hypothesis $\mu_1 > \mu_0$ and the test of the null hypothesis $\mu \leqslant \mu_0$ against the alternative $\mu > \mu_0$ apply also in this case.

To illustrate this **small-sample test**, as it is usually called, consider the following example:

EXAMPLE 13.3

Suppose that the specifications for a certain kind of ribbon call for a mean breaking strength of 185 pounds, and that five pieces randomly selected from different rolls have a mean breaking strength of 183.1 pounds with a standard deviation of 8.2 pounds. Assuming that we can look upon the data as a random sample from a normal population, test the null hypothesis $\mu = 185$ against the alternative hypothesis $\mu < 185$ at $\alpha = 0.05$.

Solution

1. H_0: $\mu = 185$
   H_1: $\mu < 185$

2. Critical region: $t \leqslant -t_{.05,4} = -2.132$, where $t = \dfrac{\bar{x} - \mu_0}{s/\sqrt{n}}$

3. Computations: $\bar{x} = 183.1$, $s = 8.2$, and $n = 5$, so that

$$t = \frac{183.1 - 185}{8.2/\sqrt{5}} = -0.49$$

4. Decision: Since $t = -0.49$ is greater than $-t_{.05,4} = -2.132$, the null hypothesis cannot be rejected. If we have to go beyond this and say that the rolls of ribbon from which the sample was selected meet specifications, we are, of course, exposed to the unknown risk of committing a type II error.

13.3 TESTS CONCERNING DIFFERENCES BETWEEN MEANS

In applied research, there are many problems in which we are interested in hypotheses concerning differences between the means of two populations. For instance, we may want to decide upon the basis of suitable samples whether men can perform a certain task as fast as women, or we may want to decide on the basis of an appropriate sample survey whether the average weekly food expenditures of families in one city exceed those of families in another city by at least \$5.00.

Let us suppose that we are dealing with independent random samples of size n_1 and n_2 from two normal populations having the means μ_1 and μ_2 and the known variances σ_1^2 and σ_2^2, and that we want to test the null hypothesis $\mu_1 - \mu_2 = \delta$, where δ is a given constant, against one of the alternatives $\mu_1 - \mu_2 \neq \delta$, $\mu_1 - \mu_2 > \delta$, or $\mu_1 - \mu_2 < \delta$. Applying the likelihood ratio technique, we will arrive at a test based on $\bar{x}_1 - \bar{x}_2$, and, referring to Exercise 4 on page 263, we find that the respective critical regions can be written as $|z| \geqslant z_{\alpha/2}$, $z \geqslant z_\alpha$, and $z \leqslant -z_\alpha$, where

$$z = \frac{\bar{x}_1 - \bar{x}_2 - \delta}{\sqrt{\dfrac{\sigma_1^2}{n_1} + \dfrac{\sigma_2^2}{n_2}}}$$

When we deal with independent random samples from populations with unknown variances which may not even be normal, we can still use the test which we have just described with s_1 substituted for σ_1 and s_2 substituted for σ_2 so long as both samples are large enough for the central limit theorem to be invoked.

EXAMPLE 13.4

Suppose that the nicotine contents of two brands of cigarettes are being measured. If in an experiment fifty cigarettes of the first brand had an average nicotine content of $\bar{x}_1 = 2.61$ milligrams with a standard deviation of $s_1 = 0.12$ milligram, while forty cigarettes of the second brand had an average nicotine content of $\bar{x}_2 = 2.38$ milligrams with a standard deviation of $s_2 = 0.14$ milligram, test the null hypothesis $\mu_1 - \mu_2 = 0.2$ against the alternative $\mu_1 - \mu_2 \neq 0.2$, using $\alpha = 0.05$.

Solution

1. H_0: $\mu_1 - \mu_2 = 0.2$
H_1: $\mu_1 - \mu_2 \neq 0.2$

2. Critical region: $|z| \geqslant z_{.025} = 1.96$, where

$$z = \frac{\bar{x}_1 - \bar{x}_2 - \delta}{\sqrt{\dfrac{\sigma_1^2}{n_1} + \dfrac{\sigma_2^2}{n_2}}}$$

3. Computations: $\bar{x}_1 = 2.61$, $s_1 = 0.12$, $n_1 = 50$, $\bar{x}_2 = 2.38$, $s_2 = 0.14$, and $n_2 = 40$, so that

$$z = \frac{2.61 - 2.38 - 0.2}{\sqrt{\dfrac{(0.12)^2}{50} + \dfrac{(0.14)^2}{40}}} = 1.08$$

4. Decision: The null hypothesis cannot be rejected. Hence, we can either accept the null hypothesis, or merely say that the difference between $2.61 - 2.38 = 0.23$ and 0.2 is not large enough to reject the null hypothesis, and let it go at that.

When n_1 and n_2 are small and σ_1 and σ_2 are unknown, the test which we have been discussing cannot be used. However, for independent random samples from two normal populations having the same unknown variance σ^2, the likelihood ratio technique yields a test based on

$$t = \frac{\bar{x}_1 - \bar{x}_2 - \delta}{s_p\sqrt{\dfrac{1}{n_1} + \dfrac{1}{n_2}}}$$

where

$$s_p^2 = \frac{(n_1 - 1)s_1^2 + (n_2 - 1)s_2^2}{n_1 + n_2 - 2}$$

From Section 11.3, we know that under the given assumptions and the null hypothesis $\mu_1 - \mu_2 = \delta$, the above expression for t is a value of a random variable having the t distribution with $n_1 + n_2 - 2$ degrees of freedom. Thus, the appropriate critical region of size α for testing the null hypothesis $\mu_1 - \mu_2 = \delta$ against the alternatives $\mu_1 - \mu_2 \neq \delta$, $\mu_1 - \mu_2 > \delta$, or $\mu_1 - \mu_2 < \delta$ under the given assumptions are, respectively, $|t| \geqslant t_{\alpha/2, n_1+n_2-2}$, $t \geqslant t_{\alpha, n_1+n_2-2}$, and $t \leqslant -t_{\alpha, n_1+n_2-2}$. To illustrate this two-sample t test, consider the following problem:

EXAMPLE 13.5

In the comparison of two kinds of paint, a consumer testing service finds that four one-gallon cans of one brand cover on the average 512 square feet with a standard deviation of 31 square feet, while four one-gallon cans of another brand cover on the average 492 square feet with a standard deviation of 26 square feet. Test the null hypothesis $\mu_1 - \mu_2 = 0$ against the alternative hypothesis $\mu_1 - \mu_2 \neq 0$ at the level of significance $\alpha = 0.05$. Assume that the two populations are normal and have equal variances.

Solution

1. H_0: $\mu_1 - \mu_2 = 0$
   H_1: $\mu_1 - \mu_2 \neq 0$
2. Critical region: $|t| \geqslant t_{.025,6} = 2.447$, where

$$t = \frac{\bar{x}_1 - \bar{x}_2 - \delta}{s_p\sqrt{\dfrac{1}{n_1} + \dfrac{1}{n_2}}}$$

3. Computations: $\bar{x}_1 = 512$, $s_1 = 31$, $n_1 = 4$, $\bar{x}_2 = 492$, $s_2 = 26$, and $n_2 = 4$, so that

$$s_p = \sqrt{\frac{3(31)^2 + 3(26)^2}{4 + 4 - 2}} = 28.609$$

and

$$t = \frac{512 - 492}{28.609\sqrt{\frac{1}{4} + \frac{1}{4}}} = 0.99$$

4. Decision: The null hypothesis cannot be rejected. Even though the difference between the two sample means seems to be fairly large, the samples are so small that the results are not conclusive; that is, the difference may well be due to chance.

If the assumption $\sigma_1 = \sigma_2$ is untenable, there are several alternative methods that can be used. A relatively simple one consists of randomly pairing the values obtained in the two samples and then looking upon their differences as a random sample of size n_1 or n_2, whichever is smaller, from a normal population which, under the null hypothesis, has the mean $\mu = \delta$. Then we test this null hypothesis against the appropriate alternative by means of the methods of Section 13.2. This is a good reason for having $n_1 = n_2$, but there exist alternative techniques for handling the case where $n_1 \neq n_2$—one of these, the *Smith–Satterthwaite* test, is referred to on page 418.

So far we have limited our discussion to random samples that are *independent*, and the methods which we have introduced in this section cannot be used, for example, to decide on the basis of weights "before and after" whether a certain diet is really effective, or whether an observed difference between the average I.Q.'s of husbands and their wives is really significant. In both of these examples the samples are not independent because the data are actually *paired.* A common way of handling this kind of problem is to proceed as in the preceding paragraph, namely, to work with the differences between the paired measurements or observations. If n is large, we can then use the test described on page 388 to test the null hypothesis $\mu_1 - \mu_2 = \delta$ against the appropriate alternative, and if n is small, we can use the t test described on page 392, provided the differences can be looked upon as a random sample from a normal population.

THEORETICAL EXERCISES

1. Given a random sample of size n from a normal population with the known variance σ^2, show that the null hypothesis $\mu = \mu_0$ can be tested against the alternative $\mu \neq \mu_0$ with the use of a one-tailed criterion based on the chi-square distribution.
2. Suppose that a random sample from a normal population with the known variance σ^2 is to be used to test the null hypothesis $\mu = \mu_0$ against the alternative hypothesis $\mu = \mu_1$, where $\mu_1 > \mu_0$, and that the probabilities of type I and type II errors are to have the preassigned values α and β. Show that the required size of the sample is given by

$$n = \frac{\sigma^2(z_\alpha + z_\beta)^2}{(\mu_1 - \mu_0)^2}$$

   Also use this formula to find n when $\sigma = 9$, $\mu_0 = 15$, $\mu_1 = 20$, $\alpha = 0.05$, and $\beta = 0.01$.
3. Suppose that independent random samples of size n from two normal populations with the known variances σ_1^2 and σ_2^2 are to be used to test the null hypothesis $\mu_1 - \mu_2 = \delta$ against the alternative hypothesis $\mu_1 - \mu_2 = \delta'$, and that the probabilities of type I and type II errors are to have the

preassigned values α and β. Show that the required size of the sample is given by

$$n = \frac{(\sigma_1^2 + \sigma_2^2)(z_\alpha + z_\beta)^2}{(\delta - \delta')^2}$$

Also use this formula to find n when $\sigma_1 = 9$, $\sigma_2 = 13$, $\delta = 80$, $\delta' = 86$, $\alpha = 0.01$, and $\beta = 0.01$.

APPLIED EXERCISES

4. According to the norms established for a reading comprehension test, eighth graders should average 84.3 with a standard deviation of 8.6. If 45 randomly selected eighth graders from a certain school district averaged 87.8, test the null hypothesis $\mu = 84.3$ against the alternative hypothesis $\mu > 84.3$, using $\alpha = 0.01$.

5. The security department of a factory wants to know whether the true average time required by the night watchman to walk his round is 30 minutes. If, in a random sample of 32 rounds, the night watchman averaged 30.8 minutes with a standard deviation of 1.5 minutes, determine at $\alpha = 0.01$ whether this is sufficient evidence to reject the null hypothesis $\mu = 30$ minutes in favor of the alternative hypothesis $\mu \neq 30$ minutes.

6. In 12 test runs over a marked course, a newly designed motorboat averaged 33.6 seconds with a standard deviation of 2.3 seconds. Assuming that it is reasonable to treat the data as a random sample from a normal population, test the null hypothesis $\mu = 35$ against the alternative $\mu < 35$ at the level of significance $\alpha = 0.05$.

7. Five measurements of the tar content of a certain kind of cigarette yielded 14.5, 14.2, 14.4, 14.3, and 14.6 mg/cig. Show that for $\alpha = 0.05$ the null hypothesis $\mu = 14.0$ must be rejected in favor of the alternative hypothesis $\mu \neq 14.0$. Assume that the data are a random sample from a normal population.

8. Suppose that in Exercise 7 the first measurement is recorded incorrectly as 16.0 instead of 14.5. Show that this will reverse the result, and explain the apparent paradox that even though the difference between the sample mean and μ_0 has increased, it is no longer significant.

9. With reference to Example 13.4, for what values of $\bar{x}_1 - \bar{x}_2$ would the null hypothesis have been rejected? Also find the probabilities of type II errors with the given criterion if (a) $\mu_1 - \mu_2 = 0.12$, (b) $\mu_1 - \mu_2 = 0.16$, (c) $\mu_1 - \mu_2 = 0.24$, and (d) $\mu_1 - \mu_2 = 0.28$.

10. A sample study was made of the number of business lunches that executives claim as deductible expenses per month. If 40 executives in the insurance industry averaged 9.1 such deductions with a standard deviation of 1.9 in a given month, while 50 bank executives averaged 8.0 with a standard deviation of 2.1, test the null hypothesis $\mu_1 - \mu_2 = 0$ against the alternative hypothesis $\mu_1 - \mu_2 \neq 0$ at $\alpha = 0.05$.

11. Sample surveys conducted in a large county in 1950 and again in 1970 showed that in 1950 the average height of 400 ten-year-old boys was 53.2 inches with a standard deviation of 2.4 inches, while in 1970 the average height of 500 ten-year-old boys was 54.5 inches with a standard deviation of 2.5 inches. Test the null hypothesis $\mu_1 - \mu_2 = -0.5$ against the alternative hypothesis $\mu_1 - \mu_2 < -0.5$ at the level of significance $\alpha = 0.05$.

12. To find out whether the inhabitants of two South Pacific islands may be regarded as having the same racial ancestry, an anthropologist determines the cephalic indices of six adult males from each island, getting $\bar{x}_1 = 77.4$, $\bar{x}_2 = 72.2$, and the corresponding standard deviations $s_1 = 3.3$ and $s_2 = 2.1$. Use $\alpha = 0.01$ to check whether the difference between the two sample means can reasonably be attributed to chance. Assume that the populations are normal and have the same variance.

13. If 8 short-range rockets of one kind have a mean target error of $\bar{x}_1 = 98$ feet with a standard deviation of $s_1 = 18$ feet while 10 short-range rockets of another kind have a mean target error of $\bar{x}_2 = 76$ feet with a standard deviation of $s_2 = 15$ feet, test the null hypothesis $\mu_1 - \mu_2 = 15$ against the alternative hypothesis $\mu_1 - \mu_2 > 15$. Let the size of the critical region be $\alpha = 0.05$ and assume that the populations are normal and have the same variance.

14. In a study of the effectiveness of certain exercises in weight reduction, a group of 16 persons engaged in these exercises for one month and showed the following results:

Weight before	*Weight after*	*Weight before*	*Weight after*
211	198	172	166
180	173	155	154
171	172	185	181
214	209	167	164
182	179	203	201
194	192	181	175
160	161	245	233
182	182	146	142

Test the null hypothesis $\mu_1 - \mu_2 = 0$ against the alternative hypothesis $\mu_1 - \mu_2 > 0$ at the level of significance $\alpha = 0.05$. Assume that the differences have a normal distribution.

15. To determine the effectiveness of an industrial safety program, the following data were collected over a period of a year on the average weekly loss of man hours due to accidents in 12 plants "before and after" the program was put into operation: 50 and 41, 87 and 75, 37 and 35, 141 and 129, 59 and 60, 65 and 53, 24 and 26, 88 and 85, 25 and 29, 36 and 31, 50 and 48, 35 and 37. Use $\alpha = 0.01$ to test the null hypothesis that the safety program is *not* effective against a suitable one-sided alternative. Assume that the differences have a normal distribution.

13.4 TESTS CONCERNING VARIANCES

There are several reasons why it is important to test hypotheses concerning the variances of populations. So far as direct applications are concerned, a manufacturer who has to meet rigid specifications will have to perform tests about the variability of his product, a teacher may want to know whether certain statements are true about the variability which he can expect in the performance of a student, and a pharmacist may have to check whether the variation in the potency of a medicine is within permissible limits. So far as indirect applications are concerned, tests about variances are often prerequisites for tests concerning other parameters. For instance, the two-sample t test described on page 394 requires that the two population variances are equal, and in practice this means that we may have to check on the reasonableness of this assumption before we perform the test concerning the means.

The tests which we shall study in this section include a test of the null hypothesis that the variance of a normal population equals a given constant, and the likelihood ratio test of the equality of the variances of two normal populations (which was referred to in Exercise 14 on page 386).

The first of these tests is essentially that of Exercise 12 on page 385. Given a random sample of size n from a normal population, we shall want to test the null hypothesis $\sigma^2 = \sigma_0^2$ against one of the alternatives $\sigma^2 \neq \sigma_0^2$, $\sigma^2 > \sigma_0^2$, or $\sigma^2 < \sigma_0^2$, and, as the reader should have discovered in Exercise 12 on page 385, the likelihood ratio technique leads to a test based on s^2, the value of the sample variance. Based on Theorem 8.9, we can thus write the critical regions for testing the null hypothesis against the two one-sided alternatives as $\chi^2 \geqslant \chi^2_{\alpha,n-1}$ and $\chi^2 \leqslant \chi^2_{1-\alpha,n-1}$, where

$$\chi^2 = \frac{(n-1)s^2}{\sigma_0^2}$$

So far as the two-sided alternative is concerned, we reject the null hypothesis if $\chi^2 \geqslant \chi^2_{\alpha/2,n-1}$ or $\chi^2 \leqslant \chi^2_{1-\alpha/2,n-1}$, and the size of all these critical regions is, of course, equal to α.

EXAMPLE 13.6

Suppose that the thickness of a part used in a semiconductor is its critical dimension and that measurements of the thickness of a random sample of 18 such parts have the variance $s^2 = 0.68$, where the measurements are in thousandths of an inch. The process is considered to be under control if the variation of the thicknesses is given by a variance not greater than 0.36. Assuming that the measurements constitute a random sample from a normal population, test the null hypothesis $\sigma^2 = 0.36$ against the alternative hypothesis $\sigma^2 > 0.36$ at $\alpha = 0.05$.

Solution

1. H_0: $\sigma^2 = 0.36$
   H_1: $\sigma^2 > 0.36$
2. Critical region: $\chi^2 \geqslant \chi^2_{.05,17} = 27.587$, where $\chi^2 = \dfrac{(n-1)s^2}{\sigma_0^2}$
3. Computations: $s^2 = 0.68$ and $n = 18$, so that

$$\chi^2 = \frac{17(0.68)}{0.36} = 32.1$$

4. Decision: Reject the null hypothesis and adjust the process used in the manufacture of these parts.

Note that if α had been 0.01 in Example 13.6, the null hypothesis could not have been rejected, since $\chi^2 = 32.1$ does not exceed $\chi^2_{.01,17} = 33.409$. This serves to indicate that the choice of α is something which should always be specified in advance, so that we will be spared the temptation of choosing a level of significance which happens to suit our purpose.

In Exercise 14 on page 386 the reader was asked to show that the likelihood ratio statistic for testing the equality of the variances of two normal populations can be expressed in terms of the ratio of the two sample variances. Given independent random samples of size n_1 and n_2 from two normal populations with the variances σ_1^2 and σ_2^2, we thus find from Theorem 8.13 that corresponding critical regions of size α for testing the null hypothesis $\sigma_1^2 = \sigma_2^2$ against the one-sided alternatives $\sigma_1^2 > \sigma_2^2$ or $\sigma_1^2 < \sigma_2^2$ are, respectively,

$$\frac{s_1^2}{s_2^2} \geqslant F_{\alpha,n_1-1,n_2-1} \quad \text{and} \quad \frac{s_2^2}{s_1^2} \geqslant F_{\alpha,n_2-1,n_1-1}$$

where F_{α,n_1-1,n_2-1} and F_{α,n_2-1,n_1-1} are as defined on page 274. The appropriate critical region for testing the null hypothesis against the two-sided alternative $\sigma_1^2 \neq \sigma_2^2$ is

$$\frac{s_1^2}{s_2^2} \geqslant F_{\alpha/2, n_1-1, n_2-1} \qquad \text{if } s_1^2 \geqslant s_2^2$$

and

$$\frac{s_2^2}{s_1^2} \geqslant F_{\alpha/2, n_2-1, n_1-1} \qquad \text{if } s_1^2 < s_2^2$$

Note that this test is based entirely on the right tail of the F distribution, which is made possible by the result of Exercise 14 on page 278, namely, by the fact that if the random variable $\mathbf{x}$ has an F distribution with ν_1 and ν_2 degrees of freedom, then $\frac{1}{\mathbf{x}}$ has an F distribution with ν_2 and ν_1 degrees of freedom.

EXAMPLE 13.7

In comparing the variability of the tensile strength of two kinds of structural steel, an experiment yielded the following results: $n_1 = 13$, $s_1^2 = 19.2$, $n_2 = 16$, and $s_2^2 = 3.5$, where the units of measurement are 1,000 pounds per square inch. Assuming that the measurements constitute independent random samples from two normal populations, test the null hypothesis $\sigma_1^2 = \sigma_2^2$ against the alternative $\sigma_1^2 \neq \sigma_2^2$ at the $\alpha = 0.02$ level of significance.

Solution

1. H_0: $\sigma_1^2 = \sigma_2^2$
H_1: $\sigma_1^2 \neq \sigma_2^2$

2. Critical region: $\frac{s_1^2}{s_2^2} \geqslant F_{.01,12,15} = 3.67$, since $s_1^2 > s_2^2$

3. Computations: $s_1^2 = 19.2$ and $s_2^2 = 3.5$, so that

$$\frac{s_1^2}{s_2^2} = \frac{19.2}{3.5} = 5.49$$

4. Decision: Reject the null hypothesis and conclude that the variability of the tensile strength of the two kinds of structural steel is not the same.

THEORETICAL EXERCISES

1. Making use of the fact that the chi-square distribution can be approximated with a normal distribution when ν, the number of degrees of freedom, is large, show that for large samples from normal populations

$$s^2 \geqslant \sigma_0^2\left[1 + z_\alpha\sqrt{\frac{2}{n-1}}\right]$$

is an approximate critical region of size α for testing the null hypothesis $\sigma^2 = \sigma_0^2$ against the alternative $\sigma^2 > \sigma_0^2$. Also construct corresponding critical regions for testing this null hypothesis against the alternatives $\sigma^2 < \sigma_0^2$ and $\sigma^2 \neq \sigma_0^2$. (See Exercise 4 on page 276.)

2. Making use of the result of Exercise 5 on page 276, show that for large random samples from normal populations, tests of the null hypothesis $\sigma^2 = \sigma_0^2$ can be based on the statistic

$$\left(\frac{s}{\sigma_0} - 1\right)\sqrt{2(n-1)}$$

which has approximately the standard normal distribution.

APPLIED EXERCISES

3. If nine determinations of the specific heat of iron had a standard deviation of 0.0086, test the null hypothesis that $\sigma = 0.0100$ for a normal population of such determinations. Use the alternative hypothesis $\sigma < 0.0100$ and the level of significance $\alpha = 0.05$.

4. In a random sample, the weights of 24 Black Angus steers of a certain age have a standard deviation of 238 pounds. Assuming that the weights constitute a random sample from a normal population, test the null hypothesis $\sigma = 250$ pounds against the two-sided alternative $\sigma \neq 250$ pounds at the level of significance $\alpha = 0.01$.

5. In a random sample, the time which 30 women took to complete the written test for their driver's license had a variance of 6.4 minutes. Assuming that the population sampled is normal, test the null hypothesis $\sigma^2 = 8$ against the alternative $\sigma^2 < 8$ at the level of significance $\alpha = 0.05$ using

(a) the method described in the text;
(b) the method of Exercise 2.

6. Test at the level of significance $\alpha = 0.02$ whether it was reasonable to assume in Example 13.5 on page 395 that the two populations have equal variances.

7. Test at the level of significance $\alpha = 0.10$ whether it was reasonable to assume in Exercise 12 on page 398 that the two populations have equal variances.

8. The following are the scores obtained in a personality test by samples of nine married women and nine unmarried women:

Unmarried	88	68	77	82	63	80	78	71	72
Married	73	77	67	74	74	64	71	71	72

Assuming that these data can be looked upon as independent random samples from two normal populations, test the null hypothesis $\sigma_1^2 = \sigma_2^2$ against the one-sided alternative $\sigma_1^2 > \sigma_2^2$ at the level of significance $\alpha = 0.05$. (σ_1^2 and σ_2^2 are, respectively, the variance of the scores of unmarried women and the variance of the scores of married women.)

13.5 TESTS CONCERNING PROPORTIONS

If an outcome of an experiment is the number of votes which a candidate receives in a poll, the number of imperfections found in a piece of cloth, the number of children who are absent from school on a given day, . . . , we refer to these data as **count data**. Appropriate models for the analysis of count data are the binomial distribution, the Poisson distribution, the multinomial distribution, and some of the other discrete distributions which we studied in Chapter 5. In this section we shall present one of the most common tests based on count data, namely, a test concerning the parameter θ of the binomial distribution.

The binomial parameter θ is the probability of a success on an individual trial and, hence, the proportion of successes one can expect in the long run. Testing on the basis of a sample whether the true proportion of cures from a certain disease is 0.90 or whether the true proportion of defectives coming off an assembly line is 0.02 is, thus, equivalent to testing hypotheses about the parameter θ of binomial populations.

In Exercise 9 on page 372 the reader was asked to show that the most powerful critical region for testing the null hypothesis $\theta = \theta_0$ against the alternative $\theta = \theta_1 < \theta_0$, where the θ is the parameter of a binomial population, is based on the value of $\mathbf{x}$, the number of "successes" obtained in n trials. When it comes to composite alternatives, the likelihood ratio technique also yields tests based on the observed number of successes (as we saw in Exercise 8 on page 384 for the special case where $\theta_0 = \frac{1}{2}$). In fact, if we want to test the null hypothesis $\theta = \theta_0$ against the one-sided alternative $\theta > \theta_0$, the critical region of size α of the likelihood ratio criterion is

$$x \geqslant k_\alpha$$

where k_α is the smallest integer for which

$$\sum_{y=k_\alpha}^{n} b(y; n, \theta_0) \leqslant \alpha$$

and $b(y; n, \theta_0)$ is the probability of getting y successes in n binomial trials when $\theta = \theta_0$. The size of this critical region, as well as the ones which follow, is thus as close as possible to α without exceeding it.

The corresponding critical region for testing the null hypothesis $\theta = \theta_0$ against the one-sided alternative $\theta < \theta_0$ is

$$x \leqslant k'_\alpha$$

where k'_α is the largest integer for which

$$\sum_{y=0}^{k'_\alpha} b(y; n, \theta_0) \leqslant \alpha$$

and, finally, the critical region for testing the null hypothesis $\theta = \theta_0$ against the two-sided alternative $\theta \neq \theta_0$ is

$$x \geqslant k_{\alpha/2} \quad \text{or} \quad x \leqslant k'_{\alpha/2}$$

EXAMPLE 13.8

If the number of successes in 20 trials of a binomial experiment is 5, test the null hypothesis $\theta = 0.50$ against the two-sided alternative $\theta \neq 0.50$ at the $\alpha = 0.05$ level of significance.

Solution

1. H_0: $\theta = 0.50$
   H_1: $\theta \neq 0.50$
2. Critical region: $x \geqslant k_{.025} = 15$ or $x \leqslant k'_{.025} = 5$ according to Table I.
3. Computation: $x = 5$
4. Decision: Reject H_0 and conclude that $\theta \neq 0.50$.

The tests which we have described require the use of a table of binomial probabilities, at least when n is small. For $n \leqslant 20$ we can use Table I at the end of this book, and for values of n up to 100 we can use the tables referred to on page

193. For larger values of n we make use of the normal approximation to the binomial distribution and treat

$$z = \frac{x - n\theta}{\sqrt{n\theta(1 - \theta)}}$$

as a value of a random variable having the standard normal distribution. For large n, we can thus test the null hypothesis $\theta = \theta_0$ against the alternatives $\theta \neq \theta_0$, $\theta > \theta_0$, or $\theta < \theta_0$ using, respectively, the critical regions $|z| \geqslant z_{\alpha/2}$, $z \geqslant z_\alpha$, and $z \leqslant -z_\alpha$, where

$$z = \frac{x - n\theta_0}{\sqrt{n\theta_0(1 - \theta_0)}}$$

or

$$z = \frac{(x \pm \frac{1}{2}) - n\theta_0}{\sqrt{n\theta_0(1 - \theta_0)}}$$

if we use the continuity correction introduced in Example 6.4 on page 215. We use the minus sign when x exceeds $n\theta_0$ and the plus sign when x is less than $n\theta_0$.

EXAMPLE 13.9

An oil company claims that at most 20 percent of all automobile owners buy brand A gasoline. Test this claim at $\alpha = 0.01$, if a random check indicates that 58 of 200 automobile owners buy brand A gasoline.

Solution

1. H_0: $\theta = 0.20$
H_1: $\theta > 0.20$

2. Critical region: $z \geqslant z_{.01} = 2.33$, where, with the continuity correction,

$$z = \frac{(x \pm \frac{1}{2}) - n\theta_0}{\sqrt{n\theta_0(1 - \theta_0)}}$$

3. Computations: $x = 58$, $n = 200$, $n\theta_0 = 200(0.20) = 40$, so that

$$z = \frac{57.5 - 40}{\sqrt{200(0.20)(0.80)}} = 3.09$$

4. Decision: Reject the null hypothesis and conclude that brand A is bought by more than 20 percent of all automobile owners. Note that if we had not used the continuity correction, we would have had $z = 3.18$, and the result would have been the same.

13.6 TESTS CONCERNING DIFFERENCES AMONG k PROPORTIONS

In applied research there are many problems in which we must decide whether observed differences among sample proportions, or percentages, are significant or whether they can be attributed to chance. For instance, if 6 percent of the frozen chicken in a sample from one supplier fails to meet certain standards and only 4 percent in a sample from another supplier fails, we may want to decide whether the difference between the two percentages is significant. Similarly, we may want to judge on the basis of sample data whether the actual proportion of voters who favor a certain candidate is the same in four different cities.

To indicate a general method for handling problems of this kind, suppose that $x_1, x_2, \ldots,$ and x_k are observed values of a set of independent random variables $\mathbf{x}_1, \mathbf{x}_2, \ldots,$ and $\mathbf{x}_k$ having binomial distributions with the respective parameters n_1 and θ_1, n_2 and $\theta_2, \ldots,$ and n_k and θ_k. If the n's are sufficiently large, we can approximate the distributions of the independent random variables

$$\mathbf{z}_i = \frac{\mathbf{x}_i - n_i\theta_i}{\sqrt{n_i\theta_i(1-\theta_i)}} \qquad \text{for } i = 1, 2, \ldots, k$$

with standard normal distributions, and, according to Theorem 8.5, we can then look upon

$$\chi^2 = \sum_{i=1}^{k} \frac{(x_i - n_i\theta_i)^2}{n_i\theta_i(1-\theta_i)}$$

as a value of a random variable having the chi-square distribution with k degrees of freedom. To test the null hypothesis $\theta_1 = \theta_2 = \cdots = \theta_k = \theta_0$ (against the alternative that at least one of the θ's does not equal θ_0) we can thus use the critical region $\chi^2 \geq \chi^2_{\alpha,k}$, where

$$\chi^2 = \sum_{i=1}^{k} \frac{(x_i - n_i\theta_0)^2}{n_i\theta_0(1-\theta_0)}$$

When θ_0 is not specified, that is, when we are interested only in the null hypothesis $\theta_1 = \theta_2 = \cdots = \theta_k$, we substitute for θ the pooled estimate

$$\hat{\theta} = \frac{x_1 + x_2 + \cdots + x_k}{n_1 + n_2 + \cdots + n_k}$$

and the critical region becomes $\chi^2 \geqslant \chi^2_{\alpha,k-1}$, where

$$\chi^2 = \sum_{i=1}^{k} \frac{(x_i - n_i\hat{\theta})^2}{n_i\hat{\theta}(1 - \hat{\theta})}$$

The loss of one degree of freedom, namely, the change in the critical region from $\chi^2_{\alpha,k}$ to $\chi^2_{\alpha,k-1}$, is due to the fact that an estimate is substituted for the unknown parameter θ; a formal discussion of this is referred to on page 418.

Let us now present an alternative form of the chi-square statistic for this kind of test which, as we shall see in Section 13.7, lends itself more readily to other applications. Arranging the data as in the following table,

	Successes	*Failures*
Sample 1	x_1	$n_1 - x_1$
Sample 2	x_2	$n_2 - x_2$
	...	...
Sample k	x_k	$n_k - x_k$

let us refer to its entries as the **observed cell frequencies** f_{ij}, where the first subscript indicates the row and the second subscript indicates the column of this $k \times 2$ table.

Under the null hypothesis $\theta_1 = \theta_2 = \cdots = \theta_k = \theta_0$ the **expected cell frequencies** for the first column are $n_i\theta_0$ for $i = 1, 2, \ldots, k$, and those for the second column are $n_i(1 - \theta_0)$. When θ_0 is not known, we substitute for it, as before, the pooled estimate $\hat{\theta}$, and estimate the expected cell frequencies as

$$e_{i1} = n_i\hat{\theta} \quad \text{and} \quad e_{i2} = n_i(1 - \hat{\theta})$$

for $i = 1, 2, \ldots,$ and k. It will be left to the reader to show in Exercise 1 on page 407 that the value of the chi-square statistic can thus be written as

$$\chi^2 = \sum_{i=1}^{k} \sum_{j=1}^{2} \frac{(f_{ij} - e_{ij})^2}{e_{ij}}$$

EXAMPLE 13.10

Determine, on the basis of the sample data shown in the following table, whether the true proportion of shoppers favoring detergent A over detergent B is the same in all three cities:

	Number favoring detergent A	*Number favoring detergent B*	
Los Angeles	232	168	$n_1 = 400$
San Diego	260	240	$n_2 = 500$
Fresno	197	203	$n_3 = 400$

Use the level of significance $\alpha = 0.05$.

Solution

1. H_0: $\theta_1 = \theta_2 = \theta_3$;
   H_1: the three θ's are not all equal.
2. Critical region: $\chi^2 \geqslant \chi^2_{.05,2} = 5.991$, where

$$\chi^2 = \sum_{i=1}^{3} \sum_{j=1}^{2} \frac{(f_{ij} - e_{ij})^2}{e_{ij}}$$

3. Computations: The pooled estimate of θ is given by

$$\hat{\theta} = \frac{232 + 260 + 197}{400 + 500 + 400} = \frac{689}{1{,}300} = 0.53$$

Thus, we estimate the expected cell frequencies as

$$e_{11} = 400(0.53) = 212 \quad \text{and} \quad e_{12} = 400(0.47) = 188$$

$$e_{21} = 500(0.53) = 265 \quad \text{and} \quad e_{22} = 500(0.47) = 235$$

$$e_{31} = 400(0.53) = 212 \quad \text{and} \quad e_{32} = 400(0.47) = 188$$

and substitution into the above formula for χ^2 yields

$$\chi^2 = \frac{(232 - 212)^2}{212} + \frac{(260 - 265)^2}{265} + \frac{(197 - 212)^2}{212} + \frac{(168 - 188)^2}{188} + \frac{(240 - 235)^2}{235} + \frac{(203 - 188)^2}{188}$$

$$= 6.48$$

4. Decision: Since $\chi^2 = 6.48$ exceeds $\chi^2_{.05,2} = 5.991$, the null hypothesis must be rejected; in other words, the true proportions of shoppers favoring detergent A over detergent B in the three cities are not the same.

THEORETICAL EXERCISES

1. Show that the two formulas for χ^2 on page 407 are equivalent.

2. Modify the criteria on pages 403 and 404 so that they can be used to test the null hypothesis $\lambda = \lambda_0$, where λ is the parameter of the Poisson distribution, on the basis of n observations. (*Hint*: Use the result of Example 7.14.) Also use Table II to find values corresponding to $k_{.025}$ and $k'_{.025}$ to test the null hypothesis $\lambda = 3.6$ against the alternative $\lambda \neq 3.6$ at $\alpha = 0.05$ on the basis of five observations.

3. For $k = 2$, show that the χ^2 formula on page 407 can be written as

$$\chi^2 = \frac{(n_1 + n_2)(n_2x_1 - n_1x_2)^2}{n_1n_2(x_1 + x_2)[(n_1 + n_2) - (x_1 + x_2)]}$$

4. Show that for $k = 2$ and large samples the null hypothesis $\theta_1 = \theta_2$, where θ_1 and θ_2 are the parameters of two binomial populations, can also be tested by looking upon

$$z = \frac{\dfrac{x_1}{n_1} - \dfrac{x_2}{n_2}}{\sqrt{\hat{\theta}(1 - \hat{\theta})\left(\dfrac{1}{n_1} + \dfrac{1}{n_2}\right)}}$$

where $\hat{\theta} = \dfrac{x_1 + x_2}{n_1 + n_2}$, as a value of a random variable having the standard normal distribution. (*Hint*: Refer to Exercise 6 on page 263.)

5. Show that the square of the expression given for z in Exercise 4 equals

$$\chi^2 = \sum_{i=1}^{2} \frac{(x_i - n_i\hat{\theta})^2}{n_i\hat{\theta}(1 - \hat{\theta})}$$

so that the two tests are actually equivalent.

APPLIED EXERCISES

6. The null hypothesis $\theta = 0.45$ is to be tested against the alternative $\theta < 0.45$ at $\alpha = 0.05$, where θ is the parameter of a binomial population with $n = 19$. Use Table I to find $k'_{.05}$ and the probabilities of committing type II errors with this criterion when $\theta = 0.35$, $\theta = 0.30$, and $\theta = 0.25$.

7. The null hypothesis $\theta = 0.25$ is to be tested against the alternative $\theta > 0.25$ at $\alpha = 0.01$, where θ is the parameter of a binomial population with $n = 20$. Use Table I to find $k_{.01}$ and the probabilities of committing type II errors with this criterion when $\theta = 0.35$, $\theta = 0.40$, and $\theta = 0.45$.

8. The null hypothesis $\theta = 0.70$ is to be tested against the alternative $\theta \neq 0.70$ at $\alpha = 0.05$, where θ is the parameter of a binomial population with $n = 18$. Use Table I to find $k_{.025}$ and $k'_{.025}$ and the probabilities of committing type II errors with this criterion when $\theta = 0.60$, $\theta = 0.65$, $\theta = 0.75$, and $\theta = 0.80$.

9. The null hypothesis $\theta = 0.40$ is to be tested against the alternative $\theta \neq 0.40$ at $\alpha = 0.01$, where θ is the parameter of a binomial population with $n = 16$. Use Table I to find $k_{.005}$ and $k'_{.005}$ and the probabilities of committing type II errors with this criterion when $\theta = 0.30$, $\theta = 0.35$, $\theta = 0.45$, and $\theta = 0.50$.

10. In a random sample of 600 cars making a right turn at a certain intersection, 157 pulled into the wrong lane. Use the level of significance $\alpha = 0.05$ to test the null hypothesis that the actual proportion of drivers who make this mistake (at the given intersection) is 0.30 against the alternative hypothesis that this figure is incorrect either way.

11. The manufacturer of a spot remover claims that his product removes at least 90 percent of all spots. What can we conclude about this claim at $\alpha = 0.05$, if the spot remover removed only 174 of 200 spots chosen at random from spots on clothes brought to a dry cleaning establishment?

12. If 74 of 250 persons who watched a certain television program in black and white and 92 of 250 persons who watched the same program in color remembered two hours later what products were advertized, test the null hypothesis that there is no difference between the corresponding population proportions at the level of significance $\alpha = 0.01$.

13. To landscape its grounds, a bank purchased 400 tulip bulbs from one nursery and 200 from another. If 46 of the 400 bulbs from the first nursery failed to bloom while 18 of the 200 bulbs from the other nursery failed to bloom, test the null hypothesis that there is no difference between the corresponding population proportions at the level of significance $\alpha = 0.05$.

14. In a random sample of 200 persons who skipped breakfast, 82 reported that they experienced midmorning fatigue, and in a random sample of 400 persons who ate breakfast, 116 reported that they experienced midmorning fatigue. Use the method of Exercise 4 and the level of significance $\alpha = 0.05$ to test the null hypothesis that there is no difference between the corresponding population proportions against the alternative that midmorning fatigue is more prevalent among persons who skip breakfast.

15. If 26 of 200 tires of Brand A failed to last 20,000 miles, while the corresponding figures for 200 tires of Brands B, C, and D were 23, 15, and 32, test the null hypothesis that there is no difference in the quality of the four kinds of tires. Use $\alpha = 0.05$.

16. In a random sample of 250 persons with low incomes 155 are for a certain piece of legislation, while in random samples of 200 persons with average incomes and 150 persons with high incomes there are, respectively, 118 and 87 who favor the legislation. Use $\alpha = 0.05$ to test the null hypothesis that the proportion of persons favoring the legislation is the same for all three groups.

13.7 CONTINGENCY TABLES

Chi-square criteria based on observed and expected cell frequencies play an important role in many problems dealing with the analysis of count data. In this section we shall use such a criterion to analyze **contingency tables** like the following 3×3 table obtained in a study of the relationship between a person's ability in mathematics and his or her interest in statistics.

		Ability in mathematics		
		Low	*Average*	*High*
Interest in statistics	*Low*	63	42	15
	Average	58	61	31
	High	14	47	29

In general, an $r \times c$ contingency table has c columns representing the different categories $A_1, A_2, \ldots,$ and A_c of one variable, r rows representing the different categories $B_1, B_2, \ldots,$ and B_r of another variable, and f_{ij} is the observed cell frequency for the cell belonging to the ith row and the jth column. Schematically,

	A_1	A_2		A_c	
B_1	f_{11}	f_{12}	$\cdots$	f_{1c}	$f_{1.}$
B_2	f_{21}	f_{22}	$\cdots$	f_{2c}	$f_{2.}$
$\cdots$	$\cdots$	$\cdots$	$\cdots$	$\cdots$	$\cdots$
B_r	f_{r1}	f_{r2}	$\cdots$	f_{rc}	$f_{r.}$
	$f_{.1}$	$f_{.2}$		$f_{.c}$	f

where the row and column totals, $f_{i.} = \sum_{j=1}^{c} f_{ij}$ and $f_{.j} = \sum_{i=1}^{r} f_{ij}$, are called the

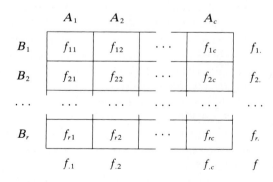

marginal frequencies, and f, the sum of all the cell frequencies, is called the **grand total**.

The null hypothesis we shall want to test is that the two variables are independent. More specifically, if θ_{ij} is the probability that an item will fall into the cell belonging to the ith row and the jth column, $\theta_{i.}$ is the probability that an item will fall into the ith row, and $\theta_{.j}$ is the probability that an item will fall into the jth column, the null hypothesis which we shall want to test is that

$$\theta_{ij} = (\theta_{i.})(\theta_{.j}) \qquad \text{for } i = 1, 2, \ldots, r \text{ and } j = 1, 2, \ldots, c$$

To test this null hypothesis against the alternative that θ_{ij} does not equal the product of $\theta_{i.}$ and $\theta_{.j}$ for at least one pair of values of i and j, we estimate the probabilities $\theta_{i.}$ and $\theta_{.j}$ as $\hat{\theta}_{i.} = \frac{f_{i.}}{f}$ and $\hat{\theta}_{.j} = \frac{f_{.j}}{f}$, and, hence, the expected cell frequencies as

$$e_{ij} = (\hat{\theta}_{i.})(\hat{\theta}_{.j}) \cdot f = \frac{f_{i.}}{f} \cdot \frac{f_{.j}}{f} \cdot f = \frac{(f_{i.})(f_{.j})}{f}$$

Then, we base our decision on

$$\chi^2 = \sum_{i=1}^{r} \sum_{j=1}^{c} \frac{(f_{ij} - e_{ij})^2}{e_{ij}}$$

and reject the null hypothesis if the value we obtain exceeds $\chi^2_{\alpha,(r-1)(c-1)}$.

The number of degrees of freedom is $(r - 1)(c - 1)$, and in connection with this let us make the following observation: Whenever expected cell frequencies in chi-square formulas are estimated on the basis of sample count data, the number of degrees of freedom is $s - t - 1$, where s is the number of terms in the summation and t is the number of independent parameters replaced by estimates. When testing for differences among k proportions with the chi-square statistic of Section 13.6, we had $s = 2k$ and $t = k$, since we had to estimate the k parameters $\theta_1, \theta_2, \ldots, \theta_k$, and the number of degrees of freedom was $2k - k - 1 = k - 1$. When testing for independence with an $r \times c$ contingency table, we have $s = rc$ and $t = r + c - 2$, since the r parameters $\theta_{i.}$ and the c parameters $\theta_{.j}$ are subject to the two restrictions that their respective sums must equal 1; hence, $s - t - 1 = rc - (r + c - 2) - 1 = (r - 1)(c - 1)$.

Since the test statistic which we have described has only approximately a chi-square distribution with $(r - 1)(c - 1)$ degrees of freedom, it is customary to use this test only when none of the e_{ij} is less than 5; sometimes this requires that we combine some of the cells with a corresponding loss in the number of degrees of freedom.

EXAMPLE 13.11

For the data shown in the table on page 411, test for independence between a person's ability in mathematics and his or her interest in statistics. Use the 0.01 level of significance.

Solution

1. H_0: Ability in mathematics and interest in statistics are independent; H_1: these two variables are not independent.
2. Critical region: $\chi^2 \geqslant \chi^2_{.01,4} = 13.277$, where

$$\chi^2 = \sum_{i=1}^{3} \sum_{j=1}^{3} \frac{(f_{ij} - e_{ij})^2}{e_{ij}}$$

3. Computations: The expected frequencies for the first row are $\frac{120 \cdot 135}{360} =$ 45, 50, and 25; those for the second row are 56, 62, and 31; and those for the third row are 34, 38, and 19. (These figures are all rounded to the nearest integer, but when such figures are small, they are often rounded to one decimal.) Then, substitution into the formula for χ^2 yields

$$\begin{aligned}\chi^2 &= \frac{(63-45)^2}{45} + \frac{(42-50)^2}{50} + \frac{(14-25)^2}{25} \\ &\quad + \frac{(58-56)^2}{56} + \frac{(61-62)^2}{62} + \frac{(31-31)^2}{31} \\ &\quad + \frac{(14-34)^2}{34} + \frac{(47-38)^2}{38} + \frac{(29-19)^2}{19} \\ &= 32.51\end{aligned}$$

4. Decision: Since $\chi^2 = 32.51$ exceeds $\chi^2_{.01,4} = 13.277$, the null hypothesis must be rejected and we conclude that there is a relationship between a person's ability in mathematics and his or her interest in statistics.

13.8 GOODNESS OF FIT

The goodness-of-fit test considered here applies to situations in which we want to determine whether a set of data may be looked upon as a random sample from a population having a given distribution. A second kind of "goodness of fit" which applies to the fitting of a curve to a set of paired data will be discussed in Chapter 14. To illustrate, suppose that we want to decide on the basis of the data (observed frequencies) shown in the following table whether the number of errors a compositor makes in setting a galley of type is a random variable having a Poisson distribution:

Number of errors	*Observed frequencies* f_i	*Poisson probabilities with* $\lambda = 3$	*Expected frequencies* e_i
0	18	0.0498	21.9
1	53	0.1494	65.7
2	103	0.2240	98.6
3	107	0.2240	98.6
4	82	0.1680	73.9
5	46	0.1008	44.0
6	18	0.0504	22.2
7	10 }	0.0216	9.5 }
8	2 } 13	0.0081	3.6 } 14.3
9	1 }	0.0027	1.2 }
	440		

To determine a corresponding set of expected frequencies for a random sample from a Poisson population, we first use the mean of the observed distribution to estimate the Poisson parameter λ, getting $\hat{\lambda} = \dfrac{1,341}{440} = 3.05$ or, approximately, $\hat{\lambda} = 3$. Then, copying the Poisson probabilities for $\lambda = 3$ from Table II and multiplying by 440, the total frequency, we get the expected frequencies shown in the right-hand column of the table. To test the null hypothesis that the observed frequencies constitute a random sample from a Poisson population, we must judge how good a fit, or how close an agreement, we have between the two sets of frequencies. In general, to test the null hypothesis H_0 that a set of observed data comes from a population having a specified distribution against the alternative that the population has some other distribution, we compute

$$\chi^2 = \sum_{i=1}^{m} \frac{(f_i - e_i)^2}{e_i}$$

and reject H_0 at the level of significance α if $\chi^2 \geqslant \chi^2_{\alpha, m-t-1}$, where m is the number of terms in the summation and t is the number of independent parameters estimated on the basis of the sample data (see discussion on page 412). In the above illustration, $t = 1$ since only one parameter is estimated on the basis of the data, and the number of degrees of freedom is $m - 2$.

EXAMPLE 13.12

For the data in the table on page 414, test whether the number of errors the compositor makes in setting a galley of type is a random variable having a Poisson distribution. Use the level of significance $\alpha = 0.05$.

Solution

1. H_0: Population has a Poisson distribution;
   H_1: population does not have a Poisson distribution.
2. Critical region: Combining the last three classes so that each expected frequency is at least 5, we get the critical region $\chi^2 \geqslant \chi^2_{.05,6} = 12.592$, where

$$\chi^2 = \sum_{i=1}^{8} \frac{(f_i - e_i)^2}{e_i}$$

3. Computations:

$$\chi^2 = \frac{(18 - 21.9)^2}{21.9} + \frac{(53 - 65.7)^2}{65.7} + \frac{(103 - 98.6)^2}{98.6} + \frac{(107 - 98.6)^2}{98.6} + \frac{(82 - 73.9)^2}{73.9} + \frac{(46 - 44.0)^2}{44.0} + \frac{(18 - 22.2)^2}{22.2} + \frac{(13 - 14.3)^2}{14.3}$$

$$= 5.95$$

4. Decision: Since $\chi^2 = 5.95$ does not exceed $\chi^2_{.05,6} = 12.592$, the null hypothesis cannot be rejected; indeed, the close agreement between the observed and expected frequencies suggests that the Poisson distribution provides a "good fit."

THEORETICAL EXERCISES

1. Show that the expected cell frequencies of an $r \times c$ contingency table satisfy the equations $\sum_{i=1}^{r} e_{ij} = f_{.j}$ and $\sum_{j=1}^{c} e_{ij} = f_{i.}$, so that once $(r - 1)(c - 1)$ of the e_{ij} have been calculated the others may be obtained by subtraction from the totals of the appropriate rows or columns.
2. If the analysis of a contingency table shows that there is a relationship between the two variables under consideration, the strength of this relationship may be measured by means of the **contingency coefficient**

$$C = \sqrt{\frac{\chi^2}{\chi^2 + f}}$$

where χ^2 is the value obtained for the test statistic and f is the grand total as defined on page 412. Show that

(a) for a 2×2 contingency table the maximum value of C is $\frac{1}{2}\sqrt{2}$;

(b) for a 3×3 contingency table the maximum value of C is $\frac{1}{3}\sqrt{6}$.

APPLIED EXERCISES

3. In a study of parents' feelings about a required course in sex education, a random sample of 360 parents are classified according to whether they have one, two, or three or more children in the school system, and also whether they feel that the course is poor, adequate, or good. Based on the results shown in the following table, test at the $\alpha = 0.05$ level of significance whether there is a relationship between parents' reaction to the course and the number of children they have in the school system:

	Number of children		
	1	*2*	*3 or more*
Poor	48	40	12
Adequate	55	53	29
Good	57	46	20

4. The following table is based on a study of the relationship between race and blood type in a Near Eastern country:

	Blood Type			
	O	*A*	*B*	*AB*
Race 1	176	148	96	72
Race 2	78	50	45	12
Race 3	15	19	8	7

Use $\alpha = 0.01$ to test the null hypothesis that there is no relationship between race and blood type in the country under consideration.

5. The following sample data pertain to the shipments received by a large firm from three different vendors:

	Number rejected	*Number imperfect but acceptable*	*Number perfect*
Vendor A	12	23	89
Vendor B	8	12	62
Vendor C	21	30	119

Test at the level of significance $\alpha = 0.01$ whether quality is independent of the source of the shipments.

6. Four coins are tossed 160 times and 0, 1, 2, 3, and 4 heads showed, respectively, 19, 54, 58, 23, and 6 times. Use the level of significance $\alpha = 0.05$ to test whether it is reasonable to suppose that the coins are balanced and randomly tossed.

7. It is desired to test whether the number of gamma rays emitted per second by a certain radioactive substance is a random variable having the Poisson distribution with $\lambda = 2.4$. Test this null hypothesis at $\alpha = 0.05$ on the basis of the following results obtained in 300 1-second intervals:

Number of gamma rays	*Frequency*
0	19
1	48
2	66
3	74
4	44
5	35
6	10
7	4

8. With reference to the data of Exercise 7, use the level of significance $\alpha = 0.05$ to test the null hypothesis that the data constitute a random sample from a population having a Poisson distribution.

9. The following is the distribution of the readings obtained with a Geiger counter of the number of particles emitted by a radioactive substance in 100 successive 40-second intervals:

Number of particles	*Frequency*
5–9	1
10–14	10
15–19	37
20–24	36
25–29	13
30–34	2
35–39	1

(a) Verify that the mean and the standard deviation of this distribution are, respectively, $\bar{x} = 20$ and $s = 5$.

(b) Find the probabilities that a random variable having a normal distribution with $\mu = 20$ and $\sigma = 5$ will take on a value less than 9.5, between 9.5 and 14.5, between 14.5 and 19.5, between 19.5 and 24.5, between 24.5 and 29.5, between 29.5 and 34.5, and greater than 34.5.

(c) Find the expected normal curve frequencies for the various classes by multiplying the probabilities obtained in part (b) by the total frequency, and then test the null hypothesis that the data constitute a random sample from a normal population. Use $\alpha = 0.05$.

References

The problem of determining the appropriate number of degrees of freedom for various uses of the chi-square statistic is discussed in

CRAMÉR, H., *Mathematical Methods of Statistics.* Princeton, N.J.: Princeton University Press, 1946.

The *Smith–Satterthwaite* test of the null hypothesis that two normal populations with unequal variances have the same mean is given in

MILLER, I., and FREUND, J. E., *Probability and Statistics for Engineers*, 2nd ed. Englewood Cliffs, N.J.: Prentice-Hall, Inc., 1977.

Additional comments relative to using the continuity correction for testing hypotheses concerning binomial parameters can be found in

BROWNLEE, K. A., *Statistical Theory and Methodology in Science and Engineering*, 2nd ed. New York: John Wiley & Sons, Inc., 1965.

regression and correlation

14

14.1 REGRESSION

A major objective of many statistical investigations is to establish relationships which make it possible to predict one or more variables in terms of others. Thus, studies are made to predict the potential sales of a new product in terms of its price, a patient's weight in terms of the number of weeks he or she has been on a diet, family expenditures on entertainment in terms of family income, the per capita consumption of certain foods in terms of their nutritional values and the amount of money spent advertising them on television, and so forth.

Although it is, of course, desirable to be able to predict one quantity exactly in terms of others, this is seldom possible, and in most instances we have to be satisfied with predicting averages or expected values. Thus, we may not be able to predict exactly how much money Mr. Brown will make ten years after graduating from college, but, given suitable data, we can predict the average income of a college graduate in terms of the number of years he has been out of college. Similarly, we can at best predict the average yield of a given variety of wheat in terms of data on the rainfall in July, and we can at best predict the average performance of students starting college in terms of their I.Q.'s.

Formally, if we are given the joint distribution of two random variables **x** and **y** and **x** is known to take on the value x, the basic problem of **bivariate regression** is that of determining the conditional mean $\mu_{\mathbf{y}|x}$, namely, the "average" value of **y** for the given value of **x**. [The term "regression," as it is used here, dates back to

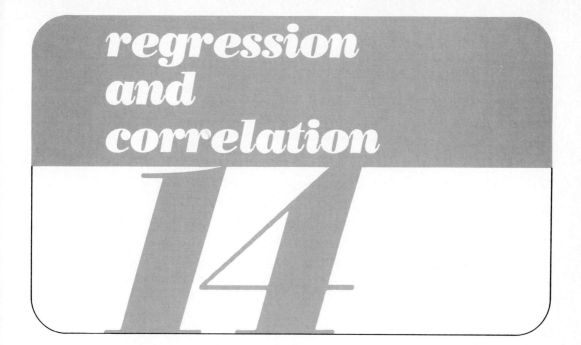

Francis Galton, who employed it first in connection with a study of the heights of fathers and sons, in which he observed a regression (a "turning back") from the heights of sons to the heights of their fathers.] In problems involving more than two random variables, that is, in **multivariate regression**, we are correspondingly concerned with quantities such as $\mu_{\mathbf{z}|x,y}$, the mean value of $\mathbf{z}$ for given values of $\mathbf{x}$ and $\mathbf{y}$, $\mu_{\mathbf{x}_4|x_1,x_2,x_3}$, the mean value of $\mathbf{x}_4$ for given values of $\mathbf{x}_1$, $\mathbf{x}_2$, and $\mathbf{x}_3$, and so on.

If $f(x, y)$ is the value of the joint density of two random variables $\mathbf{x}$ and $\mathbf{y}$ at (x, y), the problem of bivariate regression is simply that of determining the conditional density of $\mathbf{y}$ given $\mathbf{x} = x$ and then evaluating the integral

$$\mu_{\mathbf{y}|x} = E(\mathbf{y}|x) = \int_{-\infty}^{\infty} y \cdot w(y|x)\, dy$$

as outlined in Section 4.8. The resulting equation is called the **regression equation** of $\mathbf{y}$ on $\mathbf{x}$. Alternatively, we might be interested in the regression equation

$$\mu_{\mathbf{x}|y} = E(\mathbf{x}|y) = \int_{-\infty}^{\infty} x \cdot f(x|y)\, dx$$

In the discrete case, where we are dealing with probability distributions instead of probability densities, the integrals of the two preceding regression equations are simply replaced by sums.

When we do not know the joint density of the two random variables, or at least not all of its parameters, the determination of $\mu_{\mathbf{y}|x}$ and $\mu_{\mathbf{x}|y}$ becomes a problem of estimation based on sample data; this is an entirely different problem, which we shall discuss in Sections 14.3 and 14.4.

EXAMPLE 14.1

Given the random variables $\mathbf{x}$ and $\mathbf{y}$ which have the joint density

$$f(x, y) = \begin{cases} x \cdot e^{-x(1+y)} & \text{for } x > 0 \text{ and } y > 0 \\ 0 & \text{elsewhere} \end{cases}$$

find the regression equation of $\mathbf{y}$ on $\mathbf{x}$ and sketch the **regression curve**.

Solution

Integrating out y we find that the marginal density of $\mathbf{x}$ is given by

$$g(x) = \begin{cases} e^{-x} & \text{for } x > 0 \\ 0 & \text{elsewhere} \end{cases}$$

and, hence, the conditional density of $\mathbf{y}$ given $\mathbf{x} = x$ is given by

$$w(y|x) = \frac{f(x, y)}{g(x)} = \frac{x \cdot e^{-x(1+y)}}{e^{-x}} = x \cdot e^{-xy}$$

for $y > 0$ and $w(y|x) = 0$ elsewhere, which we recognize as an exponential density with $\theta = \frac{1}{x}$. Hence, by evaluating

$$\mu_{\mathbf{y}|x} = \int_0^{\infty} y \cdot x \cdot e^{-xy}\, dy$$

or by referring to Corollary 1 of Theorem 6.3, we find that the regression equation of $\mathbf{y}$ on $\mathbf{x}$ is given by

$$\mu_{\mathbf{y}|x} = \frac{1}{x}$$

The corresponding regression curve is shown in Figure 14.1.

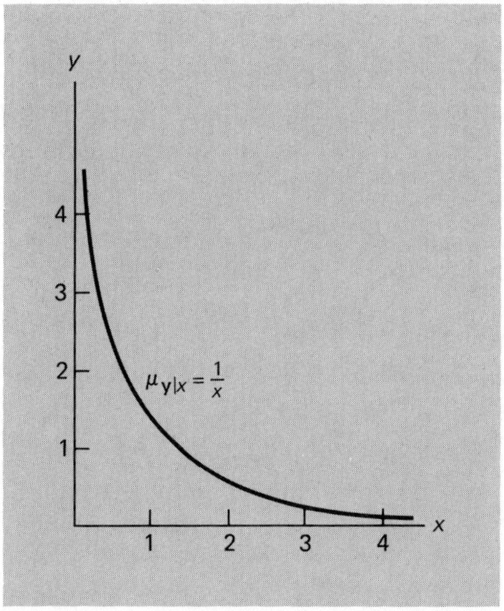

Figure 14.1 Regression curve of Example 14.1.

EXAMPLE 14.2

If **x** and **y** have the multinomial distribution

$$f(x, y) = \binom{n}{x, y, n-x-y} \cdot \theta_1^x \theta_2^y (1 - \theta_1 - \theta_2)^{n-x-y}$$

for $x = 0, 1, 2, \ldots, n$, and $y = 0, 1, 2, \ldots, n$, subject to the restriction that $x + y \leqslant n$, find the regression equation of **y** on **x**.

Solution

The marginal distribution of **x** is given by

$$\begin{aligned} g(x) &= \sum_{y=0}^{n-x} \binom{n}{x, y, n-x-y} \cdot \theta_1^x \theta_2^y (1 - \theta_1 - \theta_2)^{n-x-y} \\ &= \binom{n}{x} \theta_1^x (1 - \theta_1)^{n-x} \end{aligned}$$

for $x = 0, 1, 2, \ldots, n$, which we recognize as a binomial distribution with the parameters n and θ. Hence,

$$w(y|x) = \frac{f(x, y)}{g(x)} = \frac{\binom{n-x}{y} \theta_2^y (1 - \theta_1 - \theta_2)^{n-x-y}}{(1 - \theta_1)^{n-x}}$$

for $y = 0, 1, 2, \ldots, n - x$, and, rewriting this formula as

$$w(y|x) = \binom{n-x}{y} \left(\frac{\theta_2}{1 - \theta_1}\right)^y \left(\frac{1 - \theta_1 - \theta_2}{1 - \theta_1}\right)^{n-x-y}$$

we find by inspection that the conditional distribution of **y** given $\mathbf{x} = x$ is a binomial distribution with the parameters $n - x$ and $\frac{\theta_2}{1 - \theta_1}$, so that the regression equation of **y** on **x** is

$$\mu_{\mathbf{y}|x} = \frac{(n - x)\theta_2}{1 - \theta_1}$$

according to Theorem 5.2.

If we let **x** be the number of times an even number comes up in 30 rolls of a balanced die, and **y** the number of times the result is a five, then the regression equation becomes

$$\mu_{\mathbf{y}|x} = \frac{(30-x)\frac{1}{6}}{1-\frac{1}{2}} = \frac{1}{3}(30-x)$$

This stands to reason, because for each of the $30 - x$ outcomes that are not even, there are the three equally likely possibilities 1, 3, or 5.

EXAMPLE 14.3

If the joint density of $\mathbf{x}_1$, $\mathbf{x}_2$, and $\mathbf{x}_3$ is given by

$$f(x_1, x_2, x_3) = \begin{cases} (x_1 + x_2)\, e^{-x_3} & \text{for } 0 < x_1 < 1, 0 < x_2 < 1, x_3 > 0 \\ 0 & \text{elsewhere} \end{cases}$$

find the regression equation of $\mathbf{x}_2$ on $\mathbf{x}_1$ and $\mathbf{x}_3$.

Solution

Referring to Example 3.22 on page 115, we find that the joint marginal density of $\mathbf{x}_1$ and $\mathbf{x}_3$ is given by

$$m(x_1, x_3) = \begin{cases} (x_1 + \frac{1}{2})\, e^{-x_3} & \text{for } 0 < x_1 < 1, x_3 > 0 \\ 0 & \text{elsewhere} \end{cases}$$

Therefore,

$$\begin{aligned} \mu_{\mathbf{x}_2|x_1,x_3} &= \int_{-\infty}^{\infty} x_2 \frac{f(x_1, x_2, x_3)}{m(x_1, x_3)}\, dx_2 = \int_0^1 \frac{x_2(x_1 + x_2)}{(x_1 + \frac{1}{2})}\, dx_2 \\ &= \frac{x_1 + \frac{2}{3}}{2x_1 + 1} \end{aligned}$$

Note that this conditional expectation depends on x_1 but not on x_3, which should really have been expected since we indicated on page 120 that there is a pairwise independence between $\mathbf{x}_2$ and $\mathbf{x}_3$.

14.2 LINEAR REGRESSION

An important feature of Example 14.2 is that the regression equation is **linear**, namely, that it is of the form

$$\mu_{\mathbf{y}|x} = \alpha + \beta x$$

where α and β are constants, called the **regression coefficients**. There are several reasons why linear regression equations are of special interest: First, they lend themselves readily to further mathematical treatment; then, they often provide good approximations to otherwise complicated regression equations; and, finally, in the case of the bivariate normal distribution, which we studied in Section 6.7, the regression equations are, in fact, linear.

To simplify the study of linear regression equations, let us express the regression coefficients α and β in terms of some of the lower moments of the joint distribution of **x** and **y**, namely, in terms of $E(\mathbf{x}) = \mu_1$, $E(\mathbf{y}) = \mu_2$, $\text{var}(\mathbf{x}) = \sigma_1^2$, $\text{var}(\mathbf{y}) = \sigma_2^2$, and $\text{cov}(\mathbf{x}, \mathbf{y}) = \sigma_{12}$. Then, also using the correlation coefficient

$$\rho = \frac{\sigma_{12}}{\sigma_1 \sigma_2}$$

defined in Section 6.7, we can prove the following results:

THEOREM 14.1 If the regression of **y** on **x** is linear, then

$$\mu_{\mathbf{y}|x} = \mu_2 + \rho \frac{\sigma_2}{\sigma_1}(x - \mu_1)$$

and if the regression of **x** on **y** is linear, then

$$\mu_{\mathbf{x}|y} = \mu_1 + \rho \frac{\sigma_1}{\sigma_2}(y - \mu_2)$$

Proof. Since $\mu_{\mathbf{y}|x} = \alpha + \beta x$, it follows that

$$\int y \cdot w(y|x)\, dy = \alpha + \beta x$$

and if we multiply the expression on both sides of this equation by $g(x)$, the corresponding value of the marginal density of **x**, and integrate on x, we obtain

$$\iint y \cdot w(y|x) g(x)\, dy\, dx = \alpha \int g(x)\, dx + \beta \int x \cdot g(x)\, dx$$

or

$$\mu_2 = \alpha + \beta \mu_1$$

since $w(y|x)g(x) = f(x, y)$. If we had multiplied the equation for $\mu_{\mathbf{y}|x}$ on both sides by $x \cdot g(x)$ before integrating on x, we would have obtained

$$\iint xy \cdot f(x, y)\, dy\, dx = \alpha \int x \cdot g(x)\, dx + \beta \int x^2 \cdot g(x)\, dx$$

$$E(\mathbf{xy}) = \alpha\mu_1 + \beta E(\mathbf{x}^2)$$

Solving $\mu_2 = \alpha + \beta\mu_1$ and $E(\mathbf{xy}) = \alpha\mu_1 + \beta E(\mathbf{x}^2)$ for α and β and making use of the fact that $E(\mathbf{xy}) = \sigma_{12} + \mu_1\mu_2$ and $E(\mathbf{x}^2) = \sigma_1^2 + \mu_1^2$, we find that

$$\alpha = \mu_2 - \frac{\sigma_{12}}{\sigma_1^2} \cdot \mu_1 = \mu_2 - \rho\frac{\sigma_2}{\sigma_1} \cdot \mu_1$$

and

$$\beta = \frac{\sigma_{12}}{\sigma_1^2} = \rho\frac{\sigma_2}{\sigma_1}$$

which enables us to write the linear regression equation of **y** on **x** as

$$\mu_{\mathbf{y}|x} = \mu_2 + \rho\frac{\sigma_2}{\sigma_1}(x - \mu_1)$$

When the regression of **x** on **y** is linear, similar steps lead to the equation

$$\mu_{\mathbf{x}|y} = \mu_1 + \rho\frac{\sigma_1}{\sigma_2}(y - \mu_2)$$

It follows from Theorem 14.1 that if the regression equation is linear and $\rho = 0$, then $\mu_{\mathbf{y}|x}$ does not depend on x (or $\mu_{\mathbf{x}|y}$ does not depend on y). When $\rho = 0$ and, hence, $\sigma_{12} = 0$, the two random variables **x** and **y** are **uncorrelated**, and we can paraphrase the assertion which we made on page 155 by saying that if two random variables are independent they are also uncorrelated, but if two random variables are uncorrelated they are not necessarily independent; the latter is again illustrated in Exercise 9 on page 431.

The correlation coefficient and its estimates are of importance in many statistical investigations, and they will be discussed in some detail in Section 14.5. At this time, let us again point out that $-1 \leq \rho \leq +1$, as the reader will be asked to prove in Exercise 11 on page 431, and that the sign of ρ tells us directly whether the slope of a regression line is upward or downward.

14.3 THE METHOD OF LEAST SQUARES

In the preceding sections we have discussed the problem of regression only in connection with random variables having known joint distributions. In actual practice, there are many problems where a set of **paired data** gives the indication that the regression is linear, where we do not know the joint distribution of the random variables under consideration, but nevertheless want to estimate the regression coefficients α and β. Problems of this kind are usually handled by the **method of least squares**, a method of curve fitting suggested early in the nineteenth century by the French mathematician Adrien Legendre.

To illustrate this technique, let us consider the following data on the number of hours which ten persons studied for a French test and their scores on the test:

Hours studied x	*Test score* y
4	31
9	58
10	65
14	73
4	37
7	44
12	60
22	91
1	21
17	84

Plotting these data as in Figure 14.2, we get the impression that a straight line provides a reasonably good fit. Although the points do not all fall on a straight line, the overall pattern suggests that the average test score for a given number of hours studied may well be related to the number of hours studied by means of an equation of the form $\mu_{y|x} = \alpha + \beta x$.

Once we have decided in a given problem that the regression is approximately linear, we face the problem of estimating the coefficients α and β from the sample data. In other words, we face the problem of obtaining estimates $\hat{\alpha}$ and $\hat{\beta}$ such that the estimated regression line $\hat{y} = \hat{\alpha} + \hat{\beta}x$ in some sense provides the best possible fit to the given data.

Denoting the vertical deviation from a point to the line by e_i, as indicated in Figure 14.3, the least squares criterion on which we shall base this "goodness of fit" requires that we minimize the sum of the squares of these deviations. Thus, if we are given a set of paired data $\{(x_i, y_i);\ i = 1, 2, \ldots, n\}$, the **least squares estimates** of the regression coefficients are the values $\hat{\alpha}$ and $\hat{\beta}$ for which the quantity

$$Q = \sum_{i=1}^{n} e_i^2 = \sum_{i=1}^{n} [y_i - (\hat{\alpha} + \hat{\beta}x_i)]^2$$

is a minimum. Differentiating partially with respect to $\hat{\alpha}$ and $\hat{\beta}$, and equating

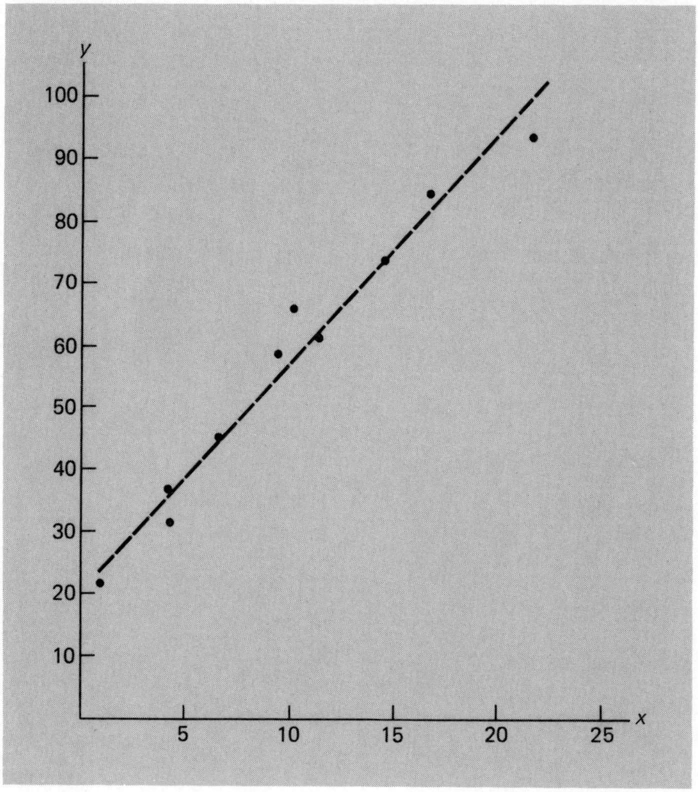

Figure 14.2 Data on test scores and number of hours studied.

these partial derivatives to zero, we obtain

$$\frac{\partial Q}{\partial \hat{\alpha}} = \sum_{i=1}^{n} (-2)[y_i - (\hat{\alpha} + \hat{\beta}x_i)] = 0$$

and

$$\frac{\partial Q}{\partial \hat{\beta}} = \sum_{i=1}^{n} (-2)x_i[y_i - (\hat{\alpha} + \hat{\beta}x_i)] = 0$$

which yield the so-called system of **normal equations**

$$\sum_{i=1}^{n} y_i = \hat{\alpha}n + \hat{\beta} \cdot \sum_{i=1}^{n} x_i$$

$$\sum_{i=1}^{n} x_i y_i = \hat{\alpha} \cdot \sum_{i=1}^{n} x_i + \hat{\beta} \cdot \sum_{i=1}^{n} x_i^2$$

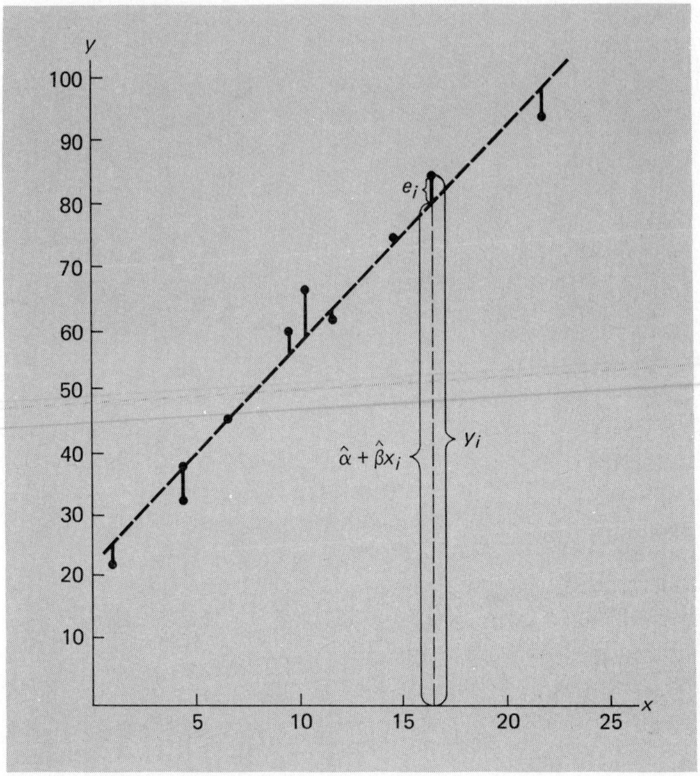

Figure 14.3 Least squares criterion.

Solving this system of equations for $\hat{\beta}$, we find that the least squares estimate of β is

$$\hat{\beta} = \frac{n\left(\sum_{i=1}^{n} x_i y_i\right) - \left(\sum_{i=1}^{n} x_i\right)\left(\sum_{i=1}^{n} y_i\right)}{n\left(\sum_{i=1}^{n} x_i^2\right) - \left(\sum_{i=1}^{n} x_i\right)^2}$$

and then, using the first of the normal equations, we obtain the least squares estimate of α by noting that

$$\hat{\alpha} = \frac{\sum_{i=1}^{n} y_i - \hat{\beta} \cdot \sum_{i=1}^{n} x_i}{n}$$

We have therefore arrived at the following result:

THEOREM 14.2 Given the sample data $\{(x_i, y_i);\ i = 1, 2, \ldots, n\}$, the coefficients of the least squares line $\hat{y} = \hat{\alpha} + \hat{\beta}x$ are

$$\hat{\beta} = \frac{n\left(\sum_{i=1}^{n} x_i y_i\right) - \left(\sum_{i=1}^{n} x_i\right)\left(\sum_{i=1}^{n} y_i\right)}{n\left(\sum_{i=1}^{n} x_i^2\right) - \left(\sum_{i=1}^{n} x_i\right)^2}$$

and

$$\hat{\alpha} = \frac{\sum_{i=1}^{n} y_i - \hat{\beta} \cdot \sum_{i=1}^{n} x_i}{n}$$

EXAMPLE 14.4

With reference to the data on page 426, find the equation of the least squares line that approximates the regression of test scores on the number of hours studied.

Solution

We find that $n = 10$, $\sum x = 100$, $\sum x^2 = 1{,}376$, $\sum y = 564$, and $\sum xy = 6{,}945$, so that

$$\hat{\beta} = \frac{10(6{,}945) - (100)(564)}{10(1{,}376) - (100)^2} = 3.471$$

and then

$$\hat{\alpha} = \frac{564 - (3.471)(100)}{10} = 21.69$$

Therefore, the equation of the least squares line is

$$\hat{y} = 21.69 + 3.471x$$

Using this equation, we can predict, for example, that a person who studies 14 hours for the test will get a score of $21.69 + 3.471(14) = 70.284$ (or 70 rounded to the nearest unit). Since we did not make any assumptions about the joint distribution of the random variables with which we are concerned, we cannot judge the "goodness" of this prediction; also, we cannot judge the "goodness" of the estimates $\hat{\alpha} = 21.69$ and $\hat{\beta} = 3.471$.

The least squares criterion, or in other words the method of least squares, is used in many problems of curve fitting which are more general than the one treated in this section. In Exercise 21 on page 434 the reader will be asked to use the method of least squares to estimate the coefficients β_0, β_1, and β_2 in the parabolic regression equation $\mu_{\mathbf{y}|x} = \beta_0 + \beta_1 x + \beta_2 x^2$, and in Exercises 19 and 20 he/she will be asked to estimate the constants of regression equations representing exponential relationships and power functions. Also, in Exercise 22 on page 435 the reader will be asked to use the method of least squares to estimate the coefficients of a **multiple regression equation** of the form $\mu_{\mathbf{z}|x,y} = \beta_0 + \beta_1 x + \beta_2 y$.

THEORETICAL EXERCISES

1. With reference to Example 14.1, show that the regression equation of **x** on **y** is

$$\mu_{\mathbf{x}|y} = \frac{2}{1+y}$$

and sketch the regression curve.

2. Given the joint density

$$f(x, y) = \begin{cases} \frac{2}{5}(2x + 3y) & \text{for } 0 < x < 1 \text{ and } 0 < y < 1 \\ 0 & \text{elsewhere} \end{cases}$$

find $\mu_{\mathbf{y}|x}$ and $\mu_{\mathbf{x}|y}$.

3. Given the joint density

$$f(x, y) = \begin{cases} 6x & \text{for } 0 < x < y < 1 \\ 0 & \text{elsewhere} \end{cases}$$

find $\mu_{\mathbf{y}|x}$ and $\mu_{\mathbf{x}|y}$.

4. Given the joint density

$$f(x, y) = \begin{cases} \dfrac{2x}{(1 + x + xy)^3} & \text{for } x > 0 \text{ and } y > 0 \\ 0 & \text{elsewhere} \end{cases}$$

show that $\mu_{\mathbf{y}|x} = 1 + \dfrac{1}{x}$ and that $\text{var}(\mathbf{y}|x)$ does not exist.

5. With reference to Exercise 2 on page 122, use the results of parts (c) and (d) to find $\mu_{\mathbf{x}|1}$ and $\mu_{\mathbf{y}|0}$.
6. With reference to Exercise 3 on page 122, find an expression for $\mu_{\mathbf{y}|x}$.
7. Given the joint density

$$f(x, y) = \begin{cases} 2 & \text{for } 0 < y < x < 1 \\ 0 & \text{elsewhere} \end{cases}$$

show that

(a) $\mu_{\mathbf{y}|x} = \dfrac{x}{2}$ and $\mu_{\mathbf{x}|y} = \dfrac{1+y}{2}$;

(b) $E(\mathbf{x}^m\mathbf{y}^n) = \dfrac{2}{(n+1)(m+n+2)}$.

Also, verify the results of part (a) by substituting the values of σ_1, σ_2, and ρ, obtained with the formula of part (b), into the formulas of Theorem 14.1.
8. Given the joint density

$$f(x, y) = \begin{cases} 24xy & \text{for } x > 0,\ y > 0, \text{ and } x + y < 1 \\ 0 & \text{elsewhere} \end{cases}$$

show that $\mu_{\mathbf{y}|x} = \frac{2}{3}(1 - x)$ and verify this result by determining the values of σ_1, σ_2, and ρ, and substituting them into the appropriate formula of Theorem 14.1.
9. Given the joint density

$$f(x, y) = \begin{cases} 1 & \text{for } -y < x < y \text{ and } 0 < y < 1 \\ 0 & \text{elsewhere} \end{cases}$$

show that the random variables **x** and **y** are uncorrelated but not independent.
10. Show that if $\mu_{\mathbf{y}|x}$ is linear in x and $\text{var}(\mathbf{y}|x)$ is constant, then $\text{var}(\mathbf{y}|x) = \sigma_2^2(1 - \rho^2)$.
11. Given a pair of random variables **x** and **y** having the variances σ_1^2 and σ_2^2 and the correlation coefficient ρ, use Theorem 4.14 to express $\text{var}\left(\dfrac{\mathbf{x}}{\sigma_1} + \dfrac{\mathbf{y}}{\sigma_2}\right)$ and $\text{var}\left(\dfrac{\mathbf{x}}{\sigma_1} - \dfrac{\mathbf{y}}{\sigma_2}\right)$ in terms of σ_1, σ_2, and ρ. Then, making use of the fact that variances cannot be negative, show that $-1 \leq \rho \leq +1$.
12. Given the random variables $\mathbf{x}_1$, $\mathbf{x}_2$, and $\mathbf{x}_3$ having the joint density $f(x_1, x_2, x_3)$, show that if the regression of $\mathbf{x}_3$ on $\mathbf{x}_1$ and $\mathbf{x}_2$ is linear and written as

$$\mu_{\mathbf{x}_3|x_1,x_2} = \alpha + \beta_1(x_1 - \mu_1) + \beta_2(x_2 - \mu_2)$$

then

$$\alpha = \mu_3$$

$$\beta_1 = \frac{\sigma_{13}\sigma_2^2 - \sigma_{12}\sigma_{23}}{\sigma_1^2\sigma_2^2 - \sigma_{12}^2}$$

$$\beta_2 = \frac{\sigma_{23}\sigma_1^2 - \sigma_{12}\sigma_{13}}{\sigma_1^2\sigma_2^2 - \sigma_{12}^2}$$

where $\mu_i = E(\mathbf{x}_i)$, $\sigma_i^2 = \text{var}(\mathbf{x}_i)$, and $\sigma_{ij} = \text{cov}(\mathbf{x}_i, \mathbf{x}_j)$. [*Hint:* Proceed as on page 425, multiplying by $(x_1 - \mu_1)$ and $(x_2 - \mu_2)$, respectively, to obtain the second and third equations.]

13. Find the least squares estimate of the parameter β in the regression equation $\mu_{\mathbf{y}|x} = \beta x$.

14. By solving the normal equations on page 427 simultaneously, show that

$$\hat{\alpha} = \frac{\left(\sum_{i=1}^{n} x_i^2\right)\left(\sum_{i=1}^{n} y_i\right) - \left(\sum_{i=1}^{n} x_i\right)\left(\sum_{i=1}^{n} x_i y_i\right)}{n\left(\sum_{i=1}^{n} x_i^2\right) - \left(\sum_{i=1}^{n} x_i\right)^2}$$

15. In some problems we estimate a regression equation of the form $\mu_{\mathbf{y}|x} = \beta_0 + \beta_1 x + \beta_2 x^2$ by fitting the parabola $\hat{y} = \hat{\beta}_0 + \hat{\beta}_1 x + \hat{\beta}_2 x^2$ to a set of paired data $\{(x_i, y_i); i = 1, 2, \ldots, n\}$. Minimizing

$$Q = \sum_{i=1}^{n} [y_i - (\hat{\beta}_0 + \hat{\beta}_1 x_i + \hat{\beta}_2 x_i^2)]^2$$

with respect to $\hat{\beta}_0$, $\hat{\beta}_1$, and $\hat{\beta}_2$, derive a set of normal equations whose solution will yield the least squares estimates $\hat{\beta}_0$, $\hat{\beta}_1$, and $\hat{\beta}_2$.

16. In some problems we estimate a regression equation of the form $\mu_{\mathbf{z}|x,y} = \beta_0 + \beta_1 x + \beta_2 y$ by fitting the plane $\hat{z} = \hat{\beta}_0 + \hat{\beta}_1 x + \hat{\beta}_2 y$ to a set of data consisting of the triplets $\{(x_i, y_i, z_i); i = 1, 2, \ldots, n\}$. Minimizing

$$Q = \sum_{i=1}^{n} [z_i - (\hat{\beta}_0 + \hat{\beta}_1 x_i + \hat{\beta}_2 y_i)]^2$$

with respect to $\hat{\beta}_0$, $\hat{\beta}_1$, and $\hat{\beta}_2$, derive a set of normal equations whose solution will yield the least squares estimates $\hat{\beta}_0$, $\hat{\beta}_1$, and $\hat{\beta}_2$.

APPLIED EXERCISES

17. Various doses of a poison were given to groups of 25 mice and the following results were observed:

Dose (mg) x	*Number of deaths* y
4	1
6	3
8	6
10	8
12	14
14	16
16	20
18	21

(a) Find the equation of the least squares line fit to these data.

(b) Estimate the number of deaths in a group of 25 mice who receive a 7-mg dose of this poison.

18. The following are the grades which 12 students obtained in the mid-term and final examinations in a course in statistics:

Mid-term examination x	*Final examination* y
71	83
49	62
80	76
73	77
93	89
85	74
58	48
82	78
64	76
32	51
87	73
80	89

(a) Find the equation of the least squares line which will enable us to predict a student's final examination grade in this course on the basis of his/her mid-term grade.

(b) Predict the final examination grade of a student who received an 84 on the mid-term examination.

19. If a set of paired data gives the indication that the regression equation is of the form $\mu_{y|x} = \alpha \cdot \beta^x$, it is customary to estimate α and β by fitting the line $\log \hat{y} = \log \hat{\alpha} + x \cdot \log \hat{\beta}$ to the points $\{(x_i, \log y_i); i = 1, 2, \ldots, n\}$ by the method of least squares. Use this technique to fit an exponential curve of the form $\hat{y} = \hat{\alpha} \cdot \hat{\beta}^x$ to the following data on the growth of cactus grafts under controlled environmental conditions:

Weeks after grafting x	*Height (inches)* y
1	2.0
2	2.4
4	5.1
5	7.3
6	9.4
8	18.3

20. If a set of paired data gives the indication that the regression equation is of the form $\mu_{y|x} = \alpha \cdot x^{\beta}$, it is customary to estimate α and β by fitting the line $\log \hat{y} = \log \hat{a} + \hat{\beta} \cdot \log x$ to the points $\{(\log x_i, \log y_i); i = 1, 2, \ldots, n\}$ by the method of least squares.

(a) Use this technique to fit a power function of the form $\hat{y} = \hat{\alpha} \cdot x^{\hat{\beta}}$ to the following data on the unit cost of producing certain electronic components and the number of units produced:

Lot size x	*Unit cost* y
50	\$108
100	\$ 53
250	\$ 24
500	\$ 9
1,000	\$ 5

(b) Use the result of part (a) to estimate the unit cost for a lot of 300 components.

21. The following data pertain to the drying time of a varnish and the amount of a certain chemical that has been added:

Amount of additive (*grams*) x	*Drying time* (*hours*) y
1	8.5
2	8.0
3	6.0
4	5.0
5	6.0
6	5.5
7	6.5
8	7.0

(a) Use the theory of Exercise 15 to fit a parabola of the form $\hat{y} = \hat{\beta}_0 + \hat{\beta}_1 x + \hat{\beta}_2 x^2$ to the data.

(b) Estimate the drying time of the varnish when 5.5 grams of the chemical are added.

22. The following data consist of the grades which 10 students obtained in an examination, their I.Q.'s, and the number of hours they spent studying for the examination:

Grade z	*I.Q.* x	*Hours studied* y
79	112	5
97	126	13
51	100	3
65	114	7
82	112	11
93	121	9
81	110	8
38	103	4
60	111	6
86	124	2

(a) Use the theory of Exercise 16 to fit a plane of the form $\hat{z} = \hat{\beta}_0 + \hat{\beta}_1 x + \hat{\beta}_2 y$ to the data.

(b) Predict the grade of a student with an I.Q. of 108 who studies 6 hours for this examination.

14.4 NORMAL REGRESSION ANALYSIS

When we analyze a set of paired data $\{(x_i, y_i); i = 1, 2, \ldots, n\}$ by **regression analysis**, we assume that the x_i are constants while the y_i are values of corresponding independent random variables $\mathbf{y}_i$. This clearly differs from **correlation analysis**, which we shall take up in Section 14.5, where the x_i and the y_i are values of corresponding random variables $\mathbf{x}_i$ and $\mathbf{y}_i$. For example, if we want to analyze data on the ages and prices of used cars, treating the ages as known constants and the prices as values of random variables, this is a problem of regression analysis. On the other hand, if we want to analyze data on the height and weight of certain animals, and height and weight are both looked upon as random variables, this is a problem of correlation analysis.

This section will be devoted to some of the basic problems of **normal regression analysis**, where it is assumed that for each fixed x_i the conditional density of the corresponding random variable $\mathbf{y}_i$ is the normal density

$$w(y_i|x_i) = \frac{1}{\sigma\sqrt{2\pi}} \cdot e^{-\frac{1}{2}\left[\frac{y_i - (\alpha + \beta x_i)}{\sigma}\right]^2} \qquad -\infty < y_i < \infty$$

where α, β, and σ are the same for each i. Given a random sample of such paired data, normal regression analysis concerns itself mainly with the estimation of σ and the regression coefficients α and β, with tests of hypotheses concerning these three parameters, and with predictions based on the estimated regression equation $\hat{y} = \hat{\alpha} + \hat{\beta}x$, where $\hat{\alpha}$ and $\hat{\beta}$ are estimates of α and β.

To obtain maximum likelihood estimates of the parameters α, β, and σ, we partially differentiate the likelihood function (or its logarithm, which is easier) with respect to α, β, and σ, equate the expressions to zero, and then solve the resulting system of equations. Thus, differentiating

$$\ln L = -n \cdot \ln \sigma - \frac{n}{2} \cdot \ln 2\pi - \frac{1}{2\sigma^2} \cdot \sum_{i=1}^{n} [y_i - (\alpha + \beta x_i)]^2$$

partially with respect to α, β, and σ, and equating the expressions which we obtain to zero, we get

$$\frac{\partial \ln L}{\partial \alpha} = \frac{1}{\sigma^2} \cdot \sum_{i=1}^{n} [y_i - (\alpha + \beta x_i)] = 0$$

$$\frac{\partial \ln L}{\partial \beta} = \frac{1}{\sigma^2} \cdot \sum_{i=1}^{n} x_i[y_i - (\alpha + \beta x_i)] = 0$$

$$\frac{\partial \ln L}{\partial \sigma} = -\frac{n}{\sigma} + \frac{1}{\sigma^3} \cdot \sum_{i=1}^{n} [y_i - (\alpha + \beta x_i)]^2 = 0$$

As the reader will be asked to verify in Exercise 1 on page 445, the first two of these equations yield the maximum likelihood estimates

$$\hat{\alpha} = \frac{\left(\sum_{i=1}^{n} x_i^2\right)\left(\sum_{i=1}^{n} y_i\right) - \left(\sum_{i=1}^{n} x_i\right)\left(\sum_{i=1}^{n} x_i y_i\right)}{n\left(\sum_{i=1}^{n} x_i^2\right) - \left(\sum_{i=1}^{n} x_i\right)^2}$$

and

$$\hat{\beta} = \frac{n\left(\sum_{i=1}^{n} x_i y_i\right) - \left(\sum_{i=1}^{n} x_i\right)\left(\sum_{i=1}^{n} y_i\right)}{n\left(\sum_{i=1}^{n} x_i^2\right) - \left(\sum_{i=1}^{n} x_i\right)^2}$$

which, according to Exercise 14 on page 432, are identical with the least squares estimates given in Theorem 14.2. Thus, to simplify the calculation of the estimated regression coefficients we first compute $\hat{\beta}$ and then determine $\hat{\alpha}$ by the formula of Theorem 14.2.

If we substitute $\hat{\alpha}$ and $\hat{\beta}$ for α and β into the equation obtained by equating $\frac{\partial \ln L}{\partial \sigma}$ to zero, it follows immediately that the maximum likelihood estimate of σ is given by

$$\hat{\sigma} = \sqrt{\frac{1}{n} \cdot \sum_{i=1}^{n} [y_i - (\hat{\alpha} + \hat{\beta} x_i)]^2}$$

In practice, it is generally easier to find $\hat{\sigma}$ by using an alternative, but equivalent, formula for $\hat{\sigma}^2$. To derive this formula, let us first introduce the following notation:

$$S_{xx} = \sum_{i=1}^{n} (x_i - \bar{x})^2 = \sum_{i=1}^{n} x_i^2 - \frac{\left(\sum_{i=1}^{n} x_i\right)^2}{n}$$

$$S_{yy} = \sum_{i=1}^{n} (y_i - \bar{y})^2 = \sum_{i=1}^{n} y_i^2 - \frac{\left(\sum_{i=1}^{n} y_i\right)^2}{n}$$

and

$$S_{xy} = \sum_{i=1}^{n} (x_i - \bar{x})(y_i - \bar{y}) = \sum_{i=1}^{n} x_i y_i - \frac{\left(\sum_{i=1}^{n} x_i\right)\left(\sum_{i=1}^{n} y_i\right)}{n}$$

Noting that $\hat{\beta} = \frac{S_{xy}}{S_{xx}}$, and using the result of Exercise 2 on page 446, it follows that

$$\hat{\sigma}^2 = \frac{S_{yy} - \hat{\beta} S_{xy}}{n}$$

Having obtained maximum likelihood estimators of the regression coefficients, let us now investigate their use in testing hypotheses concerning α and β, and in constructing confidence intervals for these two parameters. Since problems concerning β are usually of more immediate interest than problems concerning α (the null hypothesis $\beta = 0$ is equivalent to the null hypothesis $\rho = 0$), we shall discuss some of the sampling theory connected with β in the text, while leaving corresponding theory concerning α to Exercises 4 and 6 on page 446.

To study the sampling distribution of $\hat{\boldsymbol{\beta}}$, let us write

$$\hat{\boldsymbol{\beta}} = \frac{S_{x\mathbf{y}}}{S_{xx}} = \frac{\sum_{i=1}^{n} (x_i - \bar{x})(\mathbf{y}_i - \bar{\mathbf{y}})}{S_{xx}}$$

$$= \sum_{i=1}^{n} \left(\frac{x_i - \bar{x}}{S_{xx}}\right) \mathbf{y}_i$$

which is seen to be a linear combination of the n independent normal random variables $\mathbf{y}_i$. It follows from Exercise 6 on page 251 that $\hat{\boldsymbol{\beta}}$, itself, has a normal distribution with the mean

$$E(\hat{\boldsymbol{\beta}}) = \sum_{i=1}^{n} \left[\frac{x_i - \bar{x}}{S_{xx}}\right] \cdot E(\mathbf{y}_i | x_i)$$

$$= \sum_{i=1}^{n} \left[\frac{x_i - \bar{x}}{S_{xx}}\right] (\alpha + \beta x_i) = \beta$$

and the variance

$$\text{var}(\hat{\boldsymbol{\beta}}) = \sum_{i=1}^{n} \left[\frac{x_i - \bar{x}}{S_{xx}}\right]^2 \cdot \text{var}(\mathbf{y}_i | x_i)$$

$$= \sum_{i=1}^{n} \left[\frac{x_i - \bar{x}}{S_{xx}}\right]^2 \cdot \sigma^2 = \frac{\sigma^2}{S_{xx}}$$

In order to apply this theory to construct confidence intervals for β or test hypotheses concerning β, we shall have to make use of the following theorem, a proof of which is referred to on page 451:

THEOREM 14.3 Under the assumptions of normal regression analysis, $\frac{n\hat{\sigma}^2}{\sigma^2}$ has the chi-square distribution with $n-2$ degrees of freedom. Furthermore, $\frac{n\hat{\sigma}^2}{\sigma^2}$ and $\hat{\beta}$ are independent.

Making use of this theorem as well as the previously proved result that $\hat{\beta}$ has a normal distribution with the mean β and the variance $\frac{\sigma^2}{S_{xx}}$, the definition of the t distribution in Section 8.5 leads to the result that the random variable

$$\mathbf{t} = \frac{\dfrac{\hat{\beta} - \beta}{\sigma/\sqrt{S_{xx}}}}{\sqrt{\dfrac{n\hat{\sigma}^2}{\sigma^2}\Big/(n-2)}} = \frac{\hat{\beta} - \beta}{\hat{\sigma}}\sqrt{\frac{(n-2)S_{xx}}{n}}$$

has a t distribution with $n - 2$ degrees of freedom. Hence,

$$P\left(-t_{\alpha/2,n-2} < \frac{\hat{\beta} - \beta}{\hat{\sigma}}\sqrt{\frac{(n-2)S_{xx}}{n}} < t_{\alpha/2,n-2}\right) = 1 - \alpha$$

or, equivalently,

$$P\left[\hat{\beta} - t_{\alpha/2,n-2} \cdot \hat{\sigma}\sqrt{\frac{n}{(n-2)S_{xx}}} < \beta < \hat{\beta} + t_{\alpha/2,n-2} \cdot \hat{\sigma}\sqrt{\frac{n}{(n-2)S_{xx}}}\right] = 1 - \alpha$$

so that

THEOREM 14.4 Under the assumptions of normal regression analysis, a $(1-\alpha)100\%$ confidence interval for the parameter β in the regression line $\mu_{y|x} = \alpha + \beta x$ is given by

$$\hat{\beta} - t_{\alpha/2,n-2} \cdot \hat{\sigma}\sqrt{\frac{n}{(n-2)S_{xx}}} < \beta < \hat{\beta} + t_{\alpha/2,n-2} \cdot \hat{\sigma}\sqrt{\frac{n}{(n-2)S_{xx}}}$$

EXAMPLE 14.5

With reference to the test scores and hours studied data on page 426, construct a 95% confidence interval for the regression coefficient β.

Solution

In Exercise 14.4 we had $n = 10, \sum x = 100, \sum x^2 = 1{,}376, \sum y = 564, \sum xy = 6{,}945$, and we obtained $\hat{\beta} = 3.471$. Including $\sum y^2 = 36{,}562$ along with these sums, we find that

$$S_{xx} = 1{,}376 - \frac{100^2}{10} = 376$$

$$S_{yy} = 36{,}562 - \frac{564^2}{10} = 4{,}752.4$$

$$S_{xy} = 6{,}945 - \frac{100 \cdot 564}{10} = 1{,}305$$

and, hence,

$$\hat{\sigma}^2 = \frac{S_{yy} - \hat{\beta}S_{xy}}{n} = \frac{4{,}752.4 - (3.471)(1{,}305)}{10} = 22.275$$

so that $\hat{\sigma} = 4.720$. Thus, by Theorem 14.4,

$$3.471 - (2.306)(4.720)\sqrt{\frac{10}{8(376)}} < \beta < 3.471 + (2.306)(4.720)\sqrt{\frac{10}{8(376)}}$$

or

$$2.84 < \beta < 4.10$$

since $t_{.025,8} = 2.306$ according to Table IV.

14.5 NORMAL CORRELATION ANALYSIS

In normal correlation analysis we analyze a set of paired data $\{(x_i, y_i); i = 1, 2, \ldots, n\}$, where the x_i's and y_i's are values of a random sample from a bivariate normal population with the parameters $\mu_1, \mu_2, \sigma_1, \sigma_2$, and ρ. To estimate these parameters by the method of maximum likelihood, we shall have to maximize the likelihood

$$L = \prod_{i=1}^{n} f(x_i, y_i)$$

where $f(x_i, y_i)$ is given by Definition 6.8, and to this end we shall have to differentiate L, or $\ln L$, partially with respect to μ_1, μ_2, σ_1, σ_2, and ρ, equate the resulting expressions to zero, and then solve the resulting system of equations for the five parameters. Leaving the details to the reader, let us merely state that when $\frac{\partial \ln L}{\partial \mu_1}$ and $\frac{\partial \ln L}{\partial \mu_2}$ are equated to zero, we get

$$-\frac{\sum_{i=1}^{n}(x_i - \mu_1)}{\sigma_1^2} + \frac{\rho \sum_{i=1}^{n}(y_i - \mu_2)}{\sigma_1 \sigma_2} = 0$$

and

$$-\frac{\rho \sum_{i=1}^{n}(x_i - \mu_1)}{\sigma_1 \sigma_2} + \frac{\sum_{i=1}^{n}(y_i - \mu_2)}{\sigma_2^2} = 0$$

Solving these two equations for μ_1 and μ_2, we find that the maximum likelihood estimates of these two parameters are

$$\hat{\mu}_1 = \bar{x} \quad \text{and} \quad \hat{\mu}_2 = \bar{y}$$

namely, the respective sample means. Subsequently, equating $\frac{\partial \ln L}{\partial \sigma_1}$, $\frac{\partial \ln L}{\partial \sigma_2}$, and $\frac{\partial \ln L}{\partial \rho}$ to zero and substituting $\bar{x}$ and $\bar{y}$ for μ_1 and μ_2, we obtain a system of equations whose solution is

$$\hat{\sigma}_1 = \sqrt{\frac{\sum_{i=1}^{n}(x_i - \bar{x})^2}{n}}, \qquad \hat{\sigma}_2 = \sqrt{\frac{\sum_{i=1}^{n}(y_i - \bar{y})^2}{n}}$$

$$\hat{\rho} = \frac{\sum_{i=1}^{n}(x_i - \bar{x})(y_i - \bar{y})}{\sqrt{\sum_{i=1}^{n}(x_i - \bar{x})^2}\sqrt{\sum_{i=1}^{n}(y_i - \bar{y})^2}}$$

(A detailed derivation of these maximum likelihood estimates is referred to at the end of this chapter.) It is of interest to note that the maximum likelihood estimates of σ_1 and σ_2 are identical with the one obtained on page 331 for the standard deviation of the univariate normal distribution; they differ from the respective sample standard deviations s_1 and s_2 only by the factor $\sqrt{\frac{n-1}{n}}$.

The estimate $\hat{\rho}$, called the **sample correlation coefficient**, is usually denoted by the letter r, and its calculation is facilitated by using the following alternative, but equivalent, formula:

DEFINITION 14.1 If $\{(x_i, y_i); i = 1, 2, \ldots, n\}$ are the values of a random sample from a bivariate population, then

$$r = \frac{n \cdot \sum_{i=1}^{n} x_i y_i - \left(\sum_{i=1}^{n} x_i\right)\left(\sum_{i=1}^{n} y_i\right)}{\sqrt{n \cdot \sum_{i=1}^{n} x_i^2 - \left(\sum_{i=1}^{n} x_i\right)^2}\sqrt{n \cdot \sum_{i=1}^{n} y_i^2 - \left(\sum_{i=1}^{n} y_i\right)^2}}$$

is called the **sample correlation coefficient.**

It will be left to the reader in Exercise 11 on page 448 to verify that the maximum likelihood estimate $\hat{\rho}$ and the formula for r in Definition 14.1 are equivalent.

Since ρ measures the strength of the linear relationship between **x** and **y**, there are many problems in which the estimation of ρ and tests concerning ρ are of special interest. When $\rho = 0$, the two random variables are uncorrelated, and as we have already seen, in the case of the bivariate normal distribution this means that they are also independent. When ρ equals +1 or −1, it follows from the relationship

$$\sigma_{\mathbf{y}|x}^2 = \sigma^2 = \sigma_2^2(1 - \rho^2)$$

established in Theorem 6.9, that $\sigma = 0$, and this means that there is a perfect linear relationship between **x** and **y**. Using the invariance property of maximum likelihood estimators, we can write

$$\hat{\sigma}^2 = \hat{\sigma}_2^2(1 - r^2)$$

which not only provides an alternative method for finding $\hat{\sigma}^2$ in Section 14.4, but also serves to tie together the concepts of regression and correlation. From this formula for $\hat{\sigma}^2$ it is clear that when $\hat{\sigma}^2 = 0$, namely, when the set of data points $\{(x_i, y_i); i = 1, 2, \ldots, n\}$ fall on a straight line, then r will equal +1 or −1, depending on whether the line has an upward or downward slope. In order to interpret values of r between 0 and +1 or 0 and −1, we solve the preceding equation for r^2 and multiply by 100, getting

$$100r^2 = \frac{\hat{\sigma}_2^2 - \hat{\sigma}^2}{\hat{\sigma}_2^2} \cdot 100$$

where $\hat{\sigma}_2^2$ measures the total variation of the y's, $\hat{\sigma}^2$ measures the conditional variation of the y's for fixed values of x, and, hence, $\hat{\sigma}_2^2 - \hat{\sigma}^2$ measures that part of the total variation of the y's which is accounted for by the relationship with x. *Thus,* $100r^2$ *is the percentage of the total variation of the y's which is accounted for by the relationship with x.* For instance, when $r = 0.5$ then 25 percent of the variation of the y's is accounted for by the relationship with x, when $r = 0.7$ then 49 percent of the variation of the y's is accounted for by the relationship with x, and we might thus say that a correlation of $r = 0.7$ is almost "twice as strong" as a correlation of $r = 0.5$. Similarly, we might say that a correlation of $r = 0.6$ is "nine times as strong" as a correlation of $r = 0.2$.

EXAMPLE 14.6

Suppose that we want to determine on the basis of the following data whether there is a relationship between the time, in minutes, it takes a secretary to complete a certain form in the morning and in the late afternoon:

Morning x	*Afternoon* y
8.2	8.7
9.6	9.6
7.0	6.9
9.4	8.5
10.9	11.3
7.1	7.6
9.0	9.2
6.6	6.3
8.4	8.4
10.5	12.3

Compute and interpret the sample correlation coefficient.

Solution

From the data we get $n = 10$, $\sum x = 86.7$, $\sum x^2 = 771.35$, $\sum y = 88.8$, $\sum y^2 = 819.34$, and $\sum xy = 792.92$, so that substitution into the formula for r yields

$$r = \frac{10(792.92) - (86.7)(88.8)}{\sqrt{10(771.35) - (86.7)^2}\sqrt{10(819.34) - (88.8)^2}}$$

$$= 0.936$$

This is indicative of a positive association between the time it takes a secretary to perform the given task in the morning and in the late afternoon, and this is also apparent from the **scattergram** of Figure 14.4. Since $100r^2 = 100(0.936)^2 = 87.6$, we can say that almost 88% of the variation of the y's is accounted for by a linear relationship with x.

Since the sampling distribution of $\mathbf{r}$ for random samples from bivariate normal populations is rather complicated, it is common practice to base confidence intervals for ρ and tests concerning ρ on the statistic $\frac{1}{2} \cdot \ln \frac{1+\mathbf{r}}{1-\mathbf{r}}$, whose distribution is approximately normal with the mean $\frac{1}{2} \cdot \ln \frac{1+\rho}{1-\rho}$ and the variance $\frac{1}{n-3}$. Thus,

$$z = \frac{\frac{1}{2} \cdot \ln \frac{1+r}{1-r} - \frac{1}{2} \cdot \ln \frac{1+\rho}{1-\rho}}{\frac{1}{\sqrt{n-3}}}$$

$$= \frac{\sqrt{n-3}}{2} \cdot \ln \frac{(1+r)(1-\rho)}{(1-r)(1+\rho)}$$

can be looked upon as a value of a random variable having approximately the standard normal distribution. Using this approximation, we can test the null hypothesis $\rho = \rho_0$ against an appropriate alternative as illustrated in Example 14.7 below, or calculate confidence intervals for ρ by the method suggested in Exercise 13 on page 448.

EXAMPLE 14.7

With reference to Example 14.6, test the null hypothesis $\rho = 0$ against the alternative hypothesis $\rho \neq 0$ at the 0.01 level of significance.

Solution

1. H_0: $\rho = 0$
H_1: $\rho \neq 0$

2. Critical region: $|z| \geqslant z_{.005} = 2.575$, where

$$z = \frac{\sqrt{n-3}}{2} \cdot \ln \frac{(1+r)}{(1-r)}$$

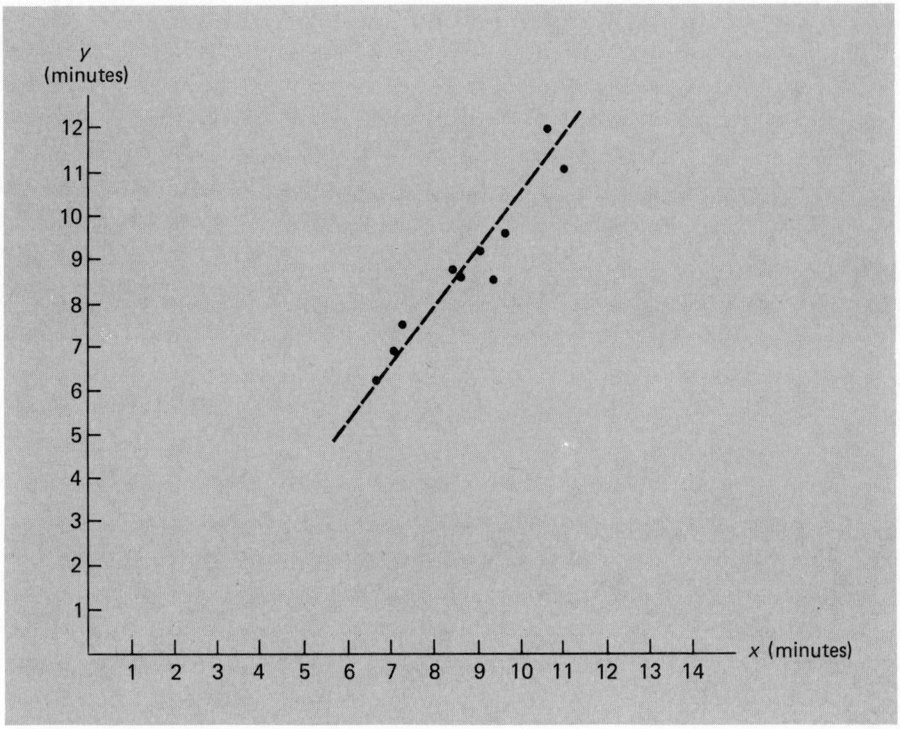

Figure 14.4 Scattergram of data of Example 14.6.

3. Computations: $r = 0.936$ and $n = 10$, so that

$$z = \frac{\sqrt{7}}{2} \cdot \ln \frac{1.936}{0.064} = 4.5$$

4. Decision: Since $z = 4.5$ exceeds $z_{.005} = 2.575$, the null hypothesis of no correlation must be rejected and we conclude that there is a relationship between the time it takes a secretary to complete the form in the morning and in the late afternoon. Of course, it must be remembered that this test is only approximate and that the data have to be looked upon as a random sample from a bivariate normal population.

THEORETICAL EXERCISES

1. Verify that the simultaneous solution of $\frac{\partial \ln L}{\partial \alpha} = 0$ and $\frac{\partial \ln L}{\partial \beta} = 0$ yields the formulas for $\hat{\alpha}$ and $\hat{\beta}$ on page 437.

2. Making use of the fact that $\hat{\alpha} = \bar{y} - \hat{\beta}\bar{x}$ and $\hat{\beta} = \dfrac{S_{xy}}{S_{xx}}$, show that

$$\sum_{i=1}^{n} [y_i - (\hat{\alpha} + \hat{\beta}x_i)]^2 = S_{yy} - \hat{\beta}S_{xy}$$

3. Show that $n\hat{\sigma}^2/(n-2)$ is an unbiased estimator of σ^2.
4. Under the assumption of normal regression analysis, show that
   (a) the least squares estimate of $\hat{\alpha}$ in Theorem 14.2 can be written in the form

$$\hat{\alpha} = \sum_{i=1}^{n} \left[\frac{S_{xx} + n\bar{x}^2 - n\bar{x}x_i}{nS_{xx}}\right] y_i$$

   (b) $\hat{\boldsymbol{\alpha}}$ has a normal distribution with

$$E(\hat{\boldsymbol{\alpha}}) = \alpha \quad \text{and} \quad \text{var}(\hat{\boldsymbol{\alpha}}) = \frac{(S_{xx} + n\bar{x}^2)\sigma^2}{nS_{xx}}$$

5. Use Theorem 4.15 to show that

$$\text{cov}(\hat{\boldsymbol{\alpha}}, \hat{\boldsymbol{\beta}}) = -\frac{\bar{x}}{S_{xx}} \cdot \sigma^2$$

6. Use the result of part (b) of Exercise 4 to show that

$$z = \frac{(\hat{\alpha} - \alpha)\sqrt{nS_{xx}}}{\sigma\sqrt{S_{xx} + n\bar{x}^2}}$$

is a value of a random variable having the standard normal distribution. Also, use the first part of Theorem 14.3 and the fact that $\hat{\boldsymbol{\alpha}}$ and $\dfrac{n\hat{\boldsymbol{\sigma}}^2}{\sigma^2}$ are independent to show that

$$t = \frac{(\hat{\alpha} - \alpha)\sqrt{(n-2)S_{xx}}}{\hat{\sigma}\sqrt{S_{xx} + n\bar{x}^2}}$$

is a value of a random variable having the t distribution with $n - 2$ degrees of freedom.
7. Use the results of Exercises 4 and 5 and the fact that $E(\hat{\boldsymbol{\beta}}) = \beta$ and $\text{var}(\hat{\boldsymbol{\beta}}) = \dfrac{\sigma^2}{S_{xx}}$ to show that $\hat{\mathbf{y}}_0 = \hat{\boldsymbol{\alpha}} + \hat{\boldsymbol{\beta}}x_0$ is a random variable having a normal

distribution with the mean

$$\alpha + \beta x_0 = \mu_{\mathbf{y}|x_0}$$

and the variance

$$\sigma^2\left[\frac{1}{n} + \frac{(x_0 - \bar{x})^2}{S_{xx}}\right]$$

Also, use the first part of Theorem 14.3 as well as the fact that $\hat{\mathbf{y}}_0$ and $\frac{n\hat{\boldsymbol{\sigma}}^2}{\sigma^2}$ are independent to show that

$$t = \frac{(\hat{y}_0 - \mu_{\mathbf{y}|x_0})\sqrt{n-2}}{\hat{\sigma}\sqrt{1 + \frac{n(x_0 - \bar{x})^2}{S_{xx}}}}$$

is a value of a random variable having the t distribution with $n - 2$ degrees of freedom.

8. Find a $(1 - \alpha)100\%$ confidence interval for $\mu_{\mathbf{y}|x_0}$, the mean of $\mathbf{y}$ at $x = x_0$, by solving the double inequality $-t_{\alpha/2,n-2} < t < t_{\alpha/2,n-2}$ with t as given in Exercise 7.

9. Use the results of Exercises 4 and 5 and the fact that $E(\hat{\boldsymbol{\beta}}) = \beta$ and $\text{var}(\hat{\boldsymbol{\beta}}) = \frac{\sigma^2}{S_{xx}}$ to show that $\mathbf{y}_0 - (\hat{\boldsymbol{\alpha}} + \hat{\boldsymbol{\beta}}x_0)$ is a random variable having a normal distribution with zero mean and the variance

$$\sigma^2\left[1 + \frac{1}{n} + \frac{(x_0 - \bar{x})^2}{S_{xx}}\right]$$

where $\mathbf{y}_0$ has a normal distribution with the mean $\alpha + \beta x_0$ and the variance σ^2; that is, $\mathbf{y}_0$ is a future observation of $\mathbf{y}$ corresponding to $\mathbf{x} = x_0$. Also, use the first part of Theorem 14.3 as well as the fact that $\mathbf{y}_0 - (\hat{\boldsymbol{\alpha}} + \hat{\boldsymbol{\beta}}x_0)$ and $\frac{n\hat{\boldsymbol{\sigma}}^2}{\sigma^2}$ are independent to show that

$$t = \frac{[y_0 - (\hat{\alpha} + \hat{\beta}x_0)]\sqrt{n-2}}{\hat{\sigma}\sqrt{1 + n + \frac{n(x_0 - \bar{x})^2}{S_{xx}}}}$$

is a value of a random variable having the t distribution with $n - 2$ degrees of freedom.

10. Solve the double inequality $-t_{\alpha/2,n-2} < t < t_{\alpha/2,n-2}$ with t as given in Exercise 9, so that the middle term is y_0 and the two limits can be calculated without knowledge of y_0. Note that although the resulting double inequality may be interpreted like a confidence interval, it is not designed to estimate a parameter; instead, it provides **limits of prediction** for a future observation of y which corresponds to the (given or observed) value x_0.

11. Show that the maximum likelihood estimate of ρ on page 441 and the formula for r in Definition 14.1 are equivalent.

12. The calculation of r can often by simplified by adding the same constant to each x, adding the same constant to each y, or by multiplying each x and/or each y by the same positive constant.

(a) Show that if r_{xy} is the correlation coefficient calculated for a set of paired data $\{(x_i, y_i); i = 1, 2, \ldots, n\}$, then r_{uv}, the correlation coefficient for $u_i = ax_i + b$ and $v_i = cy_i + d$ (with $a > 0$ and $c > 0$), is given by $r_{uv} = r_{xy}$.

(b) Recalculate the correlation coefficient for Example 14.6 by first multiplying each x and each y by 10, and then subtracting 70 from each x and 60 from each y.

13. By solving the double inequality $-z_{\alpha/2} < z < z_{\alpha/2}$ with z given by the formula on page 444, construct a $(1 - \alpha)100\%$ confidence interval for ρ. Use the result to construct a 95% confidence interval for ρ in Example 14.6.

14. By writing the sample correlation coefficient of Definition 14.1 in the form

$$r = \frac{S_{xy}}{\sqrt{S_{xx} \cdot S_{yy}}}$$

verify that values of the t statistic on page 439 can be written as

$$t = \left(1 - \frac{\beta}{\hat{\beta}}\right)\frac{r\sqrt{n-2}}{\sqrt{1-r^2}}$$

Also use this result to derive the following $(1 - \alpha)100\%$ confidence limits for β:

$$\hat{\beta}\left[1 \pm t_{\alpha/2,n-2} \cdot \frac{\sqrt{1-r^2}}{r\sqrt{n-2}}\right]$$

15. Use the first part of Exercise 14 to show that if the assumptions underlying normal regression analysis are met and $\beta = 0$, then $\mathbf{r}^2$ has a beta distribution with the mean $\frac{1}{n-1}$.

APPLIED EXERCISES

16. Referring to the hours studied and test score data on page 426 (as well as the calculations already performed in Examples 14.4 and 14.5), test the null hypothesis $\beta = 3.20$ against the alternative hypothesis $\beta \neq 3.20$ at the level of significance $\alpha = 0.05$.

17. With reference to Example 14.6, test the null hypothesis $\beta = 1.1$ against the alternative hypothesis $\beta \neq 1.1$ at the 0.05 level of significance.

18. With reference to Exercise 17 on page 433, construct a 95% confidence interval for the regression coefficient β.

19. With reference to Exercise 18 on page 433, construct a 99% confidence interval for the regression coefficient β.

20. With reference to Example 14.4, test the null hypothesis $\alpha = 21.50$ against the alternative hypothesis $\alpha \neq 21.50$ at the level of significance 0.01. Use the theory of Exercise 6.

21. Use the result of Exercise 6 to construct a 95% confidence interval for α for the data of Exercise 17 on page 433.

22. Use the result of Exercise 6 to calculate a 99% confidence interval for α for the data of Exercise 18 on page 433.

23. Use the results of Exercises 8 and 10 (as well as the calculations already performed in Examples 14.4 and 14.5) to construct

(a) a 95% confidence interval for the mean test score of a person who has studied 14 hours for the test;

(b) 95% limits of prediction for the test score of a person who has studied 14 hours for the test.

24. Use the results of Exercises 8 and 10 and the data of Exercise 17 on page 433 to find

(a) a 99% confidence interval for the mean number of deaths in a group of 25 mice when the dosage is 9 mg;

(b) 99% limits of prediction of the number of deaths in a group of 25 mice when the dosage in 9 mg.

25. An objective achievement test is said to be reliable if a student who takes the test several times will consistently get high (or low) scores. One way of checking the reliability of a test is to divide it into two parts, usually the even-numbered problems and the odd-numbered problems, and to check the correlation between the scores which students get in both halves of the test.

Thus, the following data represent the grades, x and y, which 20 students obtained for the even-numbered problems and the odd-numbered problems of a new objective test designed to test eighth grade achievement in general science:

x	y	x	y
27	29	33	42
36	44	39	31
44	49	38	38
32	27	24	22
27	35	33	34
41	33	32	37
38	29	37	38
44	40	33	35
30	27	34	32
27	38	39	43

Calculate r for these data and test its significance, namely, the null hypothesis $\rho = 0$, at $\alpha = 0.05$.

26. With reference to the data of Exercise 25, use the result of Exercise 13 to construct a 95% confidence interval for ρ.

27. The following data pertain to x, the amount of fertilizer (in pounds) which a farmer applies to his soil, and y, his yield of wheat (in bushels per acre):

x	y	x	y	x	y
112	33	88	24	37	27
92	28	44	17	23	9
72	38	132	36	77	32
66	17	23	14	142	38
112	35	57	25	37	13
88	31	111	40	127	23
42	8	69	29	88	31
126	37	19	12	48	37
72	32	103	27	61	25
52	20	141	40	71	14
28	17	77	26	113	26

Assuming that the data can be looked upon as a sample from a bivariate normal population, calculate r and test its significance at $\alpha = 0.01$. Judging from a scattergram of these paired data, does the assumption seem reasonable?

28. With reference to the data of Exercise 27, use the result of Exercise 13 to determine a 95% confidence interval for ρ.

29. Use the result of Exercise 14 to calculate a 95% confidence interval for β for the hours studied and test scores on page 426, and compare this interval with the one obtained in Example 14.5.

References

A proof of Theorem 14.3 and other mathematical details left out in the text may be found in the book by S. S. Wilks referred to on page 252, and information about the distribution of $\frac{1}{2} \cdot \ln \frac{1 + \mathbf{r}}{1 - \mathbf{r}}$ may be found in the book by M. G. Kendall and A. Stuart referred to on page 126. A derivation of the maximum likelihood estimates of σ_1, σ_2, and ρ is given in

HOEL, P., *Introduction to Mathematical Statistics*, 3rd ed. New York: John Wiley & Sons, Inc., 1962.

A more advanced treatment of regression analysis may be found in

WALPOLE, R. E., and MYERS, R. H., *Probability and Statistics for Engineers and Scientists*, 2nd ed. New York: Macmillan Publishing Co., Inc., 1978.

analysis of variance

15

15.1 INTRODUCTION

In this chapter we shall generalize the work of Section 13.3 and consider the problem of deciding whether observed differences among more than two sample means can be attributed to chance, or whether there are real differences among the means of the populations sampled. For instance, we may want to decide on the basis of sample data whether there really is a difference in the effectiveness of three methods of teaching a foreign language, we may want to compare the average yields per acre of six varieties of wheat, or we may want to see whether there really is a difference in the average mileage obtained with four kinds of gasoline.

Since observed differences can always be due to causes other than those postulated—for instance, differences in the performance of students taught a foreign language by three different methods may be due to differences in intelligence, and differences in the average mileage obtained with four kinds of gasoline may be due to differences in road conditions—we shall also discuss some questions of **experimental design**, so that, with reasonable assurance, statistically significant results can be attributed to particular causes.

15.2 ONE-WAY ANALYSIS OF VARIANCE

To give an example of a typical situation where we would perform a one-way analysis of variance, suppose that we want to compare the cleansing action of three detergents on the basis of the following whiteness readings made on 15 swatches of white cloth, which were first soiled with India ink and then washed in an agitator-type machine with the respective detergents:

Detergent A: 77, 81, 71, 76, 80
Detergent B: 72, 58, 74, 66, 70
Detergent C: 76, 85, 82, 80, 77

The means of these three samples are 77, 68, and 80, and we want to know whether the differences among them are significant or whether they can be attributed to chance.

In general, in a problem like this, we have k independent random samples of size n from k populations. The jth value from the ith population is denoted x_{ij}, that is,

$$\begin{array}{ll} \textit{Population 1}: & x_{11}, x_{12}, \ldots, x_{1n} \\ \textit{Population 2}: & x_{21}, x_{22}, \ldots, x_{2n} \\ \cdots & \cdots\cdots \\ \textit{Population k}: & x_{k1}, x_{k2}, \ldots, x_{kn} \end{array}$$

and we shall assume that the corresponding random variables $\mathbf{x}_{ij}$, which are all independent, have normal distributions with the respective means μ_i and the common variance σ^2. Stating these assumptions somewhat differently, we could say that the model for the observations is given by

$$x_{ij} = \mu_i + e_{ij}$$

for $i = 1, 2, \ldots, k$, and $j = 1, 2, \ldots, n$, where the e_{ij} are values of nk independent random variables having normal distributions with zero means and the common variance σ^2. To permit the generalization of this model to more complicated kinds of situations (see page 464), it is usually written in the form

$$x_{ij} = \mu + \alpha_i + e_{ij}$$

for $i = 1, 2, \ldots, k$, and $j = 1, 2, \ldots, n$. Here μ is referred to as the **grand mean,** and the α_i, called the **treatment effects**, are such that $\sum_{i=1}^{k} \alpha_i = 0$. Note that we have merely written the mean of the ith population as $\mu_i = \mu + \alpha_i$ and imposed the condition $\sum_{i=1}^{k} \alpha_i = 0$ so that the mean of the μ_i equals the grand mean μ. The practice of referring to the different populations as different **treatments** is due to the fact that many analysis-of-variance techniques were originally developed in connection with agricultural experiments where different fertilizers, for example, were regarded as different treatments applied to the soil. Thus, we shall refer to the three detergents of our example on page 453 as three different treatments, and in other problems we may refer to four nationalities as four different treatments, five kinds of advertising campaigns as five different treatments, and so on.

The null hypothesis we shall want to test is that the population means are all equal, namely, that $\mu_1 = \mu_2 = \cdots = \mu_k$ or equivalently that

$$H_0\colon \quad \alpha_i = 0 \qquad \text{for } i = 1, 2, \ldots, k$$

Correspondingly, the alternative hypothesis is that the population means are not all equal, namely, that

$$H_1\colon \quad \alpha_i \neq 0 \qquad \text{for at least one value of } i$$

The test, itself, is based on an analysis of the total variability of the combined data ($nk - 1$ times their variance), which is given by

$$\sum_{i=1}^{k} \sum_{j=1}^{n} (x_{ij} - \bar{x}_{..})^2 \qquad \text{where} \qquad \bar{x}_{..} = \frac{1}{nk} \cdot \sum_{i=1}^{k} \sum_{j=1}^{n} x_{ij}$$

If the null hypothesis is true, all this variability is due to chance, but if it is not true, then part of the above sum of squares is due to the differences among the population means. To isolate, or separate, these two contributions to the total variability of the data, we refer to the following theorem:

THEOREM 15.1

$$\sum_{i=1}^{k} \sum_{j=1}^{n} (x_{ij} - \bar{x}_{..})^2 = n \cdot \sum_{i=1}^{k} (\bar{x}_{i.} - \bar{x}_{..})^2 + \sum_{i=1}^{k} \sum_{j=1}^{n} (x_{ij} - \bar{x}_{i.})^2$$

where $\bar{x}_{i.}$ is the mean of the observations from the ith population and $\bar{x}_{..}$ is the mean of all nk observations.

Proof.

$$\sum_{i=1}^{k} \sum_{j=1}^{n} (x_{ij} - \bar{x}_{..})^2 = \sum_{i=1}^{k} \sum_{j=1}^{n} [(\bar{x}_{i.} - \bar{x}_{..}) + (x_{ij} - \bar{x}_{i.})]^2$$

$$= \sum_{i=1}^{k} \sum_{j=1}^{n} [(\bar{x}_{i.} - \bar{x}_{..})^2 + 2(\bar{x}_{i.} - \bar{x}_{..})(x_{ij} - \bar{x}_{i.}) + (x_{ij} - \bar{x}_{i.})^2]$$

$$= \sum_{i=1}^{k} \sum_{j=1}^{n} (\bar{x}_{i.} - \bar{x}_{..})^2 + 2\sum_{i=1}^{k} \sum_{j=1}^{n} (\bar{x}_{i.} - \bar{x}_{..})(x_{ij} - \bar{x}_{i.}) + \sum_{i=1}^{k} \sum_{j=1}^{n} (x_{ij} - \bar{x}_{i.})^2$$

$$= n \cdot \sum_{i=1}^{k} (\bar{x}_{i.} - \bar{x}_{..})^2 + \sum_{i=1}^{k} \sum_{j=1}^{n} (x_{ij} - \bar{x}_{i.})^2$$

since $\sum_{j=1}^{n} (x_{ij} - \bar{x}_{i.}) = 0$ for each value of i.

It is customary to refer to the expression on the left-hand side of the identity of Theorem 15.1 as the **total sum of squares**, to the first term of the expression on the right-hand side as the **treatment sum of squares**, and to the second term as the **error sum of squares**, where "error" denotes the **experimental error**, or chance. Correspondingly, we denote these three sums of squares by SST, SS(Tr), and SSE, and we can write

$$\text{SST} = \text{SS(Tr)} + \text{SSE}$$

Now we have accomplished what we set out to do: We have partitioned SST, a measure of the total variation of the combined data into two components—*the second component, SSE, measures chance variation (namely, the variation within the samples); the first component, SS(Tr), also measures chance variation when the null hypothesis is true, but it also reflects the variation among the population means when the null hypothesis is false.*

Since, for each value of i, the x_{ij} are values of a random sample of size n from a normal population with the variance σ^2, it follows from Theorem 8.9 that for each value of i

$$\frac{1}{\sigma^2} \cdot \sum_{j=1}^{n} (\mathbf{x}_{ij} - \bar{\mathbf{x}}_{i.})^2$$

is a random variable having the chi-square distribution with $n - 1$ degrees of freedom. Furthermore, since the k random samples are independent, it follows from Theorem 8.7 that

$$\frac{1}{\sigma^2} \cdot \sum_{i=1}^{k} \sum_{j=1}^{n} (\mathbf{x}_{ij} - \bar{\mathbf{x}}_{i.})^2$$

is a random variable having the chi-square distribution with $k(n - 1)$ degrees of freedom. Since the mean of a chi-square distribution equals its degrees of freedom, we find that $\frac{1}{\sigma^2} \cdot \text{SSE}$ is a value of a random variable having the mean $k(n-1)$, and, hence, that $\frac{\text{SSE}}{k(n-1)}$ can serve as an estimate of σ^2.

Also, since under the null hypothesis the $\bar{x}_{i.}$ are values of independent random variables having identical normal distributions with the mean μ and the variance $\frac{\sigma^2}{n}$, it follows from Theorem 8.9 that

$$\frac{n}{\sigma^2} \cdot \sum_{i=1}^{k} (\bar{\mathbf{x}}_{i.} - \bar{\mathbf{x}}_{..})^2$$

is a random variable having the chi-square distribution with $k-1$ degrees of freedom. Since the mean of this distribution is $k-1$, it follows that $\frac{\text{SS(Tr)}}{k-1}$ provides a second estimate of σ^2.

Of course, if the null hypothesis is false, then, according to Exercise 1 on page 459, $\frac{\text{SS(Tr)}}{k-1}$ provides an estimate of σ^2 plus whatever variation there may be among the population means. This suggests that we reject the null hypothesis that the population means are all equal when $\frac{\text{SS(Tr)}}{k-1}$ is appreciably greater than $\frac{\text{SSE}}{k(n-1)}$. To put this decision on a precise basis, we shall have to assume without proof that the corresponding estimators are independent, for with this assumption we can utilize Theorem 8.12, according to which

$$F = \frac{\dfrac{\text{SS(Tr)}}{(k-1)\sigma^2}}{\dfrac{\text{SSE}}{k(n-1)\sigma^2}} = \frac{k(n-1)\text{SS(Tr)}}{(k-1)\text{SSE}}$$

is a value of a random variable having the F distribution with $k-1$ and $k(n-1)$ degrees of freedom.[†] Thus, we reject the null hypothesis that the population means are all equal if the value we obtain for

$$F = \frac{k(n-1)\text{SS(Tr)}}{(k-1)\text{SSE}}$$

exceeds $F_{\alpha,k-1,k(n-1)}$, where α is the level of significance.

[†] A proof of this independence may be found in the book by H. Scheffé listed on page 473.

The procedure we have described in this section is called a **one-way analysis of variance**, and the necessary details are usually presented in the following kind of **analysis-of-variance table**:

Source of Variation	*Degrees of Freedom*	*Sum of Squares*	*Mean Square*	*F*
Treatments	$k-1$	SS(Tr)	$\text{MS(Tr)} = \dfrac{\text{SS(Tr)}}{k-1}$	$\dfrac{\text{MS(Tr)}}{\text{MSE}}$
Error	$k(n-1)$	SSE	$\text{MSE} = \dfrac{\text{SSE}}{k(n-1)}$	
Total	$kn-1$	SST		

Note that the **mean squares** are simply the sums of squares divided by the corresponding degrees of freedom.

To simplify the calculation of the various sums of squares, we usually use the following computing formulas, which the reader will be asked to derive in Exercise 2 on page 459:

THEOREM 15.2

$$\text{SST} = \sum_{i=1}^{k}\sum_{j=1}^{n} x_{ij}^2 - \frac{1}{kn}\cdot T_{..}^2$$

and

$$\text{SS(Tr)} = \frac{1}{n}\cdot\sum_{i=1}^{k} T_{i.}^2 - \frac{1}{kn}\cdot T_{..}^2$$

where $T_{i.}$ is the total of the values obtained for the ith treatment and $T_{..}$ is the grand total of all nk observations.

Then, the value of SSE can be obtained by subtracting SS(Tr) from SST.

EXAMPLE 15.1

With reference to the illustration on page 453, test at the level of significance $\alpha = 0.01$ whether the differences among the means of the whiteness readings are significant.

Solution

1. H_0: $\alpha_i = 0$ for $i = 1, 2, 3$
H_1: $\alpha_i \neq 0$ for at least one value of i

2. Critical region: $F \geqslant F_{.01,2,12} = 6.93$.

3. Computations: The required sums and sum of squares are $T_{1.} = 385$, $T_{2.} = 340$, $T_{3.} = 400$, $T_{..} = 1{,}125$, and $\sum\sum x^2 = 85{,}041$, and substitution of these values together with $k = 3$ and $n = 5$ into the formulas of Theorem 15.2 yields

$$\begin{aligned}\text{SST} &= 85{,}041 - \tfrac{1}{15}(1{,}125)^2 \\ &= 666\end{aligned}$$

and

$$\begin{aligned}\text{SS(Tr)} &= \tfrac{1}{5}(385^2 + 340^2 + 400^2) - \tfrac{1}{15}(1{,}125)^2 \\ &= 390\end{aligned}$$

Then, by subtraction, SSE $= 666 - 390 = 276$, and the remaining calculations are shown in the following analysis-of-variance table:

Source of Variation	*Degrees of Freedom*	*Sum of Squares*	*Mean Square*	*F*
Treatments	2	390	$\frac{390}{2} = 195$	$\frac{195}{23} = 8.48$
Error	12	276	$\frac{276}{12} = 23$	
Total	14	666		

4. Decision: Since $F = 8.48$ exceeds $F_{.01,2,12} = 6.93$, the null hypothesis must be rejected, and we conclude that the three detergents are not all equally effective.

The parameters of the model on page 453, namely, μ and the α_i, are usually estimated by the method of least squares. That is, their estimates are the values which minimize

$$\sum_{i=1}^{k}\sum_{j=1}^{n}[x_{ij} - (\mu + \alpha_i)]^2$$

subject to the restriction that $\sum_{i=1}^{k} \alpha_i = 0$; as the reader will be asked to verify in Exercise 6 below, these least squares estimates are $\hat{\mu} = \bar{x}_{..}$ and $\hat{\alpha}_i = \bar{x}_{i.} - \bar{x}_{..}$.

THEORETICAL EXERCISES

1. For the one-way analysis of variance with k independent samples of size n, show that

$$E\left[\frac{n \cdot \sum_{i=1}^{k} (\bar{\mathbf{X}}_{i.} - \bar{\mathbf{X}}_{..})^2}{k-1}\right] = \sigma^2 + \frac{n \cdot \sum_{i=1}^{k} \alpha_i^2}{k-1}$$

2. Prove Theorem 15.2.

3. If, in a one-way analysis of variance, the sample sizes are unequal and there are n_i observations for the ith treatment, show that

$$\sum_{i=1}^{k} \sum_{j=1}^{n_i} (x_{ij} - \bar{x}_{..})^2 = \sum_{i=1}^{k} n_i(\bar{x}_{i.} - \bar{x}_{..})^2 + \sum_{i=1}^{k} \sum_{j=1}^{n_i} (x_{ij} - \bar{x}_{i.})^2$$

analogous to the identity of Theorem 15.1. Also show that the degrees of freedom for SST, SS(Tr), and SSE are, respectively, $N - 1$, $k - 1$, and $N - k$, where $N = \sum_{i=1}^{k} n_i$.

4. With reference to Exercise 3 show that the computing formulas for the sums of squares are

$$\text{SST} = \sum_{i=1}^{k} \sum_{j=1}^{n_i} x_{ij}^2 - \frac{1}{N} \cdot T_{..}^2$$

$$\text{SS(Tr)} = \sum_{i=1}^{k} \frac{T_{i.}^2}{n_i} - \frac{1}{N} \cdot T_{..}^2$$

and

$$\text{SSE} = \text{SST} - \text{SS(Tr)}$$

5. Show that for $k = 2$ the F test of a one-way analysis of variance is equivalent to the t test of Section 13.3 with $\delta = 0$.

6. Use Lagrange multipliers to show that the least squares estimates of the parameters of the model on page 453 are $\hat{\mu} = \bar{x}_{..}$ and $\hat{\alpha}_i = \bar{x}_{i.} - \bar{x}_{..}$.

APPLIED EXERCISES

7. To compare the effectiveness of three different types of phosphorescent coatings of airplane instrument dials, eight dials each are coated with the three types. Then the dials are illuminated by an ultraviolet light, and the following are the number of minutes each glowed after the light source was shut off:

Type 1: 52.9, 62.1, 57.4, 50.0, 59.3, 61.2, 60.8, 53.1
Type 2: 58.4, 55.0, 59.8, 62.5, 64.7, 59.9, 54.7, 58.4
Type 3: 71.3, 66.6, 63.4, 64.7, 75.8, 65.6, 72.9, 67.3

Test the null hypothesis that there is no difference in the effectiveness of the three coatings at the level of significance $\alpha = 0.01$.

8. The following are the numbers of mistakes made in five successive weeks by four technicians working for a medical laboratory:

Technician I: 13, 16, 12, 14, 15
Technician II: 14, 16, 11, 19, 15
Technician III: 13, 18, 16, 14, 18
Technician IV: 18, 10, 14, 15, 12

Test at the level of significance $\alpha = 0.05$ whether the differences among the four sample means can be attributed to chance.

9. Three groups of six guinea pigs each were injected, respectively, with 0.5 mg, 1.0 mg, and 1.5 mg of a new tranquilizier, and the following are the numbers of minutes it took them to fall asleep:

0.5 mg: 21, 23, 19, 24, 25, 23
1.0 mg: 19, 21, 20, 18, 22, 20
1.5 mg: 15, 10, 13, 14, 11, 15

Test at the level of significance $\alpha = 0.05$ whether we can reject the null hypothesis that differences in dosage have no effect. Also estimate the parameters μ, α_1, α_2, and α_3 of the model used in the analysis.

10. The following are the numbers of words per minute which a secretary typed on several occasions on four different typewriters:

Typewriter C: 71, 75, 69, 77, 61, 72, 71, 78
Typewriter D: 68, 71, 74, 66, 69, 67, 70, 62
Typewriter E: 75, 70, 81, 73, 78, 72
Typewriter F: 62, 59, 71, 68, 63, 65, 72, 60, 64

Using the computing formulas of Exercise 4 to calculate the sums of squares, test at the level of significance $\alpha = 0.05$ whether the differences among the four sample means can be attributed to chance.

11. A consumer testing service, wishing to test the accuracy of the thermostats of three different kinds of electric irons, set them at 480°F and obtained the following actual temperature readings by means of a thermocouple:

Iron X:	474, 496, 467, 471
Iron Y:	492, 498
Iron Z:	460, 495, 490

Using the computing formulas of Exercise 4 to calculate the sums of squares, test at the level of significance $\alpha = 0.05$ whether the differences among the three sample means can be attributed to chance.

15.3 EXPERIMENTAL DESIGN

In Example 15.1 it may have seemed reasonable to conclude that the three detergents are not equally effective; yet, a moment's reflection will show that this conclusion is not so "reasonable" at all. For all we know, the swatches cleaned with detergent B may have been more soiled than the others, the washing times may have been longer for detergent C, there may have been differences in water hardness or water temperature, and even the instruments used to make the whiteness readings may have gone out of adjustment after the readings for detergents A and C were made.

It is entirely possible, of course, that the differences among the three sample means are due largely to differences in the effectiveness of the detergents, but we have just listed several other factors which could be held responsible. It is important to remember that *a significance test may show that differences among sample means are too large to be attributed to chance, but it cannot say why the differences occurred.*

In general, if we want to show that one factor (among various others) can be considered the cause of an observed phenomenon, we must somehow make sure that none of the other factors can reasonably be held responsible. There are various ways in which this can be done; for instances, we can conduct a rigorously **controlled experiment** in which all variables except the one of concern are held fixed. To do this in the example dealing with the three detergents, we might soil the swatches with exactly equal amounts of india ink, always use the same washing time, water of exactly the same hardness and temperature, and inspect (and, if necessary, adjust) the measuring instruments after each use. Under such rigidly controlled conditions, significant differences among the sample means cannot be due to differently soiled swatches, or differences in washing time, water temperature, water hardness, or measuring instruments. On the positive side, the differences among the means show that the detergents are not all equally effective *if they are used in this narrowly restricted way.* Of course, we cannot say whether the same differences would exist if the washing time were longer or shorter, if the water had a different temperature or hardness, and so on.

In most cases, "overcontrolled" experiments like the one just described do not really provide us with the kind of information we want. So, we look for alternatives, and at the other extreme we can conduct experiments in which none of the extraneous factors is controlled, but in which we protect ourselves against their effects by **randomization**. That is, we design, or plan, the experiments in such a way that the variations caused by extraneous factors can all be combined under the general heading of "chance." For instance, in our example we could accomplish this by randomly assigning five of the soiled swatches to each detergent, and randomly specifying the order in which they are to be washed and measured. When all the variations due to uncontrolled extraneous factors can thus be included under the heading of chance variation, we refer to the design of the experiment as a **completely randomized design**.

As should be apparent, randomization protects against the effects of the extraneous factors only in a probabilistic sort of way. For instance, in our example it is possible, though very unlikely, that detergent A will be randomly assigned to the five swatches which happen to be the least soiled, or that the water happens to be coldest when we wash the five swatches with detergent B. It is partly for this reason that we often try to control some of the factors and randomize the others, and thus use designs that are somewhere between the two extremes which we have described.

To introduce another important concept in the design of experiments, let us consider the following data on the amount of time (in minutes) it took a certain person to drive to work, Monday through Friday, along four different routes:

Route 1: 22, 26, 25, 25, 31
Route 2: 25, 27, 28, 26, 29
Route 3: 26, 29, 33, 30, 33
Route 4: 26, 28, 27, 30, 30

The means of these four samples are 25.8, 27.0, 30.2, and 28.2, and since the differences among them are fairly large, it would seem reasonable to conclude that there are some real differences in the true average time it takes the person to drive to work along the four different routes. This does not follow, however, from a one-way analysis of variance. We get $F = 2.80$, and since this does not exceed $F_{.05,3,16} = 3.24$, the null hypothesis cannot be rejected.

Of course, the null hypothesis may be true, but observe that there are not only considerable differences among the four means, but also large differences among the values within the samples. In the first sample they range from 22 to 31, in the second sample from 25 to 29, in the third sample from 26 to 33, and in the fourth sample from 26 to 30. Not only that, but in each sample the first value is the smallest and the last value is the largest. The latter suggests that the variation within the samples may well be due to differences in driving conditions on the different days of the week. If this is the case, variations due to driving conditions were included in the error sum of squares of the one-way analysis of variance, the denominator of the F statistic was "inflated," and this may be why the results were not significant.

To avoid this kind of situation, we could hold the extraneous factor fixed, but this will seldom give us the information we want. In our example, we could limit the study to driving conditions on Monday, but then we would have no assurance that the results would apply also to driving conditions on Tuesday or on any other day of the week. Another possibility is to vary the extraneous factor deliberately over as wide a range as necessary, so that the variation it causes can be measured and, hence, eliminated from the error sum of squares. This means that we must plan the experiment in such a way that we can perform a **two-way analysis of variance**, in which the total variation of the data is partitioned into three components attributed, respectively, to treatments (in our example, the four routes), the extraneous factor (in our example, driving conditions on the different days of the week), and experimental error, or chance.

What we have suggested here is called **blocking** and the different days of the week are referred to as **blocks**. In general, blocks are the levels at which we hold an extraneous factor fixed, so that we can measure its contribution to the total variation of the data. If each treatment appears the same number of times in each block (in our example, each route is used once each day of the week), we say that the design of the experiment is a **complete block design**. Furthermore, if the treatments are distributed at random within each block (in our example, we would randomly distribute the four routes among the four Mondays, the four Tuesdays, etc.), we say that the design of the experiment is a **randomized block design**.

15.4 TWO-WAY ANALYSIS OF VARIANCE

There are essentially two different ways of analyzing two-variable experiments, and they depend on whether the two variables are independent or whether they **interact**. To illustrate what we mean here by "interact," suppose that a tire manufacturer is experimenting with different treads, and that he finds that one kind is especially good for use on dirt roads while another kind is especially good for use on hard pavement. If this is the case, we say that there is an **interaction** between road conditions and tread design. In this book we shall study only the no-interaction case.

To present the theory of a two-way analysis of variance, we shall use the terminology introduced in the preceding section and refer to the two variables as treatments and blocks; alternatively, we could refer to them also as **factor A** and **factor B**, or as **rows** and **columns**. Thus, if x_{ij} for $i = 1, 2, \ldots, k$ and $j = 1, 2, \ldots,$

n are values of independent random variables having normal distributions with the respective means μ_{ij} and the common variance σ^2, we shall consider the array

	Block 1	*Block 2*	$\cdots$	*Block n*
Treatment 1	x_{11}	x_{12}	$\cdots$	x_{1n}
Treatment 2	x_{21}	x_{22}	$\cdots$	x_{2n}
$\cdots$	$\cdots$	$\cdots$	$\cdots$	$\cdots$
Treatment k	x_{k1}	x_{k2}	$\cdots$	x_{kn}

and write the model for a two-way analysis of variance (without interaction) as

$$x_{ij} = \mu + \alpha_i + \beta_j + e_{ij}$$

for $i = 1, 2, \ldots, k$ and $j = 1, 2, \ldots, n$. Here μ is the **grand mean,** the **treatment effects** α_i are such that $\sum_{i=1}^{k} \alpha_i = 0$, the **block effects** β_j are such that $\sum_{j=1}^{n} \beta_j = 0$, and the e_{ij} are values of independent random variables having normal distributions with zero means and the common variance σ^2. Note that $\mu_{ij} = \mu + \alpha_i + \beta_j$, and, as the reader will be asked to verify in Exercise 2 on page 468,

$$\frac{\sum_{i=1}^{k} \sum_{j=1}^{n} \mu_{ij}}{nk} = \mu$$

The two null hypotheses we shall want to test are that the treatment effects are all equal to zero and that the block effects are all equal to zero, namely,

$$H_0\colon \quad \alpha_i = 0 \qquad \text{for } i = 1, 2, \ldots, k$$

and

$$H'_0\colon \quad \beta_j = 0 \qquad \text{for } j = 1, 2, \ldots, n$$

The alternative to H_0 is that the treatment effects are not all equal to zero, and the alternative to H'_0 is that the block effects are not all equal to zero. Symbolically,

$$H_1\colon \quad \alpha_i \neq 0 \qquad \text{for at least one value of } i$$

and

$$H_1': \quad \beta_j \neq 0 \qquad \text{for at least one value of } j$$

The two-way analysis, itself, is based on the following generalization of Theorem 15.1, which the reader will be asked to prove in Exercise 1 on page 468:

THEOREM 15.3

$$\sum_{i=1}^{k}\sum_{j=1}^{n}(x_{ij}-\bar{x}_{..})^2 = n\cdot\sum_{i=1}^{k}(\bar{x}_{i.}-\bar{x}_{..})^2 + k\cdot\sum_{j=1}^{n}(\bar{x}_{.j}-\bar{x}_{..})^2 + \sum_{i=1}^{k}\sum_{j=1}^{n}(x_{ij}-\bar{x}_{i.}-\bar{x}_{.j}+\bar{x}_{..})^2$$

where $\bar{x}_{i.}$ is the mean of the observations for the ith treatment, $\bar{x}_{.j}$ is the mean of the observations for the jth block, and $\bar{x}_{..}$ is the mean of all nk observations.

The expression on the left-hand side of the identity of Theorem 15.3 is the total sum of squares SST as defined on page 455, and the first term on the right-hand side is the treatment sum of squares SS(Tr). Measuring the variation among the $\bar{x}_{.j}$, the second term on the right-hand side is the **block sum of squares** SSB, and the third term on the right-hand side is the *new* error sum of squares SSE. Thus, we have

$$\text{SST} = \text{SS(Tr)} + \text{SSB} + \text{SSE}$$

and it can be shown that if H_0 is true, then $\frac{\text{SS(Tr)}}{\sigma^2}$ and $\frac{\text{SSE}}{\sigma^2}$ are values of independent random variables having chi-square distributions with $k-1$ and $(n-1)(k-1)$ degrees of freedom. If H_0 is not true, then SS(Tr) will also reflect the variation among the α_i, and according to Theorem 8.12 we reject H_0 if $F_{\text{Tr}} \geqslant F_{\alpha,k-1,(n-1)(k-1)}$, where

$$F_{\text{Tr}} = \frac{\dfrac{\text{SS(Tr)}}{k-1}}{\dfrac{\text{SSE}}{(n-1)(k-1)}} = \frac{(n-1)\cdot\text{SS(Tr)}}{\text{SSE}}$$

Similarly, if H_0' is true, then $\frac{\text{SSB}}{\sigma^2}$ and $\frac{\text{SSE}}{\sigma^2}$ are values of independent random variables having chi-square distributions with $n-1$ and $(n-1)(k-1)$ degrees of freedom. If H_0' is not true, then SSB will also reflect the variation among the β_j, and according to Theorem 8.12 we reject H_0' if $F_B \geqslant F_{\alpha,n-1,(n-1)(k-1)}$, where

$$F_B = \frac{\frac{\text{SSB}}{n-1}}{\frac{\text{SSE}}{(n-1)(k-1)}} = \frac{(k-1)\text{SSB}}{\text{SSE}}$$

This kind of analysis is called a **two-way analysis of variance**, and the necessary details are usually presented in the following kind of analysis-of-variance table:

Source of Variation	*Degrees of Freedom*	*Sum of Squares*	*Mean Square*	*F*
Treatments	$k-1$	SS(Tr)	$\text{MS(Tr)} = \frac{\text{SS(Tr)}}{k-1}$	$F_{\text{Tr}} = \frac{\text{MS(Tr)}}{\text{MSE}}$
Blocks	$n-1$	SSB	$\text{MSB} = \frac{\text{SSB}}{n-1}$	$F_B = \frac{\text{MSB}}{\text{MSE}}$
Error	$(n-1)(k-1)$	SSE	$\text{MSE} = \frac{\text{SSE}}{(n-1)(k-1)}$	
Total	$nk-1$	SST		

Again, the mean squares are simply the sums of squares divided by the corresponding degrees of freedom.

To simplify the calculations, SST and SS(Tr) are usually determined by means of the formulas of Theorem 15.2, and SSB can be determined by means of the following formula, which the reader will be asked to derive in Exercise 4 on page 469:

THEOREM 15.4

$$\text{SSB} = \frac{1}{k} \cdot \sum_{j=1}^{n} T_{.j}^2 - \frac{1}{kn} \cdot T_{..}^2$$

where $T_{.j}$ is the total of the values obtained for the jth block and $T_{..}$ is the grand total of all nk observations.

Then, the value of SSE can be obtained by subtracting SS(Tr) and SSB from SSE.

EXAMPLE 15.2

With reference to the illustration on page 462, where we had

	Monday	*Tuesday*	*Wednesday*	*Thursday*	*Friday*
Route 1	22	26	25	25	31
Route 2	25	27	28	26	29
Route 3	26	29	33	30	33
Route 4	26	28	27	30	30

test at the level of significance 0.05 whether the differences among the means obtained for the different routes (treatments) are significant, and also whether the differences among the means obtained for the different days of the week (blocks) are significant.

Solution

1. H_0: $\alpha_i = 0$ for $i = 1, 2, 3, 4$
   H'_0: $\beta_j = 0$ for $j = 1, 2, 3, 4, 5$
   H_1: $\alpha_i \neq 0$ for at least one value of i
   H'_1: $\beta_j \neq 0$ for at least one value of j
2. Critical regions: $F \geqslant F_{.05,3,12} = 3.49$ for treatments, and $F \geqslant F_{.05,4,12} = 3.26$ for blocks.
3. Computations: The required sums and sum of squares are $T_{1.} = 129$, $T_{2.} = 135$, $T_{3.} = 151$, $T_{4.} = 141$, $T_{.1} = 99$, $T_{.2} = 110$, $T_{.3} = 113$, $T_{.4} = 111$, $T_{.5} = 123$, $T_{..} = 556$, and $\sum \sum x^2 = 15{,}610$, and substitution of these values together with $k = 4$ and $n = 5$ into the formulas of Theorems 15.2 and 15.4 yields

$$\begin{aligned} \text{SST} &= 15{,}610 - \tfrac{1}{20}(556)^2 \\ &= 153.2 \\ \text{SS(Tr)} &= \tfrac{1}{5}(129^2 + 135^2 + 151^2 + 141^2) - \tfrac{1}{20}(556)^2 \\ &= 52.8 \\ \text{SSB} &= \tfrac{1}{4}(99^2 + 110^2 + 113^2 + 111^2 + 123^2) - \tfrac{1}{20}(556)^2 \\ &= 73.2 \end{aligned}$$

and, hence,

$$\begin{aligned} \text{SSE} &= 153.2 - 52.8 - 73.2 \\ &= 27.2 \end{aligned}$$

The remaining calculations are shown in the following analysis-of-variance table:

Source of Variation	*Degrees of Freedom*	*Sum of Squares*	*Mean Square*	*F*
Treatments	3	52.8	$\frac{52.8}{3} = 17.6$	$\frac{17.6}{2.27} = 7.75$
Blocks	4	73.2	$\frac{73.2}{4} = 18.3$	$\frac{18.3}{2.27} = 8.06$
Error	12	27.2	$\frac{27.2}{12} = 2.27$	
Total	19	153.2		

4. Decisions: Since $F_{\text{Tr}} = 7.75$ exceeds $F_{.05,3,12} = 3.49$ and $F_B = 8.06$ exceeds $F_{.05,4,12} = 3.26$, we find that both null hypotheses must be rejected. In other words, the differences among the means obtained for the four routes are significant, and so are the differences among the means obtained for the different days of the week. Note, however, that we cannot conclude that Route 1 is necessarily fastest and that on Fridays traffic conditions are always the worst. All we have shown by means of the analysis is that differences exist, and if we want to go one step further and pinpoint the nature of the differences, we will have to use one of the so-called **multiple comparisons tests** referred to on page 473.

THEORETICAL EXERCISES

1. Make use of the identity

$$x_{ij} - \bar{x}_{..} = (\bar{x}_{i.} - \bar{x}_{..}) + (\bar{x}_{.j} - \bar{x}_{..}) + (x_{ij} - \bar{x}_{i.} - \bar{x}_{.j} + \bar{x}_{..})$$

to prove Theorem 15.3.

2. With reference to the notation on page 464 show that

$$\frac{\sum_{i=1}^{k} \sum_{j=1}^{n} \mu_{ij}}{nk} = \mu$$

3. For the two-way analysis of variance with k treatments and n blocks, show that

$$E\left[\frac{k \cdot \sum_{j=1}^{n} (\bar{\mathbf{x}}_{.j} - \bar{\mathbf{x}}_{..})^2}{n-1}\right] = \sigma^2 + \frac{k \cdot \sum_{j=1}^{n} \beta_j^2}{n-1}$$

4. Prove Theorem 15.4.

5. A **Latin square** is a square array in which each letter (or some other kind of symbol) appears exactly once in each row and once in each column. For instance,

A	*B*	*C*	*D*
B	*C*	*D*	*A*
C	*D*	*A*	*B*
D	*A*	*B*	*C*

is a 4×4 Latin square. If we look upon the m rows of a Latin square as the levels of one variable, the m columns as the levels of a second variable, and A, B, C, . . . , as m "treatments," namely, as the levels of a third variable, it is possible to test hypotheses concerning all three of these variables on the basis of as few as m^2 observations (provided there are no interactions). Letting $x_{ij(k)}$ denote the observation in the ith row and the jth column of a Latin square (so that k is determined when we give i and j), we write the model equation as

$$x_{ij(k)} = \mu + \alpha_i + \beta_j + \tau_k + e_{ij}$$

for $i = 1, 2, \ldots, m$, $j = 1, 2, \ldots, m$, and $k = 1, 2, \ldots, m$, where μ is the grand mean, the **row effects** α_i are such that $\sum_{i=1}^{m} \alpha_i = 0$, the **column effects** β_j are such that $\sum_{j=1}^{m} \beta_j = 0$, the treatment effects τ_k are such that $\sum_{k=1}^{m} \tau_k = 0$, and the e_{ij} are values of independent random variables having normal distributions with zero means and the common variance σ^2. The null hypotheses we shall want to test (against appropriate alternatives) are that the row effects are all zero, that the column effects are all zero, and that the treatment effects are all zero.

(a) Show that

$$\sum_{i=1}^{m}\sum_{j=1}^{m}(x_{ij(k)}-\bar{x}_{..})^2 =$$

$$m\cdot\sum_{i=1}^{m}(\bar{x}_{i.}-\bar{x}_{..})^2 + m\cdot\sum_{j=1}^{m}(\bar{x}_{.j}-\bar{x}_{..})^2 + m\cdot\sum_{k=1}^{m}(\bar{x}_{(k)}-\bar{x}_{..})^2$$

$$+\sum_{i=1}^{m}\sum_{j=1}^{m}(x_{ij(k)}-\bar{x}_{i.}-\bar{x}_{.j}-\bar{x}_{(k)}+2\bar{x}_{..})^2$$

where $\bar{x}_{(k)}$ is the mean of all the observations for the kth treatment and the other means are as defined in Theorem 15.3. The expression on the left-hand side of the above identity is the total sum of squares SST, while those on the right-hand side are, respectively, the **row sum of squares** SSR, the **column sum of squares** SSC, the treatment sum of squares SS(Tr), and the error sum of squares SSE.

(b) Construct an analysis-of-variance table for this kind of experiment, determining the degrees of freedom for SSE by subtracting those for SSR, SSC, and SS(Tr) from $m^2 - 1$, the degrees of freedom for SST.

APPLIED EXERCISES

6. An experiment was performed to judge the effect of four different fuels and three different types of launchers on the range of a certain rocket. Test, on the basis of the following ranges, in miles, whether there is a significant effect due to differences in fuels and whether there is a significant effect due to differences in launchers (use $\alpha = 0.01$):

	Fuel 1	*Fuel 2*	*Fuel 3*	*Fuel 4*
Launcher X	45.9	57.6	52.2	41.7
Launcher Y	46.0	51.0	50.1	38.8
Launcher Z	45.7	56.9	55.3	48.1

7. The following are the cholesterol contents, in milligrams per package, which four laboratories obtained for 6-ounce packages of three very similar diet foods:

	Diet food A	*Diet food B*	*Diet food C*
Laboratory 1	3.4	2.6	2.8
Laboratory 2	3.0	2.7	3.1
Laboratory 3	3.3	3.0	3.4
Laboratory 4	3.5	3.1	3.7

Perform a two-way analysis of variance, testing the null hypotheses concerning the laboratories and the diet foods at the level of significance $\alpha = 0.05$.

8. A laboratory technician measures the breaking strength of each of five kinds of linen threads by using four different measuring instruments, I_1, I_2, I_3, and I_4, and obtains the following results, in ounces:

	I_1	I_2	I_3	I_4
Thread 1	20.9	20.4	19.9	21.9
Thread 2	25.0	26.2	27.0	24.8
Thread 3	25.5	23.1	21.5	24.4
Thread 4	24.8	21.2	23.5	25.7
Thread 5	19.6	21.2	22.1	22.1

Perform a two-way analysis of variance, using $\alpha = 0.05$ for both tests of significance.

9. The sample data in the following Latin square (see Exercise 5) are the grades in an American history test obtained by nine college students of various ethnic backgrounds and of various professional interests, who were taught by instructors A, B, and C:

Ethnic background

	Mexican	*German*	*Polish*
Law	A 75	B 86	C 69
Medicine	B 95	C 79	A 86
Engineering	C 70	A 83	B 93

Use the method of Exercise 5 and the level of significance $\alpha = 0.05$ to test the null hypotheses that

(a) having a different instructor has no effect on the grades;
(b) differences in ethnic background have no effect on the grades;
(c) differences in professional interest have no effect on the grades.

10. Among the nine persons interviewed in a poll, three are Easterners, three are Southerners, and three are Westerners. By profession, three of them are teachers, three are lawyers, and three are doctors, and no two of the same profession come from the same part of the United States. Also, three are

Democrats, three are Republicans, and three are Independents, and no two of the same political affiliation are of the same profession or come from the same part of the United States. If one of the teachers is an Easterner and an Independent, another teacher is a Southerner and a Republican, and one of the lawyers is a Southerner and a Democrat, what is the political affiliation of the doctor who is a Westerner? [*Hint:* Construct a Latin square (see Exercise 5) with $m = 3$.] This exercise is a simplified version of a famous problem posed by R. A. Fisher in his classical work, *The Design of Experiments.*

15.5 SOME FURTHER CONSIDERATIONS

In this chapter we have presented a brief introduction to some of the basic methods and ideas of analysis of variance and experimental design. The scope of these subjects, which are closely interrelated, is vast, and new methods are constantly being developed as their need arises in experimentation.

The designs which we have discussed all had the special feature that there were observations corresponding to all possible combinations of the values (levels) of the variables under consideration. To show that this can be very impractical or even physically impossible, we have only to consider an experiment in which we want to compare the yield of 25 varieties of wheat and, at the same time, the effect of 12 different fertilizers. To perform an experiment in which each of the 25 varieties of wheat is used in conjunction with each of the 12 fertilizers, we would have to plant 300 plots, and it does not require much imagination to see how difficult it would be to find that many test plots for which soil composition, irrigation, slope, . . . , are constant or otherwise controllable. Consequently, there is a need for designs which make it possible to test hypotheses concerning the most relevant (though not all) parameters of the model on the basis of experiments which are feasible from a practical point of view. This leads to so-called **incomplete block designs**, which are discussed in the general references on experimental design listed at the end of the chapter.

Further complications arise when there are extraneous variables which can be measured but not controlled. For example, in a comparison of various kinds of "teaching machines" it may be impossible to use persons who all have the same I.Q., but at least their I.Q.'s can be determined. In a situation like that we might use an **analysis-of-covariance** model such as

$$x_{ij} = \mu + \alpha_i + \beta y_{ij} + e_{ij}$$

which differs from the one-way analysis of variance model in that we added the term βy_{ij}, where the y_{ij} are the I.Q.'s which can be determined. Note that in this model the estimation of β is essentially a problem of regression.

Other difficulties arise when the parameters α_i and β_j in an analysis-of-variance model are not constants, but values of random variables. This kind of situation would arise, for example, if there are 25 varieties of wheat and 12 kinds of fertilizers and we randomly select, say, six of the varieties of wheat and three of the fertilizers to be included in an experiment.

These are just some of the generalizations of the methods we have presented in this chapter; they are treated in detail in the general texts on analysis of variance and experimental design which are listed below.

References

A proof of the independence of the chi-square random variables whose values constitute the various sums of squares in an analysis of variance, for instance SS(Tr) and SSE in a one-way analysis, may be found in

SCHEFFÉ, H., *The Analysis of Variance.* New York: John Wiley & Sons, Inc., 1959,

and a discussion of various multiple comparisons tests is given in

FEDERER, W. T., *Experimental Design, Theory and Application.* New York: Macmillan Publishing Co., Inc., 1955.

The following are some general texts on analysis of variance and experimental design:

ANDERSON, V. L., and MCLEAN, R. A., *Design of Experiments: A Realistic Approach.* New York: Marcel Dekker, Inc., 1974,

COCHRAN, W. G., and COX, G. M., *Experimental Design,* 2nd ed. New York: John Wiley & Sons, Inc., 1957,

FINNEY, D. J., *An Introduction to the Theory of Experimental Design.* Chicago: University of Chicago Press, 1960,

GUENTHER, W. C., *Analysis of Variance.* Englewood Cliffs, N.J.: Prentice-Hall, Inc., 1964,

HICKS, C. R., *Fundamental Concepts in the Design of Experiments.* New York: Holt, Rinehart and Winston, Inc., 1964,

LI, C. C., *Introduction to Experimental Statistics.* New York: McGraw-Hill Book Company, 1964,

SNEDECOR, G. W., and COCHRAN, W. G., *Statistical Methods*, 6th ed. Ames, Iowa: Iowa University Press, 1973.

nonparametric methods

16

16.1 INTRODUCTION

Most of the tests discussed in the three preceding chapters required specific assumptions about the population, or populations, sampled. In most cases we assumed that the populations sampled are normal; sometimes we assumed that their standard deviations are known or are known to be equal; and sometimes we assumed that the samples are independent. Since there are many situations in which the required assumptions cannot be met, statisticians have developed alternative techniques which have become known as **nonparametric methods**. This term is used somewhat loosely to include **distribution-free methods** (like the tolerance limits of Exercise 10 on page 283) where we make no assumptions about the populations, except perhaps that they are continuous. It also includes methods which are nonparametric only in the sense that we are not concerned with the parameters of populations of a given kind.

Aside from the fact that nonparametric methods can be used under more general conditions than the standard techniques which they replace, they have great intuitive appeal; that is, they are easy to explain and easy to understand. Moreover, in many nonparametric methods the computational burden is so light that they come under the heading of "quick and easy" or "short-cut" techniques. For these reasons, nonparametric methods have become quite popular, and extensive literature is devoted to their theory and application.

The main disadvantage of nonparametric methods is that they may be wasteful of information, and thus less efficient than the standard techniques which they replace. It should be observed, however, that such efficiency comparisons usually assume that the conditions underlying the standard methods are met and, hence, they tend to understate the real worth of the nonparametric methods. To put this another way, it is true in general that *the less one assumes, the less one can infer from a set of data*, but it is also true that *the less one assumes, the more one broadens the applicability of one's method.*

16.2 THE SIGN TEST

The standard t test of the null hypothesis $\mu = \mu_0$ is based on the assumption that we are sampling a normal population. When this assumption is untenable, this standard test can be replaced by any one of several nonparametric alternatives, among them the **one-sample sign test**.

The one-sample sign test applies when we sample a continuous symmetrical population, so that the probability of getting a sample value exceeding the mean and the probability of getting a sample value less than the mean are both $\frac{1}{2}$.[†] To test the null hypothesis $\mu = \mu_0$ against an appropriate alternative on the basis of a random sample of size n, we replace each sample value exceeding μ_0 with a plus sign and each sample value less than μ_0 with a minus sign, and then we test the null hypothesis that the number of plus signs is the value of a random variable having a binomial distribution with the parameters n and $\theta = \frac{1}{2}$. The two-sided alternative $\mu \neq \mu_0$ thus becomes $\theta \neq \frac{1}{2}$, and the one-sided alternatives $\mu < \mu_0$ and $\mu > \mu_0$ become $\theta < \frac{1}{2}$ and $\theta > \frac{1}{2}$, respectively. If a sample value actually equals μ_0, which does not have zero probability when we deal with rounded data even though the population is continuous, we simply discard it.

To perform a one-sample sign test when the sample is very small, we refer directly to a table of binomial probabilities such as Table I; when the sample is large, we use the normal approximation to the binomial distribution.

EXAMPLE 16.1

The following are measurements of the breaking strength of a certain kind of 2-inch cotton ribbon in pounds:

163	165	160	189	161	171	158	151	169	162
163	139	172	165	148	166	172	163	187	173

Test the null hypothesis $\mu = 160$ against the alternative $\mu > 160$ at the level of significance $\alpha = 0.05$.

[†] If it cannot be assumed that the population is symmetrical, we use the same technique but apply it to the null hypothesis $\tilde{\mu} = \tilde{\mu}_0$ instead of $\mu = \mu_0$, where $\tilde{\mu}$ is the population median.

Solution

1. H_0: $\mu = 160$
   H_1: $\mu > 160$
2. Critical region: $x \geqslant k_{.05}$, where x is the number of plus signs and $k_{.05}$ is as defined on page 404.
3. Computations: Replacing each value exceeding 160 with a plus sign, each value less than 160 with a minus sign, and discarding the one value which actually equals 160, we get

   \+ \+ \+ \+ \+ − − \+ \+ \+ − \+ \+ − \+ \+ \+ \+ \+

   so that $n = 19$ and $x = 15$.
4. From Table I we find that $k_{.05} = 14$. Since this value is exceeded by $x = 15$, the null hypothesis must be rejected and we conclude that the mean breaking strength of the given kind of ribbon exceeds 160 pounds.

EXAMPLE 16.2

The following data, in tons, are the amounts of sulfur oxides emitted by a large industrial plant on 40 days:

17	15	20	29	19	18	22	25	27	9
24	20	17	6	24	14	15	23	24	26
19	23	28	19	16	22	24	17	20	13
19	10	23	18	31	13	20	17	24	14

Test the null hypothesis $\mu = 21.5$ against the alternative hypothesis $\mu < 21.5$ at the level of significance $\alpha = 0.01$.

Solution

1. H_0: $\mu = 21.5$
   H_1: $\mu < 21.5$
2. Critical region: $z \leqslant -z_{.01} = -2.33$, where $z = \dfrac{(x \pm \frac{1}{2}) - n\theta}{\sqrt{n\theta(1-\theta)}}$, $\theta = \frac{1}{2}$, and x is the number of plus signs (values exceeding 21.5).
3. Computations: $n\theta = 40 \cdot \frac{1}{2} = 20$, $\sqrt{n\theta(1-\theta)} = \sqrt{40(0.5)(0.5)} = 3.16$, and since there are 16 plus signs and 24 minus signs, we get

$$z = \frac{(16 + \frac{1}{2}) - 20}{3.16} = -1.11$$

4. Decision: Since $z = -1.11$ is not less than $-z_{.01} = -2.33$, the null hypothesis cannot be rejected.

The sign test can also be used when we deal with paired data as in Exercises 14 and 15 on pages 398 and 399. In such problems, each pair of sample values is replaced by a plus sign if the difference between the paired observations is positive (that is, if the first value exceeds the second value) and by a minus sign if the difference between the paired observations is negative (that is, if the first value is less than the second value). To test the null hypothesis that the two populations sampled are continuous, symmetrical, and have equal means, we can use the sign test, which, in connection with this kind of problem, is referred to as the **paired-sample sign test**. If the difference between a pair of observations is zero, we discard it.

EXAMPLE 16.3

To determine the effectiveness of a new traffic control system, the number of accidents that occurred at 12 dangerous intersections during four weeks before and four weeks after the installation of the new system was observed, and the following data were obtained:

3 and 1, 5 and 2, 2 and 0, 3 and 2, 3 and 2, 3 and 0

0 and 2, 4 and 3, 1 and 3, 6 and 4, 4 and 1, 1 and 0

Use the paired-sample sign test to test the null hypothesis that the new traffic control system is not effective at $\alpha = 0.05$. (In this case, the populations sampled are, of course, not continuous, but this does not matter since zero differences are discarded.)

Solution

1. H_0: $\mu_1 = \mu_2$
H_1: $\mu_1 > \mu_2$

2. Critical region: $x \geqslant k_{.05}$, where x is the number of plus signs (positive differences) and $k_{.05}$ is as defined on page 404.

3. Computations: Replacing each pair of values by the sign of their difference, we get

$$+ \quad + \quad + \quad + \quad + \quad + \quad - \quad + \quad - \quad + \quad + \quad +$$

so that $n = 12$ and $x = 10$.

4. Decision: From Table I we find that $k_{.05} = 10$. Since $x = 10$ equals $k_{.05} = 10$, the null hypothesis must be rejected and we conclude that the new traffic control system is effective in reducing the number of accidents at dangerous intersections.

16.3 THE SIGNED-RANK TEST

As we saw in the preceding section, the sign test is very easy to perform, but since we utilize only the signs of the differences between the observations and μ_0 in the one-sample case, or the signs of the differences between the pairs of observations in the paired-sample case, it tends to be wasteful of information. An alternative nonparametric test, the **Wilcoxon signed-rank test**, is less wasteful in that it takes into account also the magnitudes of the differences. In this test, we rank the differences without regard to their signs, assigning rank 1 to the smallest difference in absolute value, rank 2 to the second smallest difference in absolute value, . . . , and rank n to the largest difference in absolute value. Zero differences are again discarded, and if the absolute values of two or more differences are the same, we assign each one the mean of the ranks which they jointly occupy. Then, the signed-rank test is based on T^+, the sum of the ranks assigned to the positive differences, T^-, the sum of the ranks of the negative differences, $T^+ - T^-$, or $\min(T^+, T^-)$. Since $T^+ + T^- = \frac{n(n+1)}{2}$, the resulting tests are all equivalent.

For very small values of n, the test of the null hypothesis that we are sampling a continuous symmetrical population with the mean μ_0 in the one-sample case, or the null hypothesis that we are sampling two continuous symmetrical populations with equal means in the paired-sample case, is usually based on special tables (see references on page 499). For $n \geqslant 15$, it is considered reasonable to assume that the distribution of $\mathbf{T}^+$ is approximately normal, and to perform the signed-rank test on the basis of this assumption, we need the following results:

THEOREM 16.1 Under either of the given null hypotheses, the mean and the variance of $\mathbf{T}^+$ are

$$E(\mathbf{T}^+) = \frac{n(n+1)}{4}$$

and

$$\text{var}(\mathbf{T}^+) = \frac{n(n+1)(2n+1)}{24}$$

Proof. Expressed in terms of ranks and signed differences, the null hypotheses for the one-sample and paired-sample signed-rank tests may be stated as follows: For each rank, the probabilities that it will be assigned to a positive difference or to a negative difference are both equal to $\frac{1}{2}$. Thus, we can write

$$\mathbf{T}^+ = 1 \cdot \mathbf{x}_1 + 2 \cdot \mathbf{x}_2 + \cdots + n \cdot \mathbf{x}_n$$

where $\mathbf{x}_1, \mathbf{x}_2, \ldots,$ and $\mathbf{x}_n$ are independent random variables having the Bernoulli distribution with $\theta = \frac{1}{2}$. Since $E(\mathbf{x}_i) = \theta = \frac{1}{2}$ and $\text{var}(\mathbf{x}_i) = \theta(1 - \theta) = \frac{1}{4}$ for $i = 1, 2, \ldots, n$ by Theorem 5.2 with $n = 1$, it follows that

$$E(\mathbf{T}^+) = 1 \cdot \tfrac{1}{2} + 2 \cdot \tfrac{1}{2} + \cdots + n \cdot \tfrac{1}{2}$$
$$= \frac{1 + 2 + \cdots + n}{2}$$
$$= \frac{n(n + 1)}{4}$$

Also, according to the corollary to Theorem 4.14 on page 156, we find that

$$\text{var}(\mathbf{T}^+) = 1^2 \cdot \tfrac{1}{4} + 2^2 \cdot \tfrac{1}{4} + \cdots + n^2 \cdot \tfrac{1}{4}$$
$$= \frac{1^2 + 2^2 + \cdots + n^2}{4}$$
$$= \frac{n(n + 1)(2n + 1)}{24}$$

We made use here of the familiar formulas for the sum and the sum of the squares of the first n positive integers, which are proved in Appendix II.

EXAMPLE 16.4

The following are the weights in pounds, before and after, of 16 persons who stayed on a certain reducing diet for four weeks:

Before	*After*
147.0	137.9
183.5	176.2
232.1	219.0
161.6	163.8
197.5	193.5
206.3	201.4
177.0	180.6
215.4	203.2
147.7	149.0
208.1	195.4
166.8	158.5
131.9	134.4
150.3	149.3
197.2	189.1
159.8	159.1
171.7	173.2

Use the signed-rank test to test, at the level of significance $\alpha = 0.05$, whether the weight-reducing diet is effective.

Solution

1. H_0: $\mu_1 = \mu_2$
   H_1: $\mu_1 > \mu_2$

2. Critical region: $z \geqslant z_{.05} = 1.645$, where $z = \dfrac{T^+ - E(\mathbf{T}^+)}{\sqrt{\mathrm{var}(\mathbf{T}^+)}}$

3. Computations: The differences between the respective pairs are 9.1, 7.3, 13.1, −2.2, 4.0, 4.9, −3.6, 12.2, −1.3, 12.7, 8.3, −2.5, 1.0, 8.1, 0.7, −1.5, and if their absolute values are ranked, we find that the positive differences occupy ranks 13, 10, 16, 8, 9, 14, 15, 12, 2, 11, and 1. Thus,

$$T^+ = 13 + 10 + 16 + 8 + 9 + 14 + 15 + 12 + 2 + 11 + 1$$
$$= 111$$

Since $E(\mathbf{T}^+) = \dfrac{16 \cdot 17}{4} = 68$ and $\mathrm{var}(\mathbf{T}^+) = \dfrac{16 \cdot 17 \cdot 33}{24} = 374$, we get

$$z = \frac{111 - 68}{\sqrt{374}} = 2.22$$

4. Decision: Since $z = 2.22$ exceeds $z_{.05} = 1.645$, the null hypothesis must be rejected, and we conclude that the diet is, indeed, effective in reducing weight.

16.4 RANK-SUM TESTS: THE *U* TEST

In this section we shall present a nonparametric alternative to the two-sample t test, which is called the **U test**, the **Wilcoxon test**, or the **Mann–Whitney test**, named after the statisticians who contributed to its development. Without having to assume that the two populations sampled have normal distributions, we will be able to test the null hypothesis that we are sampling identical continuous populations against the alternative that the two populations have unequal means.

To illustrate the procedure, suppose that we want to compare two kinds of emergency flares on the basis of the following burning times (rounded to the nearest tenth of a minute):

Brand A: 14.9, 11.3, 13.2, 16.6, 17.0, 14.1, 15.4, 13.0, 16.9
Brand B: 15.2, 19.8, 14.7, 18.3, 16.2, 21.2, 18.9, 12.2, 15.3, 19.4

Arranging these values jointly (as if they were one sample) in an increasing order of magnitude and assigning them in this order the ranks 1, 2, 3, . . . , and 19, we find that the values of the first sample (Brand *A*) occupy ranks 1, 3, 4, 5, 7, 10, 12, 13, and 14, while those of the second sample ((Brand *B*), occupy ranks 2, 6, 8, 9, 11, 15, 16, 17, 18, and 19. Had there been ties, we would have assigned to each of the tied observations the mean of the ranks which they jointly occupy.

If there is an appreciable difference between the means of the two populations, most of the lower ranks are likely to go to the values of one sample, while most of the higher ranks are likely to go to the values of the other sample. As originally proposed by Wilcoxon, the test is thus based on the value of $\mathbf{W}_1$, the sum of the ranks of the values of the first sample, or $\mathbf{W}_2$, the sum of the ranks of the values of the second sample. It does not matter whether we choose $\mathbf{W}_1$ or $\mathbf{W}_2$, for if there are n_1 values in the first sample and n_2 values in the second sample, $W_1 + W_2$ is the sum of the first $n_1 + n_2$ positive integers; that is,

$$W_1 + W_2 = \frac{(n_1 + n_2)(n_1 + n_2 + 1)}{2}$$

for any pair of values of $\mathbf{W}_1$ and $\mathbf{W}_2$. Thus, tests based on $\mathbf{W}_1$ and $\mathbf{W}_2$ are equivalent.

In actual practice, we seldom base tests on the statistics $\mathbf{W}_1$ or $\mathbf{W}_2$; instead, we use the related statistics

$$\mathbf{U}_1 = \mathbf{W}_1 - \frac{n_1(n_1 + 1)}{2}$$

or

$$\mathbf{U}_2 = \mathbf{W}_2 - \frac{n_2(n_2 + 1)}{2}$$

or the statistic $\min(\mathbf{U}_1, \mathbf{U}_2)$.† The resulting tests are all equivalent to the one's

---

† In many books, $\mathbf{U}_1$ and $\mathbf{U}_2$ are given as

$$\mathbf{U}_1 = n_1 n_2 + \frac{n_2(n_2 + 1)}{2} - \mathbf{W}_2$$

and

$$\mathbf{U}_2 = n_1 n_2 + \frac{n_1(n_1 + 1)}{2} - \mathbf{W}_1$$

and in Exercise 5 on page 485 the reader will be asked to verify that these expressions are equivalent to the ones given above.

based on $\mathbf{W}_1$ or $\mathbf{W}_2$, but they have the advantage that they lend themselves more readily to the construction of tables of critical values. Not only must $\mathbf{U}_1$ and $\mathbf{U}_2$ always take on values from 0 to $n_1 n_2$, but their sampling distributions are symmetrical about $\frac{n_1 n_2}{2}$ and they remain the same if we interchange n_1 and n_2. Also, $\mathbf{U}_1$ is equivalent to the $\mathbf{U}$ statistic of the Mann–Whitney test (see Exercise 6 on page 485), which approaches the problem in a somewhat different way.

For small values of n_1 and n_2, rank-sum tests of the null hypothesis that we are sampling two identical continuous populations are usually based on special tables (see references on page 498). However, when n_1 and n_2 are both greater than 8, it is considered reasonable to assume that the distribution of $\mathbf{U}_1$ (or that of $\mathbf{U}_2$) is approximately normal, and to perform the given rank-sum test on the basis of this assumption, we need the following results:

THEOREM 16.2 Under the null hypothesis, the means and the variances of $\mathbf{U}_1$ and $\mathbf{U}_2$ are

$$E(\mathbf{U}_1) = E(\mathbf{U}_2) = \frac{n_1 n_2}{2}$$

and

$$\text{var}(\mathbf{U}_1) = \text{var}(\mathbf{U}_2) = \frac{n_1 n_2 (n_1 + n_2 + 1)}{12}$$

Proof. Under the null hypothesis that the two samples come from identical populations which are continuous (so that the probability is zero that there will be any ties), the random variable $\mathbf{W}_1$ is the sum of n_1 positive integers selected at random from among the first $n_1 + n_2$ positive integers. Making use of the results of part (c) of Exercise 13 on page 264 with $n = n_1$ and $N = n_1 + n_2$, we thus find that

$$E(\mathbf{W}_1) = \frac{n_1(n_1 + n_2 + 1)}{2}$$

and

$$\text{var}(\mathbf{W}_1) = \frac{n_1 n_2 (n_1 + n_2 + 1)}{12}$$

Since $\mathbf{U}_1 = \mathbf{W}_1 - \frac{n_1(n_1 + 1)}{2}$, it follows that

$$E(\mathbf{U}_1) = \frac{n_1(n_1 + n_2 + 1)}{2} - \frac{n_1(n_1 + 1)}{2} = \frac{n_1 n_2}{2}$$

and

$$\text{var}(\mathbf{U}_1) = \text{var}(\mathbf{W}_1) = \frac{n_1 n_2(n_1 + n_2 + 1)}{12}$$

Also, since $U_1 + U_2 = n_1 n_2$ for any pair of values of $\mathbf{U}_1$ and $\mathbf{U}_2$ (see Exercise 3 on page 485), we get $E(\mathbf{U}_2) = n_1 n_2 - E(\mathbf{U}_1) = \frac{n_1 n_2}{2}$ and $\text{var}(\boldsymbol{U}_2) = \text{var}(\mathbf{U}_1)$.

EXAMPLE 16.5

With reference to the data on page 480, use the $\mathbf{U}_1$ statistic to test the null hypothesis that the two samples come from identical continuous populations against the alternative that the average burning time of Brand *A* flares is less than that of Brand *B* flares. Use the level of significance $\alpha = 0.05$.

Solution

1. H_0: $\mu_1 = \mu_2$
H_1: $\mu_1 < \mu_2$

2. Critical region: $z \leqslant -z_{.05} = -1.645$, where $z = \frac{U_1 - E(\mathbf{U}_1)}{\sqrt{\text{var}(\mathbf{U}_1)}}$

3. Computations: $W_1 = 1 + 3 + 4 + 5 + 7 + 10 + 12 + 13 + 14 = 69$, so that $U_1 = 69 - \frac{9 \cdot 10}{2} = 24$, and since $E(\mathbf{U}_1) = \frac{9 \cdot 10}{2} = 45$ and $\text{var}(\mathbf{U}_1) = \frac{9 \cdot 10 \cdot 20}{12} = 150$, we get

$$z = \frac{24 - 45}{\sqrt{150}} = -1.71$$

4. Decision: Since $z = -1.71$ is less than $-z_{.05} = -1.645$, the null hypothesis must be rejected, and we conclude that Brand *A* flares have, indeed, a shorter burning time than Brand *B* flares.

16.5 RANK-SUM TESTS: THE *H* TEST

The **H test**, or the **Kruskal–Wallis test**, is a generalization of the rank-sum test of the preceding section to the case where we test the null hypothesis that k samples come from identical continuous population. In other words, it is a nonparametric alternative to the one-way analysis of variance.

As in the U test, the data are ranked jointly from low to high, as though they constitute one sample. Then, letting $\mathbf{R}_i$ be the sum of the ranks of the values of the ith sample, we base the test on the statistic

$$\mathbf{H} = \frac{12}{n(n+1)} \cdot \sum_{i=1}^{k} \frac{\mathbf{R}_i^2}{n_i} - 3(n+1)$$

where $n = n_1 + n_2 + \cdots + n_k$, and k is the number of populations sampled. As it can be shown (see Exercise 7 on page 486) that the **H** statistic is proportional to a weighted mean of the squared differences $\left(\frac{\mathbf{R}_i}{n_i} - \frac{n+1}{2}\right)^2$, where $\frac{\mathbf{R}_i}{n_i}$ is the mean rank of the values of the ith sample and $\frac{n+1}{2}$ is the mean rank of all the data, it follows that the null hypothesis must be rejected for large values of H.

For very small values of k and the n_i, the test of the null hypothesis may be based on special tables (see references on page 499), but since the sampling distribution of **H** depends on the values of the n_i it is impossible to tabulate it in a compact form. Hence, the test is usually based on the large-sample theory that the sampling distribution of **H** can be approximated closely with a chi-square distribution with $k - 1$ degrees of freedom. Proofs of this result may be found in the books on nonparametric statistics referred to on page 499, and they are based on the form of the **H** statistic as it is given in Exercise 7 on page 486.

EXAMPLE 16.6

The following are the final examination grades of samples from three groups of students who were taught German by three different methods (classroom instruction and language laboratory, only classroom instruction, and only self-study in language laboratory):

First method:	94, 88, 91, 74, 87, 97
Second method:	85, 82, 79, 84, 61, 72, 80
Third method:	89, 67, 72, 76, 69

Use the H test at the level of significance $\alpha = 0.05$ to test the null hypothesis that the three methods are equally effective.

Solution

1. H_0: $\mu_1 = \mu_2 = \mu_3$
   H_1: the three means are not all equal.
2. Critical region: $H \geqslant \chi^2_{.05,2} = 5.991$.
3. Computations: Ranking the grades from 1 to 18, we find that $R_1 = 6 + 13 + 14 + 16 + 17 + 18 = 84$, $R_2 = 1 + 4.5 + 8 + 9 + 10 + 11 + 12 = 55.5$, and $R_3 = 2 + 3 + 4.5 + 7 + 15 = 31.5$, where there is one tie and the tied grades are each assigned the rank 4.5. Substituting the values of R_1, R_2, and R_3 together with $n_1 = 6$, $n_2 = 7$, $n_3 = 5$, and $n = 18$ into the formula for H, we get

$$H = \frac{12}{18 \cdot 19}\left(\frac{84^2}{6} + \frac{55.5^2}{7} + \frac{31.5^2}{5}\right) - 3 \cdot 19$$
$$= 6.67$$

4. Decision: Since $H = 6.67$ exceeds $\chi^2_{.05,2} = 5.991$, the null hypothesis must be rejected, and we conclude that the three methods are not all equally effective.

THEORETICAL EXERCISES

1. Show that under the null hypotheses of Section 16.3, the distribution of $\mathbf{T}^+$ is symmetrical about $\dfrac{n(n+1)}{4}$.
2. With reference to the signed-rank test, find $E(\mathbf{T}^+ - \mathbf{T}^-)$ and $\text{var}(\mathbf{T}^+ - \mathbf{T}^-)$.
3. Show that the minimum and maximum values of $\mathbf{U}_1$ (and $\mathbf{U}_2$) are, respectively, 0 and n_1n_2, and that $U_1 + U_2 = n_1n_2$ for any pair of values of $\mathbf{U}_1$ and $\mathbf{U}_2$.
4. Show that the distribution of $\mathbf{W}_1$ is symmetrical about $\dfrac{n_1(n_1 + n_2 + 1)}{2}$ and, hence, that the distribution of $\mathbf{U}_1$ is symmetrical about $\dfrac{n_1n_2}{2}$. (*Hint:* Rank the combined data in an increasing as well as a decreasing order of magnitude.)
5. Verify that the expressions given for $\mathbf{U}_1$ and $\mathbf{U}_2$ in the footnote to page 481 are equivalent to the expressions given on that page.
6. If $\mathbf{x}_1, \mathbf{x}_2, \ldots, \mathbf{x}_{n_1}$, and $\mathbf{y}_1, \mathbf{y}_2, \ldots, \mathbf{y}_{n_2}$, are independent random samples, we can test the null hypothesis that they come from identical continuous populations on the basis of the **Mann–Whitney statistic U**, which is simply the number of pairs (x_i, y_j) for which $x_i > y_j$. Symbolically,

$$\mathbf{U} = \sum_{i=1}^{n_1} \sum_{j=1}^{n_2} \mathbf{d}_{ij}$$

where

$$d_{ij} = \begin{cases} 1 & \text{if } x_i > y_j \\ 0 & \text{if } x_i < y_j \end{cases}$$

for $i = 1, 2, \ldots, n_1$, and $j = 1, 2, \ldots, n_2$. Making use of the fact that

$$\sum_{j=1}^{n_2} d_{ij} = r_i - m_i$$

where r_i is the rank of x_i and m_i is the number of x's that are less than or equal to x_i, show that $U = U_1$, namely, that tests based on the Mann–Whitney statistic are equivalent to the rank-sum tests of Section 16.4.

7. Verify that the Kruskal–Wallis statistic on page 484 is equivalent to

$$\mathbf{H} = \frac{12}{n(n+1)} \cdot \sum_{i=1}^{k} n_i \left[\frac{\mathbf{R}_i}{n_i} - \frac{n+1}{2} \right]^2$$

8. Show that if a one-way analysis of variance is applied to the ranks of the observations rather than to the observations themselves, the F test is equivalent to a test based on the H statistic.

APPLIED EXERCISES

9. The following are the amounts of time, in minutes, which it took a random sample of 20 technicians to perform a certain task:

18.1	20.3	18.3	15.6	22.5	16.8	17.6	16.9	18.2	17.0
19.3	16.5	19.5	18.6	20.0	18.8	19.1	17.5	18.5	18.0

Use the sign test at the level of significance $\alpha = 0.01$ to test the null hypothesis that these measurements constitute a random sample from a continuous population with the mean $\mu = 19.4$ against the two-sided alternative $\mu \neq 19.4$.

10. Rework Exercise 9 using the signed-rank test instead of the sign test.

11. The following are the amounts of money (in dollars) spent by 16 persons at a certain amusement park: 10.15, 9.85, 13.75, 8.63, 11.09, 15.63, 6.65, 9.27, 8.80, 11.45, 10.29, 9.51, 13.80, 10.00, 7.48, and 9.11. Use the sign test and the level of significance $\alpha = 0.05$ to test the null hypothesis that on the average a person spends \$9.00 at the park against the alternative that this figure is too low.

12. Rework Exercise 11 using the signed-rank test instead of the sign test.

13. The following are the numbers of speeding tickets issued by two policemen on a random sample of 30 days: 7 and 10, 11 and 13, 10 and 11, 14 and 14, 11 and 15, 12 and 9, 6 and 10, 9 and 13, 8 and 11, 10 and 11, 11 and 15, 13 and 11, 7 and 10, 6 and 12, 10 and 14, 8 and 8, 11 and 12, 9 and 14, 9 and 7, 10 and 12, 6 and 7, 12 and 14, 9 and 11, 12 and 10, 11 and 13, 12 and 15, 7 and 9, 10 and 9, 11 and 13, and 8 and 10. Use the sign test at the level of significance $\alpha = 0.01$ to test the null hypothesis that on the average the two policemen issue equally many speeding tickets against the alternative hypothesis that on the average the second policeman issues more speeding tickets than the first.

14. The following are the numbers of employees absent from the two subsidiaries of a large firm on 16 days: 36 and 23, 20 and 12, 10 and 13, 15 and 11, 33 and 26, 23 and 21, 18 and 23, 25 and 24, 17 and 10, 24 and 34, 30 and 21, 18 and 24, 25 and 25, 19 and 14, 22 and 11, 28 and 16. Use the sign test at the level of significance $\alpha = 0.05$ to test the null hypothesis that on the average there are equally many absences in the two subsidiaries against the alternative that on the average there are more absences in the first subsidiary.

15. Rework Exercises 14 using the signed-rank test instead of the sign test.

16. The following are the weekly food expenditures (in dollars) of 10 households with no children chosen at random from each of two suburbs of a large city:

Suburb A: 54.78, 62.60, 51.89, 54.50, 56.00, 59.38, 48.19, 70.45, 55.15, 51.95

Suburb B: 50.12, 44.63, 65.91, 72.16, 74.59, 39.35, 52.76, 78.19, 45.75, 68.72

Use the $\mathbf{U}_1$ rank-sum statistic and the level of significance $\alpha = 0.05$ to test the null hypothesis that the two samples come from identical populations against the alternative that the populations have unequal means.

17. The following are the lifetimes, in hours, of random samples of two kinds of light bulbs in continuous use:

Brand 1: 407, 426, 453, 378, 434, 396, 441, 373, 393, 386, 415, 418

Brand 2: 403, 424, 383, 445, 439, 417, 412, 462, 439, 432, 413, 433

Use the U test (that is, the statistic $\mathbf{U}_1$) at the level of significance $\alpha = 0.01$ to determine whether it is reasonable to claim that there is no difference between the average lifetimes of the two kinds of light bulbs.

18. With reference to the illustration on page 480, calculate the value of the Mann–Whitney **U** statistic (see Exercise 6) and verify that it equals the value obtained for $\mathbf{U}_1$ in Example 16.5.

19. With reference to Exercise 16, calculate the value of the Mann–Whitney **U** statistic (see Exercise 6) and verify that it equals the value obtained for $\mathbf{U}_1$.

20. With reference to Exercise 17, calculate the value of the Mann–Whitney **U** statistic (see Exercise 6) and verify that it equals the value obtained for $\mathbf{U}_1$.

21. To compare four bowling balls, a professional bowler bowls five games with each ball and gets the following results:

Ball D: 208, 220, 247, 192, 229
Ball E: 216, 196, 189, 205, 210
Ball F: 226, 218, 252, 225, 202
Ball G: 212, 198, 207, 232, 221

Use the Kruskal–Wallis test at the level of significance $\alpha = 0.05$ to test the null hypothesis that the bowler performs equally well with the four bowling balls.

22. The following are the miles per gallon which a test driver got for ten tankfuls of each of three kinds of gasoline:

Gasoline A: 20, 31, 24, 33, 23, 24, 28, 16, 19, 26
Gasoline B: 29, 18, 29, 19, 20, 21, 34, 33, 30, 23
Gasoline C: 19, 31, 16, 26, 31, 33, 28, 28, 25, 30

Use the Kruskal–Wallis test at the level of significance $\alpha = 0.05$ to test the null hypothesis that there is no difference in the average mileage yield of the three kinds of gasoline.

16.6 TESTS BASED ON RUNS

There are several nonparametric methods for testing the randomness of observed data on the basis of the order in which they were obtained. The technique we shall describe here is based on the **theory of runs**, where a **run** is a succession of identical letters (or other kinds of symbols) which is preceded and followed by different letters or no letters at all. To illustrate, consider the following arrangement of defective, *d*, and nondefective, *n*, pieces produced in the given order by a certain machine:

$$\underbrace{n\,n\,n\,n\,n}\,\underbrace{d\,d\,d\,d}\,\underbrace{n\,n\,n\,n\,n\,n\,n\,n\,n\,n}\,\underbrace{d\,d}\,\underbrace{n\,n}\,\underbrace{d\,d\,d\,d}\,\underbrace{n}\,\underbrace{d\,d}\,\underbrace{n\,n}$$

Using braces to combine the letters which constitute a run, we find that there is first a run of five *n*'s, then a run of four *d*'s, then a run of ten *n*'s, . . . , and finally a run of two *n*'s; in all, there are nine runs of varying lengths.

The total number of runs appearing in an arrangement of this kind is often a good indication of a possible lack of randomness. If there are too few runs, we might suspect a definite grouping or clustering, or perhaps a trend; if there are too many runs, we might suspect some sort of repeated alternating pattern. In our illustration there seems to be a definite clustering, the defective pieces seem to come in groups, but it remains to be seen whether this is significant or whether it can be attributed to chance.

To find the probability that n_1 letters of one kind and n_2 letters of another kind will form u runs when each of the $\binom{n_1+n_2}{n_1}$ possible arrangements of these letters is regarded as equally likely, let us first investigate the case where u is even, namely, where $u = 2k$ and k is a positive integer. In that case there will have to be k runs of each kind alternating with one another. To find the number of ways in which n_1 letters can form k runs, let us first consider the very simple case where we have five letters c which are to be divided up into three runs. Using vertical bars to separate the five letters into three runs, we find that there are the *six* possibilities

$$c\,|c\,|c\,c\,c \qquad c\,|c\,c\,|c\,c \qquad c\,|c\,c\,c\,|c$$

$$c\,c\,|c\,|c\,c \qquad c\,c\,|c\,c\,|c \qquad c\,c\,c\,|c\,|c$$

corresponding to the $\binom{4}{2}$ ways in which we can put two vertical bars into two of the four spaces between the five c's. By the same token there are $\binom{n_1-1}{k-1}$ ways in which the n_1 letters of the first kind can form k runs, $\binom{n_2-1}{k-1}$ ways in which the n_2 letters of the second kind can form k runs, and it follows that there are altogether $2\binom{n_1-1}{k-1}\binom{n_2-1}{k-1}$ ways in which these $n_1 + n_2$ letters can form $2k$ runs. The factor 2 is accounted for by the fact that when we combine the two kinds of runs so that they alternate, we can begin either with a run of the first kind of letter or with a run of the second kind. Thus, when $u = 2k$ (where k is a positive integer), the probability of getting that many runs is

$$f(u) = \frac{2\binom{n_1-1}{k-1}\binom{n_2-1}{k-1}}{\binom{n_1+n_2}{n_1}}$$

and it will be left to the reader to show in Exercise 1 on page 495 that similar arguments lead to

$$f(u) = \frac{\binom{n_1-1}{k}\binom{n_2-1}{k-1} + \binom{n_1-1}{k-1}\binom{n_2-1}{k}}{\binom{n_1+n_2}{n_1}}$$

when $u = 2k + 1$ (where k is a positive integer).

When n_1 and n_2 are small, tests of the null hypothesis of randomness are usually based on specially constructed tables (see references on page 499); however, when n_1 and n_2 are both 10 or more, the sampling distribution of $\mathbf{u}$ can be approximated with a normal distribution. For this we require the following results:

THEOREM 16.3 Under the null hypothesis of randomness, the mean and the variance of $\mathbf{u}$ are

$$E(\mathbf{u}) = \frac{2n_1n_2}{n_1 + n_2} + 1$$

and

$$\text{var}(\mathbf{u}) = \frac{2n_1n_2(2n_1n_2 - n_1 - n_2)}{(n_1 + n_2)^2(n_1 + n_2 - 1)}$$

These results can be obtained directly with the use of the probabilities given above. The details of such a proof, as well as an alternative approach which is easier, may be found in the book by J. D. Gibbons listed on page 499.

EXAMPLE 16.7

With reference to the illustration on page 488, test the null hypothesis of randomness at the level of significance $\alpha = 0.01$.

Solution

1. H_0: Arrangement is random;
   H_1: arrangement is not random.
2. Critical region: $z \geqslant z_{.005} = 2.575$ or $z \leqslant -z_{.005} = -2.575$, where

$$z = \frac{(u \pm \frac{1}{2}) - E(\mathbf{u})}{\sqrt{\text{var}(\mathbf{u})}}$$

3. Computations: Substituting $n_1 = 20$, $n_2 = 12$, and $u = 9$, we get

$$E(\mathbf{u}) = \frac{2 \cdot 20 \cdot 12}{20 + 12} + 1 = 16$$

and

$$\text{var}(\mathbf{u}) = \frac{2 \cdot 20 \cdot 12(2 \cdot 20 \cdot 12 - 20 - 12)}{(20 + 12)^2(20 + 12 - 1)} = 6.77$$

so that

$$z = \frac{(9 + \frac{1}{2}) - 16}{\sqrt{6.77}} = -2.50$$

4. Decision: Since $z = -2.50$ falls between -2.575 and 2.575, the null hypothesis of randomness cannot be rejected. Note that if we had not used the continuity correction in this example, we would have obtained $z = -2.69$ and the decision would have been reversed.

The method we have discussed in this section is not limited to tests of the randomness of series of attributes (such as the d's and n's of our example). Any sample which consists of numerical measurements or observations can be treated similarly by using the letters a and b to denote, respectively, values falling above and below the median of the sample. (Numbers equaling the median are omitted.) The resulting series of a's and b's can then be tested for randomness on the basis of the total number of runs of a's and b's, namely, the total number of **runs above and below the median**.

EXAMPLE 16.8

The following are the speeds (in miles per hour) at which every fifth passenger car was timed at a certain checkpoint: 46, 58, 60, 56, 70, 66, 48, 54, 62, 41, 39, 52, 45, 62, 53, 69, 65, 65, 67, 76, 52, 52, 59, 59, 67, 51, 46, 61, 40, 43, 42, 77, 67, 63, 59, 63, 63, 72, 57, 59, 42, 56, 47, 62, 67, 70, 63, 66, 69, and 73. Test the null hypothesis of randomness at the level of significance $\alpha = 0.05$.

Solution

1. H_0: The sample is random;
   H_1: the sample is not random.
2. Critical region: $z \geqslant z_{.005} = 2.575$ or $z \leqslant -z_{.005} = -2.575$, where

$$z = \frac{(u \pm \frac{1}{2}) - E(\mathbf{u})}{\sqrt{\text{var}(\mathbf{u})}}$$

3. Computations: The median of the speeds is 59.5, so that the given data can be represented by the following arrangement of a's and b's:

$$b\ b\ a\ b\ a\ a\ b\ b\ a\ b\ b\ b\ b\ a\ b\ a\ a\ a\ a\ a\ b\ b\ b\ b\ a$$
$$b\ b\ a\ b\ b\ b\ a\ a\ a\ b\ a\ a\ a\ b\ b\ b\ b\ b\ a\ a\ a\ a\ a\ a\ a$$

Since $n_1 = 25$, $n_2 = 25$, and $u = 20$, we get

$$E(\mathbf{u}) = \frac{2 \cdot 25 \cdot 25}{25 + 25} + 1 = 26$$

$$\text{var}(\mathbf{u}) = \frac{2 \cdot 25 \cdot 25(2 \cdot 25 \cdot 25 - 25 - 25)}{(25 + 25)^2(25 + 25 - 1)} = 12.2$$

and

$$z = \frac{(20 + \frac{1}{2}) - 26}{\sqrt{12.2}} = -1.57$$

4. Decision: Since $z = -1.57$ falls between -2.575 and 2.575, the null hypothesis of randomness cannot be rejected.

The method of runs above and below the median is especially useful in detecting trends and cyclical patterns in economic data. If there is a trend, there will be first mostly a's and later mostly b's (or vice versa), and if there is a repeated cyclical pattern there will be a systematic alternation of a's and b's and, probably, too many runs.

16.7 THE RANK CORRELATION COEFFICIENT

Since the assumptions underlying the significance test for the correlation coefficient are rather stringent, it is sometimes preferable to use a nonparametric alternative. Most popular among the nonparametric measures of association is the **rank correlation coefficient**, also called **Spearman's rank correlation coefficient**, r_S. For a given set of paired data $\{(x_i, y_i);\ i = 1, 2, \ldots, n\}$, it is obtained by ranking the x's among themselves, and also the y's, and then substituting into the following formula:

DEFINITION 16.1 The rank correlation coefficient is given by

$$r_S = 1 - \frac{6 \cdot \sum_{i=1}^{n} d_i^2}{n(n^2 - 1)}$$

where d_i is the difference between the ranks assigned to x_i and y_i.

When there are ties in rank, we proceed as before and assign the tied observations the mean of the ranks which they jointly occupy.

When there are no ties in rank, r_S actually equals the correlation coefficient r calculated for the ranks. To verify this, let r_i and s_i be the ranks of x_i and y_i. Making use of the fact that the sum and the sum of the squares of the first n positive integers are $\frac{n(n+1)}{2}$ and $\frac{n(n+1)(2n+1)}{6}$, respectively, we find that

$$\sum_{i=1}^{n} r_i = \sum_{i=1}^{n} s_i = \frac{n(n+1)}{2}$$

$$\sum_{i=1}^{n} r_i^2 = \sum_{i=1}^{n} s_i^2 = \frac{n(n+1)(2n+1)}{6}$$

$$\sum_{i=1}^{n} r_i s_i = \frac{n(n+1)(2n+1)}{6} - \frac{1}{2} \cdot \sum_{i=1}^{n} d_i^2$$

and if we substitute these expressions into the formula for r, we get the above formula for r_S.

EXAMPLE 16.9

The following are the numbers of hours which ten students studied for an examination and the grades which they obtained:

Number of hours studied x	*Grade* y
8	56
5	44
11	79
13	72
10	70
5	54
18	94
15	85
2	33
8	65

Calculate r_S.

Solution

Ranking the x's and the y's, and proceeding as in the following table, we get

Rank of x	*Rank of y*	d	d^2
6.5	7	−0.5	0.25
8.5	9	−0.5	0.25
4	3	1.0	1.00
3	4	1.0	1.00
5	5	0.0	0.00
8.5	8	0.5	0.25
1	1	0.0	0.00
2	2	0.0	0.00
10	10	0.0	0.00
6.5	6	0.5	0.25
			3.00

and substitution into the formula for r_S yields

$$r_S = 1 - \frac{6 \cdot 3}{10(10^2 - 1)} = 0.98$$

As can be seen from this example, r_S is very easy to compute; indeed, it is sometimes used instead of r mainly because of its computational ease. If we calculated r for the data of Example 16.9, we would get 0.96, and this is very close to the value we obtained for r_S.

For small values of n ($n \leqslant 10$), the test of the null hypothesis of no correlation, indeed, the test of the null hypothesis that the x's and y's are randomly matched, may be based on special tables determined from the exact sampling distribution of $\mathbf{r}_S$ (see references on page 499). Most of the time, though, we use the fact that the distribution of $\mathbf{r}_S$ can be approximated closely with a normal distribution, and to this end we need the following results:

THEOREM 16.4 Under the null hypothesis of no correlation, the mean and the variance of $\mathbf{r}_S$ are

$$E(\mathbf{r}_S) = 0 \quad \text{and} \quad \text{var}(\mathbf{r}_S) = \frac{1}{n - 1}$$

A proof of this theorem may be found in the book by J. D. Gibbons listed on page 498. Strictly speaking, the theorem applies when there are no ties, but it can also be used when there are ties unless the number of ties is very extensive.

EXAMPLE 16.10

With reference to Example 16.9, test the significance of the value obtained for r_S at the level of significance $\alpha = 0.01$.

Solution

1. H_0: No correlation;
   H_1: there is a correlation.
2. Critical region: $z \geqslant z_{.005} = 2.575$ or $z \leqslant -z_{.005} = -2.575$, where

$$z = r_S\sqrt{n-1}$$

3. Computations: For $n = 10$ and $r_S = 0.98$, we get

$$z = 0.98\sqrt{10-1} = 2.94$$

4. Decision: Since $z = 2.94$ exceeds $z_{.005} = 2.575$, the null hypothesis must be rejected, and we conclude that there is a real (positive) relationship between study time and grades.

THEORETICAL EXERCISES

1. Verify the formula given on page 489 for the values of the probability distribution of **u** when $u = 2k + 1$, where k is a positive integer.
2. If a person gets 7 heads and 3 tails in 10 tosses of a balanced coin, find the probabilities for 2, 3, 4, 5, 6, and 7 runs.
3. Find the probability that $n_1 = 6$ letters of one kind and $n_2 = 5$ letters of another kind will form at least 8 runs.
4. If there are $n_1 = 8$ letters of one kind and $n_2 = 8$ letters of another kind, what are the values of u for which we would reject the null hypothesis of randomness at the level of significance $\alpha = 0.01$?
5. Given a set of k-tuples $(x_{11}, x_{12}, \ldots, x_{1k})$, $(x_{21}, x_{22}, \ldots, x_{2k}), \ldots,$ and $(x_{n1}, x_{n2}, \ldots, x_{nk})$, the extent of their association, or agreement, may be measured by means of the **coefficient of concordance**

$$W = \frac{12}{k^2n(n^2-1)} \cdot \sum_{i=1}^{n} \left[R_i - \frac{k(n+1)}{2}\right]^2$$

   where R_i is the sum of the ranks assigned to $x_{i1}, x_{i2}, \ldots,$ and x_{ik} when the x's with the second subscript 1 are ranked among themselves, and so are the x's with the second subscript 2, ..., and the x's with the second subscript k. What are the maximum and minimum values of W, and what do they reflect with respect to the agreement, or lack of agreement, of the values of the k random variables?

APPLIED EXERCISES

6. The following is the order in which healthy, H, and diseased, D, pine trees were observed in a survey conducted by the Forestry Service:

$$H\,H\,H\,D\,H\,H\,H\,H\,H\,D\,D\,D\,H\,H\,H\,H\,H\,H\,D\,D\,D\,D$$

(cont.) $$H\,H\,H\,H\,D\,D\,H\,H\,H\,H\,H\,H\,H\,H\,H\,H\,D\,D\,D\,D\,H\,H$$

Test for randomness at the level of significance $\alpha = 0.05$.

7. The following arrangement indicates whether 60 consecutive cars which went by the toll booth of a bridge had local plates, L, or out-of-state plates, O:

$$L\,L\,O\,L\,L\,L\,L\,O\,O\,L\,L\,L\,L\,O\,L\,O\,O\,L\,L\,L\,L\,O\,L\,O\,O\,L\,L\,L\,L\,L$$

(cont.) $$O\,L\,L\,L\,O\,L\,O\,L\,L\,L\,L\,O\,O\,L\,O\,O\,O\,O\,L\,L\,L\,L\,O\,L\,O\,O\,L\,L\,L\,O$$

Use the level of significance $\alpha = 0.05$ to test whether this arrangement of L's and O's may be regarded as random.

8. To test whether a radio signal contains a message or constitutes random noise, an interval of time is subdivided into a number of very short intervals and for each of these it is determined whether the signal strength exceeds, E, or does not exceed, N, a certain level of background noise. Test at the level of significance $\alpha = 0.01$ whether the following arrangement, thus obtained, may be regarded as random, and hence that the signal contains no message and may be regarded as random noise:

$$N\,N\,N\,E\,N\,E\,N\,E\,N\,E\,E\,E\,N\,E\,E\,E\,N\,E\,E\,N\,E\,N\,E\,E$$

(cont.) $$N\,E\,E\,N\,N\,E\,N\,E\,E\,E\,N\,E\,N\,N\,N\,E\,N\,N\,E\,N\,N\,N\,N\,E$$

9. Choose 200 random digits from an appropriate table (see reference on page 499), represent each even digit by the letter E, each odd digit by the letter O, and test for randomness at the level of significance $\alpha = 0.01$.

10. The following are the numbers of defective pieces produced by a machine on fifty consecutive days: 7, 14, 17, 10, 18, 19, 23, 19, 14, 10, 12, 18, 19, 13, 24, 26, 9, 16, 19, 14, 19, 10, 15, 22, 25, 24, 20, 9, 17, 28, 29, 19, 25, 23, 24, 28, 31, 19, 24, 30, 27, 24, 39, 35, 23, 26, 28, 31, 37, and 40. Use the method of runs above and below the median and $\alpha = 0.05$ to test the null hypothesis of randomness against the alternative that there is a trend.

11. The following are the numbers of students absent from school on 24 consecutive school days: 29, 25, 31, 28, 30, 28, 33, 31, 35, 29, 31, 33, 35, 28, 36, 30, 33, 26, 30, 28, 32, 31, 38, and 27. Test for randomness at the level of significance $\alpha = 0.01$.

12. The theory of runs may also be used as an alternative to the rank-sum test of Section 16.4, namely, the test of the null hypothesis that two independent random samples come from identical continuous populations. We simply rank the data jointly, write a 1 below each value belonging to the first sample, a 2 below each value belonging to the second sample, and then test the randomness of the resulting arrangement of 1's and 2's. If there are too few runs, this may well be accounted for by the fact that the two samples come from populations with unequal means.

(a) Apply this technique to the illustration on page 480, the one which deals with the burning times of two kinds of flares.

(b) Use this technique to rework Exercise 16 on page 487.

13. Calculate r_S for the following data representing the statistics grades, x, and psychology grades, y, of 18 students:

x	y	x	y
78	80	97	90
86	74	74	85
49	63	53	71
94	85	58	67
53	55	62	64
89	86	74	69
94	90	74	71
71	84	70	67
70	71	74	71

Also test for significance at $\alpha = 0.05$.

14. The following shows how a panel of nutrition experts and a panel of housewives ranked fifteen breakfast foods on their palatability:

Breakfast food	*Nutrition experts*	*Housewives*
A	3	5
B	7	4
C	11	8
D	9	14
E	1	2
F	4	6
G	10	12
H	8	7
I	5	1
J	13	15
K	12	9
L	2	3
M	15	10
N	6	11
O	14	13

Calculate r_S as a measure of the consistency of the two rankings.

15. Calculate r_S for the data of Exercise 18 on page 433 and test the null hypothesis of no correlation between the two variables at the level of significance $\alpha = 0.05$.

16. The following are the rankings given by three judges to the works of ten artists:

Judge A	*Judge B*	*Judge C*
6	2	7
4	5	3
2	4	1
5	8	2
9	10	10
3	1	6
1	6	4
8	9	9
10	7	8
7	3	5

Calculate the value of W, the coefficient of concordance of Exercise 5, as a measure of the agreement of the three sets of rankings.

17. With reference to Exercise 16, calculate the $k = 3$ pairwise rank correlation coefficients, and verify that the relationship between their mean, $\bar{r}_S$, and the coefficient of concordance (see Exercise 5) is given by

$$\bar{r}_S = \frac{kW - 1}{k - 1}$$

References

The tables that are needed to perform exact small-sample tests for various nonparametric methods may be found in

OWEN, D. B., *Handbook of Statistical Tables.* Reading, Mass.: Addison-Wesley Publishing Company, Inc., 1962.

In particular, tables for small-sample tests of the significance of the rank correlation coefficient are given in

KENDALL, M. G., *Rank Correlation Methods.* New York: Hafner Publishing Co., Inc., 1962.

A wealth of information about the various nonparametric tests may be found in

GIBBONS, J. D., *Nonparametric Statistical Inference.* New York: McGraw-Hill Book Company, 1971,

LEHMANN, E. L., *Nonparametrics: Statistical Methods Based on Ranks.* San Francisco: Holden-Day, Inc., 1975,

MOSTELLER, F., and ROURKE, R. E. K., *Sturdy Statistics.* Reading, Mass.: Addison-Wesley Publishing Company, Inc., 1973,

NOETHER, G. E., *Introduction to Statistics: A Nonparametric Approach,* 2nd ed. Boston: Houghton Mifflin Company, 1976,

SIEGEL, S., *Nonparametric Statistics for the Behavioral Sciences.* New York: McGraw-Hill Book Company, 1956.

Among the many published tables of random numbers, or random digits, one of the most widely used is

RAND Corporation, *A Million Random Digits with 100,000 Normal Deviates.* New York: Macmillan Publishing Co., Inc., third printing 1966.

the algebra of events

I

I.1 BOOLEAN ALGEBRA

Given two events A and B in a sample space S, we can obtain further events by forming **unions**, **intersections**, and **complements**.

DEFINITION I.1 The union of events A and B, denoted $A \cup B$, is the event in S which contains all the elements that are either in A, in B, or in both.

DEFINITION I.2 The intersection of events A and B, denoted $A \cap B$, is the event in S which contains all the elements that are both in A and in B.

DEFINITION I.3 The complement of event A, denoted A', is the event in S which contains all the elements of S that are not in A.

The formation of unions, intersections, and complements is governed by the following rules, called the postulates of **Boolean Algebra**:

POSTULATE I.1 (Closure laws) For each pair of events A and B in a sample space S there is a unique event $A \cup B$ and a unique event $A \cap B$ in S.

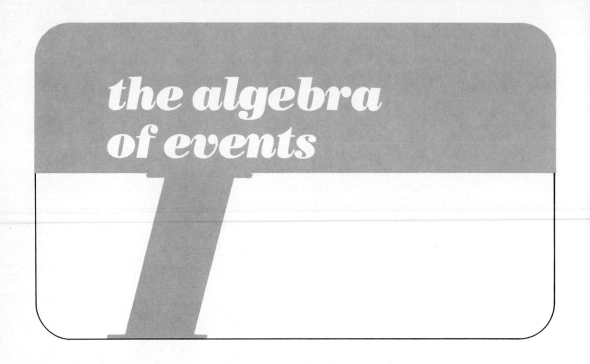

POSTULATE I.2 (Commutative laws)

$$A \cup B = B \cup A \quad \text{and} \quad A \cap B = B \cap A$$

POSTULATE I.3 (Associative laws)

$$(A \cup B) \cup C = A \cup (B \cup C) \quad \text{and} \quad (A \cap B) \cap C = A \cap (B \cap C)$$

POSTULATE I.4 (Distributive laws)

$$A \cap (B \cup C) = (A \cap B) \cup (A \cap C)$$

and

$$A \cup (B \cap C) = (A \cup B) \cap (A \cup C)$$

POSTULATE I.5 (Identity laws) $A \cap S = A$ for each event A in the sample space S; also, there exists a unique event $\varnothing$ such that $A \cup \varnothing = A$ for each event A in S.

POSTULATE I.6 (Complementation law) For each event A in a sample space S there exists a unique event A' in S such that $A \cap A' = \varnothing$ and $A \cup A' = S$.

Observe that Postulate I.5 defines, in fact, what we mean by the empty set $\varnothing$, and that Postulate I.6 defines what we mean by the complement A' of event A.

Based on these postulates, we can prove many further theorems about the manipulation of events. For instance,

THEOREM I.1 $A \cap A = A$ for any event A in a sample space S.

Proof.

$A = A \cap S$	Postulate I.5
$= A \cap (A \cup A')$	Postulate I.6
$= (A \cap A) \cup (A \cap A')$	Postulate I.4
$= (A \cap A) \cup \varnothing$	Postulate I.6
$= A \cap A$	Postulate I.5

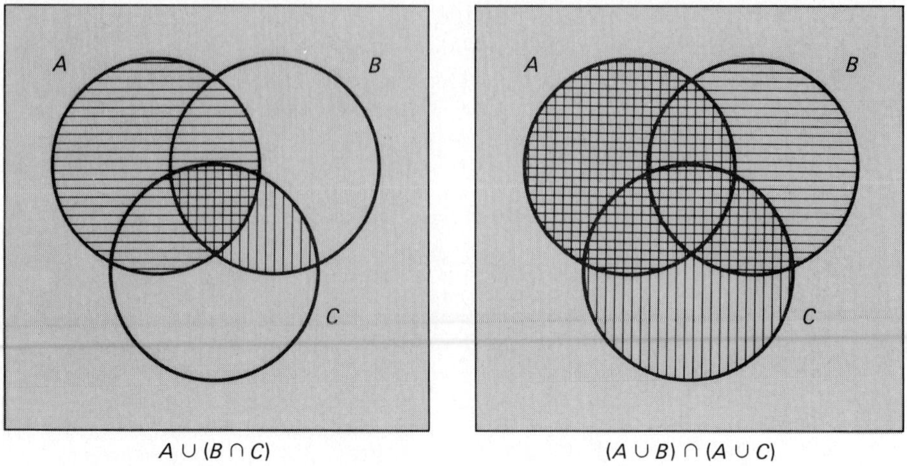

Figure I.1 Venn diagrams showing that $A \cup (B \cap C) = (A \cup B) \cap (A \cup C)$.

The following are some other theorems which can be proved in a similar way: $A \cup A = A$, $(A')' = A$, $A \cup S = S$, and $A \cap \varnothing = \varnothing$ for any event A in the sample space S, $S' = \varnothing$, and $\varnothing' = S$.

In most instances it is easiest to verify rules about events by inspection of appropriate Venn diagrams. In this way it can be shown, for example, that the second distributive law of Postulate I.4 does, indeed, hold when we represent events by means of regions of Venn diagrams. In the first Venn diagram of Figure I.1, A is ruled horizontally, $B \cap C$ is ruled vertically, and $A \cup (B \cap C)$ is represented by the region ruled horizontally and/or vertically; in the second Venn diagram of Figure I.1, $A \cup B$ is ruled horizontally, $A \cup C$ is ruled vertically, and $(A \cup B) \cap (A \cup C)$ is ruled both ways. As can be seen, the region ruled horizontally and/or vertically in the first diagram is the same as that ruled both ways in the second diagram.

EXERCISES

1. Use Venn diagrams to verify the first distributive law of Postulate I.4.
2. Use Venn diagrams to verify the two **de Morgan laws**
   (a) $(A \cap B)' = A' \cup B'$;
   (b) $(A \cup B)' = A' \cap B'$.
3. Use Venn diagrams to verify that
   (a) $A \cup (A \cap B) = A$;
   (b) $A \cap (A \cup B) = A$;
   (c) $(A \cap B) \cup (A \cap B') = A$;
   (d) $A \cup B = (A \cap B) \cup (A \cap B') \cup (A' \cap B)$;
   (e) $A \cup (A' \cap B) = A \cup B$.

4. Use Venn diagrams to verify that if $A \subset B$ (namely, A is contained in B), then $A \cap B = A$ and $A \cap B' = \varnothing$.
5. Prove that $A \cup A = A$ for any event A in a sample space S, justifying each step by means of one of the postulates.
6. Prove that $A \cap \varnothing = \varnothing$ for any event A in a sample space S, justifying each step by means of one of the postulates.

II sums and products

II.1 RULES FOR SUMS AND PRODUCTS

To simplify expressions involving sums and products, the $\sum$ and $\prod$ notations are widely used in statistics. In the usual notation we write

$$\sum_{i=a}^{b} x_i = x_a + x_{a+1} + x_{a+2} + \cdots + x_b$$

and

$$\prod_{i=a}^{b} x_i = x_a \cdot x_{a+1} \cdot x_{a+2} \cdot \ldots \cdot x_b$$

for any non-negative integers a and b with $a \leqslant b$.

When working with sums or products, it is often helpful to apply the following rules, which can all be verified by writing the respective expressions in full, that is, without the $\sum$ or $\prod$ notation:

THEOREM II.1

1. $\sum_{i=1}^{n} kx_i = k \cdot \sum_{i=1}^{n} x_i$

2. $\sum_{i=1}^{n} k = nk$

3. $\sum_{i=1}^{n} (x_i + y_i) = \sum_{i=1}^{n} x_i + \sum_{i=1}^{n} y_i$

4. $\prod_{i=1}^{n} kx_i = k^n \cdot \prod_{i=1}^{n} x_i$

5. $\prod_{i=1}^{n} k = k^n$

6. $\prod_{i=1}^{n} x_i y_i = \left(\prod_{i=1}^{n} x_i\right)\left(\prod_{i=1}^{n} y_i\right)$

7. $\ln \prod_{i=1}^{n} x_i = \sum_{i=1}^{n} \ln x_i$

Double sums, triple sums, . . . , are also widely used in statistics, and if we repeatedly apply the definition of $\sum$ given above, we have, for example,

$$\begin{aligned}\sum_{i=1}^{m} \sum_{j=1}^{n} x_{ij} &= \sum_{i=1}^{m} (x_{i1} + x_{i2} + \cdots + x_{in}) \\ &= (x_{11} + x_{12} + \cdots + x_{1n}) \\ &\quad + (x_{21} + x_{22} + \cdots + x_{2n}) \\ &\quad \cdot \quad \cdot \quad \cdot \quad \cdot \quad \cdot \\ &\quad + (x_{m1} + x_{m2} + \cdots + x_{mn})\end{aligned}$$

Note that when the x_{ij} are thus arranged in a rectangular array, the first subscript denotes the row to which a particular element belongs, and the second subscript denotes the column.

When we work with double sums, the following theorem is of special interest; it is an immediate consequence of the multinomial expansion of $(x_1 + x_2 + \cdots + x_n)^2$:

THEOREM II.2

$$\sum_{i<j} \sum x_i x_j = \frac{1}{2}\left[\left(\sum_{i=1}^{n} x_i\right)^2 - \sum_{i=1}^{n} x_i^2\right]$$

where

$$\sum_{i<j}\sum x_i x_j = \sum_{i=1}^{n-1}\sum_{j=i+1}^{n} x_i x_j$$

II.2 SPECIAL SUMS

In the theory of nonparametric statistics, particularly when we deal with rank sums, we often need expressions for the sums of powers of the first n positive integers; namely, expressions for

$$S(n, r) = 1^r + 2^r + 3^r + \cdots + n^r$$

for $r = 0, 1, 2, 3, \ldots$. The following theorem, which the reader will be asked to prove in Exercise 1 below, provides a convenient way of obtaining these sums:

THEOREM II.3

$$\sum_{r=0}^{k-1}\binom{k}{r}S(n, r) = (n + 1)^k - 1$$

for any positive integers n and k.

A disadvantage of this theorem is that we have to find the sums $S(n, r)$ one at a time, first for $r = 0$, then for $r = 1$, then for $r = 2$, and so forth. For instance, for $k = 1$ we get

$$\binom{1}{0}S(n, 0) = (n + 1) - 1 = n$$

and, hence, $S(n, 0) = 1^0 + 2^0 + \cdots + n^0 = n$. Similarly, for $k = 2$ we get

$$\binom{2}{0}S(n, 0) + \binom{2}{1}S(n, 1) = (n + 1)^2 - 1$$

$$n + 2S(n, 1) = n^2 + 2n$$

and, hence, $S(n, 1) = 1^1 + 2^1 + \cdots + n^1 = \frac{1}{2}n(n + 1)$. Using the same technique, the reader will be asked to show in Exercise 2 below that

$$S(n, 2) = \tfrac{1}{6}n(n + 1)(2n + 1) \quad \text{and} \quad S(n, 3) = \tfrac{1}{4}n^2(n + 1)^2$$

THEORETICAL EXERCISES

1. Prove Theorem II.3 by making use of the fact that

$$(m + 1)^k - m^k = \sum_{r=0}^{k-1} \binom{k}{r} m^r$$

   which follows from the binomial expansion of $(m + 1)^k$.
2. Verify the formulas for $S(n, 2)$ and $S(n, 3)$ given above, and find an expression for $S(n, 4)$.

statistical tables

TABLE I

Binomial Probabilities[†]

		θ									
n	x	.05	.10	.15	.20	.25	.30	.35	.40	.45	.50
1	0	.9500	.9000	.8500	.8000	.7500	.7000	.6500	.6000	.5500	.5000
	1	.0500	.1000	.1500	.2000	.2500	.3000	.3500	.4000	.4500	.5000
2	0	.9025	.8100	.7225	.6400	.5625	.4900	.4225	.3600	.3025	.2500
	1	.0950	.1800	.2550	.3200	.3750	.4200	.4550	.4800	.4950	.5000
	2	.0025	.0100	.0225	.0400	.0625	.0900	.1225	.1600	.2025	.2500
3	0	.8574	.7290	.6141	.5120	.4219	.3430	.2746	.2160	.1664	.1250
	1	.1354	.2430	.3251	.3840	.4219	.4410	.4436	.4320	.4084	.3750
	2	.0071	.0270	.0574	.0960	.1406	.1890	.2389	.2880	.3341	.3750
	3	.0001	.0010	.0034	.0080	.0156	.0270	.0429	.0640	.0911	.1250
4	0	.8145	.6561	.5220	.4096	.3164	.2401	.1785	.1296	.0915	.0625
	1	.1715	.2916	.3685	.4096	.4219	.4116	.3845	.3456	.2995	.2500
	2	.0135	.0486	.0975	.1536	.2109	.2646	.3105	.3456	.3675	.3750
	3	.0005	.0036	.0115	.0256	.0469	.0756	.1115	.1536	.2005	.2500
	4	.0000	.0001	.0005	.0016	.0039	.0081	.0150	.0256	.0410	.0625
5	0	.7738	.5905	.4437	.3277	.2373	.1681	.1160	.0778	.0503	.0312
	1	.2036	.3280	.3915	.4096	.3955	.3602	.3124	.2592	.2059	.1562
	2	.0214	.0729	.1382	.2048	.2637	.3087	.3364	.3456	.3369	.3125
	3	.0011	.0081	.0244	.0512	.0879	.1323	.1811	.2304	.2757	.3125
	4	.0000	.0004	.0022	.0064	.0146	.0284	.0488	.0768	.1128	.1562
	5	.0000	.0000	.0001	.0003	.0010	.0024	.0053	.0102	.0185	.0312
6	0	.7351	.5314	.3771	.2621	.1780	.1176	.0754	.0467	.0277	.0156
	1	.2321	.3543	.3993	.3932	.3560	.3025	.2437	.1866	.1359	.0938
	2	.0305	.0984	.1762	.2458	.2966	.3241	.3280	.3110	.2780	.2344
	3	.0021	.0146	.0415	.0819	.1318	.1852	.2355	.2765	.3032	.3125
	4	.0001	.0012	.0055	.0154	.0330	.0595	.0951	.1382	.1861	.2344
	5	.0000	.0001	.0004	.0015	.0044	.0102	.0205	.0369	.0609	.0938
	6	.0000	.0000	.0000	.0001	.0002	.0007	.0018	.0041	.0083	.0156
7	0	.6983	.4783	.3206	.2097	.1335	.0824	.0490	.0280	.0152	.0078
	1	.2573	.3720	.3960	.3670	.3115	.2471	.1848	.1306	.0872	.0547
	2	.0406	.1240	.2097	.2753	.3115	.3177	.2985	.2613	.2140	.1641
	3	.0036	.0230	.0617	.1147	.1730	.2269	.2679	.2903	.2918	.2734
	4	.0002	.0026	.0109	.0287	.0577	.0972	.1442	.1935	.2388	.2734
	5	.0000	.0002	.0012	.0043	.0115	.0250	.0466	.0774	.1172	.1641
	6	.0000	.0000	.0001	.0004	.0013	.0036	.0084	.0172	.0320	.0547
	7	.0000	.0000	.0000	.0000	.0001	.0002	.0006	.0016	.0037	.0078
8	0	.6634	.4305	.2725	.1678	.1001	.0576	.0319	.0168	.0084	.0039
	1	.2793	.3826	.3847	.3355	.2670	.1977	.1373	.0896	.0548	.0312
	2	.0515	.1488	.2376	.2936	.3115	.2965	.2587	.2090	.1569	.1094
	3	.0054	.0331	.0839	.1468	.2076	.2541	.2786	.2787	.2568	.2188
	4	.0004	.0046	.0185	.0459	.0865	.1361	.1875	.2322	.2627	.2734

[†] Based on *Tables of the Binomial Probability Distribution*, National Bureau of Standards Applied Mathematics Series No. 6. Washington, D.C.: U.S. Government Printing Office, 1950.

TABLE I (continued)

n	x	θ .05	.10	.15	.20	.25	.30	.35	.40	.45	.50
8	5	.0000	.0004	.0026	.0092	.0231	.0467	.0808	.1239	.1719	.2188
	6	.0000	.0000	.0002	.0011	.0038	.0100	.0217	.0413	.0703	.1094
	7	.0000	.0000	.0000	.0001	.0004	.0012	.0033	.0079	.0164	.0312
	8	.0000	.0000	.0000	.0000	.0000	.0001	.0002	.0007	.0017	.0039
9	0	.6302	.3874	.2316	.1342	.0751	.0404	.0207	.0101	.0046	.0020
	1	.2985	.3874	.3679	.3020	.2253	.1556	.1004	.0605	.0339	.0176
	2	.0629	.1722	.2597	.3020	.3003	.2668	.2162	.1612	.1110	.0703
	3	.0077	.0446	.1069	.1762	.2336	.2668	.2716	.2508	.2119	.1641
	4	.0006	.0074	.0283	.0661	.1168	.1715	.2194	.2508	.2600	.2461
	5	.0000	.0008	.0050	.0165	.0389	.0735	.1181	.1672	.2128	.2461
	6	.0000	.0001	.0006	.0028	.0087	.0210	.0424	.0743	.1160	.1641
	7	.0000	.0000	.0000	.0003	.0012	.0039	.0098	.0212	.0407	.0703
	8	.0000	.0000	.0000	.0000	.0001	.0004	.0013	.0035	.0083	.0176
	9	.0000	.0000	.0000	.0000	.0000	.0000	.0001	.0003	.0008	.0020
10	0	.5987	.3487	.1969	.1074	.0563	.0282	.0135	.0060	.0025	.0010
	1	.3151	.3874	.3474	.2684	.1877	.1211	.0725	.0403	.0207	.0098
	2	.0746	.1937	.2759	.3020	.2816	.2335	.1757	.1209	.0763	.0439
	3	.0105	.0574	.1298	.2013	.2503	.2668	.2522	.2150	.1665	.1172
	4	.0010	.0112	.0401	.0881	.1460	.2001	.2377	.2508	.2384	.2051
	5	.0001	.0015	.0085	.0264	.0584	.1029	.1536	.2007	.2340	.2461
	6	.0000	.0001	.0012	.0055	.0162	.0368	.0689	.1115	.1596	.2051
	7	.0000	.0000	.0001	.0008	.0031	.0090	.0212	.0425	.0746	.1172
	8	.0000	.0000	.0000	.0001	.0004	.0014	.0043	.0106	.0229	.0439
	9	.0000	.0000	.0000	.0000	.0000	.0001	.0005	.0016	.0042	.0098
	10	.0000	.0000	.0000	.0000	.0000	.0000	.0000	.0001	.0003	.0010
11	0	.5688	.3138	.1673	.0859	.0422	.0198	.0088	.0036	.0014	.0005
	1	.3293	.3835	.3248	.2362	.1549	.0932	.0518	.0266	.0125	.0054
	2	.0867	.2131	.2866	.2953	.2581	.1998	.1395	.0887	.0513	.0269
	3	.0137	.0710	.1517	.2215	.2581	.2568	.2254	.1774	.1259	.0806
	4	.0014	.0158	.0536	.1107	.1721	.2201	.2428	.2365	.2060	.1611
	5	.0001	.0025	.0132	.0388	.0803	.1321	.1830	.2207	.2360	.2256
	6	.0000	.0003	.0023	.0097	.0268	.0566	.0985	.1471	.1931	.2256
	7	.0000	.0000	.0003	.0017	.0064	.0173	.0379	.0701	.1128	.1611
	8	.0000	.0000	.0000	.0002	.0011	.0037	.0102	.0234	.0462	.0806
	9	.0000	.0000	.0000	.0000	.0001	.0005	.0018	.0052	.0126	.0269
	10	.0000	.0000	.0000	.0000	.0000	.0000	.0002	.0007	.0021	.0054
	11	.0000	.0000	.0000	.0000	.0000	.0000	.0000	.0000	.0002	.0005
12	0	.5404	.2824	.1422	.0687	.0317	.0138	.0057	.0022	.0008	.0002
	1	.3413	.3766	.3012	.2062	.1267	.0712	.0368	.0174	.0075	.0029
	2	.0988	.2301	.2924	.2835	.2323	.1678	.1088	.0639	.0339	.0161
	3	.0173	.0852	.1720	.2362	.2581	.2397	.1954	.1419	.0923	.0537
	4	.0021	.0213	.0683	.1329	.1936	.2311	.2367	.2128	.1700	.1208
	5	.0002	.0038	.0193	.0532	.1032	.1585	.2039	.2270	.2225	.1934
	6	.0000	.0005	.0040	.0155	.0401	.0792	.1281	.1766	.2124	.2256
	7	.0000	.0000	.0006	.0033	.0115	.0291	.0591	.1009	.1489	.1934
	8	.0000	.0000	.0001	.0005	.0024	.0078	.0199	.0420	.0762	.1208
	9	.0000	.0000	.0000	.0001	.0004	.0015	.0048	.0125	.0277	.0537

TABLE I (continued)

n	x	.05	.10	.15	.20	.25	.30	.35	.40	.45	.50
						θ					
12	10	.0000	.0000	.0000	.0000	.0000	.0002	.0008	.0025	.0068	.0161
	11	.0000	.0000	.0000	.0000	.0000	.0000	.0001	.0003	.0010	.0029
	12	.0000	.0000	.0000	.0000	.0000	.0000	.0000	.0000	.0001	.0002
13	0	.5133	.2542	.1209	.0550	.0238	.0097	.0037	.0013	.0004	.0001
	1	.3512	.3672	.2774	.1787	.1029	.0540	.0259	.0113	.0045	.0016
	2	.1109	.2448	.2937	.2680	.2059	.1388	.0836	.0453	.0220	.0095
	3	.0214	.0997	.1900	.2457	.2517	.2181	.1651	.1107	.0660	.0349
	4	.0028	.0277	.0838	.1535	.2097	.2337	.2222	.1845	.1350	.0873
	5	.0003	.0055	.0266	.0691	.1258	.1803	.2154	.2214	.1989	.1571
	6	.0000	.0008	.0063	.0230	.0559	.1030	.1546	.1968	.2169	.2095
	7	.0000	.0001	.0011	.0058	.0186	.0442	.0833	.1312	.1775	.2095
	8	.0000	.0000	.0001	.0011	.0047	.0142	.0336	.0656	.1089	.1571
	9	.0000	.0000	.0000	.0001	.0009	.0034	.0101	.0243	.0495	.0873
	10	.0000	.0000	.0000	.0000	.0001	.0006	.0022	.0065	.0162	.0349
	11	.0000	.0000	.0000	.0000	.0000	.0001	.0003	.0012	.0036	.0095
	12	.0000	.0000	.0000	.0000	.0000	.0000	.0000	.0001	.0005	.0016
	13	.0000	.0000	.0000	.0000	.0000	.0000	.0000	.0000	.0000	.0001
14	0	.4877	.2288	.1028	.0440	.0178	.0068	.0024	.0008	.0002	.0001
	1	.3593	.3559	.2539	.1539	.0832	.0407	.0181	.0073	.0027	.0009
	2	.1229	.2570	.2912	.2501	.1802	.1134	.0634	.0317	.0141	.0056
	3	.0259	.1142	.2056	.2501	.2402	.1943	.1366	.0845	.0462	.0222
	4	.0037	.0349	.0998	.1720	.2202	.2290	.2022	.1549	.1040	.0611
	5	.0004	.0078	.0352	.0860	.1468	.1963	.2178	.2066	.1701	.1222
	6	.0000	.0013	.0093	.0322	.0734	.1262	.1759	.2066	.2088	.1833
	7	.0000	.0002	.0019	.0092	.0280	.0618	.1082	.1574	.1952	.2095
	8	.0000	.0000	.0003	.0020	.0082	.0232	.0510	.0918	.1398	.1833
	9	.0000	.0000	.0000	.0003	.0018	.0066	.0183	.0408	.0762	.1222
	10	.0000	.0000	.0000	.0000	.0003	.0014	.0049	.0136	.0312	.0611
	11	.0000	.0000	.0000	.0000	.0000	.0002	.0010	.0033	.0093	.0222
	12	.0000	.0000	.0000	.0000	.0000	.0000	.0001	.0005	.0019	.0056
	13	.0000	.0000	.0000	.0000	.0000	.0000	.0000	.0001	.0002	.0009
	14	.0000	.0000	.0000	.0000	.0000	.0000	.0000	.0000	.0000	.0001
15	0	.4633	.2059	.0874	.0352	.0134	.0047	.0016	.0005	.0001	.0000
	1	.3658	.3432	.2312	.1319	.0668	.0305	.0126	.0047	.0016	.0005
	2	.1348	.2669	.2856	.2309	.1559	.0916	.0476	.0219	.0090	.0032
	3	.0307	.1285	.2184	.2501	.2252	.1700	.1110	.0634	.0318	.0139
	4	.0049	.0428	.1156	.1876	.2252	.2186	.1792	.1268	.0780	.0417
	5	.0006	.0105	.0449	.1032	.1651	.2061	.2123	.1859	.1404	.0916
	6	.0000	.0019	.0132	.0430	.0917	.1472	.1906	.2066	.1914	.1527
	7	.0000	.0003	.0030	.0138	.0393	.0811	.1319	.1771	.2013	.1964
	8	.0000	.0000	.0005	.0035	.0131	.0348	.0710	.1181	.1647	.1964
	9	.0000	.0000	.0001	.0007	.0034	.0116	.0298	.0612	.1048	.1527
	10	.0000	.0000	.0000	.0001	.0007	.0030	.0096	.0245	.0515	.0916
	11	.0000	.0000	.0000	.0000	.0001	.0006	.0024	.0074	.0191	.0417
	12	.0000	.0000	.0000	.0000	.0000	.0001	.0004	.0016	.0052	.0139
	13	.0000	.0000	.0000	.0000	.0000	.0000	.0001	.0003	.0010	.0032
	14	.0000	.0000	.0000	.0000	.0000	.0000	.0000	.0000	.0001	.0005
	15	.0000	.0000	.0000	.0000	.0000	.0000	.0000	.0000	.0000	.0000

TABLE I (continued)

						θ					
n	*x*	.05	.10	.15	.20	.25	.30	.35	.40	.45	.50
16	0	.4401	.1853	.0743	.0281	.0100	.0033	.0010	.0003	.0001	.0000
	1	.3706	.3294	.2097	.1126	.0535	.0228	.0087	.0030	.0009	.0002
	2	.1463	.2745	.2775	.2111	.1336	.0732	.0353	.0150	.0056	.0018
	3	.0359	.1423	.2285	.2463	.2079	.1465	.0888	.0468	.0215	.0085
	4	.0061	.0514	.1311	.2001	.2252	.2040	.1553	.1014	.0572	.0278
	5	.0008	.0137	.0555	.1201	.1802	.2099	.2008	.1623	.1123	.0667
	6	.0001	.0028	.0180	.0550	.1101	.1649	.1982	.1983	.1684	.1222
	7	.0000	.0004	.0045	.0197	.0524	.1010	.1524	.1889	.1969	.1746
	8	.0000	.0001	.0009	.0055	.0197	.0487	.0923	.1417	.1812	.1964
	9	.0000	.0000	.0001	.0012	.0058	.0185	.0442	.0840	.1318	.1746
	10	.0000	.0000	.0000	.0002	.0014	.0056	.0167	.0392	.0755	.1222
	11	.0000	.0000	.0000	.0000	.0002	.0013	.0049	.0142	.0337	.0667
	12	.0000	.0000	.0000	.0000	.0000	.0002	.0011	.0040	.0115	.0278
	13	.0000	.0000	.0000	.0000	.0000	.0000	.0002	.0008	.0029	.0085
	14	.0000	.0000	.0000	.0000	.0000	.0000	.0000	.0001	.0005	.0018
	15	.0000	.0000	.0000	.0000	.0000	.0000	.0000	.0000	.0001	.0002
	16	.0000	.0000	.0000	.0000	.0000	.0000	.0000	.0000	.0000	.0000
17	0	.4181	.1668	.0631	.0225	.0075	.0023	.0007	.0002	.0000	.0000
	1	.3741	.3150	.1893	.0957	.0426	.0169	.0060	.0019	.0005	.0001
	2	.1575	.2800	.2673	.1914	.1136	.0581	.0260	.0102	.0035	.0010
	3	.0415	.1556	.2359	.2393	.1893	.1245	.0701	.0341	.0144	.0052
	4	.0076	.0605	.1457	.2093	.2209	.1868	.1320	.0796	.0411	.0182
	5	.0010	.0175	.0668	.1361	.1914	.2081	.1849	.1379	.0875	.0472
	6	.0001	.0039	.0236	.0680	.1276	.1784	.1991	.1839	.1432	.0944
	7	.0000	.0007	.0065	.0267	.0668	.1201	.1685	.1927	.1841	.1484
	8	.0000	.0001	.0014	.0084	.0279	.0644	.1134	.1606	.1883	.1855
	9	.0000	.0000	.0003	.0021	.0093	.0276	.0611	.1070	.1540	.1855
	10	.0000	.0000	.0000	.0004	.0025	.0095	.0263	.0571	.1008	.1484
	11	.0000	.0000	.0000	.0001	.0005	.0026	.0090	.0242	.0525	.0944
	12	.0000	.0000	.0000	.0000	.0001	.0006	.0024	.0081	.0215	.0472
	13	.0000	.0000	.0000	.0000	.0000	.0001	.0005	.0021	.0068	.0182
	14	.0000	.0000	.0000	.0000	.0000	.0000	.0001	.0004	.0016	.0052
	15	.0000	.0000	.0000	.0000	.0000	.0000	.0000	.0001	.0003	.0010
	16	.0000	.0000	.0000	.0000	.0000	.0000	.0000	.0000	.0000	.0001
	17	.0000	.0000	.0000	.0000	.0000	.0000	.0000	.0000	.0000	.0000
18	0	.3972	.1501	.0536	.0180	.0056	.0016	.0004	.0001	.0000	.0000
	1	.3763	.3002	.1704	.0811	.0338	.0126	.0042	.0012	.0003	.0001
	2	.1683	.2835	.2556	.1723	.0958	.0458	.0190	.0069	.0022	.0006
	3	.0473	.1680	.2406	.2297	.1704	.1046	.0547	.0246	.0095	.0031
	4	.0093	.0700	.1592	.2153	.2130	.1681	.1104	.0614	.0291	.0117
	5	.0014	.0218	.0787	.1507	.1988	.2017	.1664	.1146	.0666	.0327
	6	.0002	.0052	.0301	.0816	.1436	.1873	.1941	.1655	.1181	.0708
	7	.0000	.0010	.0091	.0350	.0820	.1376	.1792	.1892	.1657	.1214
	8	.0000	.0002	.0022	.0120	.0376	.0811	.1327	.1734	.1864	.1669
	9	.0000	.0000	.0004	.0033	.0139	.0386	.0794	.1284	.1694	.1855

TABLE I (continued)

n	x	.05	.10	.15	.20	.25	.30	.35	.40	.45	.50
						θ					
18	10	.0000	.0000	.0001	.0008	.0042	.0149	.0385	.0771	.1248	.1669
	11	.0000	.0000	.0000	.0001	.0010	.0046	.0151	.0374	.0742	.1214
	12	.0000	.0000	.0000	.0000	.0002	.0012	.0047	.0145	.0354	.0708
	13	.0000	.0000	.0000	.0000	.0000	.0002	.0012	.0045	.0134	.0327
	14	.0000	.0000	.0000	.0000	.0000	.0000	.0002	.0011	.0039	.0117
	15	.0000	.0000	.0000	.0000	.0000	.0000	.0000	.0002	.0009	.0031
	16	.0000	.0000	.0000	.0000	.0000	.0000	.0000	.0000	.0001	.0006
	17	.0000	.0000	.0000	.0000	.0000	.0000	.0000	.0000	.0000	.0001
	18	.0000	.0000	.0000	.0000	.0000	.0000	.0000	.0000	.0000	.0000
19	0	.3774	.1351	.0456	.0144	.0042	.0011	.0003	.0001	.0000	.0000
	1	.3774	.2852	.1529	.0685	.0268	.0093	.0029	.0008	.0002	.0000
	2	.1787	.2852	.2428	.1540	.0803	.0358	.0138	.0046	.0013	.0003
	3	.0533	.1796	.2428	.2182	.1517	.0869	.0422	.0175	.0062	.0018
	4	.0112	.0798	.1714	.2182	.2023	.1491	.0909	.0467	.0203	.0074
	5	.0018	.0266	.0907	.1636	.2023	.1916	.1468	.0933	.0497	.0222
	6	.0002	.0069	.0374	.0955	.1574	.1916	.1844	.1451	.0949	.0518
	7	.0000	.0014	.0122	.0443	.0974	.1525	.1844	.1797	.1443	.0961
	8	.0000	.0002	.0032	.0166	.0487	.0981	.1489	.1797	.1771	.1442
	9	.0000	.0000	.0007	.0051	.0198	.0514	.0980	.1464	.1771	.1762
	10	.0000	.0000	.0001	.0013	.0066	.0220	.0528	.0976	.1449	.1762
	11	.0000	.0000	.0000	.0003	.0018	.0077	.0233	.0532	.0970	.1442
	12	.0000	.0000	.0000	.0000	.0004	.0022	.0083	.0237	.0529	.0961
	13	.0000	.0000	.0000	.0000	.0001	.0005	.0024	.0085	.0233	.0518
	14	.0000	.0000	.0000	.0000	.0000	.0001	.0006	.0024	.0082	.0222
	15	.0000	.0000	.0000	.0000	.0000	.0000	.0001	.0005	.0022	.0074
	16	.0000	.0000	.0000	.0000	.0000	.0000	.0000	.0001	.0005	.0018
	17	.0000	.0000	.0000	.0000	.0000	.0000	.0000	.0000	.0001	.0003
	18	.0000	.0000	.0000	.0000	.0000	.0000	.0000	.0000	.0000	.0000
	19	.0000	.0000	.0000	.0000	.0000	.0000	.0000	.0000	.0000	.0000
20	0	.3585	.1216	.0388	.0115	.0032	.0008	.0002	.0000	.0000	.0000
	1	.3774	.2702	.1368	.0576	.0211	.0068	.0020	.0005	.0001	.0000
	2	.1887	.2852	.2293	.1369	.0669	.0278	.0100	.0031	.0008	.0002
	3	.0596	.1901	.2428	.2054	.1339	.0716	.0323	.0123	.0040	.0011
	4	.0133	.0898	.1821	.2182	.1897	.1304	.0738	.0350	.0139	.0046
	5	.0022	.0319	.1028	.1746	.2023	.1789	.1272	.0746	.0365	.0148
	6	.0003	.0089	.0454	.1091	.1686	.1916	.1712	.1244	.0746	.0370
	7	.0000	.0020	.0160	.0545	.1124	.1643	.1844	.1659	.1221	.0739
	8	.0000	.0004	.0046	.0222	.0609	.1144	.1614	.1797	.1623	.1201
	9	.0000	.0001	.0011	.0074	.0271	.0654	.1158	.1597	.1771	.1602
	10	.0000	.0000	.0002	.0020	.0099	.0308	.0686	.1171	.1593	.1762
	11	.0000	.0000	.0000	.0005	.0030	.0120	.0336	.0710	.1185	.1602
	12	.0000	.0000	.0000	.0001	.0008	.0039	.0136	.0355	.0727	.1201
	13	.0000	.0000	.0000	.0000	.0002	.0010	.0045	.0146	.0366	.0739
	14	.0000	.0000	.0000	.0000	.0000	.0002	.0012	.0049	.0150	.0370
	15	.0000	.0000	.0000	.0000	.0000	.0000	.0003	.0013	.0049	.0148
	16	.0000	.0000	.0000	.0000	.0000	.0000	.0000	.0003	.0013	.0046
	17	.0000	.0000	.0000	.0000	.0000	.0000	.0000	.0000	.0002	.0011
	18	.0000	.0000	.0000	.0000	.0000	.0000	.0000	.0000	.0000	.0002
	19	.0000	.0000	.0000	.0000	.0000	.0000	.0000	.0000	.0000	.0000
	20	.0000	.0000	.0000	.0000	.0000	.0000	.0000	.0000	.0000	.0000

TABLE II

Poisson Probabilities[†]

	λ									
x	0.1	0.2	0.3	0.4	0.5	0.6	0.7	0.8	0.9	1.0
0	.9048	.8187	.7408	.6703	.6065	.5488	.4966	.4493	.4066	.3679
1	.0905	.1637	.2222	.2681	.3033	.3293	.3476	.3595	.3659	.3679
2	.0045	.0164	.0333	.0536	.0758	.0988	.1217	.1438	.1647	.1839
3	.0002	.0011	.0033	.0072	.0126	.0198	.0284	.0383	.0494	.0613
4	.0000	.0001	.0002	.0007	.0016	.0030	.0050	.0077	.0111	.0153
5	.0000	.0000	.0000	.0001	.0002	.0004	.0007	.0012	.0020	.0031
6	.0000	.0000	.0000	.0000	.0000	.0000	.0001	.0002	.0003	.0005
7	.0000	.0000	.0000	.0000	.0000	.0000	.0000	.0000	.0000	.0001

	λ									
x	1.1	1.2	1.3	1.4	1.5	1.6	1.7	1.8	1.9	2.0
0	.3329	.3012	.2725	.2466	.2231	.2019	.1827	.1653	.1496	.1353
1	.3662	.3614	.3543	.3452	.3347	.3230	.3106	.2975	.2842	.2707
2	.2014	.2169	.2303	.2417	.2510	.2584	.2640	.2678	.2700	.2707
3	.0738	.0867	.0998	.1128	.1255	.1378	.1496	.1607	.1710	.1804
4	.0203	.0260	.0324	.0395	.0471	.0551	.0636	.0723	.0812	.0902
5	.0045	.0062	.0084	.0111	.0141	.0176	.0216	.0260	.0309	.0361
6	.0008	.0012	.0018	.0026	.0035	.0047	.0061	.0078	.0098	.0120
7	.0001	.0002	.0003	.0005	.0008	.0011	.0015	.0020	.0027	.0034
8	.0000	.0000	.0001	.0001	.0001	.0002	.0003	.0005	.0006	.0009
9	.0000	.0000	.0000	.0000	.0000	.0000	.0001	.0001	.0001	.0002

	λ									
x	2.1	2.2	2.3	2.4	2.5	2.6	2.7	2.8	2.9	3.0
0	.1225	.1108	.1003	.0907	.0821	.0743	.0672	.0608	.0550	.0498
1	.2572	.2438	.2306	.2177	.2052	.1931	.1815	.1703	.1596	.1494
2	.2700	.2681	.2652	.2613	.2565	.2510	.2450	.2384	.2314	.2240
3	.1890	.1966	.2033	.2090	.2138	.2176	.2205	.2225	.2237	.2240
4	.0992	.1082	.1169	.1254	.1336	.1414	.1488	.1557	.1622	.1680
5	.0417	.0476	.0538	.0602	.0668	.0735	.0804	.0872	.0940	.1008
6	.0146	.0174	.0206	.0241	.0278	.0319	.0362	.0407	.0455	.0504
7	.0044	.0055	.0068	.0083	.0099	.0118	.0139	.0163	.0188	.0216
8	.0011	.0015	.0019	.0025	.0031	.0038	.0047	.0057	.0068	.0081
9	.0003	.0004	.0005	.0007	.0009	.0011	.0014	.0018	.0022	.0027
10	.0001	.0001	.0001	.0002	.0002	.0003	.0004	.0005	.0006	.0008
11	.0000	.0000	.0000	.0000	.0000	.0001	.0001	.0001	.0002	.0002
12	.0000	.0000	.0000	.0000	.0000	.0000	.0000	.0000	.0000	.0001

[†] Reproduced by permission from *Handbook of Probability and Statistics with Tables*, by R. S. Burington and D. C. May, Jr. New York: McGraw-Hill Book Company, 1953.

TABLE II (continued)

x	λ = 3.1	3.2	3.3	3.4	3.5	3.6	3.7	3.8	3.9	4.0
0	.0450	.0408	.0369	.0334	.0302	.0273	.0247	.0224	.0202	.0183
1	.1397	.1304	.1217	.1135	.1057	.0984	.0915	.0850	.0789	.0733
2	.2165	.2087	.2008	.1929	.1850	.1771	.1692	.1615	.1539	.1465
3	.2237	.2226	.2209	.2186	.2158	.2125	.2087	.2046	.2001	.1954
4	.1734	.1781	.1823	.1858	.1888	.1912	.1931	.1944	.1951	.1954
5	.1075	.1140	.1203	.1264	.1322	.1377	.1429	.1477	.1522	.1563
6	.0555	.0608	.0662	.0716	.0771	.0826	.0881	.0936	.0989	.1042
7	.0246	.0278	.0312	.0348	.0385	.0425	.0466	.0508	.0551	.0595
8	.0095	.0111	.0129	.0148	.0169	.0191	.0215	.0241	.0269	.0298
9	.0033	.0040	.0047	.0056	.0066	.0076	.0089	.0102	.0116	.0132
10	.0010	.0013	.0016	.0019	.0023	.0028	.0033	.0039	.0045	.0053
11	.0003	.0004	.0005	.0006	.0007	.0009	.0011	.0013	.0016	.0019
12	.0001	.0001	.0001	.0002	.0002	.0003	.0003	.0004	.0005	.0006
13	.0000	.0000	.0000	.0000	.0001	.0001	.0001	.0001	.0002	.0002
14	.0000	.0000	.0000	.0000	.0000	.0000	.0000	.0000	.0000	.0001

x	λ = 4.1	4.2	4.3	4.4	4.5	4.6	4.7	4.8	4.9	5.0
0	.0166	.0150	.0136	.0123	.0111	.0101	.0091	.0082	.0074	.0067
1	.0679	.0630	.0583	.0540	.0500	.0462	.0427	.0395	.0365	.0337
2	.1393	.1323	.1254	.1188	.1125	.1063	.1005	.0948	.0894	.0842
3	.1904	.1852	.1798	.1743	.1687	.1631	.1574	.1517	.1460	.1404
4	.1951	.1944	.1933	.1917	.1898	.1875	.1849	.1820	.1789	.1755
5	.1600	.1633	.1662	.1687	.1708	.1725	.1738	.1747	.1753	.1755
6	.1093	.1143	.1191	.1237	.1281	.1323	.1362	.1398	.1432	.1462
7	.0640	.0686	.0732	.0778	.0824	.0869	.0914	.0959	.1002	.1044
8	.0328	.0360	.0393	.0428	.0463	.0500	.0537	.0575	.0614	.0653
9	.0150	.0168	.0188	.0209	.0232	.0255	.0280	.0307	.0334	.0363
10	.0061	.0071	.0081	.0092	.0104	.0118	.0132	.0147	.0164	.0181
11	.0023	.0027	.0032	.0037	.0043	.0049	.0056	.0064	.0073	.0082
12	.0008	.0009	.0011	.0014	.0016	.0019	.0022	.0026	.0030	.0034
13	.0002	.0003	.0004	.0005	.0006	.0007	.0008	.0009	.0011	.0013
14	.0001	.0001	.0001	.0001	.0002	.0002	.0003	.0003	.0004	.0005
15	.0000	.0000	.0000	.0000	.0001	.0001	.0001	.0001	.0001	.0002

x	λ = 5.1	5.2	5.3	5.4	5.5	5.6	5.7	5.8	5.9	6.0
0	.0061	.0055	.0050	.0045	.0041	.0037	.0033	.0030	.0027	.0025
1	.0311	.0287	.0265	.0244	.0225	.0207	.0191	.0176	.0162	.0149
2	.0793	.0746	.0701	.0659	.0618	.0580	.0544	.0509	.0477	.0446
3	.1348	.1293	.1239	.1185	.1133	.1082	.1033	.0985	.0938	.0892
4	.1719	.1681	.1641	.1600	.1558	.1515	.1472	.1428	.1383	.1339

TABLE II (continued)

x	λ 5.1	5.2	5.3	5.4	5.5	5.6	5.7	5.8	5.9	6.0
5	.1753	.1748	.1740	.1728	.1714	.1697	.1678	.1656	.1632	.1606
6	.1490	.1515	.1537	.1555	.1571	.1584	.1594	.1601	.1605	.1606
7	.1086	.1125	.1163	.1200	.1234	.1267	.1298	.1326	.1353	.1377
8	.0692	.0731	.0771	.0810	.0849	.0887	.0925	.0962	.0998	.1033
9	.0392	.0423	.0454	.0486	.0519	.0552	.0586	.0620	.0654	.0688
10	.0200	.0220	.0241	.0262	.0285	.0309	.0334	.0359	.0386	.0413
11	.0093	.0104	.0116	.0129	.0143	.0157	.0173	.0190	.0207	.0225
12	.0039	.0045	.0051	.0058	.0065	.0073	.0082	.0092	.0102	.0113
13	.0015	.0018	.0021	.0024	.0028	.0032	.0036	.0041	.0046	.0052
14	.0006	.0007	.0008	.0009	.0011	.0013	.0015	.0017	.0019	.0022
15	.0002	.0002	.0003	.0003	.0004	.0005	.0006	.0007	.0008	.0009
16	.0001	.0001	.0001	.0001	.0001	.0002	.0002	.0002	.0003	.0003
17	.0000	.0000	.0000	.0000	.0000	.0001	.0001	.0001	.0001	.0001

x	λ 6.1	6.2	6.3	6.4	6.5	6.6	6.7	6.8	6.9	7.0
0	.0022	.0020	.0018	.0017	.0015	.0014	.0012	.0011	.0010	.0009
1	.0137	.0126	.0116	.0106	.0098	.0090	.0082	.0076	.0070	.0064
2	.0417	.0390	.0364	.0340	.0318	.0296	.0276	.0258	.0240	.0223
3	.0848	.0806	.0765	.0726	.0688	.0652	.0617	.0584	.0552	.0521
4	.1294	.1249	.1205	.1162	.1118	.1076	.1034	.0992	.0952	.0912
5	.1579	.1549	.1519	.1487	.1454	.1420	.1385	.1349	.1314	.1277
6	.1605	.1601	.1595	.1586	.1575	.1562	.1546	.1529	.1511	.1490
7	.1399	.1418	.1435	.1450	.1462	.1472	.1480	.1486	.1489	.1490
8	.1066	.1099	.1130	.1160	.1188	.1215	.1240	.1263	.1284	.1304
9	.0723	.0757	.0791	.0825	.0858	.0891	.0923	.0954	.0985	.1014
10	.0441	.0469	.0498	.0528	.0558	.0588	.0618	.0649	.0679	.0710
11	.0245	.0265	.0285	.0307	.0330	.0353	.0377	.0401	.0426	.0452
12	.0124	.0137	.0150	.0164	.0179	.0194	.0210	.0227	.0245	.0264
13	.0058	.0065	.0073	.0081	.0089	.0098	.0108	.0119	.0130	.0142
14	.0025	.0029	.0033	.0037	.0041	.0046	.0052	.0058	.0064	.0071
15	.0010	.0012	.0014	.0016	.0018	.0020	.0023	.0026	.0029	.0033
16	.0004	.0005	.0005	.0006	.0007	.0008	.0010	.0011	.0013	.0014
17	.0001	.0002	.0002	.0002	.0003	.0003	.0004	.0004	.0005	.0006
18	.0000	.0001	.0001	.0001	.0001	.0001	.0001	.0002	.0002	.0002
19	.0000	.0000	.0000	.0000	.0000	.0000	.0000	.0001	.0001	.0001

x	λ 7.1	7.2	7.3	7.4	7.5	7.6	7.7	7.8	7.9	8.0
0	.0008	.0007	.0007	.0006	.0006	.0005	.0005	.0004	.0004	.0003
1	.0059	.0054	.0049	.0045	.0041	.0038	.0035	.0032	.0029	.0027
2	.0208	.0194	.0180	.0167	.0156	.0145	.0134	.0125	.0116	.0107
3	.0492	.0464	.0438	.0413	.0389	.0366	.0345	.0324	.0305	.0286
4	.0874	.0836	.0799	.0764	.0729	.0696	.0663	.0632	.0602	.0573
5	.1241	.1204	.1167	.1130	.1094	.1057	.1021	.0986	.0951	.0916
6	.1468	.1445	.1420	.1394	.1367	.1339	.1311	.1282	.1252	.1221
7	.1489	.1486	.1481	.1474	.1465	.1454	.1442	.1428	.1413	.1396
8	.1321	.1337	.1351	.1363	.1373	.1382	.1388	.1392	.1395	.1396
9	.1042	.1070	.1096	.1121	.1144	.1167	.1187	.1207	.1224	.1241

TABLE II (continued)

x	7.1	7.2	7.3	7.4	7.5	7.6	7.7	7.8	7.9	8.0
10	.0740	.0770	.0800	.0829	.0858	.0887	.0914	.0941	.0967	.0993
11	.0478	.0504	.0531	.0558	.0585	.0613	.0640	.0667	.0695	.0722
12	.0283	.0303	.0323	.0344	.0366	.0388	.0411	.0434	.0457	.0481
13	.0154	.0168	.0181	.0196	.0211	.0227	.0243	.0260	.0278	.0296
14	.0078	.0086	.0095	.0104	.0113	.0123	.0134	.0145	.0157	.0169
15	.0037	.0041	.0046	.0051	.0057	.0062	.0069	.0075	.0083	.0090
16	.0016	.0019	.0021	.0024	.0026	.0030	.0033	.0037	.0041	.0045
17	.0007	.0008	.0009	.0010	.0012	.0013	.0015	.0017	.0019	.0021
18	.0003	.0003	.0004	.0004	.0005	.0006	.0006	.0007	.0008	.0009
19	.0001	.0001	.0001	.0002	.0002	.0002	.0003	.0003	.0003	.0004
20	.0000	.0000	.0001	.0001	.0001	.0001	.0001	.0001	.0001	.0002
21	.0000	.0000	.0000	.0000	.0000	.0000	.0000	.0000	.0001	.0001

x	8.1	8.2	8.3	8.4	8.5	8.6	8.7	8.8	8.9	9.0
0	.0003	.0003	.0002	.0002	.0002	.0002	.0002	.0002	.0001	.0001
1	.0025	.0023	.0021	.0019	.0017	.0016	.0014	.0013	.0012	.0011
2	.0100	.0092	.0086	.0079	.0074	.0068	.0063	.0058	.0054	.0050
3	.0269	.0252	.0237	.0222	.0208	.0195	.0183	.0171	.0160	.0150
4	.0544	.0517	.0491	.0466	.0443	.0420	.0398	.0377	.0357	.0337
5	.0882	.0849	.0816	.0784	.0752	.0722	.0692	.0663	.0635	.0607
6	.1191	.1160	.1128	.1097	.1066	.1034	.1003	.0972	.0941	.0911
7	.1378	.1358	.1338	.1317	.1294	.1271	.1247	.1222	.1197	.1171
8	.1395	.1392	.1388	.1382	.1375	.1366	.1356	.1344	.1332	.1318
9	.1256	.1269	.1280	.1290	.1299	.1306	.1311	.1315	.1317	.1318
10	.1017	.1040	.1063	.1084	.1104	.1123	.1140	.1157	.1172	.1186
11	.0749	.0776	.0802	.0828	.0853	.0878	.0902	.0925	.0948	.0970
12	.0505	.0530	.0555	.0579	.0604	.0629	.0654	.0679	.0703	.0728
13	.0315	.0334	.0354	.0374	.0395	.0416	.0438	.0459	.0481	.0504
14	.0182	.0196	.0210	.0225	.0240	.0256	.0272	.0289	.0306	.0324
15	.0098	.0107	.0116	.0126	.0136	.0147	.0158	.0169	.0182	.0194
16	.0050	.0055	.0060	.0066	.0072	.0079	.0086	.0093	.0101	.0109
17	.0024	.0026	.0029	.0033	.0036	.0040	.0044	.0048	.0053	.0058
18	.0011	.0012	.0014	.0015	.0017	.0019	.0021	.0024	.0026	.0029
19	.0005	.0005	.0006	.0007	.0008	.0009	.0010	.0011	.0012	.0014
20	.0002	.0002	.0002	.0003	.0003	.0004	.0004	.0005	.0005	.0006
21	.0001	.0001	.0001	.0001	.0001	.0002	.0002	.0002	.0002	.0003
22	.0000	.0000	.0000	.0000	.0001	.0001	.0001	.0001	.0001	.0001

x	9.1	9.2	9.3	9.4	9.5	9.6	9.7	9.8	9.9	10
0	.0001	.0001	.0001	.0001	.0001	.0001	.0001	.0001	.0001	.0000
1	.0010	.0009	.0009	.0008	.0007	.0007	.0006	.0005	.0005	.0005
2	.0046	.0043	.0040	.0037	.0034	.0031	.0029	.0027	.0025	.0023
3	.0140	.0131	.0123	.0115	.0107	.0100	.0093	.0087	.0081	.0076
4	.0319	.0302	.0285	.0269	.0254	.0240	.0226	.0213	.0201	.0189

TABLE II (continued)

x	λ 9.1	9.2	9.3	9.4	9.5	9.6	9.7	9.8	9.9	10
5	.0581	.0555	.0530	.0506	.0483	.0460	.0439	.0418	.0398	.0378
6	.0881	.0851	.0822	.0793	.0764	.0736	.0709	.0682	.0656	.0631
7	.1145	.1118	.1091	.1064	.1037	.1010	.0982	.0955	.0928	.0901
8	.1302	.1286	.1269	.1251	.1232	.1212	.1191	.1170	.1148	.1126
9	.1317	.1315	.1311	.1306	.1300	.1293	.1284	.1274	.1263	.1251
10	.1198	.1210	.1219	.1228	.1235	.1241	.1245	.1249	.1250	.1251
11	.0991	.1012	.1031	.1049	.1067	.1083	.1098	.1112	.1125	.1137
12	.0752	.0776	.0799	.0822	.0844	.0866	.0888	.0908	.0928	.0948
13	.0526	.0549	.0572	.0594	.0617	.0640	.0662	.0685	.0707	.0729
14	.0342	.0361	.0380	.0399	.0419	.0439	.0459	.0479	.0500	.0521
15	.0208	.0221	.0235	.0250	.0265	.0281	.0297	.0313	.0330	.0347
16	.0118	.0127	.0137	.0147	.0157	.0168	.0180	.0192	.0204	.0217
17	.0063	.0069	.0075	.0081	.0088	.0095	.0103	.0111	.0119	.0128
18	.0032	.0035	.0039	.0042	.0046	.0051	.0055	.0060	.0065	.0071
19	.0015	.0017	.0019	.0021	.0023	.0026	.0028	.0031	.0034	.0037
20	.0007	.0008	.0009	.0010	.0011	.0012	.0014	.0015	.0017	.0019
21	.0003	.0003	.0004	.0004	.0005	.0006	.0006	.0007	.0008	.0009
22	.0001	.0001	.0002	.0002	.0002	.0002	.0003	.0003	.0004	.0004
23	.0000	.0001	.0001	.0001	.0001	.0001	.0001	.0001	.0002	.0002
24	.0000	.0000	.0000	.0000	.0000	.0000	.0000	.0001	.0001	.0001

x	λ 11	12	13	14	15	16	17	18	19	20
0	.0000	.0000	.0000	.0000	.0000	.0000	.0000	.0000	.0000	.0000
1	.0002	.0001	.0000	.0000	.0000	.0000	.0000	.0000	.0000	.0000
2	.0010	.0004	.0002	.0001	.0000	.0000	.0000	.0000	.0000	.0000
3	.0037	.0018	.0008	.0004	.0002	.0001	.0000	.0000	.0000	.0000
4	.0102	.0053	.0027	.0013	.0006	.0003	.0001	.0001	.0000	.0000
5	.0224	.0127	.0070	.0037	.0019	.0010	.0005	.0002	.0001	.0001
6	.0411	.0255	.0152	.0087	.0048	.0026	.0014	.0007	.0004	.0002
7	.0646	.0437	.0281	.0174	.0104	.0060	.0034	.0018	.0010	.0005
8	.0888	.0655	.0457	.0304	.0194	.0120	.0072	.0042	.0024	.0013
9	.1085	.0874	.0661	.0473	.0324	.0213	.0135	.0083	.0050	.0029
10	.1194	.1048	.0859	.0663	.0486	.0341	.0230	.0150	.0095	.0058
11	.1194	.1144	.1015	.0844	.0663	.0496	.0355	.0245	.0164	.0106
12	.1094	.1144	.1099	.0984	.0829	.0661	.0504	.0368	.0259	.0176
13	.0926	.1056	.1099	.1060	.0956	.0814	.0658	.0509	.0378	.0271
14	.0728	.0905	.1021	.1060	.1024	.0930	.0800	.0655	.0514	.0387
15	.0534	.0724	.0885	.0989	.1024	.0992	.0906	.0786	.0650	.0516
16	.0367	.0543	.0719	.0866	.0960	.0992	.0963	.0884	.0772	.0646
17	.0237	.0383	.0550	.0713	.0847	.0934	.0963	.0936	.0863	.0760
18	.0145	.0256	.0397	.0554	.0706	.0830	.0909	.0936	.0911	.0844
19	.0084	.0161	.0272	.0409	.0557	.0699	.0814	.0887	.0911	.0888
20	.0046	.0097	.0177	.0286	.0418	.0559	.0692	.0798	.0866	.0888
21	.0024	.0055	.0109	.0191	.0299	.0426	.0560	.0684	.0783	.0846
22	.0012	.0030	.0065	.0121	.0204	.0310	.0433	.0560	.0676	.0769
23	.0006	.0016	.0037	.0074	.0133	.0216	.0320	.0438	.0559	.0669
24	.0003	.0008	.0020	.0043	.0083	.0144	.0226	.0328	.0442	.0557

TABLE II (continued)

	λ									
x	11	12	13	14	15	16	17	18	19	20
25	.0001	.0004	.0010	.0024	.0050	.0092	.0154	.0237	.0336	.0446
26	.0000	.0002	.0005	.0013	.0029	.0057	.0101	.0164	.0246	.0343
27	.0000	.0001	.0002	.0007	.0016	.0034	.0063	.0109	.0173	.0254
28	.0000	.0000	.0001	.0003	.0009	.0019	.0038	.0070	.0117	.0181
29	.0000	.0000	.0001	.0002	.0004	.0011	.0023	.0044	.0077	.0125
30	.0000	.0000	.0000	.0001	.0002	.0006	.0013	.0026	.0049	.0083
31	.0000	.0000	.0000	.0000	.0001	.0003	.0007	.0015	.0030	.0054
32	.0000	.0000	.0000	.0000	.0001	.0001	.0004	.0009	.0018	.0034
33	.0000	.0000	.0000	.0000	.0000	.0001	.0002	.0005	.0010	.0020
34	.0000	.0000	.0000	.0000	.0000	.0000	.0001	.0002	.0006	.0012
35	.0000	.0000	.0000	.0000	.0000	.0000	.0000	.0001	.0003	.0007
36	.0000	.0000	.0000	.0000	.0000	.0000	.0000	.0001	.0002	.0004
37	.0000	.0000	.0000	.0000	.0000	.0000	.0000	.0000	.0001	.0002
38	.0000	.0000	.0000	.0000	.0000	.0000	.0000	.0000	.0000	.0001
39	.0000	.0000	.0000	.0000	.0000	.0000	.0000	.0000	.0000	.0001

TABLE III

The Standard Normal Distribution

z	.00	.01	.02	.03	.04	.05	.06	.07	.08	.09
0.0	.0000	.0040	.0080	.0120	.0160	.0199	.0239	.0279	.0319	.0359
0.1	.0398	.0438	.0478	.0517	.0557	.0596	.0636	.0675	.0714	.0753
0.2	.0793	.0832	.0871	.0910	.0948	.0987	.1026	.1064	.1103	.1141
0.3	.1179	.1217	.1255	.1293	.1331	.1368	.1406	.1443	.1480	.1517
0.4	.1554	.1591	.1628	.1664	.1700	.1736	.1772	.1808	.1844	.1879
0.5	.1915	.1950	.1985	.2019	.2054	.2088	.2123	.2157	.2190	.2224
0.6	.2257	.2291	.2324	.2357	.2389	.2422	.2454	.2486	.2517	.2549
0.7	.2580	.2611	.2642	.2673	.2704	.2734	.2764	.2794	.2823	.2852
0.8	.2881	.2910	.2939	.2967	.2995	.3023	.3051	.3078	.3106	.3133
0.9	.3159	.3186	.3212	.3238	.3264	.3289	.3315	.3340	.3365	.3389
1.0	.3413	.3438	.3461	.3485	.3508	.3531	.3554	.3577	.3599	.3621
1.1	.3643	.3665	.3686	.3708	.3729	.3749	.3770	.3790	.3810	.3830
1.2	.3849	.3869	.3888	.3907	.3925	.3944	.3962	.3980	.3997	.4015
1.3	.4032	.4049	.4066	.4082	.4099	.4115	.4131	.4147	.4162	.4177
1.4	.4192	.4207	.4222	.4236	.4251	.4265	.4279	.4292	.4306	.4319
1.5	.4332	.4345	.4357	.4370	.4382	.4394	.4406	.4418	.4429	.4441
1.6	.4452	.4463	.4474	.4484	.4495	.4505	.4515	.4525	.4535	.4545
1.7	.4554	.4564	.4573	.4582	.4591	.4599	.4608	.4616	.4625	.4633
1.8	.4641	.4649	.4656	.4664	.4671	.4678	.4686	.4693	.4699	.4706
1.9	.4713	.4719	.4726	.4732	.4738	.4744	.4750	.4756	.4761	.4767
2.0	.4772	.4778	.4783	.4788	.4793	.4798	.4803	.4808	.4812	.4817
2.1	.4821	.4826	.4830	.4834	.4838	.4842	.4846	.4850	.4854	.4857
2.2	.4861	.4864	.4868	.4871	.4875	.4878	.4881	.4884	.4887	.4890
2.3	.4893	.4896	.4898	.4901	.4904	.4906	.4909	.4911	.4913	.4916
2.4	.4918	.4920	.4922	.4925	.4927	.4929	.4931	.4932	.4934	.4936
2.5	.4938	.4940	.4941	.4943	.4945	.4946	.4948	.4949	.4951	.4952
2.6	.4953	.4955	.4956	.4957	.4959	.4960	.4961	.4962	.4963	.4964
2.7	.4965	.4966	.4967	.4968	.4969	.4970	.4971	.4972	.4973	.4974
2.8	.4974	.4975	.4976	.4977	.4977	.4978	.4979	.4979	.4980	.4981
2.9	.4981	.4982	.4982	.4983	.4984	.4984	.4985	.4985	.4986	.4986
3.0	.4987	.4987	.4987	.4988	.4988	.4989	.4989	.4989	.4990	.4990

Also, for z = 4.0, 5.0, and 6.0, the probabilities are 0.49997, 0.4999997, and 0.499999999.

TABLE IV

Values of $t_{\alpha,\nu}$[†]

ν	$\alpha = .10$	$\alpha = .05$	$\alpha = .025$	$\alpha = .01$	$\alpha = .005$	ν
1	3.078	6.314	12.706	31.821	63.657	1
2	1.886	2.920	4.303	6.965	9.925	2
3	1.638	2.353	3.182	4.541	5.841	3
4	1.533	2.132	2.776	3.747	4.604	4
5	1.476	2.015	2.571	3.365	4.032	5
6	1.440	1.943	2.447	3.143	3.707	6
7	1.415	1.895	2.365	2.998	3.499	7
8	1.397	1.860	2.306	2.896	3.355	8
9	1.383	1.833	2.262	2.821	3.250	9
10	1.372	1.812	2.228	2.764	3.169	10
11	1.363	1.796	2.201	2.718	3.106	11
12	1.356	1.782	2.179	2.681	3.055	12
13	1.350	1.771	2.160	2.650	3.012	13
14	1.345	1.761	2.145	2.624	2.977	14
15	1.341	1.753	2.131	2.602	2.947	15
16	1.337	1.746	2.120	2.583	2.921	16
17	1.333	1.740	2.110	2.567	2.898	17
18	1.330	1.734	2.101	2.552	2.878	18
19	1.328	1.729	2.093	2.539	2.861	19
20	1.325	1.725	2.086	2.528	2.845	20
21	1.323	1.721	2.080	2.518	2.831	21
22	1.321	1.717	2.074	2.508	2.819	22
23	1.319	1.714	2.069	2.500	2.807	23
24	1.318	1.711	2.064	2.492	2.797	24
25	1.316	1.708	2.060	2.485	2.787	25
26	1.315	1.706	2.056	2.479	2.779	26
27	1.314	1.703	2.052	2.473	2.771	27
28	1.313	1.701	2.048	2.467	2.763	28
29	1.311	1.699	2.045	2.462	2.756	29
inf.	1.282	1.645	1.960	2.326	2.576	inf.

[†] Based on Table 12 of *Biometrika Tables for Statisticians*, Vol. I, Cambridge University Press, 1954, by permission of the *Biometrika* trustees.

TABLE V

Values of $\chi^2_{\alpha,\nu}$ †

ν	α = .995	α = .99	α = .975	α = .95	α = .05	α = .025	α = .01	α = .005	ν
1	.0000393	.000157	.000982	.00393	3.841	5.024	6.635	7.879	1
2	.0100	.0201	.0506	.103	5.991	7.378	9.210	10.597	2
3	.0717	.115	.216	.352	7.815	9.348	11.345	12.838	3
4	.207	.297	.484	.711	9.488	11.143	13.277	14.860	4
5	.412	.554	.831	1.145	11.070	12.832	15.086	16.750	5
6	.676	.872	1.237	1.635	12.592	14.449	16.812	18.548	6
7	.989	1.239	1.690	2.167	14.067	16.013	18.475	20.278	7
8	1.344	1.646	2.180	2.733	15.507	17.535	20.090	21.955	8
9	1.735	2.088	2.700	3.325	16.919	19.023	21.666	23.589	9
10	2.156	2.558	3.247	3.940	18.307	20.483	23.209	25.188	10
11	2.603	3.053	3.816	4.575	19.675	21.920	24.725	26.757	11
12	3.074	3.571	4.404	5.226	21.026	23.337	26.217	28.300	12
13	3.565	4.107	5.009	5.892	22.362	24.736	27.688	29.819	13
14	4.075	4.660	5.629	6.571	23.685	26.119	29.141	31.319	14
15	4.601	5.229	6.262	7.261	24.996	27.488	30.578	32.801	15
16	5.142	5.812	6.908	7.962	26.296	28.845	32.000	34.267	16
17	5.697	6.408	7.564	8.672	27.587	30.191	33.409	35.718	17
18	6.265	7.015	8.231	9.390	28.869	31.526	34.805	37.156	18
19	6.844	7.633	8.907	10.117	30.144	32.852	36.191	38.582	19
20	7.434	8.260	9.591	10.851	31.410	34.170	37.566	39.997	20
21	8.034	8.897	10.283	11.591	32.671	35.479	38.932	41.401	21
22	8.643	9.542	10.982	12.338	33.924	36.781	40.289	42.796	22
23	9.260	10.196	11.689	13.091	35.172	38.076	41.638	44.181	23
24	9.886	10.856	12.401	13.848	36.415	39.364	42.980	45.558	24
25	10.520	11.524	13.120	14.611	37.652	40.646	44.314	46.928	25
26	11.160	12.198	13.844	15.379	38.885	41.923	45.642	48.290	26
27	11.808	12.879	14.573	16.151	40.113	43.194	46.963	49.645	27
28	12.461	13.565	15.308	16.928	41.337	44.461	48.278	50.993	28
29	13.121	14.256	16.047	17.708	42.557	45.722	49.588	52.336	29
30	13.787	14.953	16.791	18.493	43.773	46.979	50.892	53.672	30

† Based on Table 8 of *Biometrika Tables for Statisticians*, Vol. I, Cambridge University Press, 1954, by permission of the *Biometrika* trustees.

TABLE VIa

Values of $F_{.05,\nu_1,\nu_2}$[†]

ν_1 = Degrees of freedom for numerator

ν_2 = Degrees of freedom for denominator	1	2	3	4	5	6	7	8	9	10	12	15	20	24	30	40	60	120	∞
1	161	200	216	225	230	234	237	239	241	242	244	246	248	249	250	251	252	253	254
2	18.5	19.0	19.2	19.2	19.3	19.3	19.4	19.4	19.4	19.4	19.4	19.4	19.4	19.5	19.5	19.5	19.5	19.5	19.5
3	10.1	9.55	9.28	9.12	9.01	8.94	8.89	8.85	8.81	8.79	8.74	8.70	8.66	8.64	8.62	8.59	8.57	8.55	8.53
4	7.71	6.94	6.59	6.39	6.26	6.16	6.09	6.04	6.00	5.96	5.91	5.86	5.80	5.77	5.75	5.72	5.69	5.66	5.63
5	6.61	5.79	5.41	5.19	5.05	4.95	4.88	4.82	4.77	4.74	4.68	4.62	4.56	4.53	4.50	4.46	4.43	4.40	4.37
6	5.99	5.14	4.76	4.53	4.39	4.28	4.21	4.15	4.10	4.06	4.00	3.94	3.87	3.84	3.81	3.77	3.74	3.70	3.67
7	5.59	4.74	4.35	4.12	3.97	3.87	3.79	3.73	3.68	3.64	3.57	3.51	3.44	3.41	3.38	3.34	3.30	3.27	3.23
8	5.32	4.46	4.07	3.84	3.69	3.58	3.50	3.44	3.39	3.35	3.28	3.22	3.15	3.12	3.08	3.04	3.01	2.97	2.93
9	5.12	4.26	3.86	3.63	3.48	3.37	3.29	3.23	3.18	3.14	3.07	3.01	2.94	2.90	2.86	2.83	2.79	2.75	2.71
10	4.96	4.10	3.71	3.48	3.33	3.22	3.14	3.07	3.02	2.98	2.91	2.85	2.77	2.74	2.70	2.66	2.62	2.58	2.54
11	4.84	3.98	3.59	3.36	3.20	3.09	3.01	2.95	2.90	2.85	2.79	2.72	2.65	2.61	2.57	2.53	2.49	2.45	2.40
12	4.75	3.89	3.49	3.26	3.11	3.00	2.91	2.85	2.80	2.75	2.69	2.62	2.54	2.51	2.47	2.43	2.38	2.34	2.30
13	4.67	3.81	3.41	3.18	3.03	2.92	2.83	2.77	2.71	2.67	2.60	2.53	2.46	2.42	2.38	2.34	2.30	2.25	2.21
14	4.60	3.74	3.34	3.11	2.96	2.85	2.76	2.70	2.65	2.60	2.53	2.46	2.39	2.35	2.31	2.27	2.22	2.18	2.13
15	4.54	3.68	3.29	3.06	2.90	2.79	2.71	2.64	2.59	2.54	2.48	2.40	2.33	2.29	2.25	2.20	2.16	2.11	2.07
16	4.49	3.63	3.24	3.01	2.85	2.74	2.66	2.59	2.54	2.49	2.42	2.35	2.28	2.24	2.19	2.15	2.11	2.06	2.01
17	4.45	3.59	3.20	2.96	2.81	2.70	2.61	2.55	2.49	2.45	2.38	2.31	2.23	2.19	2.15	2.10	2.06	2.01	1.96
18	4.41	3.55	3.16	2.93	2.77	2.66	2.58	2.51	2.46	2.41	2.34	2.27	2.19	2.15	2.11	2.06	2.02	1.97	1.92
19	4.38	3.52	3.13	2.90	2.74	2.63	2.54	2.48	2.42	2.38	2.31	2.23	2.16	2.11	2.07	2.03	1.98	1.93	1.88
20	4.35	3.49	3.10	2.87	2.71	2.60	2.51	2.45	2.39	2.35	2.28	2.20	2.12	2.08	2.04	1.99	1.95	1.90	1.84
21	4.32	3.47	3.07	2.84	2.68	2.57	2.49	2.42	2.37	2.32	2.25	2.18	2.10	2.05	2.01	1.96	1.92	1.87	1.81
22	4.30	3.44	3.05	2.82	2.66	2.55	2.46	2.40	2.34	2.30	2.23	2.15	2.07	2.03	1.98	1.94	1.89	1.84	1.78
23	4.28	3.42	3.03	2.80	2.64	2.53	2.44	2.37	2.32	2.27	2.20	2.13	2.05	2.01	1.96	1.91	1.86	1.81	1.76
24	4.26	3.40	3.01	2.78	2.62	2.51	2.42	2.36	2.30	2.25	2.18	2.11	2.03	1.98	1.94	1.89	1.84	1.79	1.73
25	4.24	3.39	2.99	2.76	2.60	2.49	2.40	2.34	2.28	2.24	2.16	2.09	2.01	1.96	1.92	1.87	1.82	1.77	1.71
30	4.17	3.32	2.92	2.69	2.53	2.42	2.33	2.27	2.21	2.16	2.09	2.01	1.93	1.89	1.84	1.79	1.74	1.68	1.62
40	4.08	3.23	2.84	2.61	2.45	2.34	2.25	2.18	2.12	2.08	2.00	1.92	1.84	1.79	1.74	1.69	1.64	1.58	1.51
60	4.00	3.15	2.76	2.53	2.37	2.25	2.17	2.10	2.04	1.99	1.92	1.84	1.75	1.70	1.65	1.59	1.53	1.47	1.39
120	3.92	3.07	2.68	2.45	2.29	2.18	2.09	2.02	1.96	1.91	1.83	1.75	1.66	1.61	1.55	1.50	1.43	1.35	1.25
∞	3.84	3.00	2.60	2.37	2.21	2.10	2.01	1.94	1.88	1.83	1.75	1.67	1.57	1.52	1.46	1.39	1.32	1.22	1.00

[†] Reproduced from M. Merrington and C. M. Thompson, "Tables of percentage points of the inverted beta (*F*) distribution," *Biometrika*, Vol. 33 (1943), by permission of the *Biometrika* trustees.

TABLE VIb

Values of $F_{.01,\nu_1,\nu_2}$†

ν_1 = Degrees of freedom for numerator; ν_2 = Degrees of freedom for denominator

ν_2	1	2	3	4	5	6	7	8	9	10	12	15	20	24	30	40	60	120	∞
1	4,052	5,000	5,403	5,625	5,764	5,859	5,928	5,982	6,023	6,056	6,106	6,157	6,209	6,235	6,261	6,287	6,313	6,339	6,366
2	98.5	99.0	99.2	99.2	99.3	99.3	99.4	99.4	99.4	99.4	99.4	99.4	99.4	99.5	99.5	99.5	99.5	99.5	99.5
3	34.1	30.8	29.5	28.7	28.2	27.9	27.7	27.5	27.3	27.2	27.1	26.9	26.7	26.6	26.5	26.4	26.3	26.2	26.1
4	21.2	18.0	16.7	16.0	15.5	15.2	15.0	14.8	14.7	14.5	14.4	14.2	14.0	13.9	13.8	13.7	13.7	13.6	13.5
5	16.3	13.3	12.1	11.4	11.0	10.7	10.5	10.3	10.2	10.1	9.89	9.72	9.55	9.47	9.38	9.29	9.20	9.11	9.02
6	13.7	10.9	9.78	9.15	8.75	8.47	8.26	8.10	7.98	7.87	7.72	7.56	7.40	7.31	7.23	7.14	7.06	6.97	6.88
7	12.2	9.55	8.45	7.85	7.46	7.19	6.99	6.84	6.72	6.62	6.47	6.31	6.16	6.07	5.99	5.91	5.82	5.74	5.65
8	11.3	8.65	7.59	7.01	6.63	6.37	6.18	6.03	5.91	5.81	5.67	5.52	5.36	5.28	5.20	5.12	5.03	4.95	4.86
9	10.6	8.02	6.99	6.42	6.06	5.80	5.61	5.47	5.35	5.26	5.11	4.96	4.81	4.73	4.65	4.57	4.48	4.40	4.31
10	10.0	7.56	6.55	5.99	5.64	5.39	5.20	5.06	4.94	4.85	4.71	4.56	4.41	4.33	4.25	4.17	4.08	4.00	3.91
11	9.65	7.21	6.22	5.67	5.32	5.07	4.89	4.74	4.63	4.54	4.40	4.25	4.10	4.02	3.94	3.86	3.78	3.69	3.60
12	9.33	6.93	5.95	5.41	5.06	4.82	4.64	4.50	4.39	4.30	4.16	4.01	3.86	3.78	3.70	3.62	3.54	3.45	3.36
13	9.07	6.70	5.74	5.21	4.86	4.62	4.44	4.30	4.19	4.10	3.96	3.82	3.66	3.59	3.51	3.43	3.34	3.25	3.17
14	8.86	6.51	5.56	5.04	4.70	4.46	4.28	4.14	4.03	3.94	3.80	3.66	3.51	3.43	3.35	3.27	3.18	3.09	3.00
15	8.68	6.36	5.42	4.89	4.56	4.32	4.14	4.00	3.89	3.80	3.67	3.52	3.37	3.29	3.21	3.13	3.05	2.96	2.87
16	8.53	6.23	5.29	4.77	4.44	4.20	4.03	3.89	3.78	3.69	3.55	3.41	3.26	3.18	3.10	3.02	2.93	2.84	2.75
17	8.40	6.11	5.19	4.67	4.34	4.10	3.93	3.79	3.68	3.59	3.46	3.31	3.16	3.08	3.00	2.92	2.83	2.75	2.65
18	8.29	6.01	5.09	4.58	4.25	4.01	3.84	3.71	3.60	3.51	3.37	3.23	3.08	3.00	2.92	2.84	2.75	2.66	2.57
19	8.19	5.93	5.01	4.50	4.17	3.94	3.77	3.63	3.52	3.43	3.30	3.15	3.00	2.92	2.84	2.76	2.67	2.58	2.49
20	8.10	5.85	4.94	4.43	4.10	3.87	3.70	3.56	3.46	3.37	3.23	3.09	2.94	2.86	2.78	2.69	2.61	2.52	2.42
21	8.02	5.78	4.87	4.37	4.04	3.81	3.64	3.51	3.40	3.31	3.17	3.03	2.88	2.80	2.72	2.64	2.55	2.46	2.36
22	7.95	5.72	4.82	4.31	3.99	3.76	3.59	3.45	3.35	3.26	3.12	2.98	2.83	2.75	2.67	2.58	2.50	2.40	2.31
23	7.88	5.66	4.76	4.26	3.94	3.71	3.54	3.41	3.30	3.21	3.07	2.93	2.78	2.70	2.62	2.54	2.45	2.35	2.26
24	7.82	5.61	4.72	4.22	3.90	3.67	3.50	3.36	3.26	3.17	3.03	2.89	2.74	2.66	2.58	2.49	2.40	2.31	2.21
25	7.77	5.57	4.68	4.18	3.86	3.63	3.46	3.32	3.22	3.13	2.99	2.85	2.70	2.62	2.53	2.45	2.36	2.27	2.17
30	7.56	5.39	4.51	4.02	3.70	3.47	3.30	3.17	3.07	2.98	2.84	2.70	2.55	2.47	2.39	2.30	2.21	2.11	2.01
40	7.31	5.18	4.31	3.83	3.51	3.29	3.12	2.99	2.89	2.80	2.66	2.52	2.37	2.29	2.20	2.11	2.02	1.92	1.80
60	7.08	4.98	4.13	3.65	3.34	3.12	2.95	2.82	2.72	2.63	2.50	2.35	2.20	2.12	2.03	1.94	1.84	1.73	1.60
120	6.85	4.79	3.95	3.48	3.17	2.96	2.79	2.66	2.56	2.47	2.34	2.19	2.03	1.95	1.86	1.76	1.66	1.53	1.38
∞	6.63	4.61	3.78	3.32	3.02	2.80	2.64	2.51	2.41	2.32	2.18	2.04	1.88	1.79	1.70	1.59	1.47	1.32	1.00

†Reproduced from M. Merrington and C. M. Thompson, "Tables of percentage points of the inverted beta (*F*) distribution," *Biometrika*, Vol. 33 (1943), by permission of the *Biometrika* trustees.

TABLE VII

Factorials

n	$n!$	$\log n!$
0	1	0.0000
1	1	0.0000
2	2	0.3010
3	6	0.7782
4	24	1.3802
5	120	2.0792
6	720	2.8573
7	5,040	3.7024
8	40,320	4.6055
9	362,880	5.5598
10	3,628,800	6.5598
11	39,916,800	7.6012
12	479,001,600	8.6803
13	6,227,020,800	9.7943
14	87,178,291,200	10.9404
15	1,307,674,368,000	12.1165

Binomial Coefficients

n	$\binom{n}{0}$	$\binom{n}{1}$	$\binom{n}{2}$	$\binom{n}{3}$	$\binom{n}{4}$	$\binom{n}{5}$	$\binom{n}{6}$	$\binom{n}{7}$	$\binom{n}{8}$	$\binom{n}{9}$	$\binom{n}{10}$
0	1										
1	1	1									
2	1	2	1								
3	1	3	3	1							
4	1	4	6	4	1						
5	1	5	10	10	5	1					
6	1	6	15	20	15	6	1				
7	1	7	21	35	35	21	7	1			
8	1	8	28	56	70	56	28	8	1		
9	1	9	36	84	126	126	84	36	9	1	
10	1	10	45	120	210	252	210	120	45	10	1
11	1	11	55	165	330	462	462	330	165	55	11
12	1	12	66	220	495	792	924	792	495	220	66
13	1	13	78	286	715	1287	1716	1716	1287	715	286
14	1	14	91	364	1001	2002	3003	3432	3003	2002	1001
15	1	15	105	455	1365	3003	5005	6435	6435	5005	3003
16	1	16	120	560	1820	4368	8008	11440	12870	11440	8008
17	1	17	136	680	2380	6188	12376	19448	24310	24310	19448
18	1	18	153	816	3060	8568	18564	31824	43758	48620	43758
19	1	19	171	969	3876	11628	27132	50388	75582	92378	92378
20	1	20	190	1140	4845	15504	38760	77520	125970	167960	184756

TABLE VIII

Values of e^x and e^{-x}

x	e^x	e^{-x}	x	e^x	e^{-x}
0.0	1.000	1.000	**2.5**	12.18	0.082
0.1	1.105	0.905	**2.6**	13.46	0.074
0.2	1.221	0.819	**2.7**	14.88	0.067
0.3	1.350	0.741	**2.8**	16.44	0.061
0.4	1.492	0.670	**2.9**	18.17	0.055
0.5	1.649	0.607	**3.0**	20.09	0.050
0.6	1.822	0.549	**3.1**	22.20	0.045
0.7	2.014	0.497	**3.2**	24.53	0.041
0.8	2.226	0.449	**3.3**	27.11	0.037
0.9	2.460	0.407	**3.4**	29.96	0.033
1.0	2.718	0.368	**3.5**	33.12	0.030
1.1	3.004	0.333	**3.6**	36.60	0.027
1.2	3.320	0.301	**3.7**	40.45	0.025
1.3	3.669	0.273	**3.8**	44.70	0.022
1.4	4.055	0.247	**3.9**	49.40	0.020
1.5	4.482	0.223	**4.0**	54.60	0.018
1.6	4.953	0.202	**4.1**	60.34	0.017
1.7	5.474	0.183	**4.2**	66.69	0.015
1.8	6.050	0.165	**4.3**	73.70	0.014
1.9	6.686	0.150	**4.4**	81.45	0.012
2.0	7.389	0.135	**4.5**	90.02	0.011
2.1	8.166	0.122	**4.6**	99.48	0.010
2.2	9.025	0.111	**4.7**	109.95	0.009
2.3	9.974	0.100	**4.8**	121.51	0.008
2.4	11.023	0.091	**4.9**	134.29	0.007

TABLE VIII (continued)

x	e^x	e^{-x}	x	e^x	e^{-x}
5.0	148.4	0.0067	**7.5**	1,808.0	0.00055
5.1	164.0	0.0061	**7.6**	1,998.2	0.00050
5.2	181.3	0.0055	**7.7**	2,208.3	0.00045
5.3	200.3	0.0050	**7.8**	2,440.6	0.00041
5.4	221.4	0.0045	**7.9**	2,697.3	0.00037
5.5	244.7	0.0041	**8.0**	2,981.0	0.00034
5.6	270.4	0.0037	**8.1**	3,294.5	0.00030
5.7	298.9	0.0033	**8.2**	3,641.0	0.00027
5.8	330.3	0.0030	**8.3**	4,023.9	0.00025
5.9	365.0	0.0027	**8.4**	4,447.1	0.00022
6.0	403.4	0.0025	**8.5**	4,914.8	0.00020
6.1	445.9	0.0022	**8.6**	5,431.7	0.00018
6.2	492.8	0.0020	**8.7**	6,002.9	0.00017
6.3	544.6	0.0018	**8.8**	6,634.2	0.00015
6.4	601.8	0.0017	**8.9**	7,332.0	0.00014
6.5	665.1	0.0015	**9.0**	8,103.1	0.00012
6.6	735.1	0.0014	**9.1**	8,955.3	0.00011
6.7	812.4	0.0012	**9.2**	9,897.1	0.00010
6.8	897.8	0.0011	**9.3**	10,938	0.00009
6.9	992.3	0.0010	**9.4**	12,088	0.00008
7.0	1,096.6	0.0009	**9.5**	13,360	0.00007
7.1	1,212.0	0.0008	**9.6**	14,765	0.00007
7.2	1,339.4	0.0007	**9.7**	16,318	0.00006
7.3	1,480.3	0.0007	**9.8**	18,034	0.00006
7.4	1,636.0	0.0006	**9.9**	19,930	0.00005

answers to odd-numbered exercises

Page 17

1. (b) 6, 20, and 70.
3. (a) 0.83% and 0.69%; (b) about 635 billion.
5. (b) $x^6 + 6x^5y + 15x^4y^2 + 20x^3y^3 + 15x^2y^4 + 6xy^5 + y^6$; $x^7 + 7x^6y + 21x^5y^2 + 35x^4y^3 + 35x^3y^4 + 21x^2y^5 + 7xy^6 + y^7$.
13. 560.
17. (a) 5; (b) 4.
19. (a) 20; (b) 60.
21. $3^{15} = 14{,}348{,}907$.
23. 720.
25. 120; 72.
27. 50,400; 3,360.
29. (a) 77,520; (b) 184,756; (c) 1,351.
31. 70.
33. 8,211,173,256.
35. 280.

Page 31

1. (a) {6, 8, 9}; (b) {8}; (c) {1, 2, 3, 4, 5, 8}; (d) {1, 5}; (e) {2, 4, 8}; (f) $\varnothing$.
3. (a) {Car 5, Car 6, Car 7, Car 8}; (b) {Car 2, Car 4, Car 5, Car 7}; (c) {Car 1, Car 8}; (d) {Car 3, Car 4, Car 7, Car 8}; (e) he chooses a car with air-conditioning; (f) he chooses a car which has either no power steering or it does have bucket seats; (g) he chooses a 2- or 3-year-old car with bucket seats; (h) same as part (g).
5. (a) (0, 0), (1, 0), (2, 0), (3, 0), (4, 0), (5, 0), (0, 1), (0, 2), (0, 3), (0, 4); (b) (0, 2), (1, 1), (2, 0), (0, 4), (1, 3), (2, 2), (3, 1), (4, 0), (2, 4), (3, 3), (4, 2) (5, 1);

(c) (0, 0), (1, 1), (2, 2), (3, 3), (4, 4); (d) (0, 0), (1, 0), (2, 0), (3, 0), (4, 0), (5, 0), (0, 1), (0, 2), (0, 3), (0, 4), (1, 1), (1, 3), (2, 2), (3, 1), (2, 4), (3, 3), (4, 2), (5, 1); (e) (0, 4), (0, 2), (2, 0), (4, 0); (f) (0, 2), (1, 1), (2, 0), (0, 4), (1, 3), (2, 2), (3, 1), (4, 0), (2, 4), (3, 3), (4, 2), (5, 1), (0, 0), (4, 4); (g) (0, 4), (0, 3), (0, 2), (0, 1), (0, 0), (1, 0), (2, 0), (3, 0), (4, 0), (5, 0), (1, 4), (1, 2), (2, 3), (2, 1), (3, 4), (3, 2), (4, 4), (4, 3), (4, 1), (5, 4), (5, 3), (5, 2); (h) (0, 4), (0, 3), (0, 2), (0, 1), (1, 0), (2, 0), (3, 0), (4, 0), (5, 0); (i) (1, 4), (1, 2), (2, 3), (2, 1), (3, 4), (3, 2), (4, 4), (4, 3), (4, 1), (5, 4), (5, 3), (5, 2).

7. (a) $S = \{(0,0,0), (1,0,0), (0,1,0), (0,0,1), (1,1,0), (1,0,1), (0,1,1), (1,1,1)\}$; $A = \{(1,0,1), (0,1,1), (1,1,1)\}$; $B = \{(0,1,1)\}$; $C = \{(1,0,1)\}$; (b) B and C.

9. (a) $M \cup N = \{x|3 < x < 10\}$; (b) $M \cap N = \{x|5 < x \leq 8\}$; (c) $M \cap N' = \{x|3 < x \leq 5\}$; (s) $M' \cup N = \{x|0 < x \leq 3 \text{ or } 5 < x < 10\}$.

11. 38.

13. (a) 12; (b) 6; (c) 20.

Page 44

13. (a) Permissible assignment; (b) not a permissible assignment because the sum of the probabilities exceeds 1; (c) permissible assignment; (d) not a permissible assignment because probabilities cannot be negative; (e) not a permissible assignment because the sum of the probabilities is less than 1.

15. (a) the second probability cannot be negative; (b) the third probability does not equal the sum of the first two probabilities; (c) the sum of the probabilities exceeds one; (d) the sum of the probabilities is less than one.

17. (a) Yes; (b) 9 to 11.

19. (a) 0.46; (b) 0.40; (c) 0.11; (d) 0.68.

21. (a) $\frac{3}{8}$; (b) $\frac{1}{4}$; (c) $\frac{1}{10}$; (d) $\frac{1}{10}$; (e) $\frac{11}{40}$.

23. $\frac{20}{221}$.

25. $\frac{2}{13}$.

27. (a) 0.68; (b) 0.38; (c) 0.79; (d) 0.32.

29. (a) 0.11; (b) 0.98; (c) 0.09.

31. 0.98.

Page 63

13. $\frac{17}{18}$.

15. $\frac{13}{16}$.

17. (a) $\frac{26}{37}$; (b) $\frac{17}{26}$; (c) $\frac{3}{11}$; (d) $\frac{2}{5}$.

19. 0.7685.

21. $\frac{91}{323}$.

23. (a) $\frac{1}{343}$; (b) $\frac{3}{343}$; (c) $\frac{216}{343}$.

25. (a) $\frac{64}{125}$; (b) $\frac{12}{125}$.

27. 0.8784.

29. 0.735.

31. 0.032.

33. 0.8.

35. 0.3818.

37. (a) $\frac{37}{150}$; (b) $\frac{23}{148}$; (c) $\frac{8}{33}$.

Page 81

3. $0 < k < 1$.

7. $F(x) = \begin{cases} 0 & \text{for } x < 0 \\ \frac{1}{5} & \text{for } 0 \leq x < 1 \\ \frac{4}{5} & \text{for } 1 \leq x < 2 \\ 1 & \text{for } x \geq 2 \end{cases}$

11.

y	-4	-2	0	2	4
$f(y)$	$\frac{1}{16}$	$\frac{1}{4}$	$\frac{3}{8}$	$\frac{1}{4}$	$\frac{1}{16}$

13. (a)

x	0	1	2
$f(x)$	$\frac{2}{5}$	$\frac{8}{15}$	$\frac{1}{15}$

(b) $F(x) = \begin{cases} 0 & \text{for } x < 0 \\ \frac{2}{5} & \text{for } 0 \leq x < 1 \\ \frac{14}{15} & \text{for } 1 \leq x < 2 \\ 1 & \text{for } x \geq 2 \end{cases}$

15. $F(x) = \begin{cases} 0 & \text{for } x < 0 \\ \frac{1}{64} & \text{for } 0 \leq x < 1 \\ \frac{5}{32} & \text{for } 1 \leq x < 2 \\ \frac{37}{64} & \text{for } 2 \leq x < 3 \\ 1 & \text{for } x \geq 3 \end{cases}$

(a) $\frac{27}{32}$; (b) $\frac{27}{64}$.

17. (a) 0.33; (b) 0.13; (c) 0.04; (d) 0.78.

Page 92

1. (b) $\frac{2}{5}$.

3. (a) $\frac{1}{4}$; (b) $F(x) = \begin{cases} 0 & \text{for } x \leq 0 \\ \dfrac{\sqrt{x}}{2} & \text{for } 0 < x < 4; \\ 1 & \text{for } x \geq 4 \end{cases}$ (c) $\frac{1}{2}$.

5. (a) $\frac{1}{2}$; (b) $G(x) = \begin{cases} 0 & \text{for } x \leq 0 \\ x^2(3 - 2x) & \text{for } 0 < x < 1; \\ 1 & \text{for } x \geq 1 \end{cases}$ (c) $\frac{1}{2}$.

7. $F(x) = \begin{cases} 0 & \text{for } x \leq 0 \\ \dfrac{x}{3} & \text{for } 0 < x \leq 1 \\ \frac{1}{3} & \text{for } 1 < x \leq 2 \\ \dfrac{x-1}{3} & \text{for } 2 < x < 4 \\ 1 & \text{for } x \geq 4 \end{cases}$

9. $F(x) = \begin{cases} 0 & \text{for } x \leq 0 \\ \dfrac{x^2}{4} & \text{for } 0 < x \leq 1 \\ \dfrac{2x-1}{4} & \text{for } 1 < x \leq 2 \\ \dfrac{-(x^2 - 6x + 5)}{4} & \text{for } 2 < x < 3 \\ 1 & \text{for } x \geq 3 \end{cases}$

11. (a) $\frac{16}{25}$; (b) $\frac{9}{64}$; (c) $f(y) = \begin{cases} \dfrac{18}{y^3} & \text{for } y > 3 \\ 0 & \text{elsewhere} \end{cases}$

13. (a) $\frac{1}{4}$; (b) $\frac{1}{4}$; (c) $\frac{3}{8}$; (d) $\frac{1}{2}$.

15. (a) $\frac{7}{27}$; (b) $\frac{325}{864}$; (c) $\frac{95}{432}$; (d) 0.

17. (a) 0.4692; (b) 0.0986; (c) 0.2019.

19. (a) $\frac{1}{4}$; (b) $\frac{39}{64}$; (c) $\frac{1}{16}$.

Page 108

1. (a) $\frac{1}{20}$; (b) $\frac{3}{8}$; (c) $\frac{1}{2}$; (d) $\frac{7}{30}$.

3. If $k > 0$, $P(\mathbf{x} = 3, \mathbf{y} = 1) < 0$, and if $k < 0$, all the other probabilities are negative.

5.

y \ x	0	1	2	3
0	0	$\frac{1}{30}$	$\frac{1}{10}$	$\frac{1}{5}$
1	$\frac{1}{30}$	$\frac{2}{15}$	$\frac{3}{10}$	$\frac{8}{15}$
2	$\frac{1}{10}$	$\frac{3}{10}$	$\frac{3}{5}$	1

7. 2.

9. $1 - \dfrac{\ln 2}{2} = 0.6534$.

11. $(1 - e^{-1})(1 - e^{-4}) = 0.6205$.

13. $F(b, d) - F(a, d) - F(b, c) + F(a, c)$.

15. (a) $\frac{1}{54}$; (b) $\frac{1}{18}$; (c) $\frac{7}{27}$; (d) $\frac{1}{6}$; (e) 0; (f) 1.

17. (a) 0; (b) $\frac{1}{16}$.

19.

y \ x	0	1	2
0	$\frac{3}{28}$	$\frac{9}{28}$	$\frac{3}{28}$
1	$\frac{3}{14}$	$\frac{3}{14}$	
2	$\frac{1}{28}$		

21. (a) $\frac{1}{8}$; (b) $\frac{1}{2}$.

23. (a) $\frac{5}{2}(e^{-0.4} - e^{-0.6}) = 0.3038$; (b) $5(e^{-0.30} - e^{-0.25} + 0.05) = 0.0601$.

Page 121

1. (a)

x	-1	1
$g(x)$	$\frac{1}{4}$	$\frac{3}{4}$

(b)

y	-1	0	1
$h(y)$	$\frac{5}{8}$	$\frac{1}{4}$	$\frac{1}{8}$

(c)

x	-1	1
$f(x\|-1)$	$\frac{1}{5}$	$\frac{4}{5}$

3. (a) $m(x, y) = \frac{xy}{36}$ for $x = 1, 2, 3$ and $y = 1, 2, 3$; (b) $n(x, z) = \frac{xz}{18}$ for $x = 1, 2, 3$ and $z = 1, 2$; (c) $g(x) = \frac{x}{6}$ for $x = 1, 2, 3$; (d) $\varphi(z|1, 2) = \frac{z}{3}$ for $z = 1, 2$; (e) $\psi(y, z|3) = \frac{yz}{18}$ for $y = 1, 2, 3$ and $z = 1, 2$.

5. (a) $G(x) = \begin{cases} 0 & \text{for } x < 0 \\ \frac{5}{12} & \text{for } 0 \leqslant x < 1 \\ \frac{11}{12} & \text{for } 1 \leqslant x < 2 \\ 1 & \text{for } x \geqslant 2 \end{cases}$

(b) $F(x|1) = \begin{cases} 0 & \text{for } x < 0 \\ \frac{4}{7} & \text{for } 0 \leqslant x < 1 \\ 1 & \text{for } x \geqslant 1 \end{cases}$

7. (a) $g(x) = \begin{cases} 4(1-x)^3 & \text{for } 0 < x < 1 \\ 0 & \text{elsewhere} \end{cases}$

(b) $h(y) = \begin{cases} 12y(1-y)^2 & \text{for } 0 < y < 1 \\ 0 & \text{elsewhere} \end{cases}$

Not independent.

9. (a) $p(x_2|x_1 = \frac{1}{3}, x_3 = 2) = \begin{cases} \frac{2(1+3x_2)}{5} & \text{for } 0 < x_2 < 1 \\ 0 & \text{elsewhere} \end{cases}$

(b) $q(x_2, x_3|x_1 = \frac{1}{2}) = \begin{cases} (\frac{1}{2} + x_2)\, e^{-x_3} & \text{for } 0 < x_2 < 1 \text{ and } x_3 > 0 \\ 0 & \text{elsewhere} \end{cases}$

11. $M(x_1, x_3) = \begin{cases} 0 & \text{for } x_1 \leqslant 0 \text{ or } x_3 \leqslant 0 \\ \frac{1}{2}x_1(x_1+1)(1-e^{-x_3}) & \text{for } 0 < x_1 < 1, x_3 > 0 \\ 1 - e^{-x_3} & \text{for } x_1 \geqslant 1, x_3 > 0 \end{cases}$

$G(x_1) = \begin{cases} 0 & \text{for } x_1 \leqslant 0 \\ \frac{1}{2}x_1(x_1+1) & \text{for } 0 < x_1 < 1 \\ 1 & \text{for } x_1 \geqslant 1 \end{cases}$

13. $\frac{3}{5}$.

15. (a)

w \ z	0	1
0	$\frac{188}{221}$	
1	$\frac{16}{221}$	$\frac{16}{221}$
2		$\frac{1}{221}$

(b)

z	0	1
$g(z)$	$\frac{204}{221}$	$\frac{17}{221}$

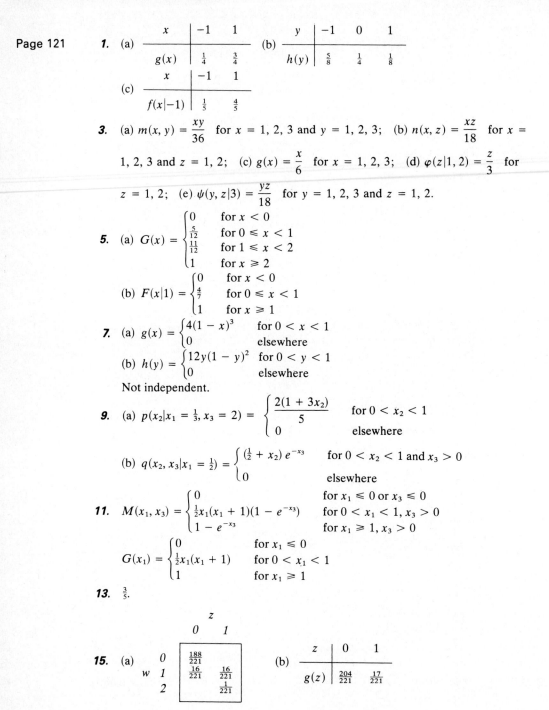

(c)

w	0	1	2
$h(w)$	$\frac{188}{221}$	$\frac{32}{221}$	$\frac{1}{221}$

(d)

w	1	2
$f(w\|1)$	$\frac{16}{17}$	$\frac{1}{17}$

17. (a) $g(x) = \begin{cases} \frac{2}{5}(x+2) & \text{for } 0 < x < 1 \\ 0 & \text{elsewhere} \end{cases}$ (b) 0.742;

(c) $w(y|x) = \begin{cases} \dfrac{x+4y}{x+2} & \text{for } 0 < y < 1 \\ 0 & \text{elsewhere} \end{cases}$ (d) 0.273.

19. (a) $g(x) = \begin{cases} \dfrac{20-x}{50} & \text{for } 10 < x < 20 \\ 0 & \text{elsewhere} \end{cases}$

$$h(y) = \begin{cases} \frac{1}{25}\left(20 \ln \dfrac{y}{5} - 2y + 10\right) & \text{for } 5 < y \leq 10 \\ \frac{1}{25}\left(20 \ln \dfrac{20}{y} - 20 + y\right) & \text{for } 10 < y < 20 \\ 0 & \text{elsewhere} \end{cases}$$

(b) $w(y|12) = \begin{cases} \frac{1}{6} & \text{for } 6 < y < 12 \\ 0 & \text{elsewhere} \end{cases}$

(c) $\frac{2}{3}$.

21. (a) $f(x_1, x_2, x_3) = \begin{cases} \dfrac{(20{,}000)^3}{(x_1+100)^3(x_2+100)^3(x_3+100)^3} & \text{for } x_1 > 0, x_2 > 0, x_3 > 0 \\ 0 & \text{elsewhere} \end{cases}$

(b) $\frac{1}{16}$.

Page 135

1. (a) $g_1 = 0$, $g_2 = 1$, $g_3 = 4$, $g_4 = 9$; (b) $P[g(\mathbf{x}) = 0] = f(0)$, $P[g(\mathbf{x}) = 1] = f(-1) + f(1)$, $P[g(\mathbf{x}) = 4] = f(-2) + f(2)$, $P[g(\mathbf{x}) = 9] = f(3)$.

5. (a) $\int_{-\infty}^{\infty}\int_{-\infty}^{\infty} xf(x, y)\,dx\,dy$; (b) $\int_{-\infty}^{\infty} xg(x)\,dx$.

7. $\frac{37}{12}$.

9. (a) $\dfrac{2}{\ln 3}, \dfrac{4}{\ln 3}, \dfrac{26}{3 \cdot \ln 3}$; (b) $\dfrac{32}{3 \ln 3} + 1$.

11. $\frac{1}{2}$.

13. $\frac{1}{12}$.

15. \$750.

17. (a) \$1.60; (b) \$1.67; (c) \$1.50.

19. 30,000 kilometers.

21. \$10,000.

Page 147

3. $\mu = \frac{4}{3}$, $\mu_2' = 2$, and $\sigma^2 = \frac{2}{9}$.

9. (a) $\alpha_4 = 3.2$; (b) $\alpha_4 = 2.6$.

13. (a) $k = \sqrt{20}$; (b) $k = 10$.

15. $M_{\mathbf{x}}(t) = \dfrac{2e^t}{3 - e^t}$, $\mu'_1 = \frac{3}{2}$, $\mu'_2 = 3$.

17. $\mu = 4$, and $\sigma^2 = 4$.

19. (a) $\sigma^2 = 2$; (b) $\sigma^2 = 2$.

21. $\mu = 0$, and $\sigma^2 = 1$.

23. $\mu = 4$, and $\sigma^2 = 16$.

25. $\mu = 1$, and $\sigma^2 = 1$.

27. At least $\frac{63}{64}$.

29. By Chebyshev's theorem, $P(\mathbf{x} < 10) \geqslant \frac{5}{9}$. The exact probability is 0.9179.

Page 161

1. 8.

3. 0.

5. $\text{cov}(\mathbf{x}, \mathbf{y}) = 0$.

7. (a) $\mu_{\mathbf{y}} = -7$, and $\sigma_{\mathbf{y}}^2 = 155$; (b) $\mu_{\mathbf{z}} = 19$, and $\sigma_{\mathbf{z}}^2 = 36$.

9. $\frac{805}{162}$.

11. $\text{var}(\mathbf{x} + \mathbf{y}) = \text{var}(\mathbf{x}) + \text{var}(\mathbf{y}) + 2\,\text{cov}(\mathbf{x}, \mathbf{y})$, $\text{var}(\mathbf{x} - \mathbf{y}) = \text{var}(\mathbf{x}) + \text{var}(\mathbf{y}) - 2\,\text{cov}(\mathbf{x}, \mathbf{y})$, $\text{cov}(\mathbf{x} + \mathbf{y}, \mathbf{x} - \mathbf{y}) = \text{var}(\mathbf{x}) - \text{var}(\mathbf{y})$.

13. -56.

15. 3.

17. $\frac{5}{12}$.

19. $\mu = 3.6$, and $\sigma = 0.022$.

21. (a) $\mu = 0.74$, and $\sigma = 0.68$; (b) $\mu = 1.91$, and $\sigma = 1.05$.

23. $\frac{1}{2}$.

Page 171

9. $\dfrac{x}{n}$.

11. (c) $F_{\mathbf{x}}(t) = [1 + \theta(t - 1)]^n$.

13. 0.1707.

15. (a) 0.2066; (b) 0.2066.

17. (a) 0.1669; (b) 0.4073; (c) 0.4073.

19. 0.9222 and 0.0778.

Page 184

1. $f(y; k, \theta) = \dbinom{y + k - 1}{k - 1}\theta^k(1 - \theta)^y$ for $y = 0, 1, 2, \cdots$;

$$\mu_{\mathbf{y}} = k\left(\frac{1}{\theta} - 1\right);\ \sigma_{\mathbf{y}}^2 = \frac{k}{\theta}\left(\frac{1}{\theta} - 1\right).$$

9. $\frac{1}{126}, \frac{10}{63}, \frac{10}{21}, \frac{20}{63}, \frac{5}{126}$.

19. $\mu_2 = \lambda$, $\mu_3 = \lambda$, $\mu_4 = \lambda + 3\lambda^2$.

21. (a) 0.1298; (b) 0.1101.

25. 0.4491.

27. (a) $\frac{1}{28}$; (b) $\frac{15}{56}$; (c) $\frac{27}{56}$; (d) $\frac{3}{14}$.

29. (a) 0.1388; (b) 0.1354.

31. 0.2700.

33. 0.2019, 0.3230, and 0.2584.

Page 192

1. 0.0840.
3. 0.0291.
5. (a) 0.1798; (b) 0.1798.

Page 202

7. (a) 0.6151; (b) 0.9197.
15. (a) $k = \alpha\beta$.
23. $\frac{1}{2}$.
25. 100.
27. (a) 0.6065; (b) 0.5276.
29. 0.2639.
31. (a) 3,200 hours; (b) 0.2057.

Page 220

11. (a) $\mu_1 = -2$, $\mu_2 = 1$, $\sigma_1 = 10$, $\sigma_2 = 5$, and $\rho = 0.7$; (b) $\mu_{y|x} = 1.70 + 0.35x$ and $\sigma^2_{y|x} = 12.75$.
13. $\rho_{uv} = \dfrac{\sigma_1^2 - \sigma_2^2}{\sqrt{(\sigma_1^2 + \sigma_2^2)^2 - 4\rho^2\sigma_1^2\sigma_2^2}}$.
15. (a) 0.1271; (b) 0.3594; (c) 0.1413; (d) 0.5876.
17. (a) 1.645; (b) 1.96; (c) 2.33; (d) 2.575.
19. 6.094 ounces.
21. 0.1446.
23. (a) 0.2990; (b) 0.1774.
25. (a) 14.5 pounds; (b) 23.625 inches.

Page 228

1. (a) $G(y) = \begin{cases} 1 - e^{-y} & \text{for } y > 0 \\ 0 & \text{elsewhere} \end{cases}$

(b) $g(y) = \begin{cases} e^{-y} & \text{for } y > 0 \\ 0 & \text{elsewhere} \end{cases}$

3. $g(y) = \begin{cases} 2y & \text{for } 0 < y < 1 \\ 0 & \text{elsewhere} \end{cases}$

5. (a) $f(y) = \begin{cases} \dfrac{1}{\theta_1 - \theta_2}[e^{-y/\theta_1} - e^{-y/\theta_2}] & \text{for } y > 0 \\ 0 & \text{elsewhere} \end{cases}$

(b) $f(y) = \begin{cases} \dfrac{1}{\theta^2} y e^{-y/\theta} & \text{for } y > 0 \\ 0 & \text{elsewhere} \end{cases}$

7. $g(z) = \begin{cases} 4z\, e^{-2z} & \text{for } z > 0 \\ 0 & \text{elsewhere} \end{cases}$

9. $f(z) = \begin{cases} -\frac{1}{25}(10 - 20 \ln 2) & \text{for } 0 < z \leq 5 \\ -\frac{1}{25}\left(20 \ln \dfrac{z}{10} - 2z + 20\right) & \text{for } 5 < z < 10 \\ 0 & \text{elsewhere} \end{cases}$

Page 244

1. $g(y) = \frac{1}{3}\left(\frac{2}{3}\right)^{-\frac{y+1}{5}}$ for $y = -1, -6, -11, -16, \ldots$

3. (a)

y	0	1	2
$g(y)$	$\frac{1}{3}$	$\frac{1}{3}$	$\frac{1}{3}$

(b)

z	-1	0	1
$g(z)$	$\frac{1}{9}$	$\frac{2}{3}$	$\frac{2}{9}$

7. $g(y) = \begin{cases} \frac{1}{6}y^{-\frac{1}{3}} & \text{for } 0 < y < 8 \\ 0 & \text{elsewhere} \end{cases}$

9. $\alpha = 1$, and $\beta = 2$.

11. (a) $g(y) = \begin{cases} \frac{1}{2} & \text{for } 0 < y < 1 \\ \frac{1}{4} & \text{for } 1 \leqslant y < 3 \\ 0 & \text{elsewhere} \end{cases}$

(b) $h(z) = \begin{cases} \frac{1}{8}z^{-3/4} & \text{for } 0 < z < 1 \\ \frac{1}{16}z^{-3/4} & \text{for } 1 \leqslant z < 81 \\ 0 & \text{elsewhere} \end{cases}$

13. (a)

u	0	1	2
$g(u)$	$\frac{1}{6}$	$\frac{5}{9}$	$\frac{5}{18}$

(b)

v	0	1
$h(v)$	$\frac{5}{6}$	$\frac{1}{6}$

(c)

w	-2	-1	0	1	2
$f(w)$	$\frac{1}{36}$	$\frac{2}{9}$	$\frac{1}{3}$	$\frac{1}{3}$	$\frac{1}{12}$

17. $h(z) = \begin{cases} 6 + 6z - 12\sqrt{z} & \text{for } 0 < z < 1 \\ 0 & \text{elsewhere} \end{cases}$

19. (a) $g(u, y) = \begin{cases} \frac{1}{2} & \text{for the region bounded by } y = 0, u = y, \text{ and } 2y - u = 2 \\ 0 & \text{elsewhere} \end{cases}$

(b) $h(u) = \begin{cases} \dfrac{2+u}{4} & \text{for } -2 < u \leqslant 0 \\ \dfrac{2-u}{4} & \text{for } 0 < u < 2 \\ 0 & \text{elsewhere} \end{cases}$

Page 250

3. $M_Y(t) = (1 - \beta t)^{-\alpha n}$; a gamma distribution with the parameters αn and β.

7. (a) 0.1781; (b) 0.9523; (c) 0.0463.

9. (a) 0.475; (b) 0.570.

Page 262

15. 0.9999994.

17. 0.025.

19. 0.2302.

21. -4.642 and 4.642.

23. (a) 0.306; (b) 0.7698.

25. (a) 0.9799; (b) 0.8361; (c) 0.7509.

Page 276

9. 0.0763.

13. $\frac{14}{27}$.

21. 0.055.

23. $t = 3.57$; the data do not support the conjecture.

Page 282

3. $\mu = \dfrac{1}{n+1}$, and $\sigma^2 = \dfrac{n}{(n+1)^2(n+2)}$.

5. (a)

y_1	1	2	3	4
$g(y_1)$	$\frac{2}{5}$	$\frac{3}{10}$	$\frac{1}{5}$	$\frac{1}{10}$

(b)

y_1	1	2	3	4	5
$g(y_1)$	$\frac{9}{25}$	$\frac{7}{25}$	$\frac{1}{5}$	$\frac{3}{25}$	$\frac{1}{25}$

7. $h(y_1, R) = n(n-1)f(y_1)f(y_1+R)\left[\int_{y_1}^{y_1+R} f(x)\,dx\right]^{n-2}$.

9. $g(R) = n(n-1)(1-R)R^{n-2}$ for $0 < R < 1$ and $g(R) = 0$ elsewhere;
$\mu = \dfrac{n-1}{n+1}$ and $\sigma^2 = \dfrac{2(n-1)}{(n+1)^2(n+2)}$.

13. 0.081.

Page 295

3. (a) Decision reversed; (b) decision is the same.

5. (a) Go to the site 33 miles away; (b) go to the site 27 miles away; (c) it does not matter.

7. (a) Expand now; (b) hotel Y; (c) go to the site 27 miles away.

9. (a) Hotel Y; (b) go to the site 27 miles away.

11. (a) Strategies I and 2, the value is 5; (b) Strategies II and 1, the value is 11; (c) Strategies I and 1, the value is −5; (d) Strategies I and 2, the value is 8.

13. (a)

		A: *NG*	*A*: *G*
B	*NSK*	0	−6
B	*SK*	8	3

(b) Give away glasses and steak knives.

15. (a) $\frac{5}{11}$ and $\frac{6}{11}$; (b) $\frac{4}{11}$ and $\frac{7}{11}$; (c) $-\frac{9}{11}$.

17. $\frac{1}{6}$ and $\frac{5}{6}$ for the defender; $\frac{5}{6}$ and $\frac{1}{6}$ for the attacker; the value is $10,333,333.

19. (a) First station owner should lower his price; (b) the owners should take turns lowering their prices on alternate days.

Page 307

3. $k = \dfrac{\theta_1\theta_2}{\sqrt{\theta_1^2 + \theta_2^2}}$.

9. (a)

	$a_1(\theta = \frac{1}{4})$	$a_2(\theta = \frac{1}{2})$
$\theta_1(\theta = \frac{1}{4})$	0	160
$\theta_2(\theta = \frac{1}{2})$	160	0

(b) $d_1(0) = \frac{1}{4}$, $d_1(1) = \frac{1}{4}$, and $d_1(2) = \frac{1}{4}$; $d_2(0) = \frac{1}{4}$, $d_2(1) = \frac{1}{4}$, and $d_2(2) = \frac{1}{2}$; $d_3(0) = \frac{1}{4}$, $d_3(1) = \frac{1}{2}$, and $d_3(2) = \frac{1}{4}$; $d_4(0) = \frac{1}{4}$, $d_4(1) = \frac{1}{2}$, and $d_4(2) = \frac{1}{2}$; $d_5(0) = \frac{1}{2}$, $d_5(1) = \frac{1}{4}$, and $d_5(2) = \frac{1}{4}$; $d_6(0) = \frac{1}{2}$, $d_6(1) = \frac{1}{4}$, and $d_6(2) = \frac{1}{2}$; $d_7(0) = \frac{1}{2}$, $d_7(1) = \frac{1}{2}$, and $d_7(2) = \frac{1}{4}$; $d_8(0) = \frac{1}{2}$, $d_8(1) = \frac{1}{2}$, and $d_8(2) = \frac{1}{2}$; (c) d_5, d_6, and d_7 are not admissible; (d) d_4; (e) d_2.

Page 323

13. $\frac{8}{9}$.
19. (a) Biased; (b) consistent.

Page 331

1. $2\mathbf{m}_1'$.
3. (a) $\bar{x}$; (b) $\bar{x}$.
5. $\sqrt{\dfrac{\sum_{i=1}^{n}(\mathbf{x}_i - \mu)^2}{n}}$.
7. The smallest sample value.
9. (a) $\dfrac{n_1 + 2n_2}{2N}$; (b) $\dfrac{n_1 + 2n_2}{2N}$.
11. The smallest and largest sample values.
13. $\frac{1}{2}(\bar{\mathbf{x}} + \bar{\mathbf{y}})$ and $\frac{1}{2}(\bar{\mathbf{x}} - \bar{\mathbf{y}})$.

Page 338

5. 0.29.
7. 0.4786.
9. (a) 100; (b) 112; (c) 108.

Page 350

1. $k = \dfrac{-1}{\ln(1 - \alpha)}$.
3. $c = \dfrac{1 + \sqrt{1 - \alpha}}{\alpha}$.
9. $27.52 < \mu < 29.48$.
11. $5.47 < \mu < 5.89$.
13. $-7.285 < \mu_1 - \mu_2 < -2.715$.
15. $-0.198 < \mu_1 - \mu_2 < 1.998$.

Page 358

3. $\hat{\theta} - \dfrac{1}{2n} - z_{\alpha/2}\sqrt{\dfrac{\hat{\theta}(1 - \hat{\theta})}{n}} < \theta < \hat{\theta} + \dfrac{1}{2n} + z_{\alpha/2}\sqrt{\dfrac{\hat{\theta}(1 - \hat{\theta})}{n}}$.
7. $0.611 < \theta < 0.749$.
9. $n = 1{,}068$.
11. $-0.372 < \theta_1 - \theta_2 < -0.204$.
13. $0.040 < \sigma^2 < 0.280$.
15. $0.165 < \dfrac{\sigma_1^2}{\sigma_2^2} < 2.752$.

Page 371

1. (a) Simple; (b) composite; (c) composite; (d) composite; (e) simple; (f) composite; (g) composite; (h) composite.
3. $\alpha = 0.1331$, and $\beta = 0.0159$.
5. $\alpha = 0.223$, and $\beta = 0.451$.
7. 0.1139.

9. $x \leqslant \dfrac{K}{\ln \dfrac{\theta_0(1-\theta_1)}{\theta_1(1-\theta_0)}}$.

11. $x \leqslant \dfrac{K}{\ln \dfrac{1-\theta_0}{1-\theta_1}} + 1$.

13. They would be committing a type I error if they erroneously rejected the hypothesis $\theta = 0.60$; they would be committing a type II error if they erroneously accepted the hypothesis $\theta = 0.60$.

15. (a) 0.034; (b) 0.045; (c) 0.052.

Page 383

1. (a) 0, 0, and $\frac{1}{21}$; (b) $\frac{6}{7}$, $\frac{5}{7}$, $\frac{11}{21}$, $\frac{2}{7}$, and 0.

3. (a) 0.852; (b) 0.016, 0.086, 0.145, 0.134, 0.122.

5. (a) 0.0375, 0.0203, 0.0107, 0.0055, 0.0027; (b) 0.9329, 0.7585, 0.3840, 0.0419.

9. (a) $\lambda = \left(\dfrac{\bar{x}}{\theta_0}\right)^n e^{-n\bar{x}\left|\frac{1}{\theta_0} - \frac{1}{\bar{x}}\right|}$.

15. $-2 \ln \lambda = 7.845$; reject the null hypothesis.

Page 396

3. $n = 151$.

5. $z = 3.02$; reject the null hypothesis.

7. $t = 5.66$; reject the null hypothesis.

9. Less than 0.145 or greater than 0.255; (a) 0.18; (b) 0.71; (c) 0.71; (d) 0.18.

11. $z = -4.88$; reject the null hypothesis.

13. t = 0.90; the null hypothesis cannot be rejected.

15. $t = 2.33$; null hypothesis cannot be rejected.

Page 402

3. $\chi^2 = 5.92$; null hypothesis cannot be rejected.

5. (a) $\chi^2 = 23$; the null hypothesis cannot be rejected; (b) $z = -0.8$; the null hypothesis cannot be rejected.

7. $F = 2.47$; assumption was reasonable.

Page 409

7. $k_{.01} = 11$; 0.95, 0.87, and 0.75.

9. $k_{.005} = 12$ and $k'_{.005} = 1$; 0.97, 0.99, 0.98, and 0.96.

11. $z = -1.30$; null hypothesis cannot be rejected.

13. $\chi^2 = 0.7$; difference is not significant.

15. $\chi^2 = 7.1$; null hypothesis cannot be rejected.

Page 415

3. $\chi^2 = 4.16$; null hypothesis cannot be rejected.

5. $\chi^2 = 1.6$; no significant difference in quality.

7. $\chi^2 = 29.2$; reject the null hypothesis.

9. (b) 0.0179, 0.1178, 0.3245, 0.3557, 0.1554, 0.0268, and 0.0019; (c) $\chi^2 = 1.44$; an excellent fit.

Page 430 **3.** $\frac{1+x}{2}, \frac{2}{3}y$. **5.** $\frac{4}{7}$, and $\frac{9}{8}$. **13.** $\hat{\beta} = \frac{\sum xy}{\sum x^2}$.

15.

$$\sum_{i=1}^{n} y_i = \hat{\beta}_0 n + \hat{\beta}_1 \sum_{i=1}^{n} x_i + \hat{\beta}_2 \sum_{i=1}^{n} x_i^2$$

$$\sum_{i=1}^{n} x_i y_i = \hat{\beta}_0 \sum_{i=1}^{n} x_i + \hat{\beta}_1 \sum_{i=1}^{n} x_i^2 + \hat{\beta}_2 \sum_{i=1}^{n} x_i^3$$

$$\sum_{i=1}^{n} x_i^2 y_i = \hat{\beta}_0 \sum_{i=1}^{n} x_i^2 + \hat{\beta}_1 \sum_{i=1}^{n} x_i^3 + \hat{\beta}_2 \sum_{i=1}^{n} x_i^4$$

17. (a) $\hat{y} = -5.96 + 1.55x$; (b) 4.89.
19. $\hat{y} = 1.37(1.38)^x$.
21. (a) $\hat{y} = 10.5 - 2.0x + 0.2x^2$; (b) 5.55.

Page 445 **13.**

$$\frac{1 + r - (1 - r)\, e^{2z_{\alpha/2}/\sqrt{n-3}}}{1 + r + (1 - r)\, e^{2z_{\alpha/2}/\sqrt{n-3}}} < \rho < \frac{1 + r - (1 - r)\, e^{-2z_{\alpha/2}/\sqrt{n-3}}}{1 + r + (1 - r)\, e^{-2z_{\alpha/2}/\sqrt{n-3}}};$$

$0.746 < \rho < 0.985$.
17. $t = 0.45$; null hypothesis cannot be rejected.
19. $0.12 < \beta < 1.04$.
21. $-8.61 < \alpha < -3.31$.
23. (a) $65.69 < \mu_{y|14} < 74.88$; (b) $57.28 < y_0 < 83.29$.
25. $r = 0.553$; significant.
27. $r = 0.727$; significant.
29. $2.84 < \beta < 4.10$.

Page 459 **7.** $F = 17.0$; differences are significant.
9. $F = 39.3$; significant, reject the null hypothesis.
$\hat{\mu} = 18.5$, $\hat{\alpha}_1 = 4$, $\hat{\alpha}_2 = 1.5$, $\hat{\alpha}_3 = -5.5$.
11. $F = 1.05$; not significant.

Page 468 **5.** (b)

Source of Variation	Degrees of Freedom	Sum of Squares	Mean Square	F
Rows	$m - 1$	SSR	$MSR = \frac{SSR}{m-1}$	$F_R = \frac{MSR}{MSE}$
Columns	$m - 1$	SSC	$MSC = \frac{SSC}{m-1}$	$F_C = \frac{MSC}{MSE}$
Treatments	$m - 1$	$SS(Tr)$	$MS(Tr) = \frac{SS(Tr)}{m-1}$	$F_{Tr} = \frac{MS(Tr)}{MSE}$
Error	$(m - 1) \times (m - 2)$	SSE	$MSE = \frac{SSE}{(m-1)(m-2)}$	
Total	$m^2 - 1$	SST		

7. For the diet foods $F = 6.64$, which is significant; for the laboratories $F = 4.91$, which is significant.

9. (a) $F = 94.24$, which is significant; (b) $F = 2.56$, which is not significant; (c) $F = 27.04$, which is significant.

Page 485

9. Cannot reject the null hypothesis.

11. Reject the null hypothesis.

13. $z = 3.21$; reject the null hypothesis.

15. $z = 1.85$; null hypothesis cannot be rejected.

17. $z = -1.33$; null hypothesis cannot be rejected.

21. $H = 4.51$; null hypothesis cannot be rejected.

Page 495

3. $\frac{11}{42}$.

5. min $W = 0$ when $R_i = \dfrac{k(n + 1)}{2}$ for all i, reflecting a complete lack of association; max $W = 1$, reflecting perfect agreement.

7. $z = -0.10$; cannot reject the null hypothesis of randomness.

11. $z = 0$; cannot reject the null hypothesis of randomness.

13. $r_s = 0.86$, $z = 3.55$; significant.

15. $z = 1.78$; cannot reject the null hypothesis.

index